Lecture Notes in Applied and Computational Mechanics

Volume 15

Series Editors

Prof. Dr.-Ing. Friedrich Pfeiffer
Prof. Dr.-Ing. Peter Wriggers

Springer

Berlin
Heidelberg
New York
Hong Kong
London
Milan
Paris
Tokyo

Micropolar Theory of Elasticity

Janusz Dyszlewicz

Springer

Professor Janusz Dyszlewicz
Institute of Mathematics
Faculty of Fundamental Problems of Technology
Wroclaw University of Technology
Wybrzeze Wyspianskiego 27
50-370 Wroclaw
Poland

With 54 Figures

Cataloging-in-Publication Data applied for
Bibliographic information published by Die Deutsche Bibliothek
Die Deutsche Bibliothek lists this publication in the Deutsche Nationalbibliografie; detailed bibliographic data is available in the Internet at <http://dnb.de>

ISBN 3-540-41835-0 Springer Verlag Berlin Heidelberg New York

Springer-Verlag is a part of Springer Science+Business Media

springeronline.com

Printed in Germany

Cover design: design & production GmbH, Heidelberg
Typesetting: Digital data supplied by author

Printed on acid-free paper 62/3020Rw -5 4 3 2 1 0

This paper is dedicated to the
memory of PROFESSOR WITOLD NOWACKI

Preface

This monograph contains the results of my research in the area of asymmetric theory of elasticity, conducted from 1969 to 1986 under the direction of PROFESSOR WITOLD NOWACKI.

I am indebted to PROFESSOR NOWACKI, thanks to whose invaluable and very kind research assistance I obtained the results which were the foundation of this monograph. Therefore, I would like to express my deepest gratitude to him and honour his memory. He will remain in my thoughts with due respect.

During my research assistantship at the Institute of Mechanics at the University of Warsaw in 1970–1973 I had the opportunity to participate in seminars and conferences, study critical reviews and carry on numerous discussions and conversations. All this resulted in many valuable remarks included in this monograph. In this connection, I would like to thank Professor Józef Ignaczak and Professor Marek Sokołowski from the Institute of Fundamental Problems of Technology at the Polish Academy of Sciences, as well as Professor Zbigniew Olesiak and Professor Adam Piskorek from the Institute of Mechanics at the University of Warsaw.

In 1974–1986 I participated in the seminar "Mathematical Models of Deformable Media" at the Institute of Mathematics of the Technical University of Wrocław (TUW). My thanks go to the organizers of the seminar, Professor Bertold Lysik from the Institute of Mathematics, Professor Otton Dąbrowski from the Institute of Civil Engineering, and Professor Igor Kisiel from the Institute of Geotechnics, for valuable discussions, in particular those concerning practical aspects of the results obtained.

I would also like to thank Professor Zbigniew Romanowicz from the Institute of Mathematics as well as Professor Wacław Kasprzak and the Senate of the Technical University of Wrocław, for granting me support for my research in 1984–1985. Thanks are also due to Professor Jarosław Stefaniak from the Institute of Applied Mechanics at the Technical University of Poznań for a number of valuable consultations in 1990 concerning my research publications.

Continuing the extensive research activity of MY ADVISOR, PROFESSOR NOWACKI, I included in this monograph the results obtained by me and my students after 1986. Thus, this work is an extended version of my earlier work entitled *Boundary and Initial-Boundary Value Problems of Equations of Micropolar Elastostatics and Elastodynamics*, published in Polish by the Publishing House of the Technical University of Wrocław in 1990 in the monograph series. It appears under [58] in the references. I would like to stress that the final form of this monograph was shaped during my stay as a research fellow at the Institute of Fundamental Problems of Technology of the Polish Academy of Sciences in the academic year 1986/1987, under the direction and with immense help of Professor Józef Ignaczak.

The manuscript of [58], listed as [49] in the references under the tentative title, *Boundary Value Problems of Micropolar Theory of Elasticity*, was highly evaluated by MY ADVISOR in a review from 1984. The main stream of this manuscript was the subject of the monograph lectures which I gave in 1980–1990 to the exclusive classes of students of mechanics at the Faculty of Fundamental Problems of Technology and the Faculty of Civil Engineering at TUW. In those classes I taught students, with whom I later wrote a number of valuable papers (quoted in the references), namely Czesław Kołodziej, Bożena Słotwińska, Monika Czub, Jacek Wytrążek, Paweł Żal, Jacek Bieńkowski and Marcin Sikora. Later, I did joint research work on the Cosserats media with my Ph.D student, Mountajab Al-Hasan, who, after graduating from Al Baath University from the Syrian Arab Republic, obtained a scholarship to the Wrocław University of Technology. I would like to add that the seminars of Professor Jan Langer from the Institute of Civil Engineering at TUW and Professor Bertold Lysik had a great influence on the high level of the research papers written in cooperation with my students.

This monograph concerns analytical methods of solving boundary and initial-boundary value problems for linear differential equations with partial derivatives describing the models of continuous media with microstructure of Cosserats type, such as the model of isotropic hemitropic medium of Aero-Kuvshinski (A-K), the model of micropolar body of Eringen-Nowacki (E-N), the model of couple-stress medium of Koiter-Mindlin (K-M), the model of hypothetical medium (HM) and the model of Hooke's medium (H). Within the frameworks of elastostatics, theromoelastostatics and elastodynamics, with the fields of body loadings and distortions taken into account, we present many methods of solving the fundamental partial differential equations of the models mentioned above, giving basic formulations of the initial-boundary problems of the dynamics and boundary value problems of the statics. These are the following methods: the method of stress equations of Beltrami-Michell type, the method of stress equations of motion of Ignaczak type (SEMP), the method of direct integration of the equations in displacements and rotations, the method of stress function, the methods of generalized representations of solutions, namely, of Galerkin, Iacovache (Cauchy-Kovalevski-Somigliana), Papkovich-Neuber, Love, then, the method of generalized potentials of Nowacki (Green-Lamé), and finally, the method of superposition by means of the generalized Schaefer vector. The methods mentioned above are used for solving boundary- and initial-boundary value problems, for plane, axially-symmetric and three-dimensional states of strain, as well as for the determination of the Green functions. In particular, we solve and discuss in detail a certain important class of boundary-value problems, containing the generalized problems of Kirsch, Goodier, Lamb, the boundary value problems for half-spaces of Dirichlet and Neumann; among others, we have here the solved problems of Boussinesq, Cerruti and Boussinesq-Mindlin. With great attention we treat the problems of stress concentration and singularities of physical fields. Of primary concern to us is micropolar theory (E-N model), which is treated not only as a limit of hemitropic theory (A-K model), but also as a theory, from which one can obtain the results of couple-stress theory (K-M model), the results of the theory of pure rotations (HM) as well as those of the classical theory of elasticity, by taking the limits of the micropolar parameters. Therefore, great emphasis is put on the original schemes of performing limits from one theory to another, within the framework of fundamental equations, general representations of solutions, singular solutions and particular boundary value problems.

The monograph consists of the Introduction and four chapters. In the Introduction we discuss the models studied in this work, namely (A-K), (E-N), (K-M), (HM) and (H) models, as well as the fundamental equations, the boundary conditions, the initial-boundary conditions, and finally, the material constants for all five models mentioned above. Chapter 1 refers to three-dimensional problems. In Chapter 2, we discuss axially-symmetric problems. In Chapter 3, we consider two-dimensional problems in the plane state of strain. In Chapter 4, we analyse hemitropic media and vector equations.

From the mathematical point of view, we use the Kupradse theorems on the existence of solutions, the Fourier and Hankel integral transformations, and the general theory of linear ordinary differential equations.

The referees viewed this work as a monograph. It contains both the results of theoretical and practial character and extends our knowledge of elastic materials with microstructure, acquired from scientific literature. This work is the first monograph on the subject that has appeared after the monograph of Professor Witold Nowacki, *Theory of Asymmetric Elasticity*, published in 1986 by Polish Scientific Publishers & Pergamon Press. In particular, it constitutes an exhaustive supplement and an extension of Professor Nowacki's work, thus it can be of interest to specialists in such areas as solid mechanics or partial differential equations, theoretical engineers and students taking courses in advanced mechanics at universities and polytechnics.

This work has had many sponsors, which has been a necessary condition for it to come out in print. The main support came from the Scientific Research Committee (KBN) in Warsaw. Also, the financial support from the Dean of the Faculty of Civil Engineering at TUW, Professor Otton Dąbrowski, and the Dean of the Faculty of Fundamental Problems of Technology at TUW, Professor Jerzy Czerwonko, is gratefully acknowledged.

Finally, I would like to warmly express my thanks to Professor Czesław Ryll-Nardzewski, Professor Zbigniew Olszak, Professor Aleksander Weron, and the Scientific and Educational Board of the Institute of Mathematics, as well as my colleagues from the Institute of Mathematics, the Institute of Civil Engineering and the Institute of Geotechnics at TUW, for their kind interest in my research and the present monograph that I experienced over the years. In particular, I am grateful to Professor Romuald Lenczewski and Dr Artur Rozwadowski for translating the manuscript from Polish, and Professor Marek Sokołowski — for verifying the text. Moreover, I would like to thank Dr Przemysław Scherwentke and Monika Kaczmarz for type-setting the text. Thanks are also due to Dr Stanisław Kroczak, who prepared the final versions of all figures and to Professor Wojciech Glabisz for consultations. Last but not least, I would like to thank Professor Wojciech Kordecki for his highly proffesional supervision of the typesetting of the manuscript.

Wrocław, September 5, 2003 Janusz Dyszlewicz

Contents

Introduction

In the classical Hooke's theory of elasticity (see Nowacki [180]) we use a theoretical, ideal and simplified model of a solid body in the form of an elastic material continuum which is regarded as a continuum in the mathematical sense. Thus, we neglect the molecular, atomic structure of the body and we treat it as a three-dimensional Euclidean space, its points being identified with the material particles of the body. The continuous distribution of matter in a given region of the body is characterized by a single scalar quantity, the density. The deformation of the body is described by the displacement vector. It is assumed that the transmission of loadings through the surface element situated inside the body is uniquely determined by the principal force vector. Adopting that assumption, we obtain the description of the body deformation in terms of symmetric tensors of strain and stress.

The adopted mathematical model of real materials complies fairly well with experimental results observed in construction materials such as steel, aluminium and concrete, with stresses remaining within the material elasticity limits. However, in the classical theory of elasticity, there are many problems in which one can hardly preserve the assumption of stress tensor symmetry. One can give here two examples in the framework of elastostatics: the problem of an unbounded plane loaded by a concentrated torque, and the two-dimensional problem of a rectangular wedge with a tangent loading applied along one of the edges. The solutions to those problems cannot be consistent with the assumption of stress tensor symmetry and they lead to singularities that have no physical explanation.

Some important differences between the classical theory and the experiment are found in those problems where considerable stress gradients occur. One should mention here the stress concentration in the neighbourhood of holes, notches or cracks, important from the point of view of safety problems in engineering structures.

The discrepancy between the classical theory of elasticity and the experiments is clearly seen in the case of dynamical problems, such as elastic vibrations characterized by high frequencies and short wavelenghts; those effects are observed whenever ultrasounds are applied. With high frequencies and short wavelengths, the body microstructure has a great impact on the actual deformation process. That is why the classical theory of elasticity fails also in the case of granular body vibrations, and this applies to most construction materials, acoustic waves in crystals, polycrystalline structures or polymers. In those media the microstructure effects become very important because some new types of waves appear that were not present in the classical theory of elasticity. Also, the classical theory cannot be applied in the case of bodies with a definite internal structure (fibrous media, surface gridwork, grates, reinforced material, see Frąckiewicz [91], Gutkowski [106], Woźniak [244]).

The first generalization of the classical theory of elasticity goes back to the work of Voigt (1887), who assumed that the transmission of interactions between two parts of a continuum across a surface element is not only assured by the force-stress vector, but also by the couple-stress vector. This assumption leads to the description of stress field by means of two asymmetric tensors: the force-stress tensor and the couple-stress tensor. A complete asymmetric elasticity theory was developed by the Cosserats brothers (1909). Their fundamental paper made an attempt to create a unified field theory, incorporating mechanics, optics and electrodynamics. However, in spite of its novelty, the theory was not appreciated for a long time. Only in the last thirty years, the research development in the area of the general theory of continuum helped the Cosserats theory to attract the attention of researchers and it became the starting point for a number of related theories in mechanics and physics of media of higher order. A lot of papers written over the years in the area of mechanics have been devoted to the media of Cosserats type; in particular, we should mention here the works of Aero & Kuvshinski [1], Burchuladse & Gegelya [19], Eringen & Suhubi [88], Günther [104], Grioli [103], Kaliski [123], Kupradse, Gegelya, Bashelishili & Burchuladse [138], Koiter [131], Kunin [135], Mindlin [157], Nowacki [173], Neuber [163], Palmov [201], Sokołowski [214], Savin [207], Schaefer [210], Bogy & Sternberg [15], Truesdell & Toupin [232], Mindlin & Tiersten [158] and Woźniak [242]. The works mentioned above deal with the mathematical and physical foundations of both linear and nonlinear continuum models with the Cosserats-type microstructures and their application to elastostatics, elastodynamics and thermoelasticity. The first well-known and widely appreciated monographs concerning the asymmetric elasticity theory have been written in Poland.

The most up-to-date monograph investigation devoted to general linear micropolar theory is the book by Nowacki (1986). The monograph discusses some general and particular problems of elastostatics, elastokinetics and thermoelasticity for the Cosserats medium. The mathematical and physical foundations of the theory, their fundamental differential equations and methods of solving those equations have been extensively presented. Some fundamental theorems have been formulated and proved, namely the energy theorem, the variational principle, the uniqueness of solutions and the works reciprocity theorem. The problems related to the generation and propagation of elastic waves with their new types occuring in the Cosserats model have been discussed. Next, the distortion problem in a micropolar medium has been presented as well as the problem of coupled fields in the Cosserats medium, in particular the field of strain with the field of temperature and electromagnetic field. Further, thermopiezoelectricity and magnetothermoelasticity have been extensively discussed. Also, a micropolar hemitropic medium has been discussed in that monograph. The work of Nowacki contains a long list of references. A more extensive reference list can be found in the monograph written by Kunin & Kosilova (1976). Due to its cognitive values, possible complete experimental verification as well as technological applications, the Cosserats theory is being developed in a great number of research centers in the world (Poland, USA, Russia, Germany, Great Britain, Sweden, Romania, Canada, Italy, Japan, Holland). One should also mention here Polish contributors to the theory of bodies with microstructure: Barański, Wilmański & Woźniak [11], Drobot [31], Gawinecki [99], Ganowicz [96], [97], Ignaczak [113]–[119], Kaliski, Kapelewski & Rymarz [125], Kurlandzki [139], Kisiel [129], Majorkowska-Knap [148], [149], Dyszlewicz & Matysiak [72]–[75], Matysiak & Wachecka-Skowron [154], Nowacki & Nowacki [194]–[197], Nowacki [165]–[168], Olesiak [198][1], Frydrychowicz, Matysiak & Olesiak [92], [93], Rymarz [204], [205], Rogula [203], Sawczuk [209], Stefaniak [217]–[220], Suchar [221]–[223], Gajl & Suchar [95], Usidus [234], [235], Wesołowski [238] and Zahorski [245]. Let us now turn to the discussion of the foundations of general linear theory of Cosserats medium.

Micropolar model of Eringen-Nowacki (E-N)

First, we collect some basic relations and equations for the Eringen-Nowacki model of a micropolar body with six degrees of freedom, subject to certain restrictions. The micropolar body is elastic, homogeneous, isotropic and centro-symmetric. We assume the right-handed orthogonal Cartesian coordinate system $Ox_1x_2x_3$ and we use the Cartesian tensor indexical notation. The partial derivatives with respect to the position variables are denoted by an apostrophe, and the derivative with respect to time t is denoted by a superscribed dot: $\dot{f} \stackrel{\text{def}}{=} \partial f/\partial t$. We apply the extended summation convention — small Latin indices i, j, k, ... assume the values 1, 2, 3, capital Latin indices K, L, M, ... — the values 2, 3, and Greek indices α, β, γ, ... — the values 1, 2. Symbol ϵ_{ijk} denotes the Levi-Civitá tensor, δ_{ij} is the Kronecker symbol, $\mathbb{R}$ denotes the set of real numbers, $\mathbb{R}^3$ is the three-dimensional Euclidean space, $\mathbf{x} \stackrel{\text{def}}{=} (x_1, x_2, x_3)$ are points in $\mathbb{R}^3$, Ω is a region in $\mathbb{R}^3$ with piecewise smooth boundary $\partial\Omega$, and n_j are the components of the unit outward normal vector to $\partial\Omega$. Further, the closed region $\overline{\Omega}$ is the union of the sets Ω and $\partial\Omega$ ($\overline{\Omega} = \Omega \cup \partial\Omega$), the time intervals are denoted as follows:

[1] Olesiak [199] and Olesiak & Wągrowska [200] consider the plasticity theory for media with microstructure.

$$\mathbf{T} \stackrel{\text{def}}{=} [0, \infty) = \{t \in \mathbb{R} : t \geq 0\}, \qquad \mathbf{T}_+ \stackrel{\text{def}}{=} (0, \infty) = \{t \in \mathbb{R} : t > 0\},$$

symbol $\overline{\Omega} \times \mathbf{T}$ denotes the Cartesian product of the sets $\overline{\Omega}$ and $\mathbf{T}$:

$$\overline{\Omega} \times \mathbf{T} = \{\mathbf{x} \in \overline{\Omega} \wedge t \in \mathbf{T}\}.$$

In a similar way we define the sets $\Omega \times \mathbf{T}_+$, $\Omega \times \mathbf{T}$ and $\partial\Omega \times \mathbf{T}$. We refer our considerations to the micropolar elastic medium $\Omega(\mu, \lambda, \alpha, \beta, \gamma, \varepsilon, \rho, J, \kappa, \eta_0, \nu_T)$ described by the set of thermo-elastodynamical real-valued parameters given in the parentheses. The initial micropolar configuration of the medium is the region Ω. All the physical fields defining the thermo-elastodynamical states of the medium are real-valued functions of the position variables $\mathbf{x}$ and the time variable t. The fact that the function f is defined in the set $\overline{\Omega} \times \mathbf{T}$ will be denoted as follows: $f : \overline{\Omega} \times \mathbf{T} \to \mathbb{R}$. A function f belongs to the class C^n in the set Ω ($f \in C^n(\Omega)$) if f and all its partial derivatives up through the n-th order are continuous in the set Ω ($n = 0, 1, 2, 3, \ldots$). Generally, thoroughout the present monograph we assume that the considered functions are sufficiently regular to make the applied procedures meaningful. The basic equations of the Eringen-Nowacki model can be divided into the following groups[2]:

— the equations of motion in $\Omega \times \mathbf{T}_+$,

$$\sigma_{ji,j} + X_i = \rho \ddot{u}_i, \qquad \epsilon_{ijk}\sigma_{jk} + \mu_{ji,j} + Y_i = J\ddot{\varphi}_i, \tag{0.1}$$

where σ_{ji}, $\mu_{ji} \in C^1(\Omega \times \mathbf{T}_+)$ are 3×3 matrices with physical components of the asymmetric force-stress tensor and the asymmetric couple-stress tensor, X_i, $Y_i \in C^0(\Omega \times \mathbf{T})$ are the fields of body loadings and body moments, respectively, u_i, $\varphi_i \in C^2(\Omega \times \mathbf{T}_+)$ are components of the displacement and the rotation vectors, respectively, ρ is the density and J is the rotational inertia of the medium;

— the equations of compatibility in $\Omega \times \mathbf{T}$,

$$\begin{aligned} &\gamma_{li,h} - \gamma_{hi,l} - \epsilon_{khi}\kappa_{lk} + \epsilon_{kli}\kappa_{hk} = 0, \\ &\kappa_{li,h} - \kappa_{hi,l} = 0, \end{aligned} \tag{0.2}$$

where γ_{ji}, $\kappa_{ji} \in C^1(\Omega \times \mathbf{T})$ — the 3×3 matrices with physical components of the asymmetric strain tensor and the asymmetric micro-strain (torsion-flexure) tensor.

The tensors γ_{ji} and κ_{ji} are defined by:

[2] The present notation is explained in and is basically the same as that used in our works [3]–[4], [24]–[25], [32]–[85], which were the basis of the present monograph. We use the vector, operator and tensor notations, with the extended summation rule.

— the geometric relations in $\Omega \times \mathbf{T}_+$,

$$\gamma_{ji} = u_{i,j} - \epsilon_{kji}\varphi_k\,, \qquad \kappa_{ji} = \varphi_{i,j}; \tag{0.3}$$

— the constitutive relations in $\Omega \times \mathbf{T}$,

$$\begin{aligned} \sigma_{ji} &= (\mu+\alpha)\gamma_{ji} + (\mu-\alpha)\gamma_{ij} + (\lambda\gamma_{kk} - \nu_T\theta)\delta_{ij}, \\ \mu_{ji} &= (\gamma+\varepsilon)\kappa_{ji} + (\gamma-\varepsilon)\kappa_{ij} + \beta\kappa_{kk}\delta_{ij}, \end{aligned} \tag{0.4}$$

where μ, λ, α, β, γ and ε are the elasticity constants of the medium, $\nu_T = (2\mu + 3\lambda)\alpha_t$ and α_t is the linear coefficient of thermal expansion of the medium.

The function $\theta = T - T_0$ describes the temperature field and it satisfies:

— the equation of heat conduction in $\Omega \times \mathbf{T}_+$,

$$\theta_{,ii} - \frac{1}{\kappa}\dot{\theta} - \eta_0 \dot{u}_{j,j} = -\frac{Q}{\kappa}, \tag{0.5}$$

where

$$Q = \frac{\kappa W}{\lambda_0}, \qquad \kappa = \frac{\lambda_0}{c_\varepsilon}, \qquad \eta_0 = \frac{\nu_T T_0}{\lambda_0}, \qquad \theta \in C^2(\Omega \times \mathbf{T}_+),$$

Q — the heat sources in the body, W — the amount of heat generated in a unit volume and unit time, λ_0 — heat conduction coefficient, T_0 — natural state temperature, c_ε — specific heat for constant deformation.

Substituting the constitutive relations (0.4) into the equations of motion (0.1) and using the geometric relations (0.3), we obtain[3]

— the equations of motion in terms of displacements and rotations in $\Omega \times \mathbf{T}_+$:

$$\begin{aligned} &(\mu+\alpha)u_{i,jj} + (\lambda+\mu-\alpha)u_{j,ji} + 2\alpha\epsilon_{ijk}\varphi_{k,j} + X_i = \rho\ddot{u}_i + \nu_T\theta_{,i}, \\ &(\gamma+\varepsilon)\varphi_{i,jj} - 4\alpha\varphi_i + (\beta+\gamma-\varepsilon)\varphi_{j,ji} + 2\alpha\epsilon_{ijk}u_{k,j} + Y_i = J\ddot{\varphi}_i. \end{aligned} \tag{0.6}$$

The principal aim of the theory is to determine the physical fields of displacements, rotations, temperature, force-stresses, couple-stresses and strains. It means that one has to determine the thermo-elastodynamical state

$$u_i,\ \varphi_i,\ \theta,\ \sigma_{ji},\ \mu_{ji},\ \gamma_{ji},\ \kappa_{ji} : \overline{\Omega} \times \mathbf{T} \to \mathbb{R} \tag{0.7}$$

of the medium $\Omega(\mu, \lambda, \alpha, \beta, \gamma, \varepsilon, \rho, J, \kappa, \eta_0, \nu_T)$ in the set $\overline{\Omega} \times \mathbf{T}$ corresponding to

---

[3] The compatibility relations (0.2) become identities.

— the equations and field relations (0.1)–(0.6) (with prescribed body forces $X_i : \Omega \times \mathbf{T} \to \mathbb{R}$, body moments $Y_i : \Omega \times \mathbf{T} \to \mathbb{R}$ and heat sources $Q : \Omega \times \mathbf{T} \to \mathbb{R}$);
— the boundary conditions (prescribed displacements and rotations on the part $\partial\Omega_u$ of the boundary, force-stresses and couple-stresses on $\partial\Omega_\sigma = \partial\Omega \setminus \partial\Omega_u$, the temperature on $\partial\Omega_\theta$ and the heat flux on $\partial\Omega_q = \partial\Omega \setminus \partial\Omega_\theta$);
— the initial conditions (at the initial instant the following fields are prescribed: displacements, rotations, temperature, velocities of displacements and rotations).

This formulation is equivalent to the following one: one has to determine a regular solution

$$u_i,\ \varphi_i,\ \theta \in C^2(\Omega \times \mathbf{T}) \cap C^1(\overline{\Omega} \times \mathbf{T}) \tag{0.8}$$

of the equations of motion in displacements and rotations (0.6) and the heat conduction equations (0.5) with the following initial-boundary conditions:

— the boundary conditions[4] on $\partial\Omega \times \mathbf{T}$[5],

$$\begin{aligned} &\sigma_{ji} n_j = p_i, \qquad \mu_{ji} n_j = m_i && \text{on } \partial\Omega_\sigma \times \mathbf{T},\\ &u_i = f_i, \qquad \varphi_i = g_i && \text{on } \partial\Omega_u \times \mathbf{T},\\ &\theta = \vartheta && \text{on } \partial\Omega_\theta \times \mathbf{T},\\ &\frac{\partial\theta}{\partial n} = q && \text{on } \partial\Omega_q \times \mathbf{T}, \end{aligned} \tag{0.9}$$

where the functions $p_i,\ m_i : \partial\Omega_\sigma \times \mathbf{T} \to \mathbb{R}$, $f_i,\ g_i : \partial\Omega_u \times \mathbf{T} \to \mathbb{R}$, $\vartheta : \partial\Omega_\theta \times \mathbf{T} \to \mathbb{R}$ and $q : \partial\Omega_q \times \mathbf{T} \to \mathbb{R}$ are given;
— the initial conditions[6] in $\Omega \times \{0\}$,

$$u_i = h_i, \qquad \varphi_i = k_i, \qquad \theta = l, \qquad \dot{u}_i = \psi_i, \qquad \dot{\varphi}_i = \chi_i, \tag{0.10}$$

where the functions $h_i,\ k_i,\ l,\ \psi_i,\ \chi_i : \Omega \to \mathbb{R}$ are given.

The micropolar theory of elasticity is of phenomenological character. The isotropic, micropolar medium is characterized by some thermo-elastodynamical parameters μ, λ, α, β, γ, ε, ρ, J, κ, η_0 and ν_T, which are to be determined

[4] The boundary conditions of mechanical type can be of three kinds, see Nowacki [192, p. 208], in the first and the second problems the pairs (u_i, φ_i), $(\sigma_{ji}n_j, \mu_{ji}n_j)$ are prescribed, the third problem refers to mixed boundary conditions. Kupradse [138, p. 57, (1976)] distinguishes two cases for the pairs $(u_i, \mu_{ji}n_j)$, $(\varphi_i, \sigma_{ji}n_j)$ in the mixed problem.

[5] Thermal conditions can be given in a more general form: the temperature gradient, or $\alpha(\partial\theta/\partial n) + \beta\theta = f$, is given on $\partial\Omega_q \times \mathbf{T}$, where $\alpha,\ \beta \in \mathbb{R}$ are some constants and f is a given function, $\partial\theta/\partial n = n_j(\partial\theta/\partial x_j)$ (see Nowacki [180, p. 76], [174]; see also Kupradse [138, p. 58, (1976)]).

[6] Ignaczak [116] analyses the initial and boundary value problems in terms of stresses for micropolar elastodynamics; in that case the initial conditions determine the fields of force-stresses and couple-stresses and their velocities in $\Omega \times \{0\}$.

by experiments. These parameters are subject to some constraints. In particular, the parameters κ, η_0, ν_T, ρ, J have to satisfy the following important inequalities:

$$\frac{\nu_T}{\eta_0} > 0, \qquad \kappa > 0, \qquad \rho > 0, \qquad J > 0. \tag{0.11}$$

Based on the thermodynamics of the irreversible processes, in the case of a homogeneous and centro-symmetric micropolar medium, we obtain the following representation of the Helmholtz free energy $F = U - ST$, where U is the internal energy, S is the entropy per unit volume, and T is the absolute temperature:

$$\begin{aligned} F = {} & \frac{\mu+\alpha}{2}\gamma_{ji}\gamma_{ji} + \frac{\mu-\alpha}{2}\gamma_{ji}\gamma_{ij} + \frac{\lambda}{2}\gamma_{kk}\gamma_{nn} + \frac{\gamma+\varepsilon}{2}\kappa_{ji}\kappa_{ji} \\ & + \frac{\gamma-\varepsilon}{2}\kappa_{ji}\kappa_{ij} + \frac{\beta}{2}\kappa_{kk}\kappa_{nn} - \nu_T\gamma_{kk}\theta + G(\theta), \qquad (\theta = T - T_0). \end{aligned} \tag{0.12}$$

Here, $G(\theta)$ denotes all the terms depending on the temperature only.

From the expression for free energy $F(\gamma_{ji}, \kappa_{ji}, \theta)$ and using the relations

$$\sigma_{ji} = \frac{\partial F}{\partial \gamma_{ji}}, \qquad \mu_{ji} = \frac{\partial F}{\partial \kappa_{ji}}, \tag{0.13}$$

we obtain the relations (0.4) between the stress fields and the strain and temperature fields. The internal energy should either be positive or equal to zero (in the case of no strain). Since the internal energy expression is a form which is quadratic with respect to the components of the strain tensor, we see that all the principal minors of the matrix associated with that quadratic form are positive. Let us represent the internal energy in the form $2U = a_{\alpha\beta}x_\alpha x_\beta$, where $\alpha = 1, 2, \ldots, 18$, $\beta = 1, 2, \ldots, 18$ and $x_1 = \gamma_{11}$, $x_2 = \gamma_{22}$, $x_3 = \gamma_{33}$, $x_4 = \kappa_{11}$, $x_5 = \kappa_{22}$, $x_6 = \kappa_{33}$, $x_7 = \gamma_{31}$, $x_8 = \gamma_{13}$, $x_9 = \kappa_{13}$, $x_{10} = \kappa_{31}$, $x_{11} = \gamma_{21}$, $x_{12} = \kappa_{12}$, $x_{13} = \gamma_{12}$, $x_{14} = \kappa_{21}$, $x_{15} = \gamma_{23}$, $x_{16} = \kappa_{23}$, $x_{17} = \gamma_{32}$, $x_{18} = \kappa_{32}$.

Thus, we obtain the following matrix of coefficients:

$$[a_{\alpha\beta}]_{18\times 18} = \begin{bmatrix} [I] & & & & \\ & [II] & & 0 & \\ & & [III] & & \\ & 0 & & [IV] & \\ & & & & [V] \end{bmatrix}, \tag{0.14}$$

where the matrices $[I]_{3\times 3}$, $[II]_{3\times 3}$, $[III]_{4\times 4}$, $[IV]_{4\times 4} \stackrel{\text{def}}{=} [V]_{4\times 4}$ take the following forms:

$$[I] = \begin{bmatrix} \lambda+2\mu & \lambda & \lambda \\ \lambda & \lambda+2\mu & \lambda \\ \lambda & \lambda & \lambda+2\mu \end{bmatrix}, \qquad [II] = \begin{bmatrix} \beta+2\gamma & \beta & \beta \\ \beta & \beta+2\gamma & \beta \\ \beta & \beta & \beta+2\gamma \end{bmatrix},$$

$$[III] = \begin{bmatrix} \mu+\alpha & \mu-\alpha & 0 & 0 \\ \mu-\alpha & \mu+\alpha & 0 & 0 \\ 0 & 0 & \gamma+\varepsilon & \gamma-\varepsilon \\ 0 & 0 & \gamma-\varepsilon & \gamma+\varepsilon \end{bmatrix}, \qquad [IV] = \begin{bmatrix} \mu+\alpha & 0 & \mu-\alpha & 0 \\ 0 & \gamma+\varepsilon & 0 & \gamma-\varepsilon \\ \mu-\alpha & 0 & \mu+\alpha & 0 \\ 0 & \gamma-\varepsilon & 0 & \gamma+\varepsilon \end{bmatrix}.$$

Applying the Sylvester theorem[7], we obtain the following set of energy inequalities:

$$\begin{aligned} &\mu > 0, \qquad \alpha > 0, \qquad \varepsilon > 0, \qquad \gamma > 0, \\ &2\mu + 3\lambda > 0, \qquad 2\gamma + 3\beta > 0, \qquad \lambda + 2\mu > 0. \end{aligned} \tag{0.15}$$

The Aero-Kuvshinski model (A-K) of isotropic hemitropic medium

In the sequel, we shall also consider a micropolar hemitropic medium. The theory of a hemitropic medium has been considered in several papers. The foundations of that theory were given by Aero & Kuvshinski [1], [2] (within the frameworks of elastostatics and elastodynamics), and by Nowacki [177] (thermoelasticity). Moreover, one should also mention here the papers by Benedict & Lakes [13], Gawinecki [99], Lenz [143], [144], Nowacki & Nowacki [169] and Nowacki [167], [168]. A hemitropic medium in an isotropic micropolar medium where the postulate of centro-symmetry is disregarded and that is why the free energy representation for a hemitropic body is more complicated than that for a micropolar centro-symmetric medium and it takes the form

$$\begin{aligned} F = {} & \frac{\mu+\alpha}{2}\gamma_{ji}\gamma_{ji} + \frac{\mu-\alpha}{2}\gamma_{ji}\gamma_{ij} + \frac{\lambda}{2}\gamma_{kk}\gamma_{nn} \\ & + \frac{\gamma+\varepsilon}{2}\kappa_{ji}\kappa_{ji} + \frac{\gamma-\varepsilon}{2}\kappa_{ji}\kappa_{ij} + \frac{\beta}{2}\kappa_{kk}\kappa_{nn} \\ & + (\chi+\nu_0)\gamma_{ji}\kappa_{ji} + (\chi-\nu_0)\gamma_{ji}\kappa_{ij} + \kappa\gamma_{kk}\kappa_{nn} \\ & - \eta\gamma_{kk}\theta - \zeta\kappa_{kk}\theta - \frac{c_\varepsilon}{T_0}\theta^2. \end{aligned} \tag{0.16}$$

One can see that apart from the material constants μ, λ, α, β, γ and ε, three additional elasticity constants χ, ν_0 and κ, as well as constants η, ζ, depending both on the mechanical and thermal properties of the body, have been introduced. The thermoelasticity parameters are subject to some constraints. Changing the notations, we give here (see Aero & Kuvshinski [1])

[7] See Mostowski & Stark [160, p. 349].

some energy inequalities which will be needed in the sequel. The free energy F (without the temperature field) is now denoted by

$$2F = a_{\alpha\beta} x_\alpha x_\beta$$

and we adopt the following notation ($\alpha, \beta = 1, 2, \ldots, 18$): $x_1 = \gamma_{11}$, $x_2 = \kappa_{11}$, $x_3 = \gamma_{22}$, $x_4 = \kappa_{22}$, $x_5 = \gamma_{33}$, $x_6 = \kappa_{33}$, $x_7 = \gamma_{13}$, $x_8 = \gamma_{31}$, $x_9 = \kappa_{13}$, $x_{10} = \kappa_{31}$, $x_{11} = \gamma_{12}$, $x_{12} = \kappa_{12}$, $x_{13} = \gamma_{21}$, $x_{14} = \kappa_{21}$, $x_{15} = \gamma_{23}$, $x_{16} = \kappa_{23}$, $x_{17} = \kappa_{32}$, $x_{18} = \gamma_{32}$.

The matrix of coefficients takes now the form (0.14) without the matrix $[V]$, where the matrices $[I]_{6\times 6}$, $[II]_{4\times 4}$, $[III]_{4\times 4}$, $[IV]_{4\times 4}$ have the form:

$$[I] = \begin{bmatrix} \lambda + 2\mu & \kappa + 2\chi & \lambda & \kappa & \lambda & \kappa \\ \kappa + 2\chi & \beta + 2\gamma & \kappa & \beta & \kappa & \beta \\ \lambda & \kappa & \lambda + 2\mu & \kappa + 2\chi & \lambda & \kappa \\ \kappa & \beta & \kappa + 2\chi & \beta + 2\gamma & \kappa & \beta \\ \lambda & \kappa & \lambda & \kappa & \lambda + 2\mu & \kappa + 2\chi \\ \kappa & \beta & \kappa & \beta & \kappa + 2\chi & \beta + 2\gamma \end{bmatrix},$$

$$[II] = \begin{bmatrix} \mu + \alpha & \mu - \alpha & \chi + \nu_0 & \chi - \nu_0 \\ \mu - \alpha & \mu + \alpha & \chi - \nu_0 & \chi + \nu_0 \\ \chi + \nu_0 & \chi - \nu_0 & \gamma + \varepsilon & \gamma - \varepsilon \\ \chi - \nu_0 & \chi + \nu_0 & \gamma - \varepsilon & \gamma + \varepsilon \end{bmatrix},$$

$$[III] = \begin{bmatrix} \mu + \alpha & \chi + \nu_0 & \mu - \alpha & \chi - \nu_0 \\ \chi + \nu_0 & \gamma + \varepsilon & \chi - \nu_0 & \gamma - \varepsilon \\ \mu - \alpha & \chi - \nu_0 & \mu + \alpha & \chi + \nu_0 \\ \chi - \nu_0 & \gamma - \varepsilon & \chi + \nu_0 & \gamma + \varepsilon \end{bmatrix},$$

$$[IV] = \begin{bmatrix} \mu + \alpha & \chi + \nu_0 & \chi - \nu_0 & \mu - \alpha \\ \chi + \nu_0 & \gamma + \varepsilon & \gamma - \varepsilon & \chi - \nu_0 \\ \chi - \nu_0 & \gamma - \varepsilon & \gamma + \varepsilon & \chi + \nu_0 \\ \mu - \alpha & \chi - \nu_0 & \chi + \nu_0 & \mu + \alpha \end{bmatrix}.$$

Now, using the Sylvester theorem, we have, for every $i = 1, 2, \ldots, 18$, that $\Delta_i > 0$. Moreover, we obtain the following inequalities:

$$
\begin{aligned}
&\alpha > 0, \qquad \mu > 0, \qquad \gamma > 0, \qquad \varepsilon > 0, \qquad \lambda + 2\mu > 0,\\
&(\lambda + 2\mu)(\beta + 2\gamma) - (\kappa + 2\chi)^2 > 0,\\
&(\lambda + \mu)[\mu(\beta + 2\gamma) - \chi^2] - 2\mu\kappa(\kappa + \chi) > 0,\\
&\alpha\mu(\gamma + \varepsilon) - \mu\nu_0^2 - \alpha\chi^2 > 0,\\
&(\alpha\varepsilon - \nu_0^2)(\mu\gamma - \chi^2) > 0,\\
&(\mu + \alpha)(\gamma + \varepsilon) - (\chi + \nu_0)^2 > 0,\\
&\gamma\varepsilon(\mu + \alpha) - \nu_0^2(\gamma + \varepsilon) > 0.
\end{aligned}
\tag{0.17}
$$

Now, the form of free energy given by (0.16) and relations (0.13) imply the constitutive relations in $\Omega \times \mathbf{T}$,

$$
\begin{aligned}
\sigma_{ji} &= (\mu + \alpha)\gamma_{ji} + (\mu - \alpha)\gamma_{ij} + (\lambda\gamma_{kk} - \eta\theta)\delta_{ji}\\
&\quad + (\chi + \nu_0)\kappa_{ji} + (\chi - \nu_0)\kappa_{ij} + \kappa\kappa_{kk}\delta_{ji},\\
\mu_{ji} &= (\chi + \nu_0)\gamma_{ji} + (\chi - \nu_0)\gamma_{ij} + \kappa\gamma_{kk}\delta_{ji}\\
&\quad + (\gamma + \varepsilon)\kappa_{ji} + (\gamma - \varepsilon)\kappa_{ij} + (\beta\kappa_{kk} - \zeta\theta)\delta_{ji}.
\end{aligned}
\tag{0.18}
$$

Notice that in contrast to a micropolar centro-symmetric medium, a hemitropic medium has the property that the fields of force-stresses and couple-stresses depend both on the strain tensor and the torsion-flexure tensor and, moreover, the couple-stress field depends on the temperature field. The heat conduction equation (0.5) loses its classical character and, in addition to the dilatation term $\gamma_{kk} = u_{i,i}$, the term $\kappa_{kk} = \varphi_{i,i}$ appears. The heat conduction equation for a hemitropic medium has the following form in $\Omega \times \mathbf{T}_+$:

$$
\theta_{,ii} - \frac{1}{\hat{\kappa}}\dot{\theta} - \eta_0\dot{u}_{j,j} - \zeta_0\dot{\varphi}_{j,j} = -\frac{Q}{\hat{\kappa}},
\tag{0.19}
$$

where

$$
Q = \frac{\hat{\kappa}W}{\lambda_0}, \qquad \hat{\kappa} = \frac{\lambda_0}{c_\varepsilon}, \qquad \eta_0 = \frac{\eta T_0}{\lambda_0}, \qquad \zeta_0 = \frac{\zeta T_0}{\lambda_0}.
$$

The momentum and angular momentum conservation principles give the same equations of motion (0.1) as in the case of a micropolar centro-symmetric medium. Similarly, the geometric relations (0.3) defining the strain tensor and the micro-strain tensor, the compatibility equations (0.2), the boundary conditions (0.9) and the initial conditions (0.10) remain unchanged. The equations of motion (0.1), expressed in terms of displacements and rotations, have a more complicated form than the analogous equations (0.6) for a micropolar centro-symmetric medium, as may be seen from the following equations in $\Omega \times \mathbf{T}_+$:

$$
\begin{aligned}
&(\mu+\alpha)u_{i,jj}+(\lambda+\mu-\alpha)u_{j,ji}+2\alpha\epsilon_{ijk}\varphi_{k,j}\\
&\qquad+\underline{(\chi+\nu_0)\varphi_{i,ll}}+\underline{(\chi+\kappa-\nu_0)\varphi_{j,ji}}+X_i=\rho\ddot{u}_i+\eta\theta_{,i},\\
&(\gamma+\varepsilon)\varphi_{i,jj}-4\alpha\varphi_i+(\beta+\gamma-\varepsilon)\varphi_{j,ji}+2\alpha\epsilon_{ijk}u_{k,j}+\underline{4\nu_0\epsilon_{ijk}\varphi_{k,j}}\\
&\qquad+\underline{(\chi+\nu_0)u_{i,ll}}+\underline{(\chi+\kappa-\nu_0)u_{j,ji}}+Y_i=J\ddot{\varphi}_i+\underline{\zeta\theta_{,i}},
\end{aligned}
\tag{0.20}
$$

where the underlined terms result from rejecting the postulate of centro-symmetry for a micropolar medium. The aim of the theory is to determine the thermo-elastodynamical state (0.7) of the medium

$$\Omega(\mu,\lambda,\alpha,\beta,\gamma,\varepsilon,\rho,J,\kappa,\chi,\nu_0,\hat{\kappa},\eta_0,\zeta_0,\eta,\zeta)$$

in the set $\overline{\Omega}\times\mathbf{T}$, corresponding to the equations and field relations (0.1)–(0.3) and (0.18)–(0.20) (with given body forces and moments and the heat sources $X_i, Y_i, Q:\Omega\times\mathbf{T}\to\mathbb{R}$), and to the given initial-boundary conditions. It means that one has to determine the regular solution (0.8) of the equations of motion in displacements-rotations (0.20), and of the heat conduction equation (0.19) with the boundary conditions (0.9) and the initial conditions (0.10). Assuming, in the theory of a hemitropic medium, that $\chi=0$, $\kappa=0$, $\nu_0=0$, $\zeta=0$, $\eta=\nu_T$, $\hat{\kappa}=\kappa$, we obtain from (0.16)–(0.20) equations and relations (0.4)–(0.6), (0.12) and (0.15) for a centro-symmetric micropolar medium (E-N).

The model of couple-stress medium of Koiter-Mindlin (K-M)

In the sequel, we shall also consider the simplified couple-stress theory, which considers the simplest model of an elastic body capable of transferring couple-stresses. In the literature of the subject, such a second order medium with constrained rotations is referred to as the Cosserats pseudo-continuum, the Koiter-Mindlin model, or the Grioli-Toupin model. Couple-stress theory is a well developed theory. A number of general theorems have been proved, integration methods of the fundamental equations have been developed, and solutions to a great number of specific problems have been given. One should mention here such authors as Günther [104], Koiter [131], Mindlin [157], Savin [207], Schaefer [210], Bogy & Sternberg [15], Muki & Sternberg [161], Sokołowski [215], Truesdell & Noll [231], Toupin [230] and Mindlin & Tiersten [158]. An essential issue of that theory is the existence of two additional elasticity constants which modify substantially the solutions of a number of basic problems of the classical theory of elasticity. We shall briefly discuss the fundamental equations of couple-stress theory (in the case of the so-called reduced formulation), arranging them in the groups described below. From the Euler equations of motion (the momentum and the angular momentum conservation principles) we obtain the equations of motion (0.1) (for $J \equiv 0$). Now, we asume that the stress tensors have the form

$$\sigma_{ji} = s_{ji} + \overset{a}{\sigma}_{ji}, \qquad \mu_{ji} = \hat{\mu}\delta_{ji} + m_{ji},$$

where

$$s_{ji} = \tfrac{1}{2}(\sigma_{ji} + \sigma_{ij}), \qquad \overset{a}{\sigma}_{ji} = \tfrac{1}{2}(\sigma_{ji} - \sigma_{ij}), \qquad \hat{\mu} = \tfrac{1}{3}\mu_{ii}$$

and we get the reduced equation of motion in $\Omega \times \mathbf{T}_+$,

$$s_{ji,j} - \tfrac{1}{2}\epsilon_{kji}(m_{lk,lj} + Y_{k,j}) + X_i = \rho\ddot{u}_i, \tag{0.21}$$

where

$$\begin{aligned} &m_{ji},\ u_i \in C^2(\Omega \times \mathbf{T}_+),\ s_{ji} \in C^1(\Omega \times \mathbf{T}_+), \\ &Y_i \in C^1(\Omega \times \mathbf{T}),\ X_i \in C^0(\Omega \times \mathbf{T}). \end{aligned}$$

In the system of equations (0.21) only the symmetric components s_{ji} of the force-stress tensor and the deviatoric components m_{ji} of the couple-stress tensor appear explicitly. In the theory, we have the symmetric strain tensor ε_{ji} and the asymmetric micro-strain tensor κ_{ji}. These tensors are determined by the geometric relations in $\Omega \times \mathbf{T}_+$,

$$\varepsilon_{ji} = \tfrac{1}{2}(u_{i,j} + u_{j,i}), \qquad \kappa_{ji} = \varphi_{j,i}. \tag{0.22}$$

The tensors ε_{ji}, κ_{ji} have to satisfy some geometric relations of strain consistency (the compatibility equations). Every point of couple-stress medium has three independent degrees of freedom, namely the three displacements u_i. The rotations φ_i depend on the displacements u_i and are given by the following classical formula:

$$\varphi_i = \tfrac{1}{2}\epsilon_{ijk}u_{k,j} \qquad \text{in } \overline{\Omega} \times \mathbf{T}. \tag{0.23}$$

The expression for the free energy has the form

$$F = \mu\varepsilon_{ij}\varepsilon_{ij} + \frac{\lambda}{2}\varepsilon_{ll}\varepsilon_{kk} + 2\mu l^{\star 2}(\kappa_{ij}\kappa_{ij} + \eta\kappa_{ij}\kappa_{ji}) - \nu_T\varepsilon_{ij}\theta + G(\theta), \tag{0.24}$$

where $G(\theta)$ denotes all the terms depending only on the temperature.

The temperature θ satisfies the classical equation of heat conduction (0.5). The form of free energy given by (0.24), and the relations between stresses and strains

$$s_{ji} = \frac{\partial F}{\partial \varepsilon_{ji}}, \qquad m_{ji} = \frac{\partial F}{\partial \kappa_{ji}} \tag{0.25}$$

imply the constitutive relations in $\Omega \times \mathbf{T}$,

$$\begin{aligned} s_{ji} &= 2\mu\varepsilon_{ji} + (\lambda\varepsilon_{kk} - \nu_T\theta)\delta_{ji}, \\ m_{ji} &= 4\mu l^{\star 2}(\kappa_{ji} + \eta\kappa_{ij}). \end{aligned} \tag{0.26}$$

The values μ, λ, $l^\star$, η are the elasticity constants of couple-stress theory satisfying the following constraints:

$$\mu > 0, \qquad \lambda > 0, \qquad l^\star \geq 0, \qquad |\eta| < 1. \tag{0.27}$$

Using the constitutive relations (0.26) and the geometric relations (0.22) and (0.23), we obtain from (0.21) the following equations of motion in terms of displacements in $\Omega \times \mathbf{T}_+$,

$$\mu u_{i,jj} + (\lambda + \mu) u_{j,ji} + \mu l^{\star 2} \epsilon_{ijk} \epsilon_{knp} u_{p,njll} + F_i = \rho \ddot{u}_i + \nu_T \theta_{,i}, \tag{0.28}$$

where $F_i = X_i + \frac{1}{2}\epsilon_{ijk} Y_{k,j}$, $u_i \in C^4(\Omega \times \mathbf{T}_+)$.

In the framework of couple-stress theory one has to determine the thermoelastodynamical state

$$u_i,\ \varphi_i,\ \theta,\ s_{ji},\ m_{ji},\ \varepsilon_{ji},\ \kappa_{ji} : \overline{\Omega} \times \mathbf{T} \to \mathbb{R}$$

of the medium $\Omega(\mu, \lambda, l^\star, \eta, \rho, \kappa, \eta_0, \nu_T)$ in the set $\overline{\Omega} \times \mathbf{T}$, which corresponds to the equations and field relations (0.21)–(0.28) (with given body loadings and heat source F_i, $Q : \Omega \times \mathbf{T} \to \mathbb{R}$), the boundary conditions on $\partial\Omega \times \mathbf{T}$, and the initial conditions in $\Omega \times \{0\}$. The presented formulation is equivalent to the one in which one has to determine the solution (u_i, θ) of the displacements equations (0.28) and the heat conduction equation (0.5) with properly formulated initial-boundary conditions. In view of the future considerations related to the thermoelastostatics of the theory, let us consider the boundary conditions of couple-stress theory.

The structure of the displacement equations (0.28) does not correspond to six but to five boundary conditions, whose final form is given below:

— the boundary conditions on $\partial\Omega \times \mathbf{T}$,

$$\begin{aligned}
&\bar{p}_i = \big[s_{ji} + \tfrac{1}{2}\epsilon_{jli}(m_{pl,p} - \overset{n}{m}_{,l} + Y_l)\big] n_j && \text{on } \partial\Omega_\sigma \times \mathbf{T},\\
&\bar{m}_i = m_{ji} n_j - \overset{n}{m}\, n_i && \text{on } \partial\Omega_\sigma \times \mathbf{T},\\
&u_i = f_i, \qquad \overset{0}{\varphi}_\alpha = g_\alpha && \text{on } \partial\Omega_u \times \mathbf{T},\\
&\theta = \vartheta && \text{on } \partial\Omega_\theta \times \mathbf{T},\\
&\frac{\partial\theta}{\partial n} = q && \text{on } \partial\Omega_q \times \mathbf{T},
\end{aligned} \tag{0.29}$$

where

$$\bar{p}_i = p_i - \tfrac{1}{2}\epsilon_{jli} \overset{n}{m}{}^0_{,l} n_j, \qquad \bar{m}_i = m_i - \overset{n}{m}{}^0 n_i$$

are appropriately reduced surface forces and moments, and

$$\overset{n}{m}{}^0 = m_i n_i, \qquad \overset{n}{m} = m_{ji} n_j n_i = m_{ll}, \qquad \overset{0}{\varphi}_\alpha \quad (\alpha = 1,\ 2)$$

are the components of the rotation vector which are tangent to $\partial\Omega$. The functions p_i, $m_i : \partial\Omega_\sigma \times \mathbf{T} \to \mathbb{R}$, f_i, $g_\alpha : \partial\Omega_u \times \mathbf{T} \to \mathbb{R}$, $\vartheta : \partial\Omega_\theta \times$

$\mathbf{T} \to \mathbb{R}$ and $q : \partial\Omega_q \times \mathbf{T} \to \mathbb{R}$ are given. The footnotes (4, 5), related to the boundary conditions (0.9), apply also to (0.29) and to the classical boundary conditions (0.48) formulated in the sequel.

Hypothetical medium (HM)

In the present work we shall also consider a hypothetical medium, in which the effects will appear in the form of the following physical fields: the rotations φ_i, the micro-strain deformation κ_{ji}, and the couple-stresses μ_{ji}. The tensors κ_{ji}, μ_{ji} are asymmetric. The deformation of a hypothetical medium is caused by the body moments Y_i, the distortions κ_{ji}^0, the surface moments m_i and the surface rotations $\hat{\varphi}_i$. Let us arrange the fundamental equations of a hypothetical medium in the following groups:

— the equations of motion in $\Omega \times \mathbf{T}_+$,

$$\mu_{ji,j} + Y_i = J\ddot{\varphi}_i, \tag{0.30}$$

where $\mu_{ji} \in C^1(\Omega \times \mathbf{T}_+)$, $Y_i \in C^0(\Omega \times \mathbf{T})$, $\varphi_i \in C^2(\Omega \times \mathbf{T}_+)$. The geometric relations in $\Omega \times \mathbf{T}_+$ determine the micro-strain tensor

$$\kappa_{ji} = \varphi_{i,j}, \tag{0.31}$$

which has to satisfy the compatibility conditions in $\Omega \times \mathbf{T}$ (one obtains them from (0.2); $\kappa_{ji} \in C^1(\Omega \times \mathbf{T})$). The expression for free energy has now the form

$$F = \frac{\gamma + \varepsilon}{2}\kappa_{ji}\kappa_{ji} + \frac{\gamma - \varepsilon}{2}\kappa_{ji}\kappa_{ij} + \frac{\beta}{2}\kappa_{kk}\kappa_{nn} \tag{0.32}$$

and, by using relation $(0.13)_2$, we get the constitutive relations in $\Omega \times \mathbf{T}$,

$$\mu_{ji} = (\gamma + \varepsilon)\kappa_{ji} + (\gamma - \varepsilon)\kappa_{ij} + \beta\kappa_{kk}\delta_{ji}, \tag{0.33}$$

where γ, ε, β are the elasticity constants satisfying the inequalities

$$\gamma > 0, \qquad \varepsilon > 0, \qquad 2\gamma + 3\beta > 0. \tag{0.34}$$

Using the constitutive relations (0.33) and the geometric relations (0.31), we obtain from the equations of motion (0.30) (the compatibility equations being identically satisfied):

— the rotation equations in $\Omega \times \mathbf{T}_+$,

$$(\gamma + \varepsilon)\varphi_{i,jj} - 4\alpha\varphi_i + (\beta + \gamma - \varepsilon)\varphi_{j,ji} + Y_i = J\ddot{\varphi}_i. \tag{0.35}$$

For a hypothetical medium one has to determine the elastodynamical state φ_i, μ_{ji}, κ_{ji} : $\overline{\Omega} \times \mathbf{T} \to \mathbb{R}$ of the medium $\Omega(\beta, \gamma, \varepsilon, J)$ in the set $\overline{\Omega} \times \mathbf{T}$, corresponding to the equations and field relations (0.30)–(0.35) (with prescribed body moments $Y_i : \Omega \times \mathbf{T} \to \mathbb{R}$) and given initial-boundary conditions. It means that one has to determine a regular solution $\varphi_i \in C^2(\Omega \times \mathbf{T}) \cap C^1(\overline{\Omega} \times \mathbf{T})$ of the rotation equation (0.35) with the following initial and boundary conditions:

— the boundary conditions on $\partial\Omega \times \mathbf{T}$,

$$\begin{aligned} \mu_{ji} n_j &= m_i &\quad& \text{on } \partial\Omega_\sigma \times \mathbf{T}, \\ \varphi_i &= g_i && \text{on } \partial\Omega_u \times \mathbf{T}, \end{aligned} \tag{0.36}$$

where $\partial\Omega = \partial\Omega_\sigma \cup \partial\Omega_u$, $\partial\Omega_\sigma \cap \partial\Omega_u = \emptyset$;

— the initial conditions in $\Omega \times \{0\}$,

$$\varphi_i = k_i, \qquad \dot{\varphi}_i = \chi_i, \tag{0.37}$$

where $m_i : \partial\Omega_\sigma \times \mathbf{T} \to \mathbb{R}$, $g_i : \partial\Omega_u \times \mathbf{T} \to \mathbb{R}$ and k_i, $\chi_i : \Omega \to \mathbb{R}$ are some given functions.

Classical model of Hooke's medium (H)

The micropolar theory of elasticity is a generalization of the classical theory, thus, in the sequel, we shall also consider basic classical relations and differential equations, arranged in the following groups. The principles of conservation of momentum and angular momentum lead to the following equations of motion in $\Omega \times \mathbf{T}_+$:

$$\sigma_{ji,j} + X_i = \rho \ddot{u}_i \tag{0.38}$$

and the symmetry of the stress tensor σ_{ji}. Here, $X_i \in C^0(\Omega \times \mathbf{T})$, $\sigma_{ji} \in C^1(\Omega \times \mathbf{T}_+)$ and $u_i \in C^2(\Omega \times \mathbf{T}_+)$.

The symmetric strain tensor is determined by the geometric relations in $\Omega \times \mathbf{T}_+$,

$$\varepsilon_{ij} = \tfrac{1}{2}(u_{i,j} + u_{j,i}) \tag{0.39}$$

and it has to satisfy the Saint-Venant compatibility conditions[8] in $\Omega \times \mathbf{T}$,

$$\varepsilon_{ij,kl} + \varepsilon_{kl,ij} - \varepsilon_{ik,jl} - \varepsilon_{jl,ik} = 0, \tag{0.40}$$

where $\varepsilon_{ij} \in C^2(\Omega \times \mathbf{T})$, $u_i \in C^3(\Omega \times \mathbf{T})$.

[8] The compatibility conditions may be derived by the method proposed by Cesáro (cf. Nowacki [180, p. 36]), or from the considerations concerning the Riemann-Christoffel tensor in the Euclidean space (cf. Lysik [147, p. 80]). Equations (0.40) may also be derived from (0.2).

In the classical theory, the field of rotations is determined from the relation in $\overline{\Omega} \times \mathbf{T}$,

$$\varphi_i = \tfrac{1}{2}\epsilon_{ijk}u_{k,j}, \tag{0.41}$$

where $\varphi_i \in C^2(\Omega \times \mathbf{T}_+)$.

The free energy in now expressed by

$$F = \mu\varepsilon_{ij}\varepsilon_{ij} + \frac{\lambda}{2}\varepsilon_{kk}\varepsilon_{nn} - \gamma\varepsilon_{kk}\theta - \frac{c_\varepsilon}{2T_0}\theta^2 \tag{0.42}$$

and, taking into account the relation

$$\sigma_{ji} = \frac{\partial F}{\partial \varepsilon_{ij}}, \tag{0.43}$$

we obtain the (Duhamel-Neumann) constitutive conditions in $\Omega \times \mathbf{T}$:

$$\sigma_{ij} = 2\mu\varepsilon_{ij} + (\lambda\varepsilon_{kk} - \gamma\theta)\delta_{ij}. \tag{0.44}$$

The temperature θ satisfies equation (0.5) (which we write here using different symbols than those usually applied in the literature):

$$\theta_{,ii} - \frac{1}{\kappa}\dot{\theta} - \eta\dot{\varepsilon}_{kk} = -\frac{Q}{\kappa}, \tag{0.45}$$

where

$$Q = \frac{\kappa W}{\lambda_0}, \qquad \kappa = \frac{\lambda_0}{c_\varepsilon}, \qquad \eta = \frac{\gamma T_0}{\lambda_0}, \qquad \theta \in C^2(\Omega \times \mathbf{T}_+),$$

λ_0 is the heat conduction coefficient, $\gamma = (2\mu + 3\lambda)\alpha_t$ and μ, λ are the Lamé constants for the isothermal process.

The following inequalities are satisfied:

$$\mu > 0, \qquad \lambda > 0, \qquad \rho > 0, \qquad \kappa > 0, \qquad \frac{\gamma}{\eta} > 0. \tag{0.46}$$

The equations of motion (0.38) with the constitutive relations (0.44) and the geometric relations (0.39) lead to the displacement equations of coupled thermoelasticity (the compatibility equations (0.40) are satisfied identically)

$$\mu u_{i,jj} + (\lambda + \mu)u_{j,ji} + X_i = \rho\ddot{u}_i + \gamma\theta_{,i} \qquad \text{in } \Omega \times \mathbf{T}_+. \tag{0.47}$$

The basic aim of classical thermoelasticity is the determination of the thermo-elastodynamical state u_i, φ_i, θ, σ_{ji}, $\varepsilon_{ji} : \overline{\Omega} \times \mathbf{T} \to \mathbb{R}$ of the medium $\Omega(\mu, \lambda, \rho, \kappa, \eta, \gamma)$ in the set $\overline{\Omega} \times \mathbf{T}$, corresponding to relations (0.38)–(0.47) (with given body forces $X_i : \Omega \times \mathbf{T} \to \mathbb{R}$ and heat sources $Q : \Omega \times \mathbf{T} \to \mathbb{R}$) and given initial-boundary conditions. It means that one has to determine a regular solution (u_i, θ) of the displacement equations of motion (0.47) and the heat conduction equation (0.45) with the following initial-boundary conditions:

— the boundary conditions on $\partial\Omega \times \mathbf{T}$,

$$\begin{aligned} &\sigma_{ji} n_j = p_i && \text{on } \partial\Omega_\sigma \times \mathbf{T}, \\ &u_i = f_i && \text{on } \partial\Omega_u \times \mathbf{T}, \\ &\theta = \vartheta, && \text{on } \partial\Omega_\theta \times \mathbf{T}, \\ &\frac{\partial\theta}{\partial n} = q, && \text{on } \partial\Omega_q \times \mathbf{T}; \end{aligned} \tag{0.48}$$

— the initial conditions in $\Omega \times \{0\}$,

$$u_i = h_i, \qquad \theta = l, \qquad \dot{u}_i = \psi_i, \tag{0.49}$$

where the functions $p_i : \partial\Omega_\sigma \times \mathbf{T} \to \mathbb{R}$, $f_i : \partial\Omega_u \times \mathbf{T} \to \mathbb{R}$, $\vartheta : \partial\Omega_\theta \times \mathbf{T} \to \mathbb{R}$, $q : \partial\Omega_q \times \mathbf{T} \to \mathbb{R}$ and h_i, ψ_i, $l : \Omega \to \mathbb{R}$ are given.

Mathematical methods

The aim of this work is to present the integration methods for the fundamental differential equations of the model of a body of Eringen-Nowacki-type (E-N) and the limiting theories (hemitropic medium of Aero-Kuvshinski (A-K), couple-stress medium of Koiter-Mindlin (K-M), classical Hookean medium (H), hypothetical medium (HM)), within the frameworks of elastostatics and elastodynamics, with the following fields taken into account: body forces, distortions and temperature. We present the methods of the stress equations of Beltrami-Michell type, the stress equations of motion of Ignaczak type, direct integration of the equations of equilibrium and the equations of motion written in terms of displacements and rotations, the stress functions, the generalized potentials (of Galerkin, Iacovache, Papkovich-Neuber, Love), the elastic potentials of Nowacki, and the superposition method based on the generalized Schaefer vector. We shall discuss the applications of those methods in solving certain two-dimensional problems (plane and axially-symmetric states of strain) and some three-dimensional problems (space, half-space, elastic layer). We shall study usefulness of the particular methods of solving the differential equations of the above-mentioned theories to the problem of determining the Green function (taking into account the effects of body forces and body moments, distortions and the temperature field on the body strain). We shall give solutions for a broad class of specific boundary value problems (the generalized problems of Lamb, Kirsch, Goodier, Boussinesq-Mindlin, the first and the second boundary value problems for a three-dimensional elastic half-space, which is hemitropic, micropolar, of couple-stress type, classical, or hypothetical). Attention is paid to the singularities of the physical fields in an elastic half-space caused by the action of a generalized concentrated loading acting on the boundary. We shall also stress the correspondences between the results obtained in various theories. We shall establish the necessary and sufficient conditions for the results of the limiting theories to be obtainable from those of more general theories. We shall also study the limit transitions from time-dependent results to the static problems of a given theory.

The present monograph deals with the differential equations in general. The fundamental equations of the micropolar theory of elasticity and the limiting theories, such as the equations of equilibrium, the equations of motion (expressed in terms of stresses, displacements or rotations), the compatibility equations, and the heat conduction equations are linear differential equations with partial derivatives. The problems of existence and uniqueness of solutions, as well as the continuous dependence of a solution on the initial-boundary conditions, belong to the most difficult ones in every theory. A great number of auxiliary mathematical tools are required, mainly from the theory of potentials and the theory of linear integral equations. The positive answer to the existence problem for the equations of the micropolar theory of elasticity was given by Kupradse [19] and [138] and his school (Burchuladse, Bashelishili, Gegelya). In the present monograph, we apply the mathematical procedure consisting in transforming partial differential equations into linear ordinary differential equations and then we use the theory of ordinary differential equations in order to construct solutions to the boundary- and initial-boundary value problems for the general and simplified micropolar media.

First of all, we shall use the Fourier integral transforms (direct and inverse) of a function $f(\mathbf{x})$,

$$\begin{aligned} \tilde{f}(\boldsymbol{\xi}) &= (2\pi)^{-\frac{n}{2}} \int_{\mathbb{R}^n} f(\mathbf{x}) e^{i\boldsymbol{\xi} \circ \mathbf{x}} d\mathbf{x}, \\ f(\mathbf{x}) &= (2\pi)^{-\frac{n}{2}} \int_{\mathbb{R}^n} \tilde{f}(\boldsymbol{\xi}) e^{-i\boldsymbol{\xi} \circ \mathbf{x}} d\boldsymbol{\xi}, \end{aligned} \tag{0.50}$$

where

$$\mathbf{x} = (x_1, \dots, x_n) \in \mathbb{R}^n, \qquad \boldsymbol{\xi} = (\xi_1, \dots, \xi_n) \in \mathbb{R}^n,$$

$$\boldsymbol{\xi} \circ \mathbf{x} = \xi_1 x_1 + \cdots + \xi_n x_n;$$

and the Hankel transforms of a function $f(r, z)$,

$$\begin{aligned} \tilde{f}_\nu(\xi, z) &= \int_0^\infty r f(r, z) \mathcal{J}_\nu(\xi r)\, dr, \\ f(r, z) &= \int_0^\infty \xi \tilde{f}_\nu(\xi, z) \mathcal{J}_\nu(\xi r)\, d\xi, \end{aligned} \tag{0.51}$$

where $\mathcal{J}_\nu(\xi r)$ is the Bessel function of the first kind and order ν.

We assume the functions f to be regular enough so that the application of the transformations given by (0.50) and (0.51) is mathematically correct (see Sneddon [211, pp. 27 and 298]).

The aim of the monograph

The results presented in this monograph, in addition to their theoretical meaning, can find applications connected with experimental design and verification of the theory. The results obtained, being a basis for some qualitative studies, may also be of fundamental importance for computer computations, provided all the material constants are known[9]. Thus, the results contained in this work may be of interest to many specialists in the area of solid mechanics, mathematicians concerned with the methods of solving partial differential equations and applied and theoretical engineers. In particular, this monograph can serve as a textbook for advanced courses in the mechanics of continuous media with microstructure, taught to graduate students at universities and polytechnics.

The numerical analysis of the obtained results has been carried out on the basis of the work of Kaliski, Kapelewski & Rymarz [125]. We assumed the following values of the material constants:

— micropolar model (E-N):

$$\left(\alpha = \gamma = \varepsilon = 0.1\mu,\ \nu = \tfrac{1}{3}\right) \Longrightarrow (a_0 = 0.067,\ l^2 = 0.55,\ \lambda = 2\mu),$$

— couple-stress model (K-M):

$$(\gamma = \varepsilon = 0.1\mu,\ \nu = \tfrac{1}{3},\ \alpha \to \infty) \Longrightarrow (a_0^\star = 0.067,\ l^{\star 2} = 0.55,\ \lambda = 2\mu,\ \eta = 0), \tag{0.52}$$

— classical model (H):

$$\nu = \tfrac{1}{3},\ \lambda = 2\mu,\ \alpha \to 0.$$

The calculations and graphs were done by means of Mathematica (see [239]).

A great number of papers have been published so far on the subject[10]. They are very important for the applications in engineering, where, in the framework of micropolar theory, some quantitative studies are conducted, concerning the stress distribution in surface girders (discs, plates, shells),

---

[9] Gauthier & Jahsman (1981) gave in [98] the values of four elasticity constants for micropolar medium. This is the most recent result in the experimental research. Micropolar waves have been identified in a specimen of epoxy resins with aluminium shot by use of Kolski's apparatus. The theoretical results of [197] were used to determine the elasticity constants.

[10] The research on that subject was led in Poland by Jemielita [122], Kaliski, Kapelewski & Rymarz [125], Majorkowska-Knap [148], [149], Frydrychowicz, Matysiak & Olesiak [92], [93], Usidus [234], [235].

beams, layers and in an elastic half-space[11]. Nowacki[12] suggested that the papers published in Poland on the subject of Cosserats media are mainly of a theoretical and cognitive character, though the role of applications of couple-stress theory should increase in the future.

[11] Naghdi, P. M., *The theory of Shells and Plates*, in: *Encyclopaedia of Physics*, Vol. VIa/2, Springer-Verlag, Berlin 1972; where on p. 633 the works of the following authors are discussed: A. E. Green, P. M. Naghdi, M. L. Wenner and others, concerning the applications of the Cosserats theory to the theory of shells and plates: (1965) [6], (1967) [4], [5], (1968) [5]–[8], (1971) [3], [5], [6]. See also the works of Woźniak quoted in the monograph [192, p. 378–9].

[12] W. Nowacki, *Autobiographical notes*, PWN, Warszawa (1985), p. 267 (in Polish).

1. Three-dimensional problems

1.1 Displacement-rotation equations of elastodynamics and coupled thermoelasticity

1.1.1 Vector equations. Superposition method

The superposition method for solving the micropolar elasticity boundary value problems was presented in [35], where the first static plane problem was considered. Starting with the problem formulated in stresses and using the equilibrium equations expressed in terms of displacements and rotations we obtain the vector $\boldsymbol{\zeta}(0,\ 0,\ \frac{1}{2}\epsilon_{\alpha\beta}u_{\beta,\alpha} - \varphi_3),\ \ \alpha,\beta = 1,2,\ \ \epsilon_{\alpha\beta}$ is the permutation symbol with two indices. Thus, we solved the boundary value problem in the framework of elastostatics and thermoelasticity and we determined the fundamental solution for the displacements and rotations due to body loadings. Further applications of the superposition method to the solutions of static boundary value problems and the determination of the fundamental solutions are given in [72]–[75] and [43]–[48], [50]–[53] and they refer to two-dimensional problems (plane and axially-symmetric states of strain) and three-dimensional problems with the temperature, body loading and distortion fields taken into account. In later works the superposition method was generalized to the case of dynamical equations of micropolar elasticity.

The vector form of displacement and rotation equations (0.6) in $\Omega \times \mathbf{T}_+$ has been considered in [49] in the form

$$\begin{aligned} &\vec{\Box}_2\mathbf{u} + (\lambda + \mu - \alpha)\,\mathrm{grad}\,\mathrm{div}\,\mathbf{u} + 2\alpha\,\mathrm{curl}\,\boldsymbol{\varphi} + \mathbf{X} = \mathbf{0},\\ &\vec{\Box}_4\boldsymbol{\varphi} + (\beta + \gamma - \varepsilon)\,\mathrm{grad}\,\mathrm{div}\,\boldsymbol{\varphi} + 2\alpha\,\mathrm{curl}\,\mathbf{u} + \mathbf{Y} = \mathbf{0}, \end{aligned} \tag{1.1}$$

where $\mathbf{u}$, $\boldsymbol{\varphi}$, $\mathbf{X}$, $\mathbf{Y}$ are the vectors of displacement, rotation, body force and body moment, respectively. By $\mathbf{0}$ we denote the zero vector. The operators are denoted as follows

$$\vec{\Box}_2 = (\mu + \alpha)\vec{\nabla}^2 - \rho\partial_t^2, \qquad \vec{\Box}_4 = (\gamma + \varepsilon)\vec{\nabla}^2 - 4\alpha - J\partial_t^2,$$

where $\vec{\nabla}^2$ is the vector Laplace operator: $\vec{\nabla}^2 \stackrel{\mathrm{def}}{=} \mathrm{grad}\,\mathrm{div} - \mathrm{curl}\,\mathrm{curl}$ and ∂_t denotes the derivative with respect to time t. In [49], we showed how to separate the system of equations (1.1) by means of the superposition method. Let us formulate the following boundary and initial conditions in displacements and rotations (Cauchy's problem) for the system of equations (1.1):

— the boundary conditions on $\partial\Omega \times \mathbf{T}$,

$$\mathbf{u} = \mathbf{f}, \qquad \boldsymbol{\varphi} = \mathbf{g}; \tag{1.2}$$

— the initial conditions in $\Omega \times \{0\}$,

$$\mathbf{u} = \mathbf{h}, \qquad \boldsymbol{\varphi} = \mathbf{k}, \qquad \dot{\mathbf{u}} = \boldsymbol{\psi}, \qquad \dot{\boldsymbol{\varphi}} = \boldsymbol{\chi}, \tag{1.3}$$

where the vectors $\mathbf{f}$, $\mathbf{g} : \partial\Omega \times \mathbf{T} \to \mathbb{R}^3$ and $\mathbf{h}$, $\mathbf{k}$, $\boldsymbol{\psi}$, $\boldsymbol{\chi} : \Omega \to \mathbb{R}^3$ are given.

Having introduced the Schaefer vector (see [192, p. 171])

$$\boldsymbol{\zeta} = \tfrac{1}{2}\operatorname{curl}\mathbf{u} - \boldsymbol{\varphi}, \tag{1.4}$$

assuming $\boldsymbol{\zeta} = \boldsymbol{\zeta}^0 + \boldsymbol{\zeta}'$, where $\boldsymbol{\zeta}^0 = \mathbf{0}$, and applying the superposition method, we obtain the initial-boundary value problem in the framework of classical elastodynamics. Thus we obtain the system of equations

$$\overset{*}{\overrightarrow{\Box}}_2\mathbf{u}^0 + (\lambda+\mu)\operatorname{grad}\operatorname{div}\mathbf{u}^0 + \mathbf{X} = \mathbf{0} \qquad \text{in } \Omega \times \mathbf{T}_+, \tag{1.5}$$

where $\mathbf{u}^0$ is the classical displacement vector, $\overset{*}{\overrightarrow{\Box}}_2 = \mu\overrightarrow{\nabla}^2 - \rho\partial_t^2$, with the following initial-boundary conditions:

— the boundary conditions on $\partial\Omega \times \mathbf{T}$,

$$\mathbf{u}^0 = \mathbf{f}; \tag{1.6}$$

— the initial conditions in $\Omega \times \{0\}$,

$$\mathbf{u}^0 = \mathbf{h}, \qquad \dot{\mathbf{u}}^0 = \boldsymbol{\psi}. \tag{1.7}$$

Moreover, we obtain the system of equations in $\Omega \times \mathbf{T}_+$,

$$\begin{aligned} &\overrightarrow{\Box}_2\mathbf{u}' + (\lambda+\mu-\alpha)\operatorname{grad}\operatorname{div}\mathbf{u}' + 2\alpha\operatorname{curl}\boldsymbol{\varphi}' + \hat{\mathbf{X}} = \mathbf{0},\\ &\overset{*}{\overrightarrow{\Box}}_2[\overrightarrow{\Box}_4\boldsymbol{\varphi}' + (\beta+\gamma-\varepsilon)\operatorname{grad}\operatorname{div}\boldsymbol{\varphi}' + 2\alpha\operatorname{curl}\mathbf{u}'] + \hat{\mathbf{Y}} = \mathbf{0}. \end{aligned} \tag{1.8}$$

The vectors $\hat{\mathbf{X}}$, $\hat{\mathbf{Y}}$ have the form

$$\hat{\mathbf{X}} = \mathbf{0}, \qquad \hat{\mathbf{Y}} = \overset{*}{\overrightarrow{\Box}}_2\mathbf{Y} - \tfrac{1}{2}\overset{*}{\overrightarrow{\Box}}_4\operatorname{curl}\mathbf{X}, \tag{1.9}$$

where $\overset{*}{\overrightarrow{\Box}}_4 = (\gamma+\varepsilon)\overrightarrow{\nabla}^2 - J\partial_t^2$.

For the system of equations (1.8), the following modified initial-boundary conditions hold:

— the boundary conditions on $\partial\Omega \times \mathbf{T}$,

$$\mathbf{u}' = \mathbf{0}, \qquad \boldsymbol{\varphi}' = \mathbf{g} - \boldsymbol{\varphi}^0; \tag{1.10}$$

— the initial conditions in $\Omega \times \{0\}$,

$$\mathbf{u}' = \mathbf{0}, \qquad \boldsymbol{\varphi}' = \mathbf{k} - \boldsymbol{\varphi}^0, \qquad \dot{\mathbf{u}}' = \mathbf{0}, \qquad \dot{\boldsymbol{\varphi}}' = \boldsymbol{\chi} - \dot{\boldsymbol{\varphi}}^0. \tag{1.11}$$

The vector $\boldsymbol{\varphi}^0$ is obtained from the classical solution of the initial-boundary value problem

$$\boldsymbol{\varphi}^0 = \tfrac{1}{2}\,\mathrm{curl}\,\mathbf{u}^0. \tag{1.12}$$

Thus, the initial-boundary value problem in the framework of micropolar theory (1.1)–(1.3) has been replaced by the classical initial-boundary value problem (1.5)–(1.7) and the complementary initial-boundary value problem (1.8)–(1.12). We also have the following relation:

$$\mathbf{u} = \mathbf{u}^0 + \mathbf{u}', \qquad \boldsymbol{\varphi} = \boldsymbol{\varphi}^0 + \boldsymbol{\varphi}'. \tag{1.13}$$

From the classical equations (1.5) and (1.12) we get

$$\begin{aligned} \overrightarrow{\Box}_1 \overset{*}{\overrightarrow{\Box}}_2 \mathbf{u}^0 &= -\overrightarrow{\Box}_1 \mathbf{X} + (\lambda + \mu)\,\mathrm{grad\,div}\,\mathbf{X}, \\ \overset{*}{\overrightarrow{\Box}}_2 \boldsymbol{\varphi}^0 &= -\tfrac{1}{2}\,\mathrm{curl}\,\mathbf{X}, \end{aligned} \tag{1.14}$$

where $\overrightarrow{\Box}_1 = (\lambda + 2\mu)\overrightarrow{\nabla}^2 - \rho\partial_t^2$. Using equations (1.8), we derive the uncoupled equations:

— for the vector $\mathbf{u}'$,

$$\begin{aligned} &\overset{*}{\overrightarrow{\Box}}_2 \overrightarrow{\Box}_1 (\overrightarrow{\Box}_2 \overrightarrow{\Box}_4 + 4\alpha^2 \overrightarrow{\nabla}^2)\mathbf{u}' \\ &\quad = \overset{*}{\overrightarrow{\Box}}_2 \Big\{ [(\lambda + \mu - \alpha)\overrightarrow{\Box}_4 - 4\alpha^2]\,\mathrm{grad\,div}\,\hat{\mathbf{X}} - \overrightarrow{\Box}_1 \overrightarrow{\Box}_4 \hat{\mathbf{X}} \Big\} \\ &\quad + 2\alpha \overrightarrow{\Box}_1\,\mathrm{curl}\,\hat{\mathbf{Y}}; \end{aligned} \tag{1.15}$$

— for the vector $\boldsymbol{\varphi}'$,

$$\begin{aligned} \overset{*}{\overrightarrow{\Box}}_2 \overrightarrow{\Box}_3 (\overrightarrow{\Box}_2 \overrightarrow{\Box}_4 + 4\alpha^2 \overrightarrow{\nabla}^2)\boldsymbol{\varphi}' &= 2\alpha \overset{*}{\overrightarrow{\Box}}_2 \overrightarrow{\Box}_3\,\mathrm{curl}\,\hat{\mathbf{X}} \\ + [(\beta + \gamma - \varepsilon)\overrightarrow{\Box}_2 &- 4\alpha^2]\,\mathrm{grad\,div}\,\hat{\mathbf{Y}} - \overrightarrow{\Box}_2 \overrightarrow{\Box}_3 \hat{\mathbf{Y}}, \end{aligned} \tag{1.16}$$

where $\overrightarrow{\Box}_3 = (\beta + 2\gamma)\overrightarrow{\nabla}^2 - 4\alpha - J\partial_t^2$.

The equation for the vector $\boldsymbol{\zeta} = \boldsymbol{\zeta}'$ has the following form:

$$\begin{aligned} &2\overset{*}{\overrightarrow{\Box}}_2 \overrightarrow{\Box}_3 (\overrightarrow{\Box}_2 \overrightarrow{\Box}_4 + 4\alpha^2 \overrightarrow{\nabla}^2)\boldsymbol{\zeta} \\ &\quad = [2\alpha \overset{*}{\overrightarrow{\Box}}_4 - 2(\beta + \gamma - \varepsilon)\overset{*}{\overrightarrow{\Box}}_2]\,\mathrm{grad\,div}\,\hat{\mathbf{Y}} \\ &\quad + \overset{*}{\overrightarrow{\Box}}_2 \overrightarrow{\Box}_3 (2\hat{\mathbf{Y}} - \overset{*}{\overrightarrow{\Box}}_4\,\mathrm{curl}\,\hat{\mathbf{X}}). \end{aligned} \tag{1.17}$$

Let us make the following important remark. Namely, having adopted the substitutions

$$(\mathbf{u}', \boldsymbol{\varphi}') \mapsto (\mathbf{u}, \boldsymbol{\varphi}), \qquad (\hat{\mathbf{X}}, \hat{\mathbf{Y}}) \mapsto (\mathbf{X}, \overset{*}{\vec{\Box}}_2 \mathbf{Y}) \tag{1.18}$$

and dividing the equations[1] $(1.8)_2$, (1.15) and (1.16) by the operator[2] $\overset{*}{\vec{\Box}}_2$ we can see that the equation system (1.8) transforms into (1.1), whereas equations (1.15) and (1.16) are transformed into the respective uncoupled equations:

— for the vector $\mathbf{u}$,

$$\begin{aligned}\vec{\Box}_1(\vec{\Box}_2\vec{\Box}_4 + 4\alpha^2\vec{\nabla}^2)\mathbf{u} = {} & [(\lambda+\mu-\alpha)\vec{\Box}_4 - 4\alpha^2]\,\mathrm{grad\,div}\,\mathbf{X} \\ & - \vec{\Box}_1\vec{\Box}_4\mathbf{X} + 2\alpha\vec{\Box}_1\,\mathrm{curl}\,\mathbf{Y};\end{aligned} \tag{1.19}$$

— for the vector $\boldsymbol{\varphi}$,

$$\begin{aligned}\vec{\Box}_3(\vec{\Box}_2\vec{\Box}_4 + 4\alpha^2\vec{\nabla}^2)\boldsymbol{\varphi} = {} & 2\alpha\vec{\Box}_3\,\mathrm{curl}\,\mathbf{X} \\ & + [(\beta+\gamma-\varepsilon)\vec{\Box}_2 - 4\alpha^2]\,\mathrm{grad\,div}\,\mathbf{Y} - \vec{\Box}_2\vec{\Box}_3\mathbf{Y}.\end{aligned} \tag{1.20}$$

We shall use the vector equations (1.1)–(1.20) when we consider the three- and two-dimensional (plane and axially-symmetric) problems. The procedure is discussed in detail in [49].

1.1.2 Fields of body loadings. Fundamental solutions and limiting cases

In [78], we determined the fundamental solution for the displacements and rotations for the case of body loadings acting in the space $\mathbb{R}^3$ and varying harmonically in time. In the rectangular coordinate system $Ox_1x_2x_3$ we have

$$(X_i, Y_i) = (P_0, M_0)e^{-i\omega t}\delta(x_1)\delta(x_2)\delta(x_3)\delta_{is}, \tag{1.21}$$

where $\delta(.)$ is the Dirac delta distribution, ω is the vibration frequency and P_0, M_0 refer to the unit force and the unit moment, respectively. Now, the equations (1.14)–(1.16), written in the system $Ox_1x_2x_3$, have the following forms:

---

1 Throughout the work, the numbers of equations and formulae are sometimes furnished with natural indices, which indicate which equation or formula from a given group is selected. Thus, for instance, $(1.236)_2$ denotes the second fromula from the group of formulae (1.236). In turn, $(1.254)_{3,4}$ denotes the third and the fourth formulae from the group (1.254).

2 Consider, for example, the equation $\vec{\mathrm{L}}_1\vec{\mathrm{L}}_2\mathbf{f} = \mathbf{0}$, where $\vec{\mathrm{L}}_1$ and $\vec{\mathrm{L}}_2$ are certain differential operators. Saying that this equation is divided by $\vec{\mathrm{L}}_1$ (or, that we remove $\vec{\mathrm{L}}_1$), we understand that we have the equation $\vec{\mathrm{L}}_2\mathbf{f} = \mathbf{0}$.

— the classical equations:

$$\square_1 \overset{*}{\square}_2 u_i^0 = -\square_1 X_i + (\lambda+\mu) X_{j,ji}, \qquad 2\overset{*}{\square}_2 \varphi_i^0 + \epsilon_{ijk} X_{k,j} = 0, \quad (1.22)$$

where

$$\square_1 = (\lambda + 2\mu)\nabla^2 - \rho\partial_t^2, \qquad \overset{*}{\square}_2 = \mu\nabla^2 - \rho\partial_t^2;$$

— the complementary equations:

$$\begin{aligned} &\square_1 \overset{*}{\square}_2 (\square_2 \square_4 + 4\alpha^2\nabla^2) u_i' \\ &\quad = \overset{*}{\square}_2 \left\{ [(\lambda+\mu-\alpha)\square_4 - 4\alpha^2] \hat{X}_{j,ji} - \square_1 \square_4 \hat{X}_i \right\} \\ &\qquad + 2\alpha \square_1 \epsilon_{ijk} \hat{Y}_{k,j}, \\ &\square_3 \overset{*}{\square}_2 (\square_2 \square_4 + 4\alpha^2\nabla^2) \varphi_i' \\ &\quad = 2\alpha \overset{*}{\square}_2 \square_3 \epsilon_{ijk} \hat{X}_{k,j} + [(\beta+\gamma-\varepsilon)\square_2 - 4\alpha^2] \hat{Y}_{j,ji} \\ &\qquad - \square_2 \square_3 \hat{Y}_i, \end{aligned} \quad (1.23)$$

where

$$\begin{aligned} &\square_2 = (\mu+\alpha)\nabla^2 - \rho\partial_t^2, \qquad \square_3 = (\beta+2\gamma)\nabla^2 - 4\alpha - J\partial_t^2, \\ &\square_4 = (\gamma+\varepsilon)\nabla^2 - 4\alpha - J\partial_t^2. \end{aligned}$$

Now, the reduced body loadings $\hat{X}_i$, $\hat{Y}_i$ take the form

$$\hat{X}_i = 0, \qquad \hat{Y}_i = \square_2 Y_i - \tfrac{1}{2} \overset{*}{\square}_4 \epsilon_{ijk} X_{k,j}, \quad (1.24)$$

where $\overset{*}{\square}_4 = (\gamma+\varepsilon)\nabla^2 - J\partial_t^2$. One has to remark that the operators $\square_i$ $(i = 1\text{–}4)$, $\overset{*}{\square}_2$ and $\overset{*}{\square}_4$ are determined by the scalar Laplace operator

$$\nabla^2(.) \overset{\text{def}}{=} \operatorname{div}\operatorname{grad}(.) = (.)_{,ii}.$$

Applying the method of integral transformations based on the Fourier transformation (0.50) (for $n = 4$)[3], and using (1.21), (1.24), as well as the formula

$$u_i = u_i^0 + u_i', \qquad \varphi_i = \varphi_i^0 + \varphi_i' \quad (1.25)$$

we obtain from equations (1.22), (1.23) the following formulae for u_i, φ_i:

[3] We use the Fourier transformation with respect to three position variables x_i and the time variable t, which, after performing the inverse transformation, gives general formulae useful for our considerations, not only in the case of harmonic time-dependence (see Nowacki [192, p. 88]).

— for the concentrated body force $(1.21)_1$ ($P_0 \neq 0$, $M_0 = 0$):

$$\begin{aligned} u_i^{(s)} &= \frac{e^{-i\omega t} P_0}{4\pi\rho\omega^2}\left[\left(\lambda_1^2 A_1 \frac{e^{i\lambda_1 R}}{R} + \lambda_2^2 A_2 \frac{e^{i\lambda_2 R}}{R}\right)\delta_{is}\right. \\ &\quad \left. + \left(A_1 \frac{e^{i\lambda_1 R}}{R} + A_2 \frac{e^{i\lambda_2 R}}{R} - \frac{e^{i\sigma_1 R}}{R}\right)_{,is}\right], \\ \varphi_i^{(s)} &= \frac{e^{-i\omega t} P_0}{4\pi\rho c_2^2} \cdot \frac{p}{\lambda_1^2 - \lambda_2^2} \epsilon_{iks} \left(\frac{e^{i\lambda_1 R} - e^{i\lambda_2 R}}{R}\right)_{,k}; \end{aligned} \tag{1.26}$$

— for the concentrated body moment $(1.21)_2$ ($P_0 = 0$, $M_0 \neq 0$) :

$$\begin{aligned} u_i^{(s)} &= \frac{e^{-i\omega t} M_0}{4\pi J c_4^2} \cdot \frac{s}{\lambda_1^2 - \lambda_2^2} \epsilon_{iks} \left(\frac{e^{i\lambda_1 R} - e^{i\lambda_2 R}}{R}\right)_{,k}, \\ \varphi_i^{(s)} &= \frac{e^{-i\omega t} M_0}{4\pi J c_4^2}\left[\left(\lambda_1^2 C_1 \frac{e^{i\lambda_1 R}}{R} + \lambda_2^2 C_2 \frac{e^{i\lambda_2 R}}{R}\right)\delta_{is}\right. \\ &\quad \left. + \left(C_1 \frac{e^{i\lambda_1 R}}{R} + C_2 \frac{e^{i\lambda_2 R}}{R} + C_3 \frac{e^{i\lambda_3 R}}{R}\right)_{,is}\right]. \end{aligned} \tag{1.27}$$

Now, let us discuss the symbols used in the final formulae (1.26), (1.27) for the pair (u_i, φ_i):

$$\sigma_i = \frac{\omega}{c_i}, \qquad c_1 = \sqrt{\frac{\lambda + 2\mu}{\rho}},\ c_2 = \sqrt{\frac{\mu + \alpha}{\rho}},\ c_3 = \sqrt{\frac{\beta + 2\gamma}{J}},\ c_4 = \sqrt{\frac{\gamma + \varepsilon}{J}},$$

$$\lambda_3 = \sqrt{\sigma_3^2 - \tau_0^2}, \qquad \tau_0^2 = \frac{4\alpha}{\beta + 2\gamma}, \qquad R = \sqrt{x_j x_j},$$

$$A_1 = \frac{\sigma_2^2 - \lambda_2^2}{\lambda_1^2 - \lambda_2^2}, \qquad A_2 = \frac{\sigma_2^2 - \lambda_1^2}{\lambda_2^2 - \lambda_1^2},$$

$$C_1 = A_2 \frac{1}{\lambda_1^2}, \qquad C_2 = A_1 \frac{1}{\lambda_2^2}, \qquad C_3 = -\frac{\sigma_2^2}{\lambda_1^2 \lambda_2^2},$$

$$\nu_0^2 = 2p = \frac{4\alpha}{\gamma + \varepsilon}, \qquad s = \frac{2\alpha}{\mu + \alpha}, \qquad \eta_0^2 = ps.$$

The values λ_1^2 and λ_2^2 are the roots of the equation

$$\lambda^4 - \lambda^2[\sigma_2^2 + \sigma_4^2 + p(s-2)] + \sigma_2^2(\sigma_4^2 - 2p) = 0$$

with the determinant

$$\Delta = (\sigma_2^2 - \sigma_4^2 - \eta_0^2 + \nu_0^2)^2 + 4\sigma_2^2\eta_0^2 > 0.$$

The roots λ_1^2, $\lambda_2^2 \in \mathbb{R}$ are different and

$$\lambda_{1,2}{}^2 = \frac{1}{2}\left[\sigma_2^2 + \sigma_4^2 + \eta_0^2 - \nu_0^2 \pm \sqrt{(\sigma_2^2 - \sigma_4^2 - \eta_0^2 + \nu_0^2)^2 + 4\sigma_2^2\eta_0^2}\right]. \quad (1.28)$$

Since we assume only real phase velocities in (1.26) and (1.27), we have to make the assumption[4] $J\omega^2 - 4\alpha > 0$ in order to satisfy the inequalities $\lambda_1^2 > 0$, $\lambda_2^2 > 0$ and $\lambda_3^2 > 0$. Identity of formulae $(1.26)_2$ and $(1.27)_1$ follows from the Betti-Maxwell[5] theorem on the works reciprocity. Formulae (1.26) and (1.27) constitute the solution of equations (0.6) for the body loadings (1.21) and for $\theta = 0$. The equation system (0.6) (for $\theta = 0$) in its uncoupled form will be obtained by use of the vector equations (1.19) and (1.20). Taking into account the fact that $\vec{\nabla}^2 \equiv (\nabla^2, \nabla^2, \nabla^2)$ in the system $Ox_1x_2x_3$, where $\nabla^2(.) = (.)_{,ii}$, we get the equations

$$\begin{aligned} \Box_1(\Box_2\Box_4 + 4\alpha^2\nabla^2)u_i &= [(\lambda + \mu - \alpha)\Box_4 - 4\alpha^2]X_{j,ji} \\ &\quad - \Box_1\Box_4 X_i + 2\alpha\Box_1\epsilon_{ijk}Y_{k,j}, \\ \Box_3(\Box_2\Box_4 + 4\alpha^2\nabla^2)\varphi_i &= 2\alpha\Box_3\epsilon_{ijk}X_{k,j} \\ &\quad + [(\beta + \gamma - \varepsilon)\Box_2 - 4\alpha^2]Y_{j,ji} - \Box_2\Box_3 Y_i, \end{aligned} \quad (1.29)$$

which imply formulae (1.26) and (1.27) for the given forces and moments. When discussing the results (1.26) and (1.27), we put the emphasis on obtaining the basic solutions for the simplified theories, as indicated on the diagram in Fig. 1.1 (double arrow denotes an analogy).

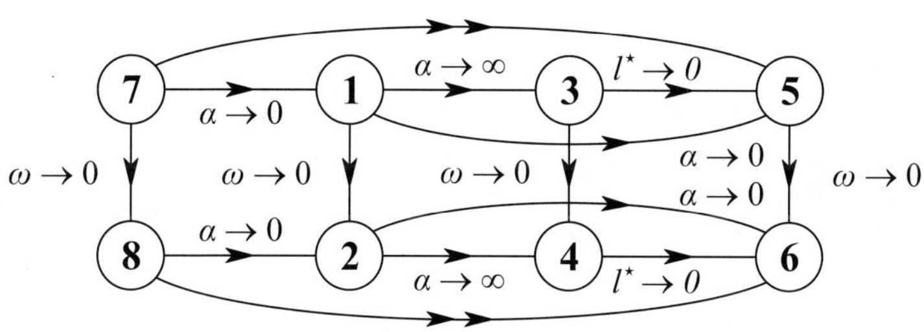

Fig. 1.1. Numbers (1), (2); (3), (4); (5), (6); (7), (8) represent, respectively, micropolar theory (E-N), couple-stress theory (K-M), the classical theory of elasticity (H), and hypothetical medium (HM), where (1), (3), (5), (7) refer to the elastodynamics and (2), (4), (6), (8) — to the elastostatics of the theories mentioned above

The fundamental solutions have been obtained by use of the corresponding equations of the theories (2)–(8), and next, starting from the results of

[4] See Nowacki [192, p. 57] and Kupradse [138, p. 71]. Ignaczak [115] considers a micropolar medium, in which $\lambda_3 = i\hat{\lambda}_3$, where $\hat{\lambda}_3 \in \mathbb{R}$, $i = \sqrt{-1}$.

[5] We assume that $M_0 = P_0 = 1$.

the theory (1), it was proved that the limiting cases of formulae (1.26) and (1.27) (as indicated in Fig. 1.1) correspond to the fundamental solutions for the limiting theories just mentioned. One has to stress that the conditions denoted by $\alpha \to 0$, $\alpha \to \infty$, $l^\star \to 0$, $\omega \to 0$ in the diagram denote limit transitions between theories and they are necessary conditions for establishing the correspondences between the results of micropolar theory (model E-N) and those of the limiting theories. However, they are not sufficient in some cases. A detailed discussion of the limit transitions is given in [78] and [52]. Explicit formulae for the fundamental solutions of the micropolar theory of elasticity were given by Kupradse [138, p. 70], Nowacki & Nowacki [192, p. 94], and Domański, Piskorek & Rojek [29].

1.1.3 Distortion fields. Fundamental solutions

In [49], we derived the uncoupled equations for the pair (u_i, φ_i) by means of the superposition method, with a distortion field taken into account. Assuming in the system $Ox_1x_2x_3$ the following relations for body forces:

$$X_i = -\overset{*}{R}{}_i^0, \qquad \hat{X}_i = -\overset{**}{R}{}_i^0, \qquad Y_i = -M_i^0, \qquad \hat{Y}_i = -\hat{R}_i^0 \tag{1.30}$$

and using equations (1.22), (1.23), we obtain the uncoupled equations for the pairs u_i^0, φ_i^0 and u_i', φ_i'. We obtain:

— the classical equations:

$$\Box_1 \overset{*}{\Box}_2 u_i^0 = \Box_1 \overset{*}{R}{}_i^0 - (\lambda+\mu)\overset{*}{R}{}^0_{j,ji}\,, \qquad 2\overset{*}{\Box}_2 \varphi_i^0 = \epsilon_{ijk}\overset{*}{R}{}^0_{k,j}; \tag{1.31}$$

— the equations for the complementary solution (u_i', φ_i'):

$$\begin{aligned} \Box_1 \overset{*}{\Box}_2(\Box_2\Box_4 + 4\alpha^2\nabla^2)u_i' &= -2\alpha\Box_1\epsilon_{ijk}\hat{R}^0_{k,j} \\ &\quad - \overset{*}{\Box}_2\left\{[(\lambda+\mu-\alpha)\Box_4 - 4\alpha^2]\overset{**}{R}{}^0_{j,ji} - \Box_1\Box_4\overset{**}{R}{}_i^0\right\}, \\ \Box_3 \overset{*}{\Box}_2(\Box_2\Box_4 + 4\alpha^2\nabla^2)\varphi_i' &= -2\alpha\overset{*}{\Box}_2\Box_3\epsilon_{ijk}\overset{**}{R}{}^0_{k,j} \\ &\quad - [(\beta+\gamma-\varepsilon)\Box_2 - 4\alpha^2]\hat{R}^0_{j,ji} + \Box_2\Box_3\hat{R}_i^0. \end{aligned} \tag{1.32}$$

Let us discuss the meaning of the symbols used in the problem with a distortion field. We have introduced the notations

$$R_i^0 = \sigma^0_{ji,j}\,, \qquad M_i^0 = \mu^0_{ji,j} + \epsilon_{ijk}\sigma^0_{jk}, \tag{1.33}$$

where σ^0_{ji}, μ^0_{ji} have the forms of the right-hand sides of the constitutive relations (0.4) (for $\theta = 0$), written for γ^0_{ji}, κ^0_{ji}, which describe the given distortion field. We have

$$R_i^0 = \overset{*}{R}{}_i^0 + \overset{**}{R}{}_i^0, \qquad \hat{R}_i^0 = \overset{*}{\Box}_2 M_i^0 - \tfrac{1}{2}\overset{*}{\Box}_4\epsilon_{ijk}\overset{*}{R}{}^0_{k,j}, \tag{1.34}$$

where

$$\begin{aligned} \overset{*}{R}{}_i^0 &= (2\mu\gamma_{(ji)}^0 + \lambda\gamma_{kk}^0\delta_{ij})_{,j}, \\ \overset{**}{R}{}_i^0 &= 2\alpha\gamma_{[ji],j}^0. \end{aligned} \tag{1.35}$$

In formula (1.35), parentheses (.) and brackets [.] denote the symmetric and antisymmetric parts of a tensor, respectively. Let us note that if we make the substitutions

$$(u_i', \varphi_i') \mapsto (u_i, \varphi_i), \qquad (\overset{**}{R}{}_i^0, \hat{R}_i^0) \mapsto (R_i^0, \overset{*}{\Box}{}_2 M_i^0) \tag{1.36}$$

in equations (1.32) and divide them by the operator $\overset{*}{\Box}{}_2$, we obtain them in the form

$$\begin{aligned} \Box_1(\Box_2\Box_4 + 4\alpha^2\nabla^2)u_i &= [(\lambda+\mu-\alpha)\Box_4 - 4\alpha^2]\sigma_{kj,kji}^0 \\ &\quad + \Box_1\Box_4\sigma_{ji,j}^0 - 2\alpha\Box_1\epsilon_{ijk}(\mu_{nk,n}^0 + \epsilon_{kml}\sigma_{ml}^0)_{,j}, \\ \Box_3(\Box_2\Box_4 + 4\alpha^2\nabla^2)\varphi_i &= [(\beta+\gamma-\varepsilon)\Box_2 - 4\alpha^2](\mu_{nj,n}^0 + \epsilon_{jml}\sigma_{ml}^0)_{,ji} \\ &\quad - 2\alpha\Box_3\epsilon_{ijk}\sigma_{lk,lj}^0 + \Box_2\Box_3(\mu_{ji,j}^0 + \epsilon_{ijk}\sigma_{jk}^0). \end{aligned} \tag{1.37}$$

Equations (1.37) constitute the uncoupled equations for the displacements u_i and the rotations φ_i, the distortion field (in the system $Ox_1x_2x_3$) being taken into account. In equations (0.6) (for $\theta = 0$) and (1.29) we should now substitute

$$X_i = -R_i^0, \qquad Y_i = -M_i^0 \tag{1.38}$$

and (1.33). Moreover, the expressions σ_{ji}, μ_{ji} should be replaced by $\sigma_{ji} + \sigma_{ji}^0$, $\mu_{ji} + \mu_{ji}^0$ in the constitutive relations (0.4) (for $\theta = 0$). In [24] and [25] Czub & Dyszlewicz determined the fundamental solutions for displacements and rotations in the case of concentrated distortions harmonically varying in time γ_{33}^0, κ_{33}^0, and acting in the space $\mathbb{R}^3$, and the limiting cases of the results concerning the limiting theories were discussed. The axially-symmetric character of the problem was also taken into account.

1.1.4 Coupled micropolar thermoelasticity. Fundamental solutions

Let us discuss the superposition method applied to the equations of coupled thermoelasticity. We start from the equation of heat conduction (0.5), the equation of motion expressed in terms of displacements and rotations (0.6), the boundary conditions (0.9) and the initial conditions (0.10). Introducing into equations (0.6) the substitution

$$\zeta_i = \tfrac{1}{2}\epsilon_{ijk}u_{k,j} - \varphi_i, \tag{1.39}$$

we obtain

$$\overset{*}{\Box}_2 u_i + (\lambda+\mu)u_{j,ji} + X_i = 2\alpha\epsilon_{ijk}\zeta_{k,j} + \nu_T\theta_{,i},$$
$$\overset{*}{\Box}_4 \epsilon_{ijk}u_{k,j} + 2Y_i = 2[\Box_4\zeta_i + (\beta+\gamma-\varepsilon)\zeta_{j,ji}]. \tag{1.40}$$

In order to write the solution (u_i, φ_i) in the form (1.25) we assume the decompositions $\zeta_i = \zeta_i^0 + \zeta_i'$, $Y_i = Y_i^0 + Y_i'$, $\theta = \theta^0 + \theta'$. Assuming $\zeta_i^0 = 0$, we arrive at the initial-boundary value problem in the framework of the classical coupled thermoelasticity. Now, using the equations $(1.40)_1$, we obtain:

— the equations of motion in terms of displacements in $\Omega \times \mathbf{T}_+$,

$$\overset{*}{\Box}_2 u_i^0 + (\lambda+\mu)u_{j,ji}^0 + X_i = \nu_T\theta_{,i}. \tag{1.41}$$

Equations (1.41) are coupled with equation (0.5), which takes here, for u_i^0, θ^0, the form of

— the equation of heat conduction in $\Omega \times \mathbf{T}_+$,

$$D\theta^0 - \eta_0\dot{u}_{j,j}^0 = -\frac{Q}{\kappa}, \tag{1.42}$$

where the operator $D \overset{\text{def}}{=} \nabla^2 - \frac{1}{\kappa}\partial_t$.

Using the conditions (0.9) and (0.10) we get:

— the boundary conditions on $\partial\Omega \times \mathbf{T}$,

$$\begin{aligned} \sigma_{ji}^0 n_j &= p_i && \text{on } \partial\Omega_\sigma \times \mathbf{T},\\ u_i^0 &= f_i && \text{on } \partial\Omega_u \times \mathbf{T},\\ \theta^0 &= \vartheta && \text{on } \partial\Omega_\theta \times \mathbf{T},\\ \frac{\partial\theta^0}{\partial n} &= q && \text{on } \partial\Omega_q \times \mathbf{T}, \end{aligned} \tag{1.43}$$

where σ_{ji}^0 is the classical stress tensor;

— the initial conditions in $\Omega \times \{0\}$,

$$u_i^0 = h_i, \qquad \theta^0 = l, \qquad \dot{u}_i^0 = \psi_i. \tag{1.44}$$

In further considerations, the system of equations (1.41) and (1.42) will be uncoupled and written in the form

$$D_2\overset{*}{\Box}_2 u_i^0 = D_1 X_{j,ji} - D_2 X_i - \frac{\nu_T}{\kappa}\overset{*}{\Box}_2 Q_{,i},$$
$$D_2\theta^0 = -\eta_0\dot{X}_{j,j} - \frac{1}{\kappa}\Box_1 Q, \tag{1.45}$$

where

$$D_1 \stackrel{\text{def}}{=} (\lambda + \mu)D - \eta_0 \nu_T \partial_t, \qquad D_2 \stackrel{\text{def}}{=} D\Box_1 - \eta_0 \nu_T \partial_t \nabla^2.$$

Next, we must apply the operator $\overset{*}{\Box}_2$ to equation $(1.40)_2$. Then we arrive at the equation

$$\overset{*}{\Box}_2(\overset{*}{\Box}_4 \epsilon_{ijk} u_{k,j} + 2Y_i) = 2\overset{*}{\Box}_2[\Box_4 \zeta_i + (\beta + \gamma - \varepsilon)\zeta_{j,ji}]. \tag{1.46}$$

Now, using (1.46) (for $\zeta_i^0 = 0$ and $Y_i^0 = 0$), we get the equation

$$\overset{*}{\Box}_4(\overset{*}{\Box}_2 \epsilon_{ijk} u^0_{k,j} + \epsilon_{ijk} X_{k,j}) = 0, \tag{1.47}$$

which will be satisfied identically in view of

$$2\overset{*}{\Box}_2 \varphi_i^0 + \epsilon_{ijk} X_{k,j} = 0, \tag{1.48}$$

implied by (1.41), where φ_i^0 are the classical rotations. From equations $(1.40)_1$ and (1.46) and the heat conduction equation (0.5) we obtain for the thermomechanical triple (u_i', ζ_i, θ') the following system of equations:

$$\begin{aligned} &\overset{*}{\Box}_2 u_i' + (\lambda + \mu) u'_{j,ji} + \hat{X}_i = 2\alpha \epsilon_{ijk} \zeta_{k,j} + \nu_T \theta'_{,i}, \\ &\overset{*}{\Box}_2 \left\{ \overset{*}{\Box}_4 \epsilon_{ijk} u'_{k,j} - 2[\Box_4 \zeta_i + (\beta + \gamma - \varepsilon)\zeta_{j,ji}] \right\} + 2\hat{Y}_i = 0, \\ &D\theta' - \eta_0 \dot{u}'_{j,j} = -\frac{\hat{Q}}{\kappa}, \end{aligned} \tag{1.49}$$

where it holds

$$\hat{X}_i = 0, \qquad \hat{Y}_i \stackrel{\text{def}}{=} \overset{*}{\Box}_2 Y_i - \tfrac{1}{2}\overset{*}{\Box}_4 \epsilon_{ijk} X_{k,j}, \qquad \hat{Q} = 0. \tag{1.50}$$

Using the system of equations (1.49), we derive a separate equation for the function ζ_i, namely

$$\begin{aligned} 2\overset{*}{\Box}_2 \Box_3 (\Box_2 \Box_4 + 4\alpha^2 \nabla^2)\zeta_i = {} & [2\alpha \overset{*}{\Box}_4 - 2(\beta + \gamma - \varepsilon)\overset{*}{\Box}_2]\hat{Y}_{j,ji} \\ & + \overset{*}{\Box}_2 \Box_3 (2\hat{Y}_i - \overset{*}{\Box}_4 \epsilon_{ijk} \hat{X}_{k,j}). \end{aligned} \tag{1.51}$$

Stating the initial-boundary value problem for the complementary solution $(u_i', \varphi_i', \theta')$, we eliminate the function ζ_i from the system of equations (1.49). Hence, we obtain

$$\begin{aligned} &\Box_2 u_i' + (\lambda + \mu - \alpha) u'_{j,ji} + 2\alpha \epsilon_{ijk} \varphi'_{k,j} + \hat{X}_i = \nu_T \theta'_{,i}, \\ &\overset{*}{\Box}_2[\Box_4 \varphi_i' + (\beta + \gamma - \varepsilon)\varphi'_{j,ji} + 2\alpha \epsilon_{ijk} u'_{k,j}] + \hat{Y}_i = 0, \\ &D\theta' - \eta_0 \dot{u}'_{j,j} = -\frac{\hat{Q}}{\kappa} \end{aligned} \tag{1.52}$$

with the following modified initial-boundary conditions:

— the boundary conditions on $\partial\Omega \times \mathbf{T}$,

$$\begin{aligned}
&\sigma'_{ji}n_j = 0, \qquad \mu'_{ji}n_j = m_i - m_i^0 && \text{on } \partial\Omega_\sigma \times \mathbf{T},\\
&u'_i = 0, \qquad \varphi'_i = g_i - \varphi_i^0 && \text{on } \partial\Omega_u \times \mathbf{T},\\
&\theta' = 0 && \text{on } \partial\Omega_\theta \times \mathbf{T},\\
&\frac{\partial\theta'}{\partial n} = 0 && \text{on } \partial\Omega_q \times \mathbf{T};
\end{aligned} \tag{1.53}$$

— the initial conditions[6] in $\Omega \times \{0\}$,

$$\begin{aligned}
&u'_i = 0, \qquad \varphi'_i = k_i - \varphi_i^0, \qquad \theta' = 0,\\
&\dot{u}'_i = 0, \qquad \dot{\varphi}'_i = \chi_i - \dot{\varphi}_i^0.
\end{aligned} \tag{1.54}$$

The quantities m_i^0, φ_i^0 result from the solution of the classical initial-boundary value problem (1.41)–(1.44):

$$\begin{aligned}
&\varphi_i^0 = \tfrac{1}{2}\epsilon_{ijk}u^0_{k,j}, \qquad m_i^0 = \mu_{ji}^0 n_j, \qquad \kappa_{ji}^0 = \varphi^0_{i,j},\\
&\mu_{ji}^0 = (\gamma+\varepsilon)\kappa_{ji}^0 + (\gamma-\varepsilon)\kappa_{ij}^0 + \beta\kappa_{kk}^0\delta_{ij}.
\end{aligned} \tag{1.55}$$

The stresses σ'_{ji}, μ'_{ji} are determined by the geometric relations (0.3) and the constitutive relations (0.4) (written for γ'_{ji}, κ'_{ji}, u'_i, φ'_i and θ'). The solution of the problem described by equations (0.5) and (0.6), as well as the initial-boundary conditions (0.9) and (0.10), takes now the form:

$$\begin{aligned}
&\theta = \theta^0 + \theta', \qquad u_i = u_i^0 + u'_i, \qquad \varphi_i = \varphi_i^0 + \varphi'_i,\\
&\sigma_{ji} = \sigma_{ji}^0 + \sigma'_{ji}, \qquad \mu_{ji} = \mu_{ji}^0 + \mu'_{ji},\\
&\gamma_{ji} = \varepsilon_{ij}^0 + \gamma'_{ji}, \qquad \kappa_{ji} = \kappa_{ji}^0 + \kappa'_{ji},
\end{aligned}$$

where $(\theta^0, u_i^0, \varphi_i^0, \sigma_{ji}^0, \varepsilon_{ij}^0)$ is the solution of the initial-boundary value problem (1.41)–(1.44) of classical thermoelasticity.

Let us now derive from the system of equations (1.52) the uncoupled equations for the triple $(u'_i, \varphi'_i, \theta')$. We obtain the following equations in $\Omega \times \mathbf{T}_+$:

— the equation for u'_i,

$$\begin{aligned}
&D_2\overset{*}{\Box}_2(\Box_2\Box_4 + 4\alpha^2\nabla^2)u'_i\\
&\quad = 2\alpha D_2\epsilon_{ijk}\hat{Y}_{k,j} - \nu_T\overset{*}{\Box}_2(\Box_2\Box_4 + 4\alpha^2\nabla^2)\frac{\hat{Q}_{,i}}{\kappa}\\
&\qquad + \overset{*}{\Box}_2\left\{D[(\lambda+\mu-\alpha)\Box_4 - 4\alpha^2] - \nu_T\eta_0\partial_t\Box_4\right\}\hat{X}_{j,ji}\\
&\qquad - D_2\Box_4\overset{*}{\Box}_2\hat{X}_i,
\end{aligned} \tag{1.56}$$

---

[6] The quantities $\dot{\varphi}_i^0$ should be understood as limits: $\dot{\varphi}_i^0 = \lim\limits_{t\to 0}\dfrac{\partial\varphi_i^0(x_i,t)}{\partial t}$.

where

$$D = \nabla^2 - \frac{1}{\kappa}\partial_t, \qquad D_2 = D\Box_1 - \eta_0\nu_T\partial_t\nabla^2,$$

— the equation for θ',

$$D_2\theta' = -\eta_0\partial_t\hat{X}_{j,j} - \Box_1\frac{\hat{Q}}{\kappa}, \tag{1.57}$$

— the equation for φ_i' (it is identical to equation $(1.23)_2$),

$$\begin{aligned}&\overset{*}{\Box}_2\Box_3(\Box_2\Box_4 + 4\alpha^2\nabla^2)\varphi_i' \\ &\quad = 2\alpha\overset{*}{\Box}_2\Box_3\epsilon_{ijk}\hat{X}_{k,j} + [(\beta+\gamma-\varepsilon)\Box_2 - 4\alpha^2]\hat{Y}_{j,ji} - \Box_2\Box_3\hat{Y}_i.\end{aligned} \tag{1.58}$$

Let us note the following. Assuming formally the substitutions

$$(u_i', \varphi_i', \theta') \mapsto (u_i, \varphi_i, \theta), \qquad (\hat{X}_i, \hat{Y}_i, \hat{Q}) \mapsto (X_i, \overset{*}{\Box}_2 Y_i, Q)$$

the system of equations (1.52) (after removing the operator $\overset{*}{\Box}_2$ from $(1.52)_2$) becomes the system of equations (0.5) and (0.6) describing micropolar coupled thermoelasticity, and should be completed by the initial-boundary conditions (0.9) and (0.10). The separated equations for the thermomechanical triple (u_i, φ_i, θ) are obtained from equations (1.56)–(1.58) (once we remove the operator $\overset{*}{\Box}_2$ from equations (1.56) and (1.58)):

$$\begin{aligned}D_2(\Box_2\Box_4 + 4\alpha^2\nabla^2)u_i &= 2\alpha D_2\epsilon_{ijk}Y_{k,j} - \frac{\nu_T}{\kappa}(\Box_2\Box_4 + 4\alpha^2\nabla^2)Q_{,i} \\ &\quad + \left\{D[(\lambda+\mu-\alpha)\Box_4 - 4\alpha^2] - \nu_T\eta_0\partial_t\Box_4\right\}X_{j,ji} - D_2\Box_4X_i, \\ \Box_3(\Box_2\Box_4 + 4\alpha^2\nabla^2)\varphi_i &= 2\alpha\Box_3\epsilon_{ijk}X_{k,j} \\ &\quad + [(\beta+\gamma-\varepsilon)\Box_2 - 4\alpha^2]Y_{j,ji} - \Box_2\Box_3Y_i, \\ D_2\theta &= -\eta_0\dot{X}_{j,j} - \Box_1\frac{Q}{\kappa}.\end{aligned} \tag{1.59}$$

Let us make another observation based on the superposition method. We neglect in our considerations the loadings X_i, Y_i and the heat sources Q. Now, if we assume in the initial-boundary conditions (1.53) and (1.54) that

$$\begin{aligned}&m_i - m_i^0 = 0 & &\text{on } \partial\Omega_\sigma \times \mathbf{T}, \\ &g_i - \varphi_i^0 = 0 & &\text{on } \partial\Omega_u \times \mathbf{T}, \\ &k_i - \varphi_i^0 = 0, \qquad \chi_i - \dot{\varphi}_i^0 = 0 & &\text{in } \Omega \times \{0\},\end{aligned}$$

then we get in $\overline{\Omega} \times \mathbf{T}$:

$$u'_i \equiv 0, \qquad \varphi'_i \equiv 0, \qquad \theta' \equiv 0, \qquad \sigma'_{ji} \equiv 0, \qquad \mu'_{ji} \equiv 0.$$

The solution of the initial-boundary value problem is now reduced to the classical solution $(u^0_i, \theta^0, \sigma^0_{ji})$. However, the body exhibits an effect of microstructure in the form of the couple-stress field μ^0_{ji} (there exist non-zero rotations φ^0_i and material constants γ, ε, β, see formula (1.55)). Nevertheless, if we model the classical initial-boundary conditions (1.43) and (1.44) for a simply connected micropolar body in such a way that there appears a displacement field u^0_i, but with the condition $\operatorname{curl}\mathbf{u}^0 = \mathbf{0}$ in $\overline{\Omega} \times \mathbf{T}$, then, assuming homogeneous non-classical initial-boundary conditions in the form

$$\begin{aligned}
&\mu_{ji}n_j = 0 && \text{on } \partial\Omega_\sigma \times \mathbf{T},\\
&\varphi_i = 0 && \text{on } \partial\Omega_u \times \mathbf{T},\\
&\varphi_i = 0 \text{ and } \dot{\varphi}_i = 0 && \text{in } \Omega \times \{0\},
\end{aligned}$$

we shall obtain the solution for the case of Hookean body (i.e. the micropolar body exhibits no microstructural effect).

The superposition method leads to an interesting conclusion concerning the effect of body forces, body moments and heat sources in an unbounded space. Let us now turn to the heat conduction equation $(1.52)_3$ and the homogeneous initial-boundary conditions (1.53) and (1.54) for the temperature θ'. Since $\hat{Q} \equiv 0$, we have $\theta' \equiv 0$ whenever the term $\eta_0 \dot{u}'_{j,j}$ vanishes, and thus the solution for the initial-boundary value problem is reduced to the determination of the pair (u'_i, φ'_i). In that case the system of equations $(1.52)_{1,2}$, uncoupled with respect to u'_i and φ'_i $(\theta' = 0)$ leads to equations (1.23). We encounter such a case in the infinite space. Since $\operatorname{div}\mathbf{u}'$ satisfies the equation

$$D_2 u'_{i,i} = -D\hat{X}_{i,i} - \frac{\nu_T}{\kappa}\nabla^2\hat{Q}, \tag{1.60}$$

we see that $u'_{i,i} \equiv 0$ $(\hat{X}_i \equiv 0,\ \hat{Q} \equiv 0)$. There will be no temperature field in the complementary solution in the case of an infinite space. The conclusions concerning the infinite space are given in Table 1.1. Moreover, the following fields: φ_i generated by X_i and u'_i, φ'_i generated by Y_i are identical to those of Subsection 1.1.2 (the problem with no temperature field).

Table 1.1.

Causes	Effects		
X_i	u_i	φ_i	θ^0
Y_i	u'_i	φ'_i	0
Q	u^0_i	0	θ^0

Let us now study the fundamental solutions[7]. Let us assume that the loadings (1.21) and a concentrated heat source

$$Q = Q_0 e^{-i\omega t}\delta(x_1)\delta(x_2)\delta(x_3), \tag{1.61}$$

act in the space $\mathbb{R}^3$, where $Q_0 = 1$. For the case of the loadings $(1.21)_2$ $(X_i = 0,\ Q = 0)$, we obtain $u_i^0 = 0$, $\theta^0 = 0$, from the classical equations (1.45). From the complementary equations (1.23) we obtain (1.27) for the pair (u_i', φ_i'), which is the final result. In the case when the heat sources (1.61) are present $(X_i = 0,\ Y_i = 0)$, we get from equations (1.23) and (1.48), respectively: $u_i' = 0$, $\varphi_i' = 0$, $\varphi_i^0 = 0$ and the final result constitutes the solution obtained from the equations

$$D_2 u_i^0 = -\frac{\nu_T}{\kappa} Q_{,i}, \qquad D_2\theta^0 = -\frac{1}{\kappa}\Box_1 Q. \tag{1.62}$$

Applying the Fourier integral transformations (0.50) (for $n = 4$) to equations (1.62) and taking into account the fact that Q varies harmonically in time with the frequency ω, we obtain from the operator D_2 the expression

$$(\lambda + 2\mu)W_4(\mu).$$

The polynomial $W_4(\mu)$ with complex coefficients has the following form:

$$W_4(\mu) = \mu^4 - [\sigma_1^2 + q(1+\varepsilon)]\mu^2 + q\sigma_1^2 = (\mu^2 - \mu_1^2)(\mu^2 - \mu_2^2),$$

where $q = \dfrac{i\omega}{\kappa}$, $\varepsilon = \kappa\eta_0 m$, $m = \dfrac{\nu_T}{2\mu + \lambda}$; μ_1^2, μ_2^2 are the roots of equation $W_4(\mu) = 0$:

$${\mu_{1,2}}^2 = \tfrac{1}{2}[\sigma_1^2 + q(1+\varepsilon) \pm \sqrt{\Delta}],$$

where $\Delta = [\sigma_1^2 + q(1+\varepsilon)]^2 - 4q\sigma_1^2$.

The roots μ_1, μ_2 are complex-valued[8]. Finally, we obtain the solution of equation (1.62):

$$\begin{aligned} u_l^0 &= \frac{e^{-i\omega t} m Q_0}{4\pi\kappa(\mu_1^2 - \mu_2^2)} \left(\frac{e^{i\mu_2 R} - e^{i\mu_1 R}}{R}\right)_{,l}, \\ \theta^0 &= -\frac{e^{-i\omega t} Q_0}{4\pi\kappa(\mu_1^2 - \mu_2^2)} (\nabla^2 + \sigma_1^2)\left(\frac{e^{i\mu_1 R} - e^{i\mu_2 R}}{R}\right). \end{aligned} \tag{1.63}$$

Now, for the case of the loading $(1.21)_1$ $(Y_i = 0,\ Q = 0)$, the classical equations (1.45) take the form

[7] Fundamental solutions of thermoelasticity were given by the method of potentials by Nowacki [192, pp. 244–250]. See also Stefaniak [218].

[8] For a detailed discussion of this problem, see Chadwick & Sneddon [21] and Nowacki [174, pp. 96–102].

$$D_2 \overset{*}{\square}_2 u_i^0 = D_1 X_{j,ji} - D_2 X_i, \qquad D_2\theta^0 = -\eta_0 \dot{X}_{j,j}. \tag{1.64}$$

From equations (1.64) we obtain the following results[9]:

$$\begin{aligned} \theta^{0(s)} &= \frac{e^{-i\omega t} q\varepsilon P_0}{4\pi\rho m c_1^2(\mu_1^2-\mu_2^2)} \left(\frac{e^{i\mu_1 R}-e^{i\mu_2 R}}{R}\right)_{,s}, \\ u_l^{0(s)} &= \frac{e^{-i\omega t} P_0}{4\pi\rho\omega^2}\left[\hat{\sigma}_2^2 \frac{e^{i\hat{\sigma}_2 R}}{R}\delta_{ls} + \left(\frac{e^{i\hat{\sigma}_2 R}}{R} - E_1\frac{e^{i\mu_1 R}}{R} + E_2\frac{e^{i\mu_2 R}}{R}\right)_{,ls}\right], \end{aligned} \tag{1.65}$$

where

$$\hat{\sigma}_2 = \frac{\omega}{\hat{c}_2}, \qquad \hat{c}_2 = \sqrt{\frac{\mu}{\rho}}, \qquad E_1 = \frac{(\mu_1^2-q)\sigma_1^2}{\mu_1^2(\mu_1^2-\mu_2^2)}, \qquad E_2 = \frac{(\mu_2^2-q)\sigma_1^2}{\mu_2^2(\mu_1^2-\mu_2^2)}.$$

The SI units are used in this work. In order to make the presented forms uniform and to preserve the physical units at both sides of formulae, the coefficients P_0, Q_0, M, M_0, M_1, γ_0^{mn}, κ_0^{mn} are preserved. The dimension of any quantity will be denoted by the Maxwell symbol [ ]. It holds

$$[\delta(A)] = [A^{-1}], \ [dY/dX] = [Y \cdot X^{-1}]$$

and $[\int Y\, dX] = [Y \cdot X]$ (see [128, p. 16 & p. 32] and also Kasprzak & Lysik, *Dimensional Analysis*, WNT, 1988). Throughout the work it holds $[L_{(.)}] = [R_{(.)}]$, where $L_{(.)}$ and $R_{(.)}$ denote the left and the right hand sides, respectively, of the formula (.). For example, let us check that the dimensions in formulae (1.65) agree. We have

$$\begin{aligned} &[P_0] = [\mathrm{N}],\ [\rho] = [\mathrm{N}\cdot\mathrm{s}^2\cdot\mathrm{m}^{-4}],\ [\omega] = [\mathrm{s}^{-1}],\ [c_1] = [c_2] = [\mathrm{m}\cdot\mathrm{s}^{-1}], \\ &[\hat{\sigma}_2] = [\mathrm{m}^{-1}],\ [\mu_1] = [\mu_2] = [\mathrm{m}^{-1}],\ [q] = [\mathrm{m}^{-2}],\ [E_1] = [E_2] = [1], \\ &[\varepsilon] = [1],\ [\mu] = [\lambda] = [\mathrm{N}\cdot\mathrm{m}^{-2}],\ [\nu_T] = [\mathrm{N}\cdot\mathrm{m}^{-2}\cdot\mathrm{K}^{-1}], \\ &[R] = [\mathrm{m}],\ [t] = [\mathrm{s}]. \end{aligned}$$

We obtain

$$[L_{(1.65)1}] = [R_{(1.65)1}] = [\mathrm{K}], \qquad [L_{(1.65)2}] = [R_{(1.65)2}] = [\mathrm{m}].$$

From equation (1.48) we get φ_i^0:

$$\varphi_l^{0(s)} = \frac{e^{-i\omega t} P_0}{8\pi\mu}\epsilon_{lks}\left(\frac{e^{i\hat{\sigma}_2 R}}{R}\right)_{,k}. \tag{1.66}$$

Using the complementary equations (1.23), we get the formulae (see [78]):

---
9 See Nowacki [170] and [172].

$$u_l'^{(s)} = \frac{e^{-i\omega t}P_0}{4\pi\rho\omega^2}\left[\left(A_1\frac{e^{i\lambda_1 R}}{R} + A_2\frac{e^{i\lambda_2 R}}{R} - \frac{e^{i\hat{\sigma}_2 R}}{R}\right)_{,ls}\right.$$
$$\left. + \left(\lambda_1^2 A_1\frac{e^{i\lambda_1 R}}{R} + \lambda_2^2 A_2\frac{e^{i\lambda_2 R}}{R} - \hat{\sigma}_2^2\frac{e^{i\hat{\sigma}_2 R}}{R}\right)\delta_{ls}\right], \tag{1.67}$$

$$\varphi_l'^{(s)} = -\varphi_l^{0(s)} + \frac{e^{-i\omega t}P_0 p}{4\pi\rho c_2^2(\lambda_1^2-\lambda_2^2)}\epsilon_{lks}\left(\frac{e^{i\lambda_1 R} - e^{i\lambda_2 R}}{R}\right)_{,k}.$$

The notations used in formulae (1.67) are the same as those in Subsection 1.1.2. Let us note that from formulae $(1.63)_1$ and $(1.65)_1$ it follows that

$$-\frac{i\omega\kappa\eta_0 P_0}{Q_0\nu_T}u_l^0 = \theta^{0(l)}$$

which is a consequence of the theorem of reciprocity of works.

1.1.5 Stress-temperature equations of motion of Ignaczak type. Fundamental solutions

In linear classical elastodynamics, the mixed initial-boundary value problem may be reduced to solving one tensorial stress equation with certain initial-boundary conditions expressed in stresses[10].

Ignaczak [116] proved that solving the stress initial-boundary value problem (SIBVP) in linear micropolar elastodynamics is equivalent to solving the system of two coupled tensor stress equations with appropriately formulated initial-boundary conditions in terms of stresses.

Al-Hasan & Dyszlewicz [3] have considered a stress-temperature initial-boundary value problem (STIBVP) of Ignaczak type for the coupled dynamical thermoelasticity (micropolar model of E-N). We shall briefly present here the results contained in [3]. Let us turn to the basic relations and equations of the E-N model: the equations of motion (0.1), the equations of compatibility (0.2), the geometric relations (0.3), the constitutive relations (0.4), the equation of heat conduction (0.5), the boundary conditions $(0.9)_{1,3}$ on $\partial\Omega \times T$, the initial conditions (0.10).

Definition 1.1.1. *By stress and temperature fields corresponding to the solution of the problem given by* (0.1)–(0.5), $(0.9)_{1,3}$ *and* (0.10) *we understand the triple* $(\underset{\sim}{\sigma}, \underset{\sim}{\mu}, \theta)$ *with such properties that there exists a quadruple* $(\mathbf{u}, \boldsymbol{\varphi}, \underset{\sim}{\gamma}, \underset{\sim}{\kappa})$ *such that the system of functions* $(\mathbf{u}, \boldsymbol{\varphi}, \underset{\sim}{\kappa}, \underset{\sim}{\gamma}, \underset{\sim}{\sigma}, \underset{\sim}{\mu}, \theta)$ *is a solution of* (0.1)–(0.5), $(0.9)_{1,3}$ *and* (0.10).

Let us assume that we know the triple $(\underset{\sim}{\sigma}, \underset{\sim}{\mu}, \theta)$ and the pair $(\underset{\sim}{\gamma}, \underset{\sim}{\kappa})$. Then the pair $(\mathbf{u}, \boldsymbol{\varphi})$ can be obtained from the pair $(\underset{\sim}{\gamma}, \underset{\sim}{\kappa})$ by integrating the geometric relations (0.3) with respect to the position coordinates x_i, and arbitrary integration functions $A_{ij}(x_k, t)$, $B_{ij}(x_k, t)$ $(j \neq k)$ can be determined

[10] See Nowacki [180, p. 511]; Gurtin [105, p. 212]; Ignaczak [113], [114].

by using the compatibility equations (0.2). In [3] we offer an alternative method of recovering $(\mathbf{u}, \varphi)$ from $(\underset{\sim}{\gamma}, \underset{\sim}{\kappa})$. Let us derive stress-temperature equations of motion of Ignaczak type. Using the geometric relations (0.3), we obtain from the equations of motion (0.1) the following equations in $\Omega \times T_+$:

$$\rho^{-1} R_{i,j} - J^{-1} \epsilon_{kji} M_k = \ddot{\gamma}_{ji}, \qquad J^{-1} M_{i,j} = \ddot{\kappa}_{ji}, \tag{1.68}$$

where

$$R_i = \sigma_{ji,j} + X_i, \qquad M_i = \epsilon_{ijk}\sigma_{jk} + \mu_{ji,j}.$$

In view of equations (0.3), the heat conduction equation (0.5) takes the form in $\Omega \times T_+$

$$\theta_{,ii} - \frac{1}{\kappa}\dot{\theta} - \eta_0 \dot{\gamma}_{kk} = -\frac{Q}{\kappa}. \tag{1.69}$$

The constitutive relations (0.4), when solved for strains, take the following equivalent form in $\Omega \times T$:

$$\begin{aligned} \gamma_{ji} &= \frac{1}{2\mu}\sigma_{(ji)} + \frac{1}{2\alpha}\sigma_{[ji]} - \frac{1}{2\mu}(\lambda e - \nu_T \theta)\delta_{ij}, \\ \kappa_{ji} &= \frac{1}{2\gamma}\mu_{(ji)} + \frac{1}{2\varepsilon}\mu_{[ji]} - \frac{\beta}{2\gamma(2\gamma + 3\beta)}\mu_{kk}\delta_{ij}, \end{aligned} \tag{1.70}$$

where

$$e = \frac{1}{2\mu + 3\lambda}(\sigma_{kk} + 3\nu_T \theta)$$

and the parentheses (.) and brackets [.] denote the symmetric part and the antisymmetric part of a tensor, respectively. By using relations (1.70), we reduce equations (1.68) and (1.69) to the form in $\Omega \times T_+$:

$$\begin{aligned} &\rho^{-1} R_{i,j} + J^{-1}\epsilon_{ijk} M_k - \frac{1}{2\mu}\ddot{\sigma}_{(ji)} - \frac{1}{2\alpha}\ddot{\sigma}_{[ji]} + \frac{1}{2\mu}(\lambda \ddot{e} - \nu_T \ddot{\theta})\delta_{ij} = 0, \\ &J^{-1} M_{i,j} - \frac{1}{2\gamma}\ddot{\mu}_{(ji)} - \frac{1}{2\varepsilon}\ddot{\mu}_{[ji]} + \frac{\beta}{2\gamma(2\gamma+3\beta)}\ddot{\mu}_{kk}\delta_{ij} = 0, \\ &\theta_{,ii} - \frac{1}{\kappa}\dot{\theta} - \eta_0 \dot{e} = -\frac{Q}{\kappa}, \end{aligned} \tag{1.71}$$

where

$$\dot{e} = \frac{1}{2\mu + 3\lambda}(\dot{\sigma}_{kk} + 3\nu_T \dot{\theta}), \qquad \ddot{e} = \frac{1}{2\mu + 3\lambda}(\ddot{\sigma}_{kk} + 3\nu_T \ddot{\theta}).$$

The initial conditions for the triple $(\underset{\sim}{\sigma}, \underset{\sim}{\mu}, \theta)$ which satisfies the field equations (1.71) follow from the initial conditions (0.10), the constitutive relations (0.4), the geometric relations (0.3) and the heat conduction equation (0.5). We obtain the following matrices of initial values of stresses $\underset{\sim}{\sigma}^0$, $\underset{\sim}{\mu}^0$ and stress velocities $\dot{\underset{\sim}{\sigma}}^0, \dot{\underset{\sim}{\mu}}^0$:

$$\underset{\sim}{\sigma}^0 \equiv \begin{bmatrix} \sigma^0_{11} & \sigma^0_{12} & \sigma^0_{13} \\ \sigma^0_{21} & \sigma^0_{22} & \sigma^0_{23} \\ \sigma^0_{31} & \sigma^0_{32} & \sigma^0_{33} \end{bmatrix}, \qquad \underset{\sim}{\mu}^0 \equiv \begin{bmatrix} \mu^0_{11} & \mu^0_{12} & \mu^0_{13} \\ \mu^0_{21} & \mu^0_{22} & \mu^0_{23} \\ \mu^0_{31} & \mu^0_{32} & \mu^0_{33} \end{bmatrix}, \tag{1.72}$$

$$\dot{\underset{\sim}{\sigma}}^0 \equiv \begin{bmatrix} \dot\sigma^0_{11} & \dot\sigma^0_{12} & \dot\sigma^0_{13} \\ \dot\sigma^0_{21} & \dot\sigma^0_{22} & \dot\sigma^0_{23} \\ \dot\sigma^0_{31} & \dot\sigma^0_{32} & \dot\sigma^0_{33} \end{bmatrix}, \qquad \dot{\underset{\sim}{\mu}}^0 \equiv \begin{bmatrix} \dot\mu^0_{11} & \dot\mu^0_{12} & \dot\mu^0_{13} \\ \dot\mu^0_{21} & \dot\mu^0_{22} & \dot\mu^0_{23} \\ \dot\mu^0_{31} & \dot\mu^0_{32} & \dot\mu^0_{33} \end{bmatrix}, \tag{1.73}$$

where

$$\begin{aligned} \sigma^0_{ji} &= (\mu+\alpha)\gamma^0_{ji} + (\mu-\alpha)\gamma^0_{ij} \\ &\qquad + (\lambda e^0 - \nu_T l)\delta_{ij}, \\ \mu^0_{ji} &= (\gamma+\varepsilon)k_{i,j} + (\gamma-\varepsilon)k_{j,i} + \beta k_{k,k}\delta_{ij}, \\ \gamma^0_{ji} &= h_{i,j} - \epsilon_{kji}k_k, \qquad e^0 = h_{k,k} \end{aligned} \tag{1.74}$$

and

$$\begin{aligned} \dot\sigma^0_{ji} &= (\mu+\alpha)\dot\gamma^0_{ji} + (\mu-\alpha)\dot\gamma^0_{ij} \\ &\qquad + \left[(\lambda + \nu_T\kappa\eta_0)\dot e^0 - \nu_T(\kappa\, l_{,kk} + Q^0)\right]\delta_{ij}, \\ \dot\mu^0_{ji} &= (\gamma+\varepsilon)\chi_{i,j} + (\gamma-\varepsilon)\chi_{j,i} + \beta\chi_{k,k}\delta_{ij}, \\ \dot\gamma^0_{ji} &= \psi_{i,j} - \epsilon_{kji}\chi_k, \qquad \dot e^0 = \psi_{k,k}, \end{aligned} \tag{1.75}$$

where Q^0 is the initial value of the heat sources. Let us formulate the following useful lemma [111].

Lemma 1.1.1. *Assume that the fields* $\mathbf{u}, \boldsymbol{\varphi}$, $\underset{\sim}{\sigma}$ *and* $\underset{\sim}{\mu}$ *are sufficiently smooth on* $\overline{\Omega} \times T$. *Then the fields* $\mathbf{u}, \boldsymbol{\varphi}$, $\underset{\sim}{\sigma}$ *and* $\underset{\sim}{\mu}$ *satisfy the equations of motion* (0.1) *and the initial conditions* (0.10) *if and only if the following relations hold in* $\overline{\Omega} \times T$

$$u_i = \psi_i t + h_i + \rho^{-1}(t * R_i), \qquad \varphi_i = \chi_i t + k_i + J^{-1}(t * M_i), \tag{1.76}$$

where $*$ *denotes the convolution* [156]*:*

$$t * f(x_1, x_2, x_3, t) = \int_0^t (t-\tau) f(x_1, x_2, x_3, \tau)\, d\tau.$$

Definition 1.1.2. (Stress-temperature equations of motion problem STEMP). *By the stress-temperature equations of motion problem of linear,* *homogeneous, isotropic, coupled, dynamical, micropolar thermoelasticity for a three-dimensional body we understand the initial-boundary valued problem, in which the field equations* (1.71) *are satisfied, together with the initial conditions in* $\Omega \times \{0\}$

$$\begin{aligned} \underset{\sim}{\sigma} &= \underset{\sim}{\sigma}^0, \qquad \underset{\sim}{\mu} = \underset{\sim}{\mu}^0, \qquad \theta = l, \\ \dot{\underset{\sim}{\sigma}} &= \dot{\underset{\sim}{\sigma}}^0, \qquad \dot{\underset{\sim}{\mu}} = \dot{\underset{\sim}{\mu}}^0 \end{aligned} \tag{1.77}$$

and the boundary conditions $(0.9)_{1,3}$ *on* $\partial\Omega \times T$.

Theorem 1.1.1. *(Formulation of the problem in terms of stresses and temperature). Let the systems of functions* $(\mathbf{u}, \boldsymbol{\varphi})$, $(\underset{\sim}{\sigma}, \underset{\sim}{\mu}, \theta)$ *and* $(\underset{\sim}{\gamma}, \underset{\sim}{\kappa})$ *be sufficiently smooth in* $\overline{\Omega} \times T$. *Then* $\underset{\sim}{\sigma}, \underset{\sim}{\mu}$ *and* θ *are the stress fields and the temperature field, respectively, corresponding to the solution of the initial-boundary value problem* (0.1)–(0.5), $(0.9)_{1,3}$, (0.10) *if and only if equations* (1.71) *hold and the initial conditions* (1.77) *and the boundary conditions* $(0.9)_{1,3}$ *are satisfied.*

Proof. Necessity. It has been shown that the relations (0.1) and (0.3)–(0.5), $(0.9)_{1,3}$, (0.10) give the field equations (1.71), the boundary conditions $(0.9)_{1,3}$ and the initial conditions (1.77).
Sufficiency. Let us assume that the stress and temperature fields $\underset{\sim}{\sigma}, \underset{\sim}{\mu}$ and θ correspond to the solution of STEMP. Therefore, by using (1.70), we determine the strain fields $\underset{\sim}{\gamma}$ and $\underset{\sim}{\kappa}$. Next, we determine the displacement field $\mathbf{u}$ and the rotation field $\boldsymbol{\varphi}$ by means of the relations (1.76). Let us notice that the conditions (1.76), (1.70) and (1.71) imply the relations (0.3). The system of functions $(\mathbf{u}, \boldsymbol{\varphi}, \underset{\sim}{\sigma}, \underset{\sim}{\mu})$ satisfies the equations of motion (0.1) and the initial conditions $(0.10)_{1,2,4,5}$ if and only if the relations (1.76) hold. Since it is easy to notice that $(1.71)_3$ and $(1.77)_3$ imply the heat conduction equation (0.5) and the initial condition $(0.10)_3$, we conclude that the system of functions $(\mathbf{u}, \boldsymbol{\varphi}, \underset{\sim}{\gamma}, \underset{\sim}{\kappa}, \underset{\sim}{\sigma}, \underset{\sim}{\mu}, \theta)$ is a solution of the problem (0.1)–(0.5), $(0.9)_{1,3}$, (0.10). This completes the proof.

Remark 1. When solving a STEMP, one does not have to refer to the compatibility equations (0.2) since equations (0.2) are implied by equations (1.71), (1.77) and (1.72)–(1.75).

Remark 2. A STEMP makes sense when the initial stress, stress-rate, and temperature are presented in an arbitrary way. However, in such a general case, the compatibility equations (0.2) will not be satisfied at $t = 0$, and the problem does not belong to classical micropolar dynamical thermoelasticity.

Let us consider harmonic vibrations of an infinite micropolar thermoelastic space which correspond to the following concentrated loads: (1.21) and (1.61).

A singular solution of STEMP for an infinite space corresponding to the loads X_i, Y_i and Q given by (1.21) and (1.61), respectively, will be obtained with the help of differential equations for the stress components and for the temperature, which are obtained from the field equations (1.71) in the form:

— the equation for force-stresses

$$
\begin{aligned}
D_2\Box_3(\Box_2\Box_4 + 4\alpha^2\nabla^2)\sigma_{ij} &= -2\alpha\epsilon_{ijk}D_2L_1Y_{s,sk} \\
&+ 2\alpha D_2\Box_3[(\mu+\alpha)\epsilon_{jsk}Y_{k,si} + (\mu-\alpha)\epsilon_{isk}Y_{k,sj} + \epsilon_{ijk}\Box_2 Y_k] \\
&+ 2\mu\Box_3(DL_2 - \nu_T\eta_0\partial_t\Box_4)X_{k,kij} \\
&- D_2\Box_3\{\Box_4[(\mu+\alpha)X_{j,i} + (\mu-\alpha)X_{i,j}] + 4\alpha^2(X_{j,i} - X_{i,j})\} \\
&- \Box_3(\Box_2\Box_4 + 4\alpha^2\nabla^2)(\lambda D - \nu_T\eta_0\partial_t)X_{k,k}\delta_{ij} \\
&- \frac{\nu_T}{\kappa}\Box_3(\Box_2\Box_4 + 4\alpha^2\nabla^2)[2\mu Q_{,ij} + (\lambda\nabla^2 - \Box_1)Q\delta_{ij}],
\end{aligned}
\tag{1.78}
$$

where

$$
L_1 = (\beta+\gamma-\varepsilon)\Box_2 - 4\alpha^2, \qquad L_2 = (\lambda+\mu-\alpha)\Box_4 - 4\alpha^2;
$$

— the equation for couple-stresses

$$
\begin{aligned}
&\Box_3(\Box_2\Box_4 + 4\alpha^2\nabla^2)\mu_{ij} \\
&\quad = 2\alpha\Box_3[(\gamma+\varepsilon)\epsilon_{jsk}X_{k,si} + (\gamma-\varepsilon)\epsilon_{isk}X_{k,sj}] \\
&\quad - \Box_2\Box_3[(\gamma+\varepsilon)Y_{j,i} + (\gamma-\varepsilon)Y_{i,j}] \\
&\quad + 2\gamma L_1 Y_{k,kij} - \beta(\Box_2\Box_4 + 4\alpha^2\nabla^2)Y_{k,k}\delta_{ij};
\end{aligned}
\tag{1.79}
$$

— the equation for temperature

$$
D_2\theta = -\left(\eta_0\partial_t X_{k,k} + \frac{1}{\kappa}\Box_1 Q\right). \tag{1.80}
$$

First, a singular solution to STEMP will be obtained for the case $P_0 \neq 0$, $M_0 = 0$, $Q_0 = 0$. Applying the metod of integral transformations based on the Fourier transformation (0.50) (for $n = 4$), and using $(1.21)_1$ we obtain from equations (1.78)–(1.80) the following formulae:

— for the force-stresses

$$
\begin{aligned}
\sigma_{ij}^{(s)} = \frac{P_0 e^{-i\omega t}}{4\pi\rho\omega^2}\Bigg[&\frac{\rho\omega^2\alpha}{\mu l^2(\lambda_1^2 - \lambda_2^2)}(\delta_{is}F_{1,j} - \delta_{js}F_{1,i}) \\
&+ 2\mu(K_1 - K_2)_{,ijs} + \left(\lambda G_2 - \frac{\rho\omega^2 q\varepsilon}{\mu_1^2 - \mu_2^2}\Gamma_1\right)_{,s}\delta_{ij} \\
&+ (\mu+\alpha)\delta_{js}G_{1,i} + (\mu-\alpha)\delta_{is}G_{1,j}\Bigg],
\end{aligned}
\tag{1.81}
$$

where

$$K_1 = \frac{1}{R}\left(A_1 e^{i\lambda_1 R} + A_2 e^{i\lambda_2 R}\right), \qquad K_2 = \frac{1}{R}\left(E_1 e^{i\mu_1 R} - E_2 e^{i\mu_2 R}\right),$$

$$G_1 = \frac{1}{R}\left(A_1 \lambda_1^2 e^{i\lambda_1 R} + A_2 \lambda_2^2 e^{i\lambda_2 R}\right), \quad G_2 = \frac{1}{R}\left(E_1 \mu_1^2 e^{i\mu_1 R} - E_2 \mu_2^2 e^{i\mu_2 R}\right),$$

$$F_1 = \frac{1}{R}\left(e^{i\lambda_1 R} - e^{i\lambda_2 R}\right), \qquad \Gamma_1 = \frac{1}{R}\left(e^{i\mu_1 R} - e^{i\mu_2 R}\right);$$

— for the couple-stresses

$$\mu_{ij}^{(s)} = \frac{P_0 e^{-i\omega t}}{8\pi\mu l^2} \frac{1}{(\lambda_1^2 - \lambda_2^2)} \left[(\gamma + \varepsilon)\epsilon_{jks} F_{1,ik} + (\gamma - \varepsilon)\epsilon_{iks} F_{1,jk}\right]; \tag{1.82}$$

— for the temperature

$$\theta^{(s)} = \frac{P_0 e^{-i\omega t}}{4\pi\rho\omega^2} \frac{q\varepsilon\sigma_1^2}{m(\mu_1^2 - \mu_2^2)} \Gamma_{1,s}. \tag{1.83}$$

Formulas (1.81), (1.82) and (1.83) represent a singular solution

$$(\underset{\sim}{\sigma}^{(s)}, \underset{\sim}{\mu}^{(s)}, \theta^{(s)}),$$

which corresponds to the load $(1.21)_1$. In order to obtain displacement and rotation fields associated with the solution, we use relations (1.76). For this purpose, we need to specify the initial values of displacements, rotations and their velocities. It follows from equations (1.74), (1.75) and (1.81), (1.82), that:

$$\psi_i^{(s)} = -i\omega h_i^{(s)}, \qquad \chi_i^{(s)} = -i\omega k_i^{(s)} \tag{1.84}$$

and

$$\begin{aligned} h_i^{(s)} &= A_0[(K_1 - K_2)_{,is} + \delta_{is} G_1], \\ k_i^{(s)} &= \frac{\rho\omega^2}{2\mu l^2} \frac{1}{\lambda_1^2 - \lambda_2^2} A_0 \epsilon_{iks} F_{1,k}, \end{aligned} \tag{1.85}$$

where $A_0 = \dfrac{P_0}{4\pi\rho\omega^2}$. Hence, from equations (1.76), (1.81), (1.82), (1.84) and (1.85) and the relations

$$\begin{aligned} &\int_0^t (t-\tau) e^{-i\omega\tau} d\tau = \omega^{-2}\left(1 - i\omega t - e^{-i\omega t}\right), \\ &A_1 + A_2 = 1, \qquad E_1 - E_2 = 1, \\ &\nabla^2\left(\frac{e^{i\lambda_1 R}}{R}\right) = \nabla^2\left(\frac{1}{R}\right) - \lambda_1^2 \frac{e^{i\lambda_1 R}}{R} \end{aligned} \tag{1.86}$$

we obtain

$$
\begin{aligned}
u_i^{(s)} &= A(t)\left[(K_1 - K_2)_{,is} + \delta_{is} G_1\right], \\
\varphi_i^{(s)} &= \frac{\rho\omega^2}{2\mu l^2} \frac{1}{\lambda_1^2 - \lambda_2^2} A(t) \epsilon_{iks} F_{1,k},
\end{aligned}
\tag{1.87}
$$

where $A(t) = A_0 e^{-i\omega t}$. One can check that when $R \to \infty$, the stress fields (1.81) and (1.82), the temperature field (1.83), as well as the displacement and rotations fields (1.87), all tend to zero. Also, by taking into account relations $(1.21)_1$, (1.81)–(1.83), (1.87) and the formula

$$
\frac{1}{4\pi}\nabla^2\left(\frac{1}{R}\right) = -\delta(x_1)\delta(x_2)\delta(x_3) \tag{1.88}
$$

we note that the equations of motion (0.1) and the heat conduction equation (0.5) are satisfied identically.

Next, a singular solution to STEMP is obtained for the case $P_0 = 0$, $M_0 \neq 0$, $Q_0 = 0$. Proceeding in a way similar to that of nonzero mass forces we receive:

— the force-stresses

$$
\begin{aligned}
\sigma_{ij}^{(s)} &= \frac{M_0 e^{-i\omega t}}{4\pi}\Bigg\{ - p(\epsilon_{ijk} K_{3,ks} + \epsilon_{ijs} F_2) \\
&+ \frac{1}{2\mu l^2(\lambda_1^2 - \lambda_2^2)}\left[(\mu + \alpha)\epsilon_{jks} F_{1,ik} + (\mu - \alpha)\epsilon_{iks} F_{1,jk}\right]\Bigg\},
\end{aligned}
\tag{1.89}
$$

where

$$
\begin{aligned}
K_3 &= \frac{1}{R}\left(C_1 e^{i\lambda_1 R} + C_2 e^{i\lambda_2 R} + C_3 e^{i\lambda_3 R}\right), \\
F_2 &= \frac{1}{R}\left(A_1 e^{i\lambda_2 R} + A_2 e^{i\lambda_1 R}\right);
\end{aligned}
$$

— the couple-stresses

$$
\begin{aligned}
\mu_{ij}^{(s)} &= \frac{M_0 e^{-i\omega t}}{4\pi J c_4^2}\Bigg[2\gamma K_{3,ijs} + \frac{\beta(\gamma + \varepsilon)}{\beta + 2\gamma}\left(\frac{e^{i\lambda_3 R}}{R}\right)_{,s}\delta_{ij} \\
&+ (\gamma + \varepsilon)\delta_{js} F_{2,i} + (\gamma - \varepsilon)\delta_{is} F_{2,j}\Bigg].
\end{aligned}
\tag{1.90}
$$

In this case, the formula for the temperature $\theta^{(s)}$ reads

$$
\theta^{(s)} \equiv 0. \tag{1.91}
$$

Formulas (1.89), (1.90) and (1.91) represent a singular solution $(\underset{\sim}{\sigma}^{(s)}, \underset{\sim}{\mu}^{(s)}, \theta^{(s)})$ of STEMP for an infinite space corresponding to the load $(1.21)_2$. Furthermore, initial values of displacements, rotations and their velocities are given by the formulas

$$\psi_i^{(s)} = -i\omega h_i^{(s)}, \qquad \chi_i^{(s)} = -i\omega k_i^{(s)}, \tag{1.92}$$

and

$$\begin{aligned} h_i^{(s)} &= \frac{Jc_4^2}{2\mu l^2} \frac{1}{\lambda_1^2 - \lambda_2^2} C_0 \epsilon_{iks} F_{1,k}, \\ k_i^{(s)} &= C_0(K_{3,is} + \delta_{is} F_2), \qquad C_0 = \frac{M_0}{4\pi J c_4^2}. \end{aligned} \tag{1.93}$$

Finally, the formulas for the displacements and the rotations read

$$\begin{aligned} u_i^{(s)} &= \frac{Jc_4^2}{2\mu l^2} \frac{1}{\lambda_1^2 - \lambda_2^2} C(t) \epsilon_{iks} F_{1,k}, \\ \varphi_i^{(s)} &= C(t)(K_{3,is} + \delta_{is} F_2), \end{aligned} \tag{1.94}$$

where $C(t) = C_0 e^{-i\omega t}$. One can check that when $R \to \infty$, the stress fields (1.89), (1.90), the temperature field (1.91), as well as the displacement and rotation fields (1.94), all tend to zero. Also, by taking into account relations $(1.21)_2$, (1.89)–(1.91), (1.94) and (1.88), we note that the equations of motion (0.1) and the heat conduction equation (0.5) are satisfied identically.

Let us note that in this case a temperature field vanishes. Finally, in case $P_0 = 0$, $M_0 = 0$, $Q_0 \neq 0$, from equations (1.78) and (1.80) we obtain

$$D_2 \sigma_{ij} = -\frac{\nu_T}{\kappa} [2\mu Q_{,ij} + (\lambda \nabla^2 - \Box_1) Q \delta_{ij}], \tag{1.95}$$

$$D_2 \theta = -\frac{1}{\kappa} \Box_1 Q. \tag{1.96}$$

Applying to equations (1.95), (1.96) and (1.61) the same integration method based on the four-dimensional Fourier transform (0.50) as in the previous cases, we obtain the following results in a closed form:

— the force-stresses

$$\sigma_{ij} = -\frac{Q_0 e^{-i\omega t}}{4\pi} \frac{m}{\kappa} \left[\frac{2\mu}{\mu_1^2 - \mu_2^2} \Gamma_{1,ij} + \left(2\mu \Gamma_2 - \frac{\lambda \sigma_1^2}{\mu_1^2 - \mu_2^2} \Gamma_1 \right) \delta_{ij} \right], \tag{1.97}$$

— the temperature

$$\theta = \frac{Q_0 e^{-i\omega t}}{4\pi} \frac{1}{\kappa} \Gamma_2, \tag{1.98}$$

where $\Gamma_2 = \dfrac{1}{R} \left(E_1 e^{i\mu_2 R} - E_2 e^{i\mu_1 R} \right)$.

By using equation (1.79), in this case, for the couple-stresses we obtain the formula

$$\mu_{ij} \equiv 0. \tag{1.99}$$

The pair $(\mathbf{u}, \varphi)$ given by formula (1.76) is determined by a method described in the previous cases. The initial values of displacements, rotations and their velocities take the form:

$$\psi_i = -i\omega h_i, \qquad \chi_i = -i\omega k_i, \tag{1.100}$$

and

$$h_i = -B_0 \frac{1}{\mu_1^2 - \mu_2^2} \Gamma_{1,i}, \qquad k_i = 0, \tag{1.101}$$

where $B_0 = \dfrac{mQ_0}{4\pi\kappa}$. Using these formulas as well as relations (1.76) and (1.86), we obtain results for u_i and φ_i in the following form:

$$u_i = -B(t) \frac{1}{\mu_1^2 - \mu_2^2} \Gamma_{1,i}, \qquad \varphi_i = 0, \tag{1.102}$$

where $B(t) = B_0 e^{-i\omega t}$. When $R \to \infty$, all physical fields obtained above tend to zero. Besides, taking into account relations (1.97)–(1.99), (1.102), (1.61) and (1.88), we note that the equations of motion (0.1) and the heat conduction equation (0.5) are satisfied identically. Formulas (1.97), $(1.102)_1$, and (1.98) determine the fields of stresses $\underset{\sim}{\sigma}$, displacements $\mathbf{u}$ and temperature θ corresponding to the heat source (1.61) ($\boldsymbol{\varphi} \equiv \mathbf{0}$, $\underset{\sim}{\mu} \equiv 0$).

The notations in this Subsection are the same as in the description of the E-N model and in Subsection 1.1.2 and 1.1.4.

Remark 3. The results for the physical fields produced by the loads (1.21), (1.61) which were obtained by the STEMP method correspond to the results given in Subsections 1.1.2 and 1.1.4, where the superposition method was used.

1.1.6 Radiation conditions of Sommerfeld type

The Sommerfeld radiation conditions[11] for a classical linear, homogeneous, isotropic elastodynamics were given by Kupradse [137, pp. 62-70]. For a linear coupled dynamic thermoelasticity, the radiation conditions were given by Ignaczak & Nowacki [120] and Kupradse and co-authors [138] (the work from 1968). While, the radiation conditions for a linear, homogeneous, dynamical, micropolar isothermal elasticity were given by Ignaczak [115] and [117]. Following the results in [117] Al-Hasan & Dyszlewicz [4] discussed the radiation conditions for a linear, homogeneous, isotropic, dynamical, coupled, micropolar thermoelasticity. Let us briefly present here the results given in [4].

We rewrite equations (0.5)–(0.6) for the E-N model in the vector form, assuming that mass forces and mass moments as well as heat sources vanish:

— displacement-rotation-temperature equations:

$$\vec{\Box}_2 \hat{\mathbf{u}} + (\lambda + \mu - \alpha)\,\mathrm{grad\,div}\,\hat{\mathbf{u}} + 2\alpha\,\mathrm{curl}\,\hat{\boldsymbol{\varphi}} = \nu_T\,\mathrm{grad}\,\hat{\theta}, \tag{1.103}$$

$$\vec{\Box}_4 \hat{\boldsymbol{\varphi}} + (\beta + \gamma - \varepsilon)\,\mathrm{grad\,div}\,\hat{\boldsymbol{\varphi}} + 2\alpha\,\mathrm{curl}\,\hat{\mathbf{u}} = \mathbf{0}; \tag{1.104}$$

[11] See [216].

— temperature-displacement equation

$$\nabla^2\hat{\theta} - \frac{1}{\kappa}\dot{\hat{\theta}} - \eta_0 \operatorname{div} \dot{\hat{\mathbf{u}}} = 0, \tag{1.105}$$

where

$$\hat{\mathbf{u}} \equiv \hat{\mathbf{u}}(\mathbf{x},\ t),\ \hat{\boldsymbol{\varphi}} \equiv \hat{\boldsymbol{\varphi}}(\mathbf{x},\ t),\ \text{and } \hat{\theta} \equiv \hat{\theta}(\mathbf{x},\ t)$$

denote the displacement vector, the rotation vector and the temperature field, respectively.

Using the Stokes-Helmholtz representation (1.147) and assuming $\hat{\mathbf{u}}$ and $\hat{\boldsymbol{\varphi}}$ in the form

$$\hat{\mathbf{u}} = \hat{\mathbf{u}}_{(p)} + \hat{\mathbf{u}}_{(s)}, \qquad \hat{\boldsymbol{\varphi}} = \hat{\boldsymbol{\varphi}}_{(p)} + \hat{\boldsymbol{\varphi}}_{(s)}, \tag{1.106}$$

and

$$\operatorname{curl} \hat{\mathbf{u}}_{(p)} = \mathbf{0}, \qquad \operatorname{curl} \hat{\boldsymbol{\varphi}}_{(p)} = \mathbf{0}, \tag{1.107}$$

$$\operatorname{div} \hat{\mathbf{u}}_{(s)} = 0, \qquad \operatorname{div} \hat{\boldsymbol{\varphi}}_{(s)} = 0. \tag{1.108}$$

Let us assume now that a micropolar thermoelastic infinite body $\Omega_e = E \setminus \Omega_i$ (E is the whole space) has a cavity Ω_i bounded by a regular surface S with the unit normal vector from Ω_i to Ω_e. Let $\boldsymbol{\xi} \in \Omega_e$ and let Σ_r be the surface of a sphere with the center at $\boldsymbol{\xi}$ and of radius r large enough so that the surface Σ_r encloses the region Ω_i. In the region Ω_e we consider equations (1.103)–(1.105) with the conditions (1.106)–(1.108), assuming the harmonic vibrations with frequency $\omega > 0$:

$$\hat{\mathbf{u}}_{(p)} = \mathbf{u}_{(p)}(\mathbf{x})e^{-i\omega t}, \qquad \hat{\boldsymbol{\varphi}}_{(p)} = \boldsymbol{\varphi}_{(p)}(\mathbf{x})e^{-i\omega t}, \tag{1.109}$$

$$\hat{\mathbf{u}}_{(s)} = \mathbf{u}_{(s)}(\mathbf{x})e^{-i\omega t}, \qquad \hat{\boldsymbol{\varphi}}_{(s)} = \boldsymbol{\varphi}_{(s)}(\mathbf{x})e^{-i\omega t}, \tag{1.110}$$

$$\hat{\theta} = \theta(\mathbf{x})e^{-i\omega t}. \tag{1.111}$$

Now, we can present the results given in [4] in the form of the following theorem[12].

Theorem 1.1.2.

1. If $\mathbf{u}_{(p)}$ and θ satisfy in the neighborhood of infinity the asymptotic conditions

$$e^{-\beta_1 r}\left(\mathbf{u}_{(p)} + m_1 \operatorname{grad}\theta\right) = \mathbf{0}(1), \tag{1.112}$$

$$e^{-\beta_2 r}\left(\mathbf{u}_{(p)} + m_2 \operatorname{grad}\theta\right) = \mathbf{0}(1), \tag{1.113}$$

$$e^{-\beta_1 r}\left(\theta - \hat{h}_1 \operatorname{div} \mathbf{u}_{(p)}\right) = 0(1), \tag{1.114}$$

[12] Al-Hasan in his dissertation „Equations of Ignaczak type in the micropolar E-N model" (2002, Institute of Mathematics, Wrocław University of Technology, in Polish) gave an even more general theorem for the E-N model, using appropriate Green functions and the Green Formula for the Laplace operator, generalized to the case of a product of three Helmholtz operators.

$$e^{-\beta_2 r}\left(\theta-\hat{h}_2 \operatorname{div}\mathbf{u}_{(p)}\right)=0(1), \tag{1.115}$$

$$re^{-\beta_1 r}\left(\partial_r-i\mu_1\right)\left(\mathbf{u}_{(p)}+m_1 \operatorname{grad}\theta\right)=\mathbf{0}(1), \tag{1.116}$$

$$re^{-\beta_2 r}\left(\partial_r-i\mu_2\right)\left(\mathbf{u}_{(p)}+m_2 \operatorname{grad}\theta\right)=\mathbf{0}(1), \tag{1.117}$$

$$re^{-\beta_1 r}\left(\partial_r-i\mu_1\right)\left(\theta-\hat{h}_1 \operatorname{div}\mathbf{u}_{(p)}\right)=0(1), \tag{1.118}$$

$$re^{-\beta_2 r}\left(\partial_r-i\mu_2\right)\left(\theta-\hat{h}_2 \operatorname{div}\mathbf{u}_{(p)}\right)=0(1) \tag{1.119}$$

then the following integral equations hold:

$$\begin{aligned}4\pi(\mu_1^2-\mu_2^2)\mathbf{u}_{(p)}(\boldsymbol{\xi}) &= (\mu_1^2-\sigma_1^2)\int_S\left[\frac{e^{i\mu_2 R}}{R}\frac{\partial}{\partial n}\left(\mathbf{u}_{(p)}+m_2 \operatorname{grad}\theta\right)\right.\\ &\quad\left.-\left(\mathbf{u}_{(p)}+m_2\operatorname{grad}\theta\right)\frac{\partial}{\partial n}\left(\frac{e^{i\mu_2 R}}{R}\right)\right]dS\\ &\quad-(\mu_2^2-\sigma_1^2)\int_S\left[\frac{e^{i\mu_1 R}}{R}\frac{\partial}{\partial n}\left(\mathbf{u}_{(p)}+m_1\operatorname{grad}\theta\right)\right.\\ &\quad\left.-\left(\mathbf{u}_{(p)}+m_1\operatorname{grad}\theta\right)\frac{\partial}{\partial n}\left(\frac{e^{i\mu_1 R}}{R}\right)\right]dS, \qquad \boldsymbol{\xi}\in\Omega_e,\end{aligned} \tag{1.120}$$

$$\begin{aligned}4\pi(\mu_1^2-\mu_2^2)\theta(\boldsymbol{\xi}) &= (\mu_1^2-q)\int_S\left[\frac{e^{i\mu_2 R}}{R}\frac{\partial}{\partial n}\left(\theta-\hat{h}_2\operatorname{div}\mathbf{u}_{(p)}\right)\right.\\ &\quad\left.-\left(\theta-\hat{h}_2\operatorname{div}\mathbf{u}_{(p)}\right)\frac{\partial}{\partial n}\left(\frac{e^{i\mu_2 R}}{R}\right)\right]dS\\ &\quad-(\mu_2^2-q)\int_S\left[\frac{e^{i\mu_1 R}}{R}\frac{\partial}{\partial n}\left(\theta-\hat{h}_1\operatorname{div}\mathbf{u}_{(p)}\right)\right.\\ &\quad\left.-\left(\theta-\hat{h}_1\operatorname{div}\mathbf{u}_{(p)}\right)\frac{\partial}{\partial n}\left(\frac{e^{i\mu_1 R}}{R}\right)\right]dS, \qquad \boldsymbol{\xi}\in\Omega_e,\end{aligned} \tag{1.121}$$

where $f(r)=0(1)$ *means that* $f(r)\to 0$ *as* $r\to\infty$, $\dfrac{\partial}{\partial n}$ *is the derivative in the direction of the unit normal vector:* $\dfrac{\partial f}{\partial n}=f_{,i}n_i$, ∂_r *denote partial derivative with respect to* r: $\partial_r f\equiv\dfrac{\partial f}{\partial r}$, $R=|\mathbf{x}-\boldsymbol{\xi}|$. *Recall from Subsection* 1.1.4 *that the polynomial* $W_4(\mu)$ *has the following form:*

$$W_4(\mu)=\mu^4-[\sigma_1^2+q(1+\varepsilon)]\mu^2+q\sigma_1^2=(\mu^2-\mu_1^2)(\mu^2-\mu_2^2).$$

The roots μ_1, μ_2 *of the equation* $W_4(\mu)=0$ *are complex numbers of the form* $\mu_1=\alpha_1+i\beta_1$, $\mu_2=\alpha_2+i\beta_2$ $(\beta_1\geqslant 0,\ \beta_2\geqslant 0)$ *and their squares are given by* $\mu_{1,2}^2=\frac{1}{2}[\sigma_1^2+q(1+\varepsilon)\pm\sqrt{\Delta}]$, *where* $\Delta=[\sigma_1^2+q(1+\varepsilon)]^2-4q\sigma_1^2$. *Besides, we have introduced the following new notations:*

$$m_1 = \frac{m}{\mu_2^2 - \sigma_1^2}, \qquad m_2 = \frac{m}{\mu_1^2 - \sigma_1^2},$$

$$\hat{h}_1 = \frac{q\varepsilon}{m(\mu_2^2 - q)}, \qquad \hat{h}_2 = \frac{q\varepsilon}{m(\mu_1^2 - q)}.$$

2. If $\boldsymbol{\varphi}_{(p)}$ satisfies the conditions

$$\boldsymbol{\varphi}_{(p)} = \mathbf{0}(1), \qquad (\partial_r - i\lambda_3)\,\boldsymbol{\varphi}_{(p)} = \mathbf{0}(r^{-1}), \tag{1.122}$$

then the following integral equation holds:

$$\boldsymbol{\varphi}_{(p)}(\boldsymbol{\xi}) = \frac{1}{4\pi}\int_S \left[\frac{e^{i\lambda_3 R}}{R}\frac{\partial}{\partial n}\boldsymbol{\varphi}_{(p)} - \boldsymbol{\varphi}_{(p)}\frac{\partial}{\partial n}\left(\frac{e^{i\lambda_3 R}}{R}\right)\right] dS, \qquad \boldsymbol{\xi} \in \Omega_e, \tag{1.123}$$

where $f(r) = 0(r^{-1}) \Longrightarrow rf(r) = 0(1)$, λ_3 is the same as in the formula $(1.27)_2$ and satisfies the condition:

$$\lambda_3^2 = \sigma_3^2 - \frac{4\alpha}{\beta + 2\gamma} > 0.$$

3. If $\mathbf{u}_{(s)}$ and $\boldsymbol{\varphi}_{(s)}$ satisfy in the neighborhood of infinity the following asymptotic conditions

$$\mathbf{u}_{(s)} = \mathbf{0}(1), \qquad \operatorname{curl}\mathbf{u}_{(s)} = \mathbf{0}(1), \tag{1.124}$$

$$\boldsymbol{\varphi}_{(s)} = \mathbf{0}(1), \qquad \operatorname{curl}\boldsymbol{\varphi}_{(s)} = \mathbf{0}(1), \tag{1.125}$$

$$(\partial_r - i\lambda_1)\left(\mathbf{u}_{(s)} - s_1 \operatorname{curl}\boldsymbol{\varphi}_{(s)}\right) = \mathbf{0}(r^{-1}), \tag{1.126}$$

$$(\partial_r - i\lambda_2)\left(\mathbf{u}_{(s)} - s_2 \operatorname{curl}\boldsymbol{\varphi}_{(s)}\right) = \mathbf{0}(r^{-1}), \tag{1.127}$$

$$(\partial_r - i\lambda_1)\left(\boldsymbol{\varphi}_{(s)} - p_1 \operatorname{curl}\mathbf{u}_{(s)}\right) = \mathbf{0}(r^{-1}), \tag{1.128}$$

$$(\partial_r - i\lambda_2)\left(\boldsymbol{\varphi}_{(s)} - p_2 \operatorname{curl}\mathbf{u}_{(s)}\right) = \mathbf{0}(r^{-1}) \tag{1.129}$$

then the following integral equations hold:

$$\begin{aligned} 4\pi(\lambda_1^2 - \lambda_2^2)\mathbf{u}_{(s)}(\boldsymbol{\xi}) &= (\lambda_1^2 - \sigma_2^2)\int_S \left[\frac{e^{i\lambda_2 R}}{R}\frac{\partial}{\partial n}\left(\mathbf{u}_{(s)} - s_2 \operatorname{curl}\boldsymbol{\varphi}_{(s)}\right)\right. \\ &\quad \left. - \left(\mathbf{u}_{(s)} - s_2 \operatorname{curl}\boldsymbol{\varphi}_{(s)}\right)\frac{\partial}{\partial n}\left(\frac{e^{i\lambda_2 R}}{R}\right)\right] dS \\ &\quad - (\lambda_2^2 - \sigma_2^2)\int_S \left[\frac{e^{i\lambda_1 R}}{R}\frac{\partial}{\partial n}\left(\mathbf{u}_{(s)} - s_1 \operatorname{curl}\boldsymbol{\varphi}_{(s)}\right)\right. \\ &\quad \left. - \left(\mathbf{u}_{(s)} - s_1 \operatorname{curl}\boldsymbol{\varphi}_{(s)}\right)\frac{\partial}{\partial n}\left(\frac{e^{i\lambda_1 R}}{R}\right)\right] dS, \qquad \boldsymbol{\xi} \in \Omega_e, \end{aligned} \tag{1.130}$$

$$
\begin{aligned}
4\pi(\lambda_1^2-\lambda_2^2)\boldsymbol{\varphi}_{(s)}(\boldsymbol{\xi}) &= (\lambda_1^2-\hat{\sigma}_4^2)\int_S\left[\frac{e^{i\lambda_2 R}}{R}\frac{\partial}{\partial n}\left(\boldsymbol{\varphi}_{(s)}-p_2\operatorname{curl}\mathbf{u}_{(s)}\right)\right.\\
&\quad\left.-\left(\boldsymbol{\varphi}_{(s)}-p_2\operatorname{curl}\mathbf{u}_{(s)}\right)\frac{\partial}{\partial n}\left(\frac{e^{i\lambda_2 R}}{R}\right)\right]dS\\
&\quad-(\lambda_2^2-\hat{\sigma}_4^2)\int_S\left[\frac{e^{i\lambda_1 R}}{R}\frac{\partial}{\partial n}\left(\boldsymbol{\varphi}_{(s)}-p_1\operatorname{curl}\mathbf{u}_{(s)}\right)\right.\\
&\quad\left.-\left(\boldsymbol{\varphi}_{(s)}-p_1\operatorname{curl}\mathbf{u}_{(s)}\right)\frac{\partial}{\partial n}\left(\frac{e^{i\lambda_1 R}}{R}\right)\right]dS,\qquad \boldsymbol{\xi}\in\Omega_e.
\end{aligned}
\tag{1.131}
$$

The values λ_1^2 and λ_2^2 are the roots of the equation

$$\lambda^4-(\sigma_2^2+\hat{\sigma}_4^2+ps)\lambda^2+\sigma_2^2\hat{\sigma}_4^2=0$$

with the determinant $\Delta>0$. The roots λ_1^2, $\lambda_2^2\in\mathbb{R}$ are different and given by the formula (1.28) *in Subsection* 1.1.2, *which can be transformed to the form*

$$\lambda_{1,2}=\frac{1}{\sqrt{2}}\left(\sigma_2^2+\hat{\sigma}_4^2+ps\pm\sqrt{\Delta}\right)^{\frac{1}{2}},$$

where

$$\sqrt{\Delta}=\left[(\hat{\sigma}_4-\sigma_2)^2+ps\right]^{\frac{1}{2}}\left[(\hat{\sigma}_4+\sigma_2)^2+ps\right]^{\frac{1}{2}}.$$

Besides, we have introduced the following new notations:

$$\hat{\sigma}_4^2=\sigma_4^2-2p>0,$$

$$s_1=\frac{s}{\lambda_2^2-\sigma_2^2},\qquad s_2=\frac{s}{\lambda_1^2-\sigma_2^2},$$

$$p_1=\frac{p}{\lambda_2^2-\hat{\sigma}_4^2},\qquad p_2=\frac{p}{\lambda_1^2-\hat{\sigma}_4^2}.$$

The radiation conditions (1.112)–(1.119), (1.122), (1.124)–(1.129), together with some smoothness conditions on $\mathbf{u}_{(p)}$, $\boldsymbol{\varphi}_{(p)}$, $\mathbf{u}_{(s)}$, $\boldsymbol{\varphi}_{(s)}$ and θ determine a regular solution $(\mathbf{u},\,\boldsymbol{\varphi},\,\theta)$ of the linear, homogeneous, isotropic, dynamical, coupled, micropolar thermoelasticity (micropolar model of (E-N)) in the infinite region Ω_e.

1.1.7 Generalized Galerkin vector. Representation of Iacovache type. Micropolar theory and couple-stress theory

By means of the vectorial differential equations (1.14)–(1.17) we introduced in [49] the Galerkin vector, generalized to a micropolar medium. Assuming

$$\mathbf{u}^0=[\overrightarrow{\Box}_1-(\lambda+\mu)\operatorname{grad}\operatorname{div}]\mathbf{F}^0,\qquad 2\boldsymbol{\varphi}^0=\operatorname{curl}\mathbf{M}^0 \tag{1.132}$$

and using equations (1.14), we write the following equations:

$$\overrightarrow{\Box}_1 \overset{*}{\overrightarrow{\Box}}_2 \mathbf{F}^0 + \mathbf{X} = \mathbf{0}, \qquad \overset{*}{\overrightarrow{\Box}}_2 \mathbf{M}^0 + \mathbf{X} = \mathbf{0}. \tag{1.133}$$

Let us rewrite equations (1.15)–(1.17) for the vectors $\mathbf{u}'$, $\boldsymbol{\zeta}$, $\boldsymbol{\varphi}'$ in the operator form:

$$\begin{aligned}
&\overset{*}{\overrightarrow{\Box}}_2 \overrightarrow{\Box}_1 (\overrightarrow{\Box}_2 \overrightarrow{\Box}_4 + 4\alpha^2 \overrightarrow{\nabla}^2)\mathbf{u}' = \overrightarrow{\mathrm{N}}_1 \hat{\mathbf{X}} + \overrightarrow{\mathrm{N}}_2 \hat{\mathbf{Y}},\\
&2\overset{*}{\overrightarrow{\Box}}_2 \overrightarrow{\Box}_3 (\overrightarrow{\Box}_2 \overrightarrow{\Box}_4 + 4\alpha^2 \overrightarrow{\nabla}^2)\boldsymbol{\zeta} = \overrightarrow{\mathrm{N}}_3 \hat{\mathbf{Y}} + \overrightarrow{\mathrm{N}}_4 \hat{\mathbf{X}},\\
&\overset{*}{\overrightarrow{\Box}}_2 \overrightarrow{\Box}_3 (\overrightarrow{\Box}_2 \overrightarrow{\Box}_4 + 4\alpha^2 \overrightarrow{\nabla}^2)\boldsymbol{\varphi}' = \overrightarrow{\mathrm{N}}_5 \hat{\mathbf{Y}} + \overrightarrow{\mathrm{N}}_6 \hat{\mathbf{X}},
\end{aligned} \tag{1.134}$$

where the operators $\overrightarrow{\mathrm{N}}_i$ $(i = 1, \ldots, 6)$ are given by

$$\overrightarrow{\mathrm{N}}_1 = \overset{*}{\overrightarrow{\Box}}_2 \Big\{ [(\lambda + \mu - \alpha)\overrightarrow{\Box}_4 - 4\alpha^2] \operatorname{grad}\operatorname{div} - \overrightarrow{\Box}_1 \overrightarrow{\Box}_4 \Big\},$$

$$\overrightarrow{\mathrm{N}}_2 = 2\alpha \overrightarrow{\Box}_1 \operatorname{curl},$$

$$\overrightarrow{\mathrm{N}}_3 = [2\alpha \overset{*}{\overrightarrow{\Box}}_4 - 2(\beta + \gamma - \varepsilon)\overset{*}{\overrightarrow{\Box}}_2] \operatorname{grad}\operatorname{div} + 2\overset{*}{\overrightarrow{\Box}}_2 \overrightarrow{\Box}_3,$$

$$\overrightarrow{\mathrm{N}}_4 = -\overset{*}{\overrightarrow{\Box}}_2 \overrightarrow{\Box}_3 \overset{*}{\overrightarrow{\Box}}_4 \operatorname{curl},$$

$$\overrightarrow{\mathrm{N}}_5 = [(\beta + \gamma - \varepsilon)\overrightarrow{\Box}_2 - 4\alpha^2] \operatorname{grad}\operatorname{div} - \overrightarrow{\Box}_2 \overrightarrow{\Box}_3,$$

$$\overrightarrow{\mathrm{N}}_6 = 2\alpha \overset{*}{\overrightarrow{\Box}}_2 \overrightarrow{\Box}_3 \operatorname{curl}.$$

Let us introduce the vectors $\mathbf{F}'$ and $\mathbf{M}'$, assuming the representation

$$\begin{aligned}
&\mathbf{u}' = -\overrightarrow{\mathrm{N}}{}^0_1 \mathbf{F}' + \overrightarrow{\mathrm{N}}{}^0_4 \mathbf{M}', \qquad 2\boldsymbol{\zeta} = -\overrightarrow{\mathrm{N}}_3 \mathbf{M}' + \overrightarrow{\mathrm{N}}{}^0_2 \mathbf{F}',\\
&\boldsymbol{\varphi}' = -\overrightarrow{\mathrm{N}}_5 \mathbf{M}' - \overrightarrow{\mathrm{N}}_2 \mathbf{F}',
\end{aligned} \tag{1.135}$$

where the operators $\overrightarrow{\mathrm{N}}{}^0_1$, $\overrightarrow{\mathrm{N}}{}^0_2$ and $\overrightarrow{\mathrm{N}}{}^0_4$ have the forms

$$\overrightarrow{\mathrm{N}}_1 = \overset{*}{\overrightarrow{\Box}}_2 \overrightarrow{\mathrm{N}}{}^0_1, \qquad \overrightarrow{\mathrm{N}}{}^0_2 = \overrightarrow{\Box}_1 \overset{*}{\overrightarrow{\Box}}_4 \operatorname{curl}, \qquad \overrightarrow{\mathrm{N}}{}^0_4 = -2\alpha \overrightarrow{\Box}_3 \operatorname{curl}.$$

Substituting (1.135) to (1.134), we obtain the equations for the vectors $\mathbf{F}'$ and $\mathbf{M}'$ in the form

$$\begin{aligned}
&\overrightarrow{\Box}_1 (\overrightarrow{\Box}_2 \overrightarrow{\Box}_4 + 4\alpha^2 \overrightarrow{\nabla}^2)\mathbf{F}' + \hat{\mathbf{X}} = \mathbf{0},\\
&\overset{*}{\overrightarrow{\Box}}_2 \overrightarrow{\Box}_3 (\overrightarrow{\Box}_2 \overrightarrow{\Box}_4 + 4\alpha^2 \overrightarrow{\nabla}^2)\mathbf{M}' + \hat{\mathbf{Y}} = \mathbf{0}.
\end{aligned} \tag{1.136}$$

Let us recall that the vectors $\hat{\mathbf{X}}$, $\hat{\mathbf{Y}}$ are given by formulae (1.9). Take into account the following observation: assuming (1.18) and the substitution $(\mathbf{F}', \mathbf{M}') \mapsto (\mathbf{F}, \mathbf{M})$, we obtain the representation for the vectors $\mathbf{u}$, $\boldsymbol{\varphi}$ in view of $(1.135)_{1,3}$ and (1.136),

$$\mathbf{u} = -\overrightarrow{\mathrm{N}}_1^0\mathbf{F} + \overrightarrow{\mathrm{N}}_4^0\mathbf{M}, \qquad \boldsymbol{\varphi} = -\overrightarrow{\mathrm{N}}_2\mathbf{F} - \overrightarrow{\mathrm{N}}_5\mathbf{M} \tag{1.137}$$

and the field equations for the vectors $\mathbf{F}$, $\mathbf{M}$,

$$\begin{aligned} &\vec{\Box}_1(\vec{\Box}_2\vec{\Box}_4 + 4\alpha^2\vec{\nabla}^2)\mathbf{F} + \mathbf{X} = \mathbf{0},\\ &\vec{\Box}_3(\vec{\Box}_2\vec{\Box}_4 + 4\alpha^2\vec{\nabla}^2)\mathbf{M} + \mathbf{Y} = \mathbf{0}. \end{aligned} \tag{1.138}$$

Another representation for $\mathbf{u}'$, $\boldsymbol{\zeta}$, $\boldsymbol{\varphi}'$ obtained by the superposition method is different from (1.135) and has the form

$$\begin{aligned} \mathbf{u}' &= \alpha\overset{*}{\vec{\Box}}_4 \operatorname{curl}\operatorname{curl}\mathbf{F}' + \overrightarrow{\mathrm{N}}_4^0\mathbf{M}',\\ 2\boldsymbol{\zeta} &= \overset{*}{\vec{\Box}}_2\overset{*}{\vec{\Box}}_4 \operatorname{curl}\mathbf{F}' - \overrightarrow{\mathrm{N}}_3\mathbf{M}',\\ \boldsymbol{\varphi}' &= -\tfrac{1}{2}\vec{\Box}_2\overset{*}{\vec{\Box}}_4 \operatorname{curl}\mathbf{F}' - \overrightarrow{\mathrm{N}}_5\mathbf{M}'. \end{aligned} \tag{1.139}$$

Equations (1.134) lead to the field equations

$$\begin{aligned} &\overset{*}{\vec{\Box}}_2(\vec{\Box}_2\vec{\Box}_4 + 4\alpha^2\vec{\nabla}^2)\mathbf{F}' + \mathbf{X} = \mathbf{0},\\ &\vec{\Box}_3(\vec{\Box}_2\vec{\Box}_4 + 4\alpha^2\vec{\nabla}^2)\mathbf{M}' + \mathbf{Y} = \mathbf{0}. \end{aligned} \tag{1.140}$$

We also have relation (1.13): $\mathbf{u} = \mathbf{u}^0 + \mathbf{u}'$, $\boldsymbol{\varphi} = \boldsymbol{\varphi}^0 + \boldsymbol{\varphi}'$.

Formulae $(1.132)_1$ and $(1.133)_1$ constitute the Iacovache[13] representation (see [180, p. 504]), which generalizes the Galerkin vector in the classical theory of elasticity to the problem of elastokinetics. The proof of completeness for this representation was given by Sternberg & Eubanks [180, p. 507]. The representations (1.132) and (1.135), or (1.132) and (1.139), were suitably related to the field equations (1.136) and (1.140), and they constitute a generalization of the Iacovache representation of the (1.137)-type with the field equations (1.138) to the micropolar theory of elasticity, which was obtained by use of various methods in the papers by Nowacki [192, p. 35] (where the completeness proof was given for this representation with no body forces), Sandru [206] and Stefaniak [217].

Let us consider the equation for the displacements in couple-stress theory (0.28) in its vector form (for $\theta = 0$ and with the vector $\boldsymbol{\varphi}$):

$$\overset{*}{\vec{\Box}}_2\mathbf{u} + (\lambda + \mu)\operatorname{grad}\operatorname{div}\mathbf{u} + 2\mu l^{\star 2}\vec{\nabla}^2\operatorname{curl}\boldsymbol{\varphi} + \mathbf{F} = \mathbf{0}, \tag{1.141}$$

where we have

$$\boldsymbol{\varphi} = \tfrac{1}{2}\operatorname{curl}\mathbf{u}, \qquad \mathbf{F} = \mathbf{X} + \tfrac{1}{2}\operatorname{curl}\mathbf{Y}. \tag{1.142}$$

From (1.141) and (1.142) we derive the following separate equations for the vectors $\mathbf{u}$, $\boldsymbol{\varphi}$:

[13] See Iacovache [109]. Gurtin [105, p. 235] calls $(1.132)_1$ the representation of Cauchy-Kovalevski-Somigliana.

$$\begin{aligned}
\vec{\Box}_1\vec{\Box}_0\mathbf{u} &= \frac{1}{\mu}\vec{\Box}_1\mathbf{F} + \left(\frac{1}{2\nu-1} - l^{\star 2}\vec{\nabla}^2\right)\operatorname{grad}\operatorname{div}\mathbf{X},\\
\vec{\Box}_0\boldsymbol{\varphi} &= \frac{1}{2\mu}\operatorname{curl}\mathbf{F},
\end{aligned} \tag{1.143}$$

where

$$\vec{\Box}_0 = \vec{\nabla}^2(l^{\star 2}\vec{\nabla}^2 - 1) + \hat{c}_2^{-2}\partial_t^2, \qquad \hat{c}_2 = \sqrt{\frac{\mu}{\rho}}.$$

Let us write equations (1.143) in the operator form:

$$\vec{\Box}_1\vec{\Box}_0\mathbf{u} = \vec{\mathrm{L}}_1\mathbf{X} + \vec{\mathrm{L}}_2\mathbf{Y}, \qquad \vec{\Box}_0\boldsymbol{\varphi} = \vec{\mathrm{L}}_3\mathbf{Y} + \vec{\mathrm{L}}_4\mathbf{X}. \tag{1.144}$$

The operators $\vec{\mathrm{L}}_i$ $(i = 1, \ldots, 4)$ are given by

$$\begin{aligned}
\vec{\mathrm{L}}_1 &= \frac{1}{\mu}\vec{\Box}_1 - \left(\frac{1}{1-2\nu} + l^{\star 2}\vec{\nabla}^2\right)\operatorname{grad}\operatorname{div},\\
\vec{\mathrm{L}}_2 &= \vec{\Box}_1\vec{\mathrm{L}}_4,\\
\vec{\mathrm{L}}_3 &= \tfrac{1}{2}\operatorname{curl}\vec{\mathrm{L}}_4,\\
\vec{\mathrm{L}}_4 &= \frac{1}{2\mu}\operatorname{curl}.
\end{aligned}$$

Let us introduce the following representation in terms of the vectors $\hat{\mathbf{F}}$, $\mathbf{M}$:

$$\mathbf{u} = -\vec{\mathrm{L}}_1\hat{\mathbf{F}} - \vec{\mathrm{L}}_4\mathbf{M}, \qquad \boldsymbol{\varphi} = -\vec{\mathrm{L}}_3\mathbf{M} - \vec{\mathrm{L}}_2\hat{\mathbf{F}}. \tag{1.145}$$

The field equations for vectors $\hat{\mathbf{F}}$, $\mathbf{M}$ take the form:

$$\vec{\Box}_1\vec{\Box}_0\hat{\mathbf{F}} + \mathbf{X} = \mathbf{0}, \qquad \vec{\Box}_0\mathbf{M} + \mathbf{Y} = \mathbf{0}. \tag{1.146}$$

Representation (1.145), together with the field equations (1.146), can be treated as still another generalization of the Galerkin vector to couple-stress elastodynamics (cf. Sokołowski [215, p. 24]). For $l^\star \to 0$ (we assume $\mathbf{Y} = \mathbf{0}$, $\mathbf{M} = \mathbf{0}$), we have

$$\begin{aligned}
&\mu\vec{\Box}_0 \to -\overset{*}{\vec{\Box}}_2, \qquad \mu\vec{\mathrm{L}}_1 \to [\vec{\Box}_1 - (\mu+\lambda)\operatorname{grad}\operatorname{div}],\\
&-\hat{\mathbf{F}} \to \mu\mathbf{F}^0, \qquad \mathbf{u} \to \mathbf{u}^0
\end{aligned}$$

and from $(1.145)_1$ and $(1.146)_1$ we obtain the representation $(1.132)_1$ and the Iacovache equation $(1.133)_1$, respectively.

1.1.8 Generalized representation of Green-Lamé. The method of Nowacki's potentials

The method of potentials is based on the decomposition of the displacement vector $\mathbf{u}$ and the rotation vector $\boldsymbol{\varphi}$ into their potential and solenoidal parts

$$\begin{aligned} \mathbf{u} &= \operatorname{grad}\Phi + \operatorname{curl}\boldsymbol{\Psi}, \qquad \operatorname{div}\boldsymbol{\Psi} = 0,\\ \boldsymbol{\varphi} &= \operatorname{grad}\Sigma + \operatorname{curl}\mathbf{H}, \qquad \operatorname{div}\mathbf{H} = 0 \end{aligned} \tag{1.147}$$

and it is the generalization of the Lamé method [142] of uncoupling the equations of classical elastokinetics.

Clebsch (1863), Duhem (1898), Sternberg (1960) (see Gurtin [105, p. 233], and Nowacki [180, p. 500]) have proved that the Green-Lamé representation, expressed in terms of the dynamical potentials Φ and $\boldsymbol{\Psi}$, is sufficient for a complete determination of the displacement field $\mathbf{u}$. Now, returning to micropolar theory, we make use of the Helmholtz decomposition for the vectors of body loadings:

$$\begin{aligned} \mathbf{X} &= \rho(\operatorname{grad}\vartheta + \operatorname{curl}\boldsymbol{\chi}), \qquad \operatorname{div}\boldsymbol{\chi} = 0,\\ \mathbf{Y} &= J(\operatorname{grad}\sigma + \operatorname{curl}\boldsymbol{\eta}), \qquad \operatorname{div}\boldsymbol{\eta} = 0. \end{aligned} \tag{1.148}$$

Introducing relations (1.147) and (1.148) into the system of equations (1.1), we obtain the equations

$$\begin{aligned} &\operatorname{grad}(\Box_1\Phi + \rho\vartheta) + \operatorname{curl}(\vec{\Box}_2\boldsymbol{\Psi} + 2\alpha\operatorname{curl}\mathbf{H} + \rho\boldsymbol{\chi}) = \mathbf{0},\\ &\operatorname{grad}(\Box_3\Sigma + J\sigma) + \operatorname{curl}(\vec{\Box}_4\mathbf{H} + 2\alpha\operatorname{curl}\boldsymbol{\Psi} + J\boldsymbol{\eta}) = \mathbf{0}. \end{aligned} \tag{1.149}$$

Equations (1.149) are satisfied in a particular way; namely, we assume that the expressions under the gradient and curl operators vanish independently. Thus, we obtain the scalar and the vector wave equations:

$$\begin{aligned} &\Box_1\Phi + \rho\vartheta = 0, \qquad \Box_3\Sigma + J\sigma = 0,\\ &\vec{\Box}_2\boldsymbol{\Psi} + 2\alpha\operatorname{curl}\mathbf{H} + \rho\boldsymbol{\chi} = \mathbf{0}, \qquad \vec{\Box}_4\mathbf{H} + 2\alpha\operatorname{curl}\boldsymbol{\Psi} + J\boldsymbol{\eta} = \mathbf{0}, \end{aligned} \tag{1.150}$$

which were given by Nowacki [192, p. 34]. The same author has also proved the completeness of the representations (1.147), (1.148) and (1.150).

In [49], the following (in fact, very instructive) observation was made. Namely, the following relations between the potentials Φ, Σ, $\boldsymbol{\Psi}$, $\mathbf{H}$:

$$\begin{aligned} &\vec{\Box}_1\operatorname{grad}\Phi + \vec{\Box}_2\operatorname{curl}\boldsymbol{\Psi} - 2\alpha\vec{\nabla}^2\mathbf{H} + \mathbf{X} = \mathbf{0},\\ &\vec{\Box}_3\operatorname{grad}\Sigma + \vec{\Box}_4\operatorname{curl}\mathbf{H} - 2\alpha\vec{\nabla}^2\boldsymbol{\Psi} + \mathbf{Y} = \mathbf{0}, \end{aligned} \tag{1.151}$$

were derived from equations (1.1) and (1.147).

The system of equations (1.151) leads to the uncoupled equations of higher orders, namely:

— the scalar equations for the potentials Φ and Σ,

$$\nabla^2\Box_1\Phi = -\operatorname{div}\mathbf{X}, \qquad \nabla^2\Box_3\Sigma = -\operatorname{div}\mathbf{Y}, \tag{1.152}$$

where $\nabla^2 \overset{\text{def}}{=} \operatorname{div}\operatorname{grad}$;

— the vector equations for the potentials $\boldsymbol{\Psi}$ and $\mathbf{H}$,

$$\begin{aligned} \vec{\nabla}^2(\vec{\Box}_2\vec{\Box}_4 + 4\alpha^2\vec{\nabla}^2)\boldsymbol{\Psi} &= \vec{\Box}_4\operatorname{curl}\mathbf{X} - 2\alpha\operatorname{curl}\operatorname{curl}\mathbf{Y}, \\ \vec{\nabla}^2(\vec{\Box}_2\vec{\Box}_4 + 4\alpha^2\vec{\nabla}^2)\mathbf{H} &= \vec{\Box}_2\operatorname{curl}\mathbf{Y} - 2\alpha\operatorname{curl}\operatorname{curl}\mathbf{X}. \end{aligned} \tag{1.153}$$

Recall once again that $\vec{\nabla}^2 \overset{\text{def}}{=} \operatorname{grad}\operatorname{div} - \operatorname{curl}\operatorname{curl}$.

Assuming the decomposition (1.148) and removing the operator ∇^2, we obtain from (1.152) the scalar equations $(1.150)_{1,2}$.

In turn, assuming (1.148) and dividing by the operator $\vec{\nabla}^2$, we obtain from (1.153) the vector equations

$$\begin{aligned} (\vec{\Box}_2\vec{\Box}_4 + 4\alpha^2\vec{\nabla}^2)\boldsymbol{\Psi} &= 2\alpha J\operatorname{curl}\boldsymbol{\eta} - \rho\vec{\Box}_4\boldsymbol{\chi}, \\ (\vec{\Box}_2\vec{\Box}_4 + 4\alpha^2\vec{\nabla}^2)\mathbf{H} &= 2\alpha\rho\operatorname{curl}\boldsymbol{\chi} - J\vec{\Box}_2\boldsymbol{\eta}, \end{aligned} \tag{1.154}$$

which follow from the wave equations $(1.150)_{3,4}$.

In the sequel, we will often return to the representation (1.147) and the field equations (1.151)–(1.153).

1.2 Displacement-rotation and stress equations of elastostatics and thermoelastostatics

1.2.1 Vector equations and superposition method

Let us consider the vector equations in displacements and rotations of micropolar elastostatics in Ω:

$$\begin{aligned} &(\mu+\alpha)\vec{\nabla}^2\mathbf{u} + (\lambda+\mu-\alpha)\operatorname{grad}\operatorname{div}\mathbf{u} + 2\alpha\operatorname{curl}\boldsymbol{\varphi} + \mathbf{X} = \mathbf{0}, \\ &[(\gamma+\varepsilon)\vec{\nabla}^2 - 4\alpha]\boldsymbol{\varphi} + (\beta+\gamma-\varepsilon)\operatorname{grad}\operatorname{div}\boldsymbol{\varphi} + 2\alpha\operatorname{curl}\mathbf{u} + \mathbf{Y} = \mathbf{0}, \end{aligned} \tag{1.155}$$

where $\vec{\nabla}^2$ is the vector Laplace operator. Next, let us formulate the boundary conditions in terms of displacements and rotations

$$\mathbf{u} = \mathbf{f}, \qquad \boldsymbol{\varphi} = \mathbf{g} \qquad \text{on } \partial\Omega, \tag{1.156}$$

where the vectors $\mathbf{f}$, $\mathbf{g} : \partial\Omega \to \mathbb{R}^3$ are given.

In [49], we showed a method of uncoupling the system of equations (1.155) with the boundary conditions[14] (1.156). Let us introduce the Schaefer vector

14 In [49] we considered the boundary conditions in displacements, rotations, force-stresses and couple-stresses with a temperature field.

(1.4) in the system of equations (1.155). Thus, (1.155) becomes a new system of equations, namely

$$\begin{aligned}&\mu\vec{\nabla}^2\mathbf{u}+(\lambda+\mu)\,\mathrm{grad\,div}\,\mathbf{u}+\mathbf{X}=2\alpha\,\mathrm{curl}\,\boldsymbol{\zeta},\\&(\gamma+\varepsilon)\vec{\nabla}^2\,\mathrm{curl}\,\mathbf{u}+2\mathbf{Y}\\&\qquad=2\Big\{[(\gamma+\varepsilon)\vec{\nabla}^2-4\alpha]\boldsymbol{\zeta}+(\beta+\gamma-\varepsilon)\,\mathrm{grad\,div}\,\boldsymbol{\zeta}\Big\}.\end{aligned}\tag{1.157}$$

We assume the following form of the solution of the boundary value problem:

$$\mathbf{u}=\mathbf{u}^0+\mathbf{u}',\qquad \boldsymbol{\varphi}=\boldsymbol{\varphi}^0+\boldsymbol{\varphi}'.\tag{1.158}$$

To this end, the vectors $\boldsymbol{\zeta}$, $\mathbf{Y}$ are assumed to have the forms $\boldsymbol{\zeta}=\boldsymbol{\zeta}^0+\boldsymbol{\zeta}'$ and $\mathbf{Y}=\mathbf{Y}^0+\mathbf{Y}'$. For $\boldsymbol{\zeta}^0=0$ ($\mathbf{Y}^0=\mathbf{0}$) we obtain from $(1.157)_1$ and (1.156) the boundary value problem in the framework of classical elasticity. Thus, we obtain

— the equation in terms of displacements in Ω,

$$\mu\vec{\nabla}^2\mathbf{u}^0+(\lambda+\mu)\,\mathrm{grad\,div}\,\mathbf{u}^0+\mathbf{X}=\mathbf{0};\tag{1.159}$$

— the boundary conditions on $\partial\Omega$,

$$\mathbf{u}^0=\mathbf{f}.\tag{1.160}$$

Moreover, from equation $(1.157)_2$ we obtain in Ω the equation

$$2\mu\vec{\nabla}^2\boldsymbol{\varphi}^0+\mathrm{curl}\,\mathbf{X}=\mathbf{0},\tag{1.161}$$

where $\boldsymbol{\varphi}^0=\frac{1}{2}\,\mathrm{curl}\,\mathbf{u}^0$, which will be an identity in view of equation (1.159). Next, for the vector $\mathbf{u}^0$ we obtain the equation

$$\mu\vec{\nabla}^2\vec{\nabla}^2\mathbf{u}^0=\frac{\lambda+\mu}{2\mu+\lambda}\,\mathrm{grad\,div}\,\mathbf{X}-\vec{\nabla}^2\mathbf{X}.\tag{1.162}$$

Moreover, assuming $\boldsymbol{\zeta}^0=0$ and taking into account (1.157), we obtain the following system of equations with the vectors $\mathbf{u}'$, $\boldsymbol{\zeta}$:

$$\begin{aligned}&\mu\vec{\nabla}^2\mathbf{u}'+(\lambda+\mu)\,\mathrm{grad\,div}\,\mathbf{u}'+\hat{\mathbf{X}}=2\alpha\,\mathrm{curl}\,\boldsymbol{\zeta},\\&(\gamma+\varepsilon)\vec{\nabla}^2\,\mathrm{curl}\,\mathbf{u}'+2\overset{*}{\mathbf{Y}}\\&\qquad=2\Big\{[(\gamma+\varepsilon)\vec{\nabla}^2-4\alpha]\boldsymbol{\zeta}+(\beta+\gamma-\varepsilon)\,\mathrm{grad\,div}\,\boldsymbol{\zeta}\Big\},\end{aligned}\tag{1.163}$$

where

$$\hat{\mathbf{X}}=\mathbf{0},\qquad \overset{*}{\mathbf{Y}}=\mathbf{Y}-\frac{\gamma+\varepsilon}{2\mu}\,\mathrm{curl}\,\mathbf{X}.\tag{1.164}$$

From the system of equations (1.163) we now derive a separate equation for the vector $\boldsymbol{\zeta}$:

$$
\begin{aligned}
4\alpha(\nu_1^2\overrightarrow{\nabla}^2-1)(l^2\overrightarrow{\nabla}^2-1)\boldsymbol{\zeta} &= -\mu_0^{-1}(\nu_1^2\overrightarrow{\nabla}^2-1)\,\mathrm{curl}\,\hat{\mathbf{X}} \\
&\quad + (\nu_1^2\overrightarrow{\nabla}^2-1)\overset{*}{\mathbf{Y}} + \left[\tfrac{1}{2}\mu_0^{-1}-(\nu_1^2-k^2)\right]\mathrm{grad\,div}\,\overset{*}{\mathbf{Y}},
\end{aligned}
\tag{1.165}
$$

where

$$
l^2 = \frac{(\gamma+\varepsilon)(\mu+\alpha)}{4\alpha\mu}, \qquad k^2 = \frac{\gamma+\varepsilon}{4\alpha}, \qquad \mu_0 = \frac{2\mu}{\gamma+\varepsilon}, \qquad \nu_1^2 = \frac{2\gamma+\beta}{4\alpha}.
$$

Still, it remains to formulate the boundary value problem for the complementary solution $\mathbf{u}'$, $\boldsymbol{\varphi}'$. To do this, we eliminate the vector $\boldsymbol{\zeta}$ from the system of equations (1.163) and obtain

$$
\begin{aligned}
&(\mu+\alpha)\overrightarrow{\nabla}^2\mathbf{u}' + (\lambda+\mu-\alpha)\,\mathrm{grad\,div}\,\mathbf{u}' + 2\alpha\,\mathrm{curl}\,\boldsymbol{\varphi}' + \hat{\mathbf{X}} = \mathbf{0}, \\
&[(\gamma+\varepsilon)\overrightarrow{\nabla}^2 - 4\alpha]\boldsymbol{\varphi}' + (\beta+\gamma-\varepsilon)\,\mathrm{grad\,div}\,\boldsymbol{\varphi}' + 2\alpha\,\mathrm{curl}\,\mathbf{u}' + \overset{*}{\mathbf{Y}} = \mathbf{0}
\end{aligned}
\tag{1.166}
$$

with the modified boundary conditions

$$
\mathbf{u}' = \mathbf{0}, \qquad \boldsymbol{\varphi}' = \mathbf{g} - \boldsymbol{\varphi}^0 \qquad \text{on } \partial\Omega,
\tag{1.167}
$$

where the vector $\boldsymbol{\varphi}^0$ is obtained by solving the classical problem, (1.159) and (1.160).

Now, we shall derive some separate equations from the system of equations (1.166):

— for the vector $\mathbf{u}'$,

$$
\begin{aligned}
2\mu\overrightarrow{\nabla}^2\overrightarrow{\nabla}^2(l^2\overrightarrow{\nabla}^2-1)\mathbf{u}' &= \overrightarrow{\nabla}^2\,\mathrm{curl}\,\overset{*}{\mathbf{Y}} - 2\overrightarrow{\nabla}^2(l^2\overrightarrow{\nabla}^2-1)\hat{\mathbf{X}} \\
&\quad - \mu_0^{-1}\overrightarrow{\nabla}^2\,\mathrm{curl\,curl}\,\hat{\mathbf{X}} + (1-\nu)^{-1}(l^2\overrightarrow{\nabla}^2-1)\,\mathrm{grad\,div}\,\hat{\mathbf{X}};
\end{aligned}
\tag{1.168}
$$

— for the vector $\boldsymbol{\varphi}'$,

$$
\begin{aligned}
&2\mu\overrightarrow{\nabla}^2(\nu_1^2\overrightarrow{\nabla}^2-1)(l^2\overrightarrow{\nabla}^2-1)\boldsymbol{\varphi}' \\
&\quad = -(2\alpha)^{-1}\mu(l^2\overrightarrow{\nabla}^2-1)\,\mathrm{grad\,div}\,\overset{*}{\mathbf{Y}} \\
&\qquad + (\nu_1^2\overrightarrow{\nabla}^2-1)\,\mathrm{curl}\,\hat{\mathbf{X}} + \alpha_0^{-1}(\nu_1^2\overrightarrow{\nabla}^2-1)\,\mathrm{curl\,curl}\,\overset{*}{\mathbf{Y}},
\end{aligned}
\tag{1.169}
$$

where $\alpha_0 = s = \dfrac{2\alpha}{\mu+\alpha}$ and $\nu = \dfrac{\lambda}{2(\mu+\lambda)}$ is the Poisson ratio. Note that taking into account the substitutions

$$
(\mathbf{u}', \boldsymbol{\varphi}') \mapsto (\mathbf{u}, \boldsymbol{\varphi}), \qquad (\hat{\mathbf{X}}, \overset{*}{\mathbf{Y}}) \mapsto (\mathbf{X}, \mathbf{Y}),
\tag{1.170}
$$

we see that the system of equations (1.166) is transformed into (1.155), whereas equations (1.168) and (1.169) take the form

$$
\begin{aligned}
2\mu\overrightarrow{\nabla}^2\overrightarrow{\nabla}^2(l^2\overrightarrow{\nabla}^2-1)\mathbf{u} &= \overrightarrow{\nabla}^2\,\mathrm{curl}\,\mathbf{Y} - 2\overrightarrow{\nabla}^2(l^2\overrightarrow{\nabla}^2-1)\mathbf{X} \\
&\quad - \mu_0^{-1}\overrightarrow{\nabla}^2\,\mathrm{curl\,curl}\,\mathbf{X} + (1-\nu)^{-1}(l^2\overrightarrow{\nabla}^2-1)\,\mathrm{grad\,div}\,\mathbf{X};
\end{aligned}
\tag{1.171}
$$

$$\begin{aligned}
&2\mu\overrightarrow{\nabla}^2(\nu_1^2\overrightarrow{\nabla}^2-1)(l^2\overrightarrow{\nabla}^2-1)\boldsymbol{\varphi}\\
&\quad=-(2\alpha)^{-1}\mu(l^2\overrightarrow{\nabla}^2-1)\,\mathrm{grad}\,\mathrm{div}\,\mathbf{Y}\\
&\qquad+(\nu_1^2\overrightarrow{\nabla}^2-1)\,\mathrm{curl}\,\mathbf{X}+\alpha_0^{-1}(\nu_1^2\overrightarrow{\nabla}^2-1)\,\mathrm{curl}\,\mathrm{curl}\,\mathbf{Y}.
\end{aligned}\tag{1.172}$$

If, in equations (1.155), one assumes the relations (called analogies of body loadings)

$$\mathbf{X}=-\nu_T\,\mathrm{grad}\,\theta,\qquad \mathbf{Y}=\mathbf{0}\tag{1.173}$$

then one obtains the equations in terms of displacements and rotations with a temperature field θ. The function θ satisfies the heat conduction equation

$$\mathrm{div}\,\mathrm{grad}\,\theta=-\frac{W}{\lambda_0}\qquad\text{in }\Omega.\tag{1.174}$$

Within the framework of thermoelastostatics, the uncoupled equations for the vectors $\mathbf{u}$, $\boldsymbol{\varphi}$ are obtained from equations (1.171), (1.172) by use of (1.173), and thus

$$\begin{aligned}
&\overrightarrow{\nabla}^2\overrightarrow{\nabla}^2(l^2\overrightarrow{\nabla}^2-1)\mathbf{u}=m\overrightarrow{\nabla}^2(l^2\overrightarrow{\nabla}^2-1)\,\mathrm{grad}\,\theta,\\
&\overrightarrow{\nabla}^2(\nu_1^2\overrightarrow{\nabla}^2-1)(l^2\overrightarrow{\nabla}^2-1)\boldsymbol{\varphi}=\mathbf{0},
\end{aligned}\tag{1.175}$$

where $m=\dfrac{\nu_T}{2\mu+\lambda}$.

Now, let us formulate the boundary value problem within the framework of thermoelastostatics. Combining the system of equations (1.155) and equations (1.173), (1.174), we have on $\partial\Omega$ the boundary conditions

$$\mathbf{u}=\mathbf{f},\qquad \boldsymbol{\varphi}=\mathbf{g},\qquad \theta=\vartheta,\tag{1.176}$$

where the function $\vartheta:\partial\Omega\to\mathbb{R}$ is given.

Using the superposition method, we get in Ω the equations of classical thermoelastostatics, i.e. equation (1.174) and the equation

$$\mu\overrightarrow{\nabla}^2\mathbf{u}^0+(\lambda+\mu)\,\mathrm{grad}\,\mathrm{div}\,\mathbf{u}^0=\nu_T\,\mathrm{grad}\,\theta\tag{1.177}$$

with the boundary conditions

$$\mathbf{u}^0=\mathbf{f},\qquad \theta=\vartheta\qquad\text{on }\partial\Omega.\tag{1.178}$$

The vectors $\mathbf{u}^0$, $\boldsymbol{\varphi}^0$ satisfy the equations

$$\overrightarrow{\nabla}^2\overrightarrow{\nabla}^2\mathbf{u}^0=m\overrightarrow{\nabla}^2\,\mathrm{grad}\,\theta,\qquad \overrightarrow{\nabla}^2\boldsymbol{\varphi}^0=\mathbf{0}.\tag{1.179}$$

The complementary solution, which takes into account the boundary effect, is obtained from the homogeneous system of equations (1.166) (without the temperature field: $\hat{\mathbf{X}}=\mathbf{0}$, $\overset{*}{\mathbf{Y}}=\mathbf{0}$) with the boundary conditions (1.167). The vectors $\mathbf{u}'$, $\boldsymbol{\varphi}'$ satisfy the equations:

$$\begin{aligned}&\vec{\nabla}^2\vec{\nabla}^2(l^2\vec{\nabla}^2-1)\mathbf{u}'=\mathbf{0},\\&\vec{\nabla}^2(\nu_1^2\vec{\nabla}^2-1)(l^2\vec{\nabla}^2-1)\boldsymbol{\varphi}'=\mathbf{0}.\end{aligned} \tag{1.180}$$

Also, the vector $\boldsymbol{\zeta}$ satisfies the equation

$$(\nu_1^2\vec{\nabla}^2-1)(l^2\vec{\nabla}^2-1)\boldsymbol{\zeta}=\mathbf{0}. \tag{1.181}$$

In [49], we showed the usefulness of the vector equations (1.155)–(1.181) in solving the boundary value problems of elastostatics and thermoelastostatics (three- and two-dimensional: plane and axially-symmetric ones), as well as in finding fundamental solutions.

1.2.2 Fields of body loadings. Fundamental solutions

In [52] and [49], we determined the fundamental solutions for the displacements and rotations in the case of body loadings acting in the space $\mathbb{R}^3$. In the rectangular coordinate system $Ox_1x_2x_3$ we have

$$(X_i, Y_i)=(P_0, M_0)\delta(x_1)\delta(x_2)\delta(x_3)\delta_{is}. \tag{1.182}$$

The symbols used here are the same as those in Subsection 1.1.2. Equations (1.161), (1.162), (1.168) and (1.169) take now the forms:

— the classical equations:

$$\begin{aligned}&2\mu\nabla^2\nabla^2u_i^0=(1-\nu)^{-1}X_{j,ji}-2\nabla^2X_i,\\&2\mu\nabla^2\varphi_i^0+\epsilon_{ijk}X_{k,j}=0,\end{aligned} \tag{1.183}$$

where $\nabla^2(.)=(.)_{,ii}$;

— the complementary equations:

$$\begin{aligned}2\mu\nabla^2\nabla^2(l^2\nabla^2-1)u_i'&=\nabla^2\epsilon_{ijk}\overset{*}{Y}_{k,j}-2\nabla^2(l^2\nabla^2-1)\hat{X}_i\\&-\mu_0^{-1}\nabla^2\epsilon_{imp}\epsilon_{pjk}\hat{X}_{k,jm}+(1-\nu)^{-1}(l^2\nabla^2-1)\hat{X}_{j,ji},\\2\mu\nabla^2(\nu_1^2\nabla^2-1)(l^2\nabla^2-1)\varphi_i'&=-(2\alpha)^{-1}\mu(l^2\nabla^2-1)\overset{*}{Y}_{j,ji}\\&+(\nu_1^2\nabla^2-1)\epsilon_{ijk}\hat{X}_{k,j}+\alpha_0^{-1}(\nu_1^2\nabla^2-1)\epsilon_{imp}\epsilon_{pjk}\overset{*}{Y}_{k,jm},\end{aligned} \tag{1.184}$$

where

$$\hat{X}_i=0, \qquad \overset{*}{Y}_i=Y_i-\mu_0^{-1}\epsilon_{ijk}X_{k,j}. \tag{1.185}$$

Applying the method of integral Fourier transforms (0.50) (for $n=3$) to equations (1.183) and (1.184) and taking into account (1.182) and (1.185), we obtain, by means of the formula

$$u_i=u_i^0+u_i', \qquad \varphi_i=\varphi_i^0+\varphi_i' \tag{1.186}$$

the following formulae:

— for the body force $(1.182)_1$ ($P_0 \neq 0$, $M_0 = 0$),

$$8\pi\mu P_0^{-1} u_i^{(s)} = -\frac{1}{2(1-\nu)} R_{,is} + 2\left(\frac{1}{R}\right)\delta_{is} - \mu_0^{-1}\epsilon_{imr}\epsilon_{rjs}\left(\frac{1-e^{-R/l}}{R}\right)_{,jm}, \tag{1.187}$$

$$8\pi\mu P_0^{-1} \varphi_i^{(s)} = \epsilon_{ijs}\left(\frac{1-e^{-R/l}}{R}\right)_{,j},$$

where

$$R = \sqrt{x_i x_i}, \qquad \nu = \frac{\lambda}{2(\mu+\lambda)}, \qquad \mu_0 = \frac{2\mu}{\gamma+\varepsilon}, \qquad l = \sqrt{\frac{(\gamma+\varepsilon)(\mu+\alpha)}{4\alpha\mu}};$$

— for the body moment $(1.182)_2$ ($P_0 = 0$, $M_0 \neq 0$)[15],

$$8\pi\mu M_0^{-1} u_i^{(s)} = \epsilon_{ijs}\left(\frac{1-e^{-R/l}}{R}\right)_{,j},$$

$$8\pi\mu M_0^{-1} \varphi_i^{(s)} = -(2\alpha)^{-1}\mu\left(\frac{1-e^{-R/\nu_1}}{R}\right)_{,is} + \alpha_0^{-1}\epsilon_{imr}\epsilon_{rjs}\left(\frac{1-e^{-R/l}}{R}\right)_{,jm}, \tag{1.188}$$

where $\alpha_0 = \dfrac{2\alpha}{\mu+\alpha}$.

Moreover, we have

$$\nu_1 = \sqrt{\frac{2\gamma+\beta}{4\alpha}}. \tag{1.189}$$

The formulae (1.187) and (1.188) constitute the solution of the equations

$$\begin{aligned}(\mu+\alpha)\nabla^2 u_i + (\lambda+\mu-\alpha)u_{j,ji} + 2\alpha\epsilon_{ijk}\varphi_{k,j} + X_i = 0,\\ [(\gamma+\varepsilon)\nabla^2 - 4\alpha]\varphi_i + (\beta+\gamma-\varepsilon)\varphi_{j,ji} + 2\alpha\epsilon_{ijk}u_{k,j} + Y_i = 0\end{aligned} \tag{1.190}$$

which, when uncoupled, take the form

[15] $[Y_i] = [\mathrm{N}\cdot\mathrm{m}^{-2}]$, $[\delta(\mathbf{x})] = [\mathrm{m}^{-3}]$, $[M_0] = [\mathrm{N}\cdot\mathrm{m}]$, $[\mu] = [\alpha] = [\mathrm{N}\cdot\mathrm{m}^{-2}]$, $[\alpha_0] = [1]$, $[\gamma] = [\beta] = [\mathrm{N}]$, $[\nu_1] = [R] = [l] = [\mathrm{m}]$, $[u_i] = [\mathrm{m}]$, $[\varphi_i] = [1]$, $[L_{(1.188)1}] = [R_{(1.188)1}] = [\mathrm{m}^{-2}]$, $[L_{(1.188)2}] = [R_{(1.188)2}] = [\mathrm{m}^{-3}]$.

$$
\begin{aligned}
2\mu\nabla^2\nabla^2(l^2\nabla^2-1)u_i &= \nabla^2\epsilon_{ijk}Y_{k,j} - 2\nabla^2(l^2\nabla^2-1)X_i \\
&\quad - \mu_0^{-1}\nabla^2\epsilon_{imp}\epsilon_{pjk}X_{k,jm} + (1-\nu)^{-1}(l^2\nabla^2-1)X_{j,ji}, \\
2\mu\nabla^2(\nu_1^2\nabla^2-1)(l^2\nabla^2-1)\varphi_i &= -(2\alpha)^{-1}\mu(l^2\nabla^2-1)Y_{j,ji} \\
&\quad + (\nu_1^2\nabla^2-1)\epsilon_{ijk}X_{k,j} + \alpha_0^{-1}(\nu_1^2\nabla^2-1)\epsilon_{imp}\epsilon_{rjk}Y_{k,jm}.
\end{aligned}
\tag{1.191}
$$

One has to replace X_i, Y_i with (1.182).

Formulae (1.187) and (1.188) provide a basis for the study of certain limiting cases. In [52] we proved that, using those formulae, one can obtain the fundamental solutions for the limiting theories (4), (6), and (8), as given in the diagram in Fig. 1.1. Moreover, we gave and discussed the solution of equations (1.190) and (1.191) with the body loadings of the form:

$$
Y_i \to 0, \qquad X_i \to \tfrac{1}{2}\epsilon_{ijk}Y_{k,j}. \tag{1.192}
$$

In particular, for the loading $(1.182)_1$, we obtain the formulae[16]

$$
\begin{aligned}
8\pi\mu M_0^{-1}u_i^{(s)} &= \epsilon_{ijs}\left(\frac{1-e^{-R/l}}{R}\right)_{,j} + \frac{\mu}{\mu+\alpha}\epsilon_{ijs}\left(\frac{e^{-R/l}}{R}\right)_{,j}, \\
8\pi\mu M_0^{-1}\varphi_i^{(s)} &= \tfrac{1}{2}\epsilon_{imr}\epsilon_{rjs}\left(\frac{1-e^{-R/l}}{R}\right)_{,jm},
\end{aligned}
\tag{1.193}
$$

which represent the result for the action of the classical unit moment M_1 in a micropolar medium. Formulae (1.187) and (1.188) were independently obtained by Nowacki [186, p. 160] and Sandru [206].

1.2.3 Distortion fields. Fundamental solutions and limiting cases

In [53] and [49], we considered the equations in displacements and rotations (1.190) with a distortion field and we derived the uncoupled equations for the pair (u_i, φ_i) by means of the superposition method. We shall give here these equations assuming the substitutions:

$$
X_i = -\overset{*}{R}{}_i^0, \qquad \hat{X}_i = -\overset{**}{R}{}_i^0, \qquad Y_i = -M_i^0, \qquad \overset{*}{Y}_i = -\hat{\overset{*}{R}}{}_i^0. \tag{1.194}
$$

From equations (1.183) and (1.184), we obtain, respectively:

— the classical equations:

$$
\begin{aligned}
2\mu\nabla^2\nabla^2u_i^0 &= 2\nabla^2\overset{*}{R}{}_i^0 - (1-\nu)^{-1}\overset{*}{R}{}_{j,ji}^0, \\
2\mu\nabla^2\varphi_i^0 &= \epsilon_{ijk}\overset{*}{R}{}_{k,j}^0;
\end{aligned}
\tag{1.195}
$$

[16] $[M_0] = [\mathrm{N}\cdot\mathrm{m}]$.

— the equations for the complementary solution:

$$\begin{aligned}
2\mu\nabla^2\nabla^2(l^2\nabla^2-1)u_i' &= -\nabla^2\epsilon_{ijk}\hat{\hat{R}}{}^0_{k,j} + 2\nabla^2(l^2\nabla^2-1)\overset{**}{R}{}^0_i \\
&\quad + \mu_0^{-1}\nabla^2\epsilon_{imp}\epsilon_{pjk}\overset{**}{R}{}^0_{k,jm} - (1-\nu)^{-1}(l^2\nabla^2-1)\overset{**}{R}{}^0_{j,ji}, \\
2\mu\nabla^2(\nu_1^2\nabla^2-1)(l^2\nabla^2-1)\varphi_i' &= (2\alpha)^{-1}\mu(l^2\nabla^2-1)\hat{\hat{R}}{}^0_{j,ji} \\
&\quad - (\nu_1^2\nabla^2-1)\epsilon_{ijk}\overset{**}{R}{}^0_{k,j} - \alpha_0^{-1}(\nu_1^2\nabla^2-1)\epsilon_{imp}\epsilon_{pjk}\hat{\hat{R}}{}^0_{k,jm},
\end{aligned} \tag{1.196}$$

where $\hat{\hat{R}}{}^0_i = M_i - \mu_0^{-1}\epsilon_{ijk}\overset{*}{R}{}^0_{k,j}$.

Other symbols referring to distortions have the same meaning as in Subsection 1.1.3. Moreover, we have

$$\nu = \frac{\lambda}{2(\mu+\lambda)}, \qquad \nu_1^2 = \frac{2\gamma+\beta}{4\alpha},$$

$$\mu_0 = \frac{2\mu}{\gamma+\varepsilon}, \qquad \alpha_0 = \frac{2\alpha}{\mu+\alpha}, \qquad l^2 = \frac{(\gamma+\varepsilon)(\mu+\alpha)}{4\alpha\mu}.$$

The following observation is of importance; namely, under the substitutions

$$(u_i', \varphi_i') \mapsto (u_i, \varphi_i), \qquad (\overset{**}{R}{}^0_i, \hat{\hat{R}}{}^0_i) \mapsto (R_i^0, M_i^0) \tag{1.197}$$

equations (1.196) are transformed into

$$\begin{aligned}
2\mu\nabla^2\nabla^2(l^2\nabla^2-1)u_i &= -\nabla^2\epsilon_{ijk}(\mu^0_{lk,l} + \epsilon_{krs}\sigma^0_{rs})_{,j} \\
&\quad + 2\nabla^2(l^2\nabla^2-1)\sigma^0_{ji,j} + \mu_0^{-1}\nabla^2\epsilon_{imp}\epsilon_{pjk}\sigma^0_{lk,ljm} \\
&\quad - (1-\nu)^{-1}(l^2\nabla^2-1)\sigma^0_{lk,lki}, \\
2\mu\nabla^2(\nu_1^2\nabla^2-1)(l^2\nabla^2-1)\varphi_i &= (2\alpha)^{-1}\mu(l^2\nabla^2-1) \\
&\quad \times (\mu^0_{lk,l} + \epsilon_{krs}\sigma^0_{rs})_{,ki} - (\nu_1^2\nabla^2-1)\epsilon_{ijk}\sigma^0_{lk,lj} \\
&\quad - \alpha_0^{-1}(\nu_1^2\nabla^2-1)\epsilon_{imp}\epsilon_{pjk}(\mu^0_{lk,l} + \epsilon_{krs}\sigma^0_{rs})_{,jm},
\end{aligned} \tag{1.198}$$

which are obtained from equations (1.190) by means of equations (1.191) and the substitution (1.38) (the analogy of body forces). Applying the integral Fourier transformation (0.50) (for $n=3$) to equations (1.195), (1.196) and (1.198), we have determined the fundamental solutions for the displacements u_i and the rotations φ_i. The distortion field is given by

$$\gamma^0_{ij} = \gamma_0^{ij}\delta(x_1)\delta(x_2)\delta(x_3), \qquad \kappa^0_{ij} = \kappa_0^{ij}\delta(x_1)\delta(x_2)\delta(x_3), \tag{1.199}$$

where γ_0^{ij}, κ_0^{ij} are of unit values[17]. The results are given below in a compact form:

— for the distortion $(1.199)_1$,

$$
\begin{aligned}
4\pi(\gamma_0^{mn})^{-1}u^{\gamma_0}_{lmn} &= -\left[\mu_0^{-1}\nabla^2\left(\frac{1-e^{-R/l}}{R}\right)+\frac{1}{R}\right]_{,n}\delta_{lm} \\
&+\left[\mu_0^{-1}\left(\frac{1-e^{-R/l}}{R}\right)+\frac{1}{2}(1-\nu)^{-1}R\right]_{,lmn}-\left(\frac{1}{R}\right)_{,m}\delta_{ln} \\
&-\left[\frac{1}{2}(1-\nu)^{-1}\nabla^2 R-\frac{1}{R}\right]_{,l}\delta_{mn},
\end{aligned} \tag{1.200}
$$

$$
\begin{aligned}
4\pi(\gamma_0^{mn})^{-1}\varphi^{\gamma_0}_{imn} &= \tfrac{1}{2}\epsilon_{mnp}\left(\frac{1-e^{-R/\nu_1}}{R}\right)_{,ip} \\
&-\epsilon_{mij}\left(\frac{1-e^{-R/l}}{R}\right)_{,nj};
\end{aligned}
$$

— for the distortion $(1.199)_2$,

$$
\begin{aligned}
4\pi(\kappa_0^{mn})^{-1}u^{\kappa_0}_{imn} &= -\mu_0^{-1}\epsilon_{nij}\left(\frac{1-e^{-R/l}}{R}\right)_{,mj} \\
&-\frac{\gamma-\varepsilon}{\gamma+\varepsilon}\mu_0^{-1}\epsilon_{mij}\left(\frac{1-e^{-R/l}}{R}\right)_{,nj},
\end{aligned}
$$

$$
\begin{aligned}
4\pi(\kappa_0^{mn})^{-1}\varphi^{\kappa_0}_{imn} &= \frac{\gamma}{2\alpha}\left[\left(\frac{1-e^{-R/\nu_1}}{R}\right)-2\mu_1^2\left(\frac{1-e^{-R/l}}{R}\right)\right]_{,imn} \\
&+\nabla^2\left[\frac{\beta}{4\alpha}\left(\frac{1-e^{-R/\nu_1}}{R}\right)_{,i}\delta_{mn}+l^2\left(\frac{1-e^{-R/l}}{R}\right)_{,m}\delta_{in}\right. \\
&\left.+\frac{\gamma-\varepsilon}{\gamma+\varepsilon}l^2\left(\frac{1-e^{-R/l}}{R}\right)_{,n}\delta_{im}\right],
\end{aligned} \tag{1.201}
$$

where we add the new notation $\mu_1^2 = \dfrac{\mu+\alpha}{2\mu}$.

It was assumed in (1.200) and (1.201) that $\gamma_0^{ij} \neq 0$, $\kappa_0^{ij} \neq 0$ for every i, j. The symbols $u^{\gamma_0,\kappa_0}_{imn}$, $\varphi^{\gamma_0,\kappa_0}_{imn}$ denote, respectively, the displacements u_i and the rotations φ_i generated by the distortions γ^0_{mn}, κ^0_{mn}. Certain limiting cases of

---

17

$$[dY/dX] = [Y\cdot X^{-1}],\ [u_i] = [\mathrm{m}],\ [\varphi_i] = [1],\ [\gamma^0_{ij}] = [\gamma_{ij}] = [u_{i,j}-\epsilon_{kji}\varphi_k] = [1],$$

$$[\kappa^0_{ji}] = [\kappa_{ji}] = [\varphi_{i,j}] = [\mathrm{m}^{-1}],\ [\delta(\mathbf{x})] = [m^{-3}],\ [\gamma_0^{ij}] = [\mathrm{m}^3],\ [\kappa_0^{ij}] = [\mathrm{m}^2].$$

the results (1.200) and (1.201) are interesting. We discuss here, as an example, the limiting case $\alpha \to \infty$. One has to assume that

$$\sqrt{\frac{\gamma+\varepsilon}{4\mu}} = l^\star, \qquad \frac{\gamma-\varepsilon}{\gamma+\varepsilon} = \eta, \tag{1.202}$$

where $l^\star$, η are the constants of couple-stress theory. Now, formulae (1.201), for instance, take the following form:

$$\begin{aligned} 4\pi(\kappa_0^{mn})^{-1} u_{imn}^{\kappa_0} &= -2l^{\star 2}\epsilon_{nij}\left(\frac{1-e^{-R/l^\star}}{R}\right)_{,mj} \\ &\quad - 2\eta l^{\star 2}\epsilon_{mij}\left(\frac{1-e^{-R/l^\star}}{R}\right)_{,nj}, \\ 4\pi(\kappa_0^{mn})^{-1} \varphi_{imn}^{\kappa_0} &= -(1+\eta) l^{\star 2}\left(\frac{1-e^{-R/l^\star}}{R}\right)_{,imn} \\ &\quad + l^{\star 2}\nabla^2\left[\left(\frac{1-e^{-R/l^\star}}{R}\right)_{,m}\delta_{in} + \eta\left(\frac{1-e^{-R/l^\star}}{R}\right)_{,n}\delta_{im}\right]. \end{aligned} \tag{1.203}$$

It was shown that formulae (1.203) constitute the fundamental solution in the framework of couple-stress theory. This follows from the fact that those formulae are obtained from the equations

$$\begin{aligned} 2\mu\nabla^2(l^{\star 2}\nabla^2 - 1)\varphi_i &= -\epsilon_{ijk}(s_{lk,l}^0 + \tfrac{1}{2}\epsilon_{krp}m_{qp,rq}^0)_{,j}, \\ 2\mu\nabla^2\nabla^2(l^{\star 2}\nabla^2 - 1)u_i &= -(1-\nu)^{-1}(l^{\star 2}\nabla^2 - 1)s_{pj,pij}^0 \\ &\quad + 2l^{\star 2}\nabla^2\epsilon_{imp}\epsilon_{pjk}(s_{lk,l}^0 + \tfrac{1}{2}\epsilon_{krs}m_{qs,rq}^0)_{,jm} \\ &\quad + 2\nabla^2(l^{\star 2}\nabla^2 - 1)(s_{ji,j}^0 + \tfrac{1}{2}\epsilon_{ijk}m_{lk,jl}^0), \end{aligned} \tag{1.204}$$

where

$$s_{ij}^0 = 2\mu\varepsilon_{ij}^0 + \lambda\varepsilon_{kk}^0\delta_{ij}, \qquad m_{ij}^0 = 4\mu l^{\star 2}(\kappa_{ji}^0 + \eta\kappa_{ij}^0),$$

assuming that the distortion field ε_{ij}^0, κ_{ij}^0 is given by formula $(1.199)_2$. The displacement equations of couple-stress theory with a distortion field in the form

$$\mu\nabla^2 u_i + (\lambda+\mu)u_{j,ji} + \mu l^{\star 2}\nabla^2\epsilon_{ilk}\epsilon_{kpq}u_{q,pl} = s_{ji,j}^0 + \tfrac{1}{2}\epsilon_{ijk}m_{lk,jl}^0 \tag{1.205}$$

become identities after substituting the displacements $(1.203)_1$ and the distortion $(1.199)_2$. The derivative is understood here in the distributive sense:

$$\nabla^2\left(\frac{1}{R}\right) = -4\pi\delta(x_1)\delta(x_2)\delta(x_3). \tag{1.206}$$

Moreover, using formula (1.203), we obtain

$$\varphi^{\kappa_0}_{imn} = \tfrac{1}{2}\epsilon_{irk} u^{\kappa_0}_{kmn,r} \tag{1.207}$$

relating the displacements $u^{\kappa_0}_{imn}$ and the rotations $\varphi^{\kappa_0}_{imn}$. Formula (1.207) is valid in couple-stress theory. The conditions $\alpha \to \infty$ and (1.202), which are necessary and sufficient for the discussed limiting case with a distortion field given by κ^0_{ij}, do not suffice any more if we want to obtain the formulae of couple-stress theory from formula (1.200) for the indices $m \neq n$. In the present case, one has to construct the expressions

$$2u^{\gamma_0}_{l(mn)} = u^{\gamma_0}_{lmn} + u^{\gamma_0}_{lnm}, \qquad 2\varphi^{\gamma_0}_{l(mn)} = \varphi^{\gamma_0}_{lmn} + \varphi^{\gamma_0}_{lnm} \tag{1.208}$$

and then perform the limit. In [53] we discussed in detail the relations between the results of the theories (2), (4), (6) and (8) (see Fig. 1.1). It was pointed out that formulae (1.200) and (1.201) lead to the fundamental solutions for the limiting theories, which are denoted by the symbols (4), (6) and (8).

1.2.4 Problem of elastic half-space

In [41], [45], [46] and [48], we considered the problem of the half-space

$$\Omega = \{(x_1, x_2, x_3) : x_1 > 0 \wedge |x_2| < \infty \wedge |x_3| < \infty\} \tag{1.209}$$

within the framework of elastostatics and thermoelastostatics. Using the method of direct integration of the fundamental equations in displacements and rotations, as well as the superposition method, the fields of displacements, rotations, (force- and couple-) stresses have been determined for the following boundary value problems:

— in displacements and rotations:

$$u_i = U_i, \qquad \varphi_i = \Phi_i \qquad \text{on } \partial\Omega; \tag{1.210}$$

— in stresses[18]:

$$\sigma_{1i} = -p_i, \qquad \mu_{1i} = -m_i \qquad \text{on } \partial\Omega; \tag{1.211}$$

— the thermoelastic boundary value problem:

$$\sigma_{1i} = \mu_{1i} = 0, \qquad \theta = f \qquad \text{on } \partial\Omega; \tag{1.212}$$

where U_i, Φ_i, p_i, m_i, $f : \partial\Omega \to \mathbb{R}$ are some given functions. Moreover, in the problems with the conditions (1.210)–(1.212) on $\partial\Omega$, we assume the conditions

$$u_i \to 0, \qquad \varphi_i \to 0, \qquad \sigma_{ji} \to 0, \qquad \mu_{ji} \to 0 \qquad \text{for } R = \sqrt{x_i x_i} \to \infty. \tag{1.213}$$

[18] Boundary value problems with conditions (1.210) and (1.211) are called first- and second boundary value problems, respectively, or the problem of Dirichlet- and Neumann type, respectively.

In the case of the boundary conditions (1.210), (1.211), the solution of the problem was obtained by applying two methods: by means of the equations

$$\nabla^2\nabla^2(l^2\nabla^2 - 1)u_i = 0, \qquad \nabla^2(\nu_1^2\nabla^2 - 1)(l^2\nabla^2 - 1)\varphi_i = 0 \tag{1.214}$$

and

$$\nabla^2\nabla^2 u_i^0 = 0, \qquad \nabla^2\varphi_i^0 = 0 \tag{1.215}$$

as well as

$$\begin{aligned} &\nabla^2\nabla^2(l^2\nabla^2 - 1)u_i' = 0, \\ &\nabla^2(\nu_1^2\nabla^2 - 1)(l^2\nabla^2 - 1)\varphi_i' = 0, \\ &(\nu_1^2\nabla^2 - 1)(l^2\nabla^2 - 1)\zeta_i = 0 \qquad \text{in } \Omega. \end{aligned} \tag{1.216}$$

In the thermoelastic problem with the boundary conditions (1.212), it is additionally assumed that there are no heat sources in the half-space and thus the heat conduction equation (1.174) is transformed into the Laplace equation [19]

$$\nabla^2\theta = 0 \qquad \text{in } \Omega \tag{1.217}$$

with the condition $\theta \to 0$ for $R = \sqrt{x_i x_i} \to \infty$.

The solution of the thermoelastic boundary value problem was derived by means of two methods. In the method of direct integration, we integrate the equations

$$\begin{aligned} &\nabla^2\nabla^2(l^2\nabla^2 - 1)u_i = m\nabla^2(l^2\nabla^2 - 1)\theta_{,i}, \\ &\nabla^2(\nu_1^2\nabla^2 - 1)(l^2\nabla^2 - 1)\varphi_i = 0. \end{aligned} \tag{1.218}$$

In turn, in the superposition method, we integrate

$$\nabla^2\nabla^2 u_i^0 = m\nabla^2\theta_{,i}, \qquad \nabla^2\varphi_i^0 = 0 \tag{1.219}$$

as well as equations (1.216).

In the papers mentioned above, the solution of the three-dimensional problem with boundary conditions (1.210)–(1.212) and the conclusions concerning the stress fields were presented. It was also proved that the results obtained for the problems with the boundary conditions (1.210) comply with the vector result of Nowacki [189]. Below we shall also give the solution of equations (1.214) in the form that can be applied to the discussion of three-dimensional and two-dimensional boundary value problems formulated for a layer or a half-space, in the framework of elastostatics and micropolar thermoelastostatics:

---

[19] The half-space problem with a heat source inside was solved in [43].
Stefaniak [218] gives on p. 37 a solution of the half-space problem with a heat source inside that harmonically changes in time, and a concentrated force (the quasi-static problem).

— the displacements u_i,

$$u_1(x_i) = \mathcal{F}^{\star}_{(2)} \left[\!\left[(A_1 + B_1\xi x_1)e^{-\xi x_1} + C_1 e^{-\rho x_1} \right.\right.$$
$$\left.\left. +(a_1 + b_1\xi x_1)e^{\xi x_1} + d_1 e^{\rho x_1}; \xi_2 \to x_2, \xi_3 \to x_3 \right]\!\right],$$
$$u_K(x_i) = \mathcal{F}^{\star}_{(2)} \left[\!\left[i\left\{\xi_K\xi^{-1}[A_1 - (3-4\nu)B_1] - 2\epsilon_{1KL}\xi_L\xi^{-2}\bar{A}_1\right\}e^{-\xi x_1} \right.\right.$$
$$+ i\xi_K B_1 x_1 e^{-\xi x_1} + \xi^{-2}(i\rho\xi_K C_1 - i\alpha_0\epsilon_{1KL}\xi_L\bar{C}_1)e^{-\rho x_1}$$
$$+ i\left\{\xi_K\xi^{-1}[-a_1 - (3-4\nu)b_1] - 2\epsilon_{1KL}\xi_L\xi^{-2}\bar{a}_1\right\}e^{\xi x_1}$$
$$- i\xi_K b_1 x_1 e^{\xi x_1} - \xi^{-2}(i\rho\xi_K d_1 + i\alpha_0\epsilon_{1KL}\xi_L\bar{d}_1)e^{\rho x_1};$$
$$\left.\left. \xi_2 \to x_2, \xi_3 \to x_3 \right]\!\right]; \tag{1.220}$$

— the rotations φ_i,

$$\varphi_1(x_i) = \mathcal{F}^{\star}_{(2)} \left[\!\left[\bar{A}_1 e^{-\xi x_1} + \bar{B}_1 e^{-\eta x_1} + \bar{C}_1 e^{-\rho x_1} \right.\right.$$
$$\left.\left. + \bar{a}_1 e^{\xi x_1} + \bar{b}_1 e^{\eta x_1} + \bar{d}_1 e^{\rho x_1}; \xi_2 \to x_2, \xi_3 \to x_3 \right]\!\right],$$
$$\varphi_K(x_i) = \mathcal{F}^{\star}_{(2)} \left[\!\left[i[\xi_K\xi^{-1}\bar{A}_1 - 2(1-\nu)\epsilon_{1KL}\xi_L B_1]e^{-\xi x_1} \right.\right.$$
$$+ i\xi_K\eta^{-1}\bar{B}_1 e^{-\eta x_1} + \xi^{-2}(i\rho\xi_K\bar{C}_1 + i\mu_0\epsilon_{1KL}\xi_L C_1)e^{-\rho x_1} \tag{1.221}$$
$$+ i[-\xi_K\xi^{-1}\bar{a}_1 + 2(1-\nu)\epsilon_{1KL}\xi_L b_1]e^{\xi x_1} - i\xi_K\eta^{-1}\bar{b}_1 e^{\eta x_i}$$
$$\left.\left. + \xi^{-2}(-i\rho\xi_K\bar{d}_1 + i\mu_0\epsilon_{1KL}\xi_L d_1)e^{\rho x_1}; \xi_2 \to x_2, \xi_3 \to x_3 \right]\!\right],$$

where

$$\rho = \sqrt{\xi^2 + l^{-2}},\ \eta = \sqrt{\xi^2 + \nu_1^{-2}},\ \xi^2 = \xi_L\xi_L,\ \alpha_0 = \frac{2\alpha}{\mu + \alpha},\ \mu_0 = \frac{2\mu}{\gamma + \varepsilon}.$$

The constant coefficients A_1, B_1, C_1, $\bar{A}_1$, $\bar{B}_1$, $\bar{C}_1$, a_1, b_1, d_1, $\bar{a}_1$, $\bar{b}_1$, $\bar{d}_1$ are determined from the boundary conditions. For the half-space, from the conditions $u_i \to 0$, $\varphi_i \to 0$ for $R \to \infty$, it folows that one has to assume $a_1 = b_1 = d_1 = \bar{a}_1 = \bar{b}_1 = \bar{d}_1 = 0$.

Further, formulae (1.220) and (1.221) also lead to the formulae for displacements and rotations within the framework of two-dimensional problems, namely the so-called first and second problems of plane strain which are represented, respectively, by the triples (u_1, u_2, φ_3), and $(\varphi_1, \varphi_2, u_3)$ (see Nowacki [192, p. 102]). In order to obtain such formulae for two-dimensional problems, it is sufficient to replace in equations (1.220) and (1.221) the double inverse

Fourier transform $\mathcal{F}^{\star}_{(2)} [\![\cdot]\!]$ with respect to the variables x_2 and x_3 by a single Fourier transform $\mathcal{F}^{\star} [\![\cdot]\!]$ with respect to the variable x_2, and assume that

$$\xi_3 = 0, \qquad \xi_2 = \xi, \qquad \xi = \sqrt{\xi_L \xi_L} = |\xi|.$$

1.2.5 Galerkin vector. Micropolar theory and couple-stress theory

In [49], we introduced the Galerkin vector, generalized to micropolar elastostatics by means of the vector equations (1.161), (1.162), (1.165), (1.168) and (1.169) from the superposition method. Assuming, in Ω,

$$\mu \mathbf{u}^0 = \left(\overrightarrow{\nabla}^2 - \frac{\lambda + \mu}{2\mu + \lambda} \operatorname{grad} \operatorname{div} \right) \mathbf{F}^0, \qquad 2\mu \boldsymbol{\varphi}^0 = \operatorname{curl} \mathbf{M}^0, \tag{1.222}$$

we get from equations (1.161) and (1.162) the equations for the vectors $\mathbf{F}^0$, $\mathbf{M}^0$:

$$\overrightarrow{\nabla}^2 \overrightarrow{\nabla}^2 \mathbf{F}^0 + \mathbf{X} = \mathbf{0}, \qquad \overrightarrow{\nabla}^2 \mathbf{M}^0 + \mathbf{X} = \mathbf{0}. \tag{1.223}$$

Let us write equations (1.165), (1.168) and (1.169) in the operator form, namely

$$\begin{aligned} &4\alpha(\nu_1^2 \overrightarrow{\nabla}^2 - 1)(l^2 \overrightarrow{\nabla}^2 - 1)\boldsymbol{\zeta} = \overrightarrow{\mathrm{L}}_3 \overset{*}{\mathbf{Y}} + \overrightarrow{\mathrm{L}}_4 \hat{\mathbf{X}}, \\ &2\mu \overrightarrow{\nabla}^2 \overrightarrow{\nabla}^2 (l^2 \overrightarrow{\nabla}^2 - 1)\mathbf{u}' = \overrightarrow{\mathrm{L}}_1 \hat{\mathbf{X}} + \overrightarrow{\mathrm{L}}_2 \overset{*}{\mathbf{Y}}, \\ &2\mu \overrightarrow{\nabla}^2 (\nu_1^2 \overrightarrow{\nabla}^2 - 1)(l^2 \overrightarrow{\nabla}^2 - 1)\boldsymbol{\varphi}' = \overrightarrow{\mathrm{L}}_5 \overset{*}{\mathbf{Y}} + \overrightarrow{\mathrm{L}}_6 \hat{\mathbf{X}}, \end{aligned} \tag{1.224}$$

where the operators $\overrightarrow{\mathrm{L}}_i$ $(i = 1, \ldots, 6)$ are given by

$$\begin{aligned} \overrightarrow{\mathrm{L}}_1 &= (1-\nu)^{-1}(l^2 \overrightarrow{\nabla}^2 - 1) \operatorname{grad} \operatorname{div} - \mu_0^{-1} \overrightarrow{\nabla}^2 \operatorname{curl} \operatorname{curl} - 2\overrightarrow{\nabla}^2 (l^2 \overrightarrow{\nabla}^2 - 1), \\ \overrightarrow{\mathrm{L}}_2 &= \overrightarrow{\nabla}^2 \operatorname{curl}, \\ \overrightarrow{\mathrm{L}}_3 &= (\nu_1^2 \overrightarrow{\nabla}^2 - 1) + \left[\tfrac{1}{2}\mu_0^{-1} - (\nu_1^2 - k^2)\right] \operatorname{grad} \operatorname{div}, \\ \overrightarrow{\mathrm{L}}_4 &= -\mu_0^{-1}(\nu_1^2 \overrightarrow{\nabla}^2 - 1) \operatorname{curl}, \\ \overrightarrow{\mathrm{L}}_5 &= -(2\alpha)^{-1}\mu(l^2 \overrightarrow{\nabla}^2 - 1) \operatorname{grad} \operatorname{div} + \alpha_0^{-1}(\nu_1^2 \overrightarrow{\nabla}^2 - 1) \operatorname{curl} \operatorname{curl}, \\ \overrightarrow{\mathrm{L}}_6 &= -\mu_0 \overrightarrow{\mathrm{L}}_4. \end{aligned}$$

Let us now introduce two new vectors $\mathbf{F}'$, $\mathbf{M}'$, by the following substitutions:

$$\begin{aligned} 2\mu \mathbf{u}' &= -\overrightarrow{\mathrm{L}}_1 \mathbf{F}' - \overrightarrow{\mathrm{L}}_6 \mathbf{M}', \\ 4\alpha \boldsymbol{\zeta} &= -\overrightarrow{\nabla}^2 \overrightarrow{\mathrm{L}}_3 \mathbf{M}' + \mu_0^{-1} \overrightarrow{\nabla}^2 \overrightarrow{\mathrm{L}}_2 \mathbf{F}', \\ 2\mu \boldsymbol{\varphi}' &= -\overrightarrow{\mathrm{L}}_5 \mathbf{M}' - \overrightarrow{\mathrm{L}}_2 \mathbf{F}'. \end{aligned} \tag{1.225}$$

Using (1.225), we obtain from (1.224) the equations for the vectors $\mathbf{F}'$, $\mathbf{M}'$:

$$\begin{aligned}&\overrightarrow{\nabla}^2\overrightarrow{\nabla}^2(l^2\overrightarrow{\nabla}^2-1)\mathbf{F}'+\hat{\mathbf{X}}=\mathbf{0},\\&\overrightarrow{\nabla}^2(l^2\overrightarrow{\nabla}^2-1)(\nu_1^2\overrightarrow{\nabla}^2-1)\mathbf{M}'+\overset{*}{\mathbf{Y}}=\mathbf{0},\end{aligned}\tag{1.226}$$

where $\hat{\mathbf{X}}$, $\overset{*}{\mathbf{Y}}$ are given by formulae (1.164). Let us note that assuming substitutions (1.170) and $(\mathbf{F}',\mathbf{M}')\mapsto(\mathbf{F},\mathbf{M})$, we get from $(1.225)_{1,3}$ and (1.226) the representation

$$\begin{aligned}2\mu\mathbf{u}&=-\overrightarrow{\mathrm{L}}_1\mathbf{F}-\overrightarrow{\mathrm{L}}_6\mathbf{M},\\2\mu\boldsymbol{\varphi}&=-\overrightarrow{\mathrm{L}}_5\mathbf{M}-\overrightarrow{\mathrm{L}}_2\mathbf{F},\end{aligned}\tag{1.227}$$

and the field equations[20]

$$\begin{aligned}&\overrightarrow{\nabla}^2\overrightarrow{\nabla}^2(l^2\overrightarrow{\nabla}^2-1)\mathbf{F}+\mathbf{X}=\mathbf{0},\\&\overrightarrow{\nabla}^2(l^2\overrightarrow{\nabla}^2-1)(\nu_1^2\overrightarrow{\nabla}^2-1)\mathbf{M}+\mathbf{Y}=\mathbf{0}.\end{aligned}\tag{1.228}$$

Let us introduce another representation for the pair $(\mathbf{u}',\boldsymbol{\varphi}')$, different from that in (1.225); namely

$$\begin{aligned}2\mu\mathbf{u}'&=\mu_0^{-1}\,\mathrm{curl}\,\overrightarrow{\mathrm{L}}_2\mathbf{F}'+\mu_0\overrightarrow{\mathrm{L}}_4\mathbf{M}',\\2\mu\boldsymbol{\varphi}'&=-\overrightarrow{\mathrm{L}}_5\mathbf{M}'-(\mu_0\alpha_0)^{-1}\overrightarrow{\nabla}^2\overrightarrow{\mathrm{L}}_2\mathbf{F}'.\end{aligned}\tag{1.229}$$

Now, taking into account relations (1.164) and (1.229) in equations $(1.224)_{2,3}$, we obtain the equations for the vectors $\mathbf{F}'$, $\mathbf{M}'$:

$$\begin{aligned}&\overrightarrow{\nabla}^2\overrightarrow{\nabla}^2(l^2\overrightarrow{\nabla}^2-1)\mathbf{F}'+\mathbf{X}=\mathbf{0},\\&\overrightarrow{\nabla}^2(l^2\overrightarrow{\nabla}^2-1)(\nu_1^2\overrightarrow{\nabla}^2-1)\mathbf{M}'+\mathbf{Y}=\mathbf{0}.\end{aligned}\tag{1.230}$$

The representation $(1.222)_1$ and equation $(1.223)_1$ correspond to the Galerkin vector of the classical theory of elasticity. In micropolar theory, the solution of the boundary value problem will be obtained in the form:

$$\mathbf{u}=\mathbf{u}^0+\mathbf{u}',\qquad\boldsymbol{\varphi}=\boldsymbol{\varphi}^0+\boldsymbol{\varphi}',$$

where the pair $(\mathbf{u}^0,\boldsymbol{\varphi}^0)$ is related to the vector $\mathbf{F}^0$ and corresponds to the classical theory of elasticity ($\mathbf{M}^0=\overrightarrow{\nabla}^2\mathbf{F}^0$, $\boldsymbol{\varphi}^0=(2\mu)^{-1}\,\mathrm{curl}\,\overrightarrow{\nabla}^2\mathbf{F}^0$), whereas the pair $(\mathbf{u}',\boldsymbol{\varphi}')$ constitutes the complementary solution and is related to the vectors $\mathbf{F}'$, $\mathbf{M}'$ (relations (1.229)). The vectors $\mathbf{F}$, $\mathbf{M}$ (relations (1.227) and

[20] The representation (1.227) and equations (1.228) are also obtained from the dynamical problem in Subsection 1.1.7. In Section 4.4 we will discuss, in general terms, the problem of going from dynamical representations to representations in elastostatics.

equations (1.228)) constitute the generalized Galerkin vector, which has been introduced to micropolar theory by means of different methods by Nowacki [192, p. 196], Sandru [206] and Stefaniak [217].

Let us recall once more that in the superposition method, the representation for the pair $(\mathbf{u}', \boldsymbol{\varphi}')$ can be assumed in the form

$$\begin{aligned} 2\mu\mathbf{u}' &= \mu_0^{-1}\,\mathrm{curl}\,\mathrm{curl}\,\mathbf{F}' - \overrightarrow{\mathrm{L}}_6\mathbf{M}', \\ 2\mu\boldsymbol{\varphi}' &= -\overrightarrow{\mathrm{L}}_5\mathbf{M}' - l^2\overrightarrow{\mathrm{L}}_2\mathbf{F}'. \end{aligned} \tag{1.231}$$

Now, the equations for the vectors $\mathbf{F}'$, $\mathbf{M}'$, obtained from equations $(1.224)_{2,3}$, (1.164) and (1.231), take the form

$$\begin{aligned} &\overrightarrow{\nabla}^2(l^2\overrightarrow{\nabla}^2 - 1)\mathbf{F}' + \mathbf{X} = \mathbf{0}, \\ &\overrightarrow{\nabla}^2(l^2\overrightarrow{\nabla}^2 - 1)(\nu_1^2\overrightarrow{\nabla}^2 - 1)\mathbf{M}' + \mathbf{Y} = \mathbf{0}. \end{aligned} \tag{1.232}$$

We can see that the equation for the vector $\mathbf{F}'$ is of order four.

Let us now turn to couple-stress theory. Here, the equations in displacements take the form

$$\mu\overrightarrow{\nabla}^2\mathbf{u} + (\lambda + \mu)\,\mathrm{grad}\,\mathrm{div}\,\mathbf{u} + 2\mu l^{\star 2}\overrightarrow{\nabla}^2\,\mathrm{curl}\,\boldsymbol{\varphi} + \mathbf{F} = \mathbf{0}, \tag{1.233}$$

where $\boldsymbol{\varphi} = \frac{1}{2}\,\mathrm{curl}\,\mathbf{u}$ and $\mathbf{F}$ denotes the introduced vector of body loadings, namely $\mathbf{F} = \mathbf{X} + \frac{1}{2}\,\mathrm{curl}\,\mathbf{Y}$. Now, using equations (1.233), we derive the uncoupled equations for $\mathbf{u}$, $\boldsymbol{\varphi}$:

$$\begin{aligned} 2\mu\overrightarrow{\nabla}^2\overrightarrow{\nabla}^2(l^{\star 2}\overrightarrow{\nabla}^2 - 1)\mathbf{u} &= \overrightarrow{\mathrm{N}}_1\mathbf{X} + \overrightarrow{\mathrm{N}}_2\mathbf{Y}, \\ 2\mu\overrightarrow{\nabla}^2(l^{\star 2}\overrightarrow{\nabla}^2 - 1)\boldsymbol{\varphi} &= \overrightarrow{\mathrm{N}}_3\mathbf{Y} + \overrightarrow{\mathrm{N}}_4\mathbf{X}, \end{aligned} \tag{1.234}$$

where the operators $\overrightarrow{\mathrm{N}}_i$ $(i = 1, \ldots, 4)$ are given by

$$\begin{aligned} &\overrightarrow{\mathrm{N}}_1 = 2\overrightarrow{\nabla}^2 - \frac{1-2\nu}{1-\nu}\left(\frac{1}{1-2\nu} + l^{\star 2}\overrightarrow{\nabla}^2\right)\mathrm{grad}\,\mathrm{div}, \\ &\overrightarrow{\mathrm{N}}_2 = \overrightarrow{\nabla}^2\,\mathrm{curl}, \qquad \overrightarrow{\mathrm{N}}_3 = \tfrac{1}{2}\,\mathrm{curl}\,\mathrm{curl}, \qquad \overrightarrow{\mathrm{N}}_4 = \mathrm{curl}\,. \end{aligned}$$

Let us now introduce the representation

$$2\mu\mathbf{u} = -\overrightarrow{\mathrm{N}}_1\hat{\mathbf{F}} - \overrightarrow{\mathrm{N}}_4\mathbf{M}, \qquad 2\mu\boldsymbol{\varphi} = -\overrightarrow{\mathrm{N}}_3\mathbf{M} - \overrightarrow{\mathrm{N}}_2\hat{\mathbf{F}}. \tag{1.235}$$

Using (1.234) and (1.235), we obtain the field equations for the vectors $\hat{\mathbf{F}}$, $\mathbf{M}$:

$$\begin{aligned} &\overrightarrow{\nabla}^2\overrightarrow{\nabla}^2(l^{\star 2}\overrightarrow{\nabla}^2 - 1)\hat{\mathbf{F}} + \mathbf{X} = \mathbf{0}, \\ &\overrightarrow{\nabla}^2(l^{\star 2}\overrightarrow{\nabla}^2 - 1)\mathbf{M} + \mathbf{Y} = \mathbf{0}. \end{aligned} \tag{1.236}$$

The vectors $\hat{\mathbf{F}}$, $\mathbf{M}$ can be treated as the Galerkin vector, generalized to couple-stress elastostatics. For $l^\star \to 0$ $(\mathbf{Y} = \mathbf{0},\ \mathbf{M} = \mathbf{0})$,

$$\overrightarrow{\mathrm{N}}_1 \to [2\overrightarrow{\nabla}^2 - (1-\nu)^{-1}\,\mathrm{grad}\,\mathrm{div}], \qquad -\hat{\mathbf{F}} \to \mathbf{F}^0, \qquad \mathbf{u} \to \mathbf{u}^0$$

and from $(1.235)_1$ and $(1.236)_1$ we obtain, respectively, the classical representation $(1.222)_1$ and the equation for the Galerkin vector $(1.223)_1$.

In the classical theory of elasticity (similarly as in couple-stress theory) it is proved[21] that the representation $(1.222)_1$ for the Galerkin vector is complete. In the micropolar theory of elasticity, the completeness problem of the representation (1.227) for the generalized Galerkin vector remains open.

1.2.6 Method of potentials. Micropolar theory and limiting theories. Superposition method

Nowacki [192, p. 33] has developed the method of potentials within the framework of micropolar theory, which has been extensively used to determine the fundamental solutions in the theory, to study wave propagation and to solve certain two-dimensional boundary value problems (see also [188]). The method has not been applied to the three-dimensional boundary value problems of elastostatics. In [49], we derived equations for the potentials and applied them to determine the fundamental solutions and, mainly, to solve certain three-dimensional boundary value problems of elastostatics. Such a vector representation will be given below. The starting point is the Stokes-Helmholtz decomposition of the vector fields $\mathbf{u}$, $\boldsymbol{\varphi}$ (1.147), which, when applied in equations (1.155), leads to equations coupling the scalar potentials Φ, Σ and the vector potentials $\boldsymbol{\Psi}$, $\mathbf{H}$ in Ω, namely:

$$\begin{aligned}
&(2\mu+\lambda)\overrightarrow{\nabla}^2\,\mathrm{grad}\,\Phi + (\mu+\alpha)\overrightarrow{\nabla}^2\,\mathrm{curl}\,\boldsymbol{\Psi} - 2\alpha\overrightarrow{\nabla}^2\mathbf{H} + \mathbf{X} = \mathbf{0},\\
&[(2\gamma+\beta)\overrightarrow{\nabla}^2 - 4\alpha]\,\mathrm{grad}\,\Sigma + [(\gamma+\varepsilon)\overrightarrow{\nabla}^2 - 4\alpha]\,\mathrm{curl}\,\mathbf{H}\\
&\qquad - 2\alpha\overrightarrow{\nabla}^2\boldsymbol{\Psi} + \mathbf{Y} = \mathbf{0}.
\end{aligned} \tag{1.237}$$

The scalar potentials Φ, Σ satisfy the equations

$$\begin{aligned}
&\mathrm{div}\,\mathrm{grad}(\mathrm{div}\,\mathrm{grad})\Phi = -\frac{1}{2\mu+\lambda}\,\mathrm{div}\,\mathbf{X},\\
&\mathrm{div}\,\mathrm{grad}(\nu_1^2\,\mathrm{div}\,\mathrm{grad} - 1)\Sigma = -\frac{1}{4\alpha}\,\mathrm{div}\,\mathbf{Y};
\end{aligned} \tag{1.238}$$

whereas the equations for the vector potentials $\mathbf{H}$, $\boldsymbol{\Psi}$ read:

$$\begin{aligned}
4\alpha\mu\overrightarrow{\nabla}^2\overrightarrow{\nabla}^2(l^2\overrightarrow{\nabla}^2 - 1)\boldsymbol{\Psi} &= [(\gamma+\varepsilon)\overrightarrow{\nabla}^2 - 4\alpha]\,\mathrm{curl}\,\mathbf{X}\\
&\quad - 2\alpha\,\mathrm{curl}\,\mathrm{curl}\,\mathbf{Y},\\
4\alpha\mu\overrightarrow{\nabla}^2\overrightarrow{\nabla}^2(l^2\overrightarrow{\nabla}^2 - 1)\mathbf{H} &= (\mu+\alpha)\overrightarrow{\nabla}^2\,\mathrm{curl}\,\mathbf{Y} - 2\alpha\,\mathrm{curl}\,\mathrm{curl}\,\mathbf{X}.
\end{aligned} \tag{1.239}$$

[21] This problem is discussed in detail by Gurtin [105, p. 141]. For couple-stress theory, see Sokołowski [215, p. 25].

The behaviour of the vectors $\mathbf{X}$, $\mathbf{Y}$ in formulae (1.237)–(1.239) allows us to extend our considerations to the temperature and distortion fields.

Let us recall Subsection 1.1.8 and make the following important remark, namely that equations (1.237)–(1.239) follow from the dynamical equations (1.151)–(1.153). One should also stress the fact that by substituting the decompositions (1.147) and (1.148) into equations (1.155), we arrive at

$$\begin{aligned} &\operatorname{grad}\Big[(2\mu+\lambda)\nabla^2\Phi+\rho\vartheta\Big] \\ &\qquad+\operatorname{curl}\Big[(\mu+\alpha)\vec{\nabla}^2\boldsymbol{\Psi}+2\alpha\operatorname{curl}\mathbf{H}+\rho\boldsymbol{\chi}\Big]=\mathbf{0}, \\ &\operatorname{grad}\big\{[(2\gamma+\beta)\nabla^2-4\alpha]\Sigma+J\sigma\big\} \\ &\qquad+\operatorname{curl}\{[(\gamma+\varepsilon)\vec{\nabla}^2-4\alpha]\mathbf{H}+2\alpha\operatorname{curl}\boldsymbol{\Psi}+J\boldsymbol{\eta}\}=\mathbf{0}. \end{aligned} \tag{1.240}$$

Now, assuming that the expressions under the gradient and curl symbols vanish independently, we obtain the system of four uncoupled equations for the potentials Φ, Σ and $\boldsymbol{\Psi}$, $\mathbf{H}$. Such a method of solving the system of equations (1.155), being a generalization of the Kelvin method, known from the classical theory of elasticity, to micropolar elastostatics, is justified only in the case of an infinite space[22].

Let us now discuss the application of the method of potentials to the limiting theories:

— the classical theory of elasticity (model H). Here, using the representation

$$\mathbf{u}^0=\operatorname{grad}\Phi^0+\operatorname{curl}\boldsymbol{\Psi}^0,\qquad \boldsymbol{\varphi}^0=\tfrac{1}{2}\operatorname{curl}\operatorname{curl}\boldsymbol{\Psi}^0,\qquad \operatorname{div}\boldsymbol{\Psi}^0=0, \tag{1.241}$$

we obtain from equations (1.159) the relations

$$(2\mu+\lambda)\vec{\nabla}^2\operatorname{grad}\Phi^0+\mu\vec{\nabla}^2\operatorname{curl}\boldsymbol{\Psi}^0+\mathbf{X}=\mathbf{0} \tag{1.242}$$

[22] See Nowacki [186, p. 161]. The following equations are obtained:

$$\begin{aligned} &(2\mu+\lambda)\nabla^2\Phi+\rho\vartheta=0, \\ &[(2\gamma+\beta)\nabla^2-4\alpha]\Sigma+J\sigma=0, \\ &(\mu+\alpha)\vec{\nabla}^2\boldsymbol{\Psi}+2\alpha\operatorname{curl}\mathbf{H}+\rho\boldsymbol{\chi}=\mathbf{0}, \\ &[(\gamma+\varepsilon)\vec{\nabla}^2-4\alpha]\mathbf{H}+2\alpha\operatorname{curl}\boldsymbol{\Psi}+J\boldsymbol{\eta}=\mathbf{0}. \end{aligned}$$

They are useful when finding fundamental solutions in the infinite space. Disregarding body loadings, we obtain in the rectangular system $Ox_1x_2x_3$ the equations:

$$\nabla^2\Phi=0,\qquad (\nu_1^2\nabla^2-1)\Sigma=0,\qquad \nabla^2(l^2\nabla^2-1)(\Psi_i,H_i)=(0,0),$$

which do not suffice to solve the boundary value problem for the half-space Ω (1.209).

coupling the potentials Φ^0 and $\boldsymbol{\Psi}^0$ in Ω, where Φ^0, $\boldsymbol{\Psi}^0$ satisfy, respectively, non-homogeneous biharmonic equations: the scalar one $(1.238)_1$ and the vector one:

$$\overrightarrow{\nabla}^2\overrightarrow{\nabla}^2\boldsymbol{\Psi}^0 = -\frac{1}{\mu}\operatorname{curl}\mathbf{X}; \tag{1.243}$$

— couple-stress theory (model K-M). Taking into account the representation

$$\mathbf{u} = \operatorname{grad}\Phi + \operatorname{curl}\boldsymbol{\Psi}, \qquad \boldsymbol{\varphi} = \tfrac{1}{2}\operatorname{curl}\operatorname{curl}\boldsymbol{\Psi}, \qquad \operatorname{div}\boldsymbol{\Psi} = 0 \tag{1.244}$$

and using equations (1.233), we obtain

$$(2\mu + \lambda)\overrightarrow{\nabla}^2\operatorname{grad}\Phi - \mu\overrightarrow{\nabla}^2(l^{\star 2}\overrightarrow{\nabla}^2 - 1)\operatorname{curl}\boldsymbol{\Psi} + \mathbf{F} = \mathbf{0}, \tag{1.245}$$

where the scalar potential Φ satisfies $(1.238)_1$, whereas for the vector potential $\boldsymbol{\Psi}$ it holds

$$\overrightarrow{\nabla}^2\overrightarrow{\nabla}^2(l^{\star 2}\overrightarrow{\nabla}^2 - 1)\boldsymbol{\Psi} = -\frac{1}{\mu}\operatorname{curl}\mathbf{F}; \tag{1.246}$$

— hypothetical medium (model HM). Assuming the decomposition $(1.147)_2$ and using the vector equations in Ω,

$$(\gamma + \varepsilon)\overrightarrow{\nabla}^2\boldsymbol{\varphi} + (\beta + \gamma - \varepsilon)\operatorname{grad}\operatorname{div}\boldsymbol{\varphi} + \mathbf{Y} = \mathbf{0} \tag{1.247}$$

we obtain

$$(2\gamma + \beta)\overrightarrow{\nabla}^2\operatorname{grad}\Sigma + (\gamma + \varepsilon)\overrightarrow{\nabla}^2\operatorname{curl}\mathbf{H} + \mathbf{Y} = \mathbf{0}, \tag{1.248}$$

where the potentials Σ, $\mathbf{H}$ satisfy the relations

$$\begin{aligned} \operatorname{div}\operatorname{grad}(\operatorname{div}\operatorname{grad})\Sigma &= -\frac{1}{2\gamma + \beta}\operatorname{div}\mathbf{Y}, \\ \overrightarrow{\nabla}^2\overrightarrow{\nabla}^2\mathbf{H} &= \frac{1}{\gamma + \varepsilon}\operatorname{curl}\mathbf{Y}. \end{aligned} \tag{1.249}$$

Now, let us return to micropolar theory and the superposition method. We assume that the boundary conditions (1.156) have been formulated for equations (1.155). To solve this problem, we use the potentials and the superposition method. For the case of the classical equations (1.159) we obtain the representation (1.241) with the field equations $(1.238)_1$ (for Φ^0), (1.242), (1.243), and with the classical boundary conditions (1.160). For the complementary solution, we assume the decomposition

$$\begin{aligned} \mathbf{u}' &= \operatorname{grad}\Phi' + \operatorname{curl}\boldsymbol{\Psi}', & \operatorname{div}\boldsymbol{\Psi}' &= 0, \\ \boldsymbol{\varphi}' &= \operatorname{grad}\Sigma' + \operatorname{curl}\mathbf{H}', & \operatorname{div}\mathbf{H}' &= 0. \end{aligned} \tag{1.250}$$

Using (1.250) in the system of equations (1.166), we obtain field equations which are valid for Φ', Σ' and $\boldsymbol{\Psi}'$, $\mathbf{H}'$, namely equations (1.237)–(1.239), in which one has to assume

$$(\Phi, \Sigma, \boldsymbol{\Psi}, \mathbf{H}, \mathbf{X}, \mathbf{Y}) \mapsto (\Phi', \Sigma', \boldsymbol{\Psi}', \mathbf{H}', \hat{\mathbf{X}}, \overset{*}{\mathbf{Y}})$$

with the boundary conditions taking into account the boundary effect for the micropolar body (1.167). The final result takes the form (1.147), where

$$\begin{aligned} \Phi &= \Phi^0 + \Phi', \qquad & \Sigma &= \Sigma', \\ \boldsymbol{\Psi} &= \boldsymbol{\Psi}^0 + \boldsymbol{\Psi}', \qquad & \mathbf{H} &= \tfrac{1}{2}\operatorname{curl}\boldsymbol{\Psi}^0 + \mathbf{H}'. \end{aligned} \tag{1.251}$$

Let us finally present certain relations between the potentials Φ, Σ, $\boldsymbol{\Psi}$, $\mathbf{H}$ and the vectors $\mathbf{F}$, $\mathbf{M}$ representing the Galerkin vector in micropolar theory. Assuming $\mathbf{X} = \mathbf{0}$, $\mathbf{Y} = \mathbf{0}$, we derive from (1.147), (1.227) the relations

$$\begin{aligned} 2\mu\Phi &= -\operatorname{div}[2 + (1+\nu)^{-1}(l^2\overrightarrow{\nabla}^2 - 1)]\mathbf{F}^1, \\ 4\alpha\Sigma &= \operatorname{div}[(l^2\overrightarrow{\nabla}^2 - 1)\mathbf{M}^3 - \mathbf{M}^1], \end{aligned} \tag{1.252}$$

and the formulae

$$\begin{aligned} 2\mu\boldsymbol{\Psi} &= \mu_0^{-1}\overrightarrow{\nabla}^2\operatorname{curl}\mathbf{F} + 2\operatorname{curl}\mathbf{F}^1 + \mathbf{M}^1 - (\nu_1^2\overrightarrow{\nabla}^2 - 1)\mathbf{M}^2, \\ 2\mu\mathbf{H} &= -\overrightarrow{\nabla}^2\mathbf{F} + \alpha_0^{-1}\operatorname{curl}[\mathbf{M}^1 - (\nu_1^2\overrightarrow{\nabla}^2 - 1)\mathbf{M}^2], \end{aligned} \tag{1.253}$$

in which $\mathbf{F} = \mathbf{F}^1 + \mathbf{F}^2$, $\mathbf{M} = \mathbf{M}^1 + \mathbf{M}^2 + \mathbf{M}^3$, and functions $\mathbf{F}^\alpha$, $\mathbf{M}^i$ satisfy the equations

$$\begin{aligned} &\overrightarrow{\nabla}^2\overrightarrow{\nabla}^2\mathbf{F}^1 = \mathbf{0}, \qquad (l^2\overrightarrow{\nabla}^2 - 1)\mathbf{F}^2 = \mathbf{0}, \\ &\overrightarrow{\nabla}^2\mathbf{M}^1 = \mathbf{0}, \qquad (l^2\overrightarrow{\nabla}^2 - 1)\mathbf{M}^2 = \mathbf{0}, \qquad (\nu_1^2\overrightarrow{\nabla}^2 - 1)\mathbf{M}^3 = \mathbf{0}. \end{aligned} \tag{1.254}$$

The (homogeneous) equations (1.238), (1.239) for (1.252), (1.253) are satisfied in view of equations (1.254) and (1.228). Additionally, we have

$$\nabla^2\operatorname{div}\mathbf{F} = 0, \qquad \operatorname{div}[\mathbf{M}^1 - (\nu_1^2\overrightarrow{\nabla}^2 - 1)\mathbf{M}^2] = 0.$$

1.2.7 Method of potentials. Fundamental solutions. Half-space problem

In [49], we considered the action of the body loadings (1.182) in the space $\mathbb{R}^3$ by use of the equations with potentials (1.238) and (1.239), written in the rectangular system of coordinates $Ox_1x_2x_3$. Applying the Fourier transformation (0.50) (for $n = 3$) to those equations, we obtain the following results:

— for the loading given by $(1.182)_1$ ($P_0 \neq 0$, $M_0 = 0$),

$$\begin{aligned} \Phi^s &= \frac{P_0}{8\pi(2\mu+\lambda)}R_{,s}, \qquad \Sigma^s = 0, \\ 8\pi\mu P_0^{-1}H_i^s &= -\tfrac{1}{2}R_{,is} - l^2\left(\frac{1-e^{-R/l}}{R}\right)_{,is} + \left(\frac{1-e^{-R/l}}{R}\right)\delta_{is}, \\ 8\pi\mu P_0^{-1}\Psi_i^s &= -\epsilon_{ijs}\left[\mu_0^{-1}\left(\frac{1-e^{-R/l}}{R}\right) + R\right]_{,j}; \end{aligned} \tag{1.255}$$

— for the loading given by $(1.182)_2$ ($P_0 = 0$, $M_0 \neq 0$),

$$\Phi^s = 0, \qquad \Sigma^s = -\frac{M_0}{16\pi\alpha}\left(\frac{1-e^{-R/\nu_1}}{R}\right)_{,s},$$

$$8\pi\mu M_0^{-1} H_i^s = \alpha_0^{-1}\epsilon_{ijs}\left(\frac{1-e^{-R/l}}{R}\right)_{,j}, \tag{1.256}$$

$$8\pi\mu M_0^{-1}\Psi_i^s = -\tfrac{1}{2}R_{,is} - l^2\left(\frac{1-e^{-R/l}}{R}\right)_{,is} + \left(\frac{1-e^{-R/l}}{R}\right)\delta_{is},$$

where $R = \sqrt{x_i x_i}$ and the constant ν_1 in the expression $\exp(-R/\nu_1)$ is given by formula (1.189),

$$l^2 = \frac{(\gamma+\varepsilon)(\mu+\alpha)}{4\alpha\mu}, \qquad \mu_0 = \frac{2\mu}{\gamma+\varepsilon}, \qquad \alpha_0 = \frac{2\alpha}{\mu+\alpha}.$$

Formulae (1.255) and (1.256) lead to formulae (1.187) and (1.188), respectively, by use of the relations:

$$u_i^s = \Phi^s_{,i} + \epsilon_{ijk}\Psi^s_{k,j}, \qquad \varphi_i^s = \Sigma^s_{,i} + \epsilon_{ijk}H^s_{k,j}. \tag{1.257}$$

The field conditions

$$\Psi^s_{i,i} = 0, \qquad H^s_{i,i} = 0 \tag{1.258}$$

become satisfied identically.

In [49] and [80], we applied the method of potentials in order to solve the micropolar half-space problem (1.209) with boundary conditions of type (1.211) and (1.213). The starting point for our considerations is given by the homogeneous field equations (1.238) and (1.239) written in the rectangular system of coordinates $Ox_1x_2x_3$ in Ω:

$$\begin{aligned} &\nabla^2\nabla^2\Phi = 0, \qquad \nabla^2(\nu_1^2\nabla^2 - 1)\Sigma = 0,\\ &\nabla^2\nabla^2(l^2\nabla^2 - 1)\Psi_i = 0 \qquad \nabla^2\nabla^2(l^2\nabla^2 - 1)H_i = 0. \end{aligned} \tag{1.259}$$

The potentials Φ, Σ and Ψ_i, H_i must satisfy in Ω the differential field conditions (1.258) and the equations

$$\begin{aligned} &(2\mu+\lambda)\nabla^2\Phi_{,i} + (\mu+\alpha)\nabla^2\epsilon_{ijk}\Psi_{k,j} - 2\alpha\nabla^2 H_i = 0,\\ &[(2\gamma+\beta)\nabla^2 - 4\alpha]\Sigma_{,i} + [(\gamma+\varepsilon)\nabla^2 - 4\alpha]\epsilon_{ijk}H_{k,j} - 2\alpha\nabla^2\Psi_i = 0. \end{aligned} \tag{1.260}$$

Applying the Fourier transformation (0.50) (for $n = 2$), we find that the partial differential equations (1.259) go over into the ordinary differential equations (with respect to the variable x_1):

$$\begin{aligned} &(\mathbb{D}_1^2 - \xi^2)^2\tilde{\Phi} = 0, \qquad (\mathbb{D}_1^2 - \eta^2)(\mathbb{D}_1^2 - \xi^2)\tilde{\Sigma} = 0,\\ &(\mathbb{D}_1^2 - \xi^2)^2(\mathbb{D}_1^2 - \rho^2)\tilde{\Psi}_i = 0, \qquad (\mathbb{D}_1^2 - \xi^2)^2(\mathbb{D}_1^2 - \rho^2)\tilde{H}_i = 0 \end{aligned} \tag{1.261}$$

for the transforms $\tilde{\Phi}$, $\tilde{\Sigma}$, $\tilde{\Psi}_i$, $\tilde{H}_i$, where $\mathbb{D}_1 \stackrel{\text{def}}{=} d/dx_1$. Recall that

$$\rho^2 = \xi^2 + l^{-2}, \qquad \eta^2 = \xi^2 + \nu_1^{-2}, \qquad \xi^2 = \xi_L \xi_L,$$
$$l^2 = \frac{(\gamma+\varepsilon)(\mu+\alpha)}{4\alpha\mu}, \qquad \nu_1^2 = \frac{2\gamma+\beta}{4\alpha}.$$

Having taken into account condition (1.213) (at infinity), one should assume the general solutions of equations (1.261) in the form:

$$\begin{aligned} &\tilde{\Phi} = (A + B\xi x_1)e^{-\xi x_1}, \qquad \tilde{\Sigma} = Ce^{-\rho x_1} + De^{-\eta x_1}, \\ &\tilde{\Psi}_i = (E_i + F_i \xi x_1)e^{-\xi x_1} + G_i e^{-\rho x_1}, \\ &\tilde{H}_i = (J_i + K_i \xi x_1)e^{-\xi x_1} + M_i e^{-\rho x_1}. \end{aligned} \tag{1.262}$$

The following observation is crucial. In formulae (1.262) there appear 22 integration constants A, B, C, D, E_i, F_i, G_i, J_i, K_i, M_i (depending on the parameter ξ), which, having satisfied the transformed field conditions (equations) (1.258) and (1.260), are expressed in terms of 10 fundamental (basic) constants. Since we still have to satisfy 6 boundary conditions of the type (1.210) or (1.211), it means that the potentials Φ, Σ, Ψ_i and H_i cannot be uniquely determined. However, it has been shown that the field relations (1.257) lead to unique fields of displacements and rotations given by formulae (1.220) and (1.221) (in which, for the half-space problem, one should set $a_1 = b_1 = d_1 = \bar{a}_1 = \bar{b}_1 = \bar{d}_1 = 0$); this result had been expected by the author. The six (basic) constants A_1, B_1, C_1, $\bar{A}_1$, $\bar{B}_1$, $\bar{C}_1$ are uniquely determined from six boundary conditions. In [80] we also discussed the half-space problem (1.209) for the limiting theories, using the method of potentials, and obtained unique expressions for displacements and rotations. Below we shall sketch (for a micropolar half-space) the procedure of unique determination of the pair (u_i, φ_i). Substituting formulae (1.262) to the transformed field conditions (1.258), (1.260), and equating to zero the expressions multiplying the exponential functions $e^{-\xi x_1}$, $x_1 e^{-\xi x_1}$, $e^{-\rho x_1}$, $e^{-\eta x_1}$, we eliminate 12 out of 22 constants, keeping 10 constants A, D, E_2, E_3, F_1, G_1, J_1, J_2, K_1, M_1, the former being expressed in terms of the latter as follows:

$$\begin{aligned} &B = \frac{2\mu}{2\mu+\lambda}\cdot\frac{1}{\xi}K_1, \qquad C = \frac{i\xi J_2 + \xi_2 J_1 - \xi_2 K_1 - \xi\xi_3 F_1}{\xi_3}, \\ &E_1 = \frac{F_1\xi - i\xi_2 E_2 - i\xi_3 E_3}{\xi}, \qquad F_K = i\left(\frac{\xi_K}{\xi}F_1 - \epsilon_{1KL}\frac{\xi_L}{\xi^2}2K_1\right), \\ &G_K = \frac{i}{\xi^2}(\rho\xi_K G_1 - \epsilon_{1KL}\xi_L\alpha_0 M_1), \qquad J_3 = \frac{i\xi J_1 - i\xi K_1 - \xi_2 J_2}{\xi_3}, \\ &K_K = \frac{i\xi_K}{\xi}K_1, \qquad M_K = \frac{i}{\xi^2}(\rho\xi_K M_1 + \epsilon_{1KL}\xi_L\mu_0 G_1), \end{aligned} \tag{1.263}$$

where ξ_2, ξ_3 are parameters resulting from the application of the Fourier transformation (with respect to the variables x_2, x_3) and $i = \sqrt{-1}$ is the imaginary unit.

Relations (1.263) enable us to express the transforms $\tilde{\Phi}$, $\tilde{\Sigma}$, $\tilde{\Psi}_i$ and $\tilde{H}_i$ in terms of 10 constants. The last step consists in substituting such transforms from (1.262) into the transformed conditions (1.257) and noticing that 8 constants A, D, E_2, E_3, F_1, G_1, K_1, M_1 appear in combinations which enable us to introduce 6 new (basic) constants A_1, B_1, C_1, $\bar{A}_1$, $\bar{B}_1$, $\bar{C}_1$:

$$\begin{aligned} &A_1 = -\xi A + \frac{2\mu}{2\mu+\lambda} K_1 - i\xi_2 E_3 + i\xi_3 E_2, \qquad B_1 = \frac{2(\mu+\lambda)}{2\mu+\lambda} K_1, \\ &C_1 = \alpha_0 M_1, \qquad \bar{A}_1 = \xi^2 F_1, \qquad \bar{B}_1 = -\eta D, \qquad \bar{C}_1 = -\mu_0 G_1. \end{aligned} \tag{1.264}$$

The constants J_1, J_2 do not appear in the transformed conditions (1.257).

1.2.8 Micropolar half-space. Problem of singularities of physical fields. Three-dimensional problem

In [79]–[82] and [49], Dyszlewicz and Słotwińska studied the problem of singularities of the fields of displacements, rotations, force-stresses and couple-stresses in the half-space (1.209), loaded on the boundary by concentrated forces and moments. The method of investigating the singularities of physical fields is due to Bogy & Sternberg [15] and Muki & Sternberg [161] (who studied physical fields in the couple-stress half-space for the plane state of strain). The starting point of our considerations will be given by general formulae derived for the fields of displacements, rotations, force-stresses and couple-stresses, in the form of Fourier integrals for the half-space in the spatial state of strain. The improper integrals appearing in those formulae are convergent on the set

$$\overline{\Omega} = \Omega \cup \partial\Omega = \{x_i : x_1 \geq 0, |x_2| < \infty, |x_3| < \infty\},$$

except, perhaps, at the point where a concentrated loading is applied (for $x_1 = 0$, $x_2 = 0$, $x_3 = 0$). However, their determination in a closed form encounters significant difficulties of mathematical nature. In [49] and [79]–[82] we found the singular parts of the solution in an explicit form, which are a basis for determining the character of the singularities of physical fields in the neighbourhood of the point of application of a concentrated loading. For that purpose, from the formal results for physical fields in the form of Fourier integrals (which are complex-variable functions), we take the real part of a solution and, making use of certain asymptotic expansions for the integrands (for $\xi = \sqrt{\xi_L \xi_L} \to \infty$), we take into account expressions which influence its singular parts. This procedure is presented in detail in the references mentioned above. The following cases of concentrated loadings were considered:

— the boundary value problem of Boussinesq type,

$$(1^o) \quad \sigma_{11} = -P_0\delta(x_2)\delta(x_3), \qquad \sigma_{1K} = \mu_{1i} = 0 \qquad \text{on } \partial\Omega;$$

— the boundary value problem of Cerruti type,

$$(2^o) \quad \sigma_{12} = -S_0\delta(x_2)\delta(x_3), \qquad \sigma_{11} = \sigma_{13} = \mu_{1i} = 0 \qquad \text{on } \partial\Omega;$$

(3°) $\sigma_{1i} = \mu_{1L} = 0, \qquad \mu_{11} = -M_\mu \delta(x_2)\delta(x_3) \qquad$ on $\partial\Omega$[23];

(4°) the bending moment with a pair of normal forces

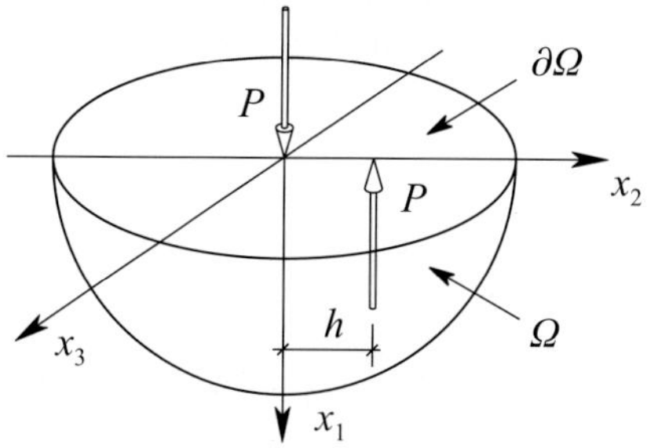

$$\lim_{h\to 0} Ph = M_0;$$

Fig. 1.2.

(5°) the twisting moment with a pair of tangent forces

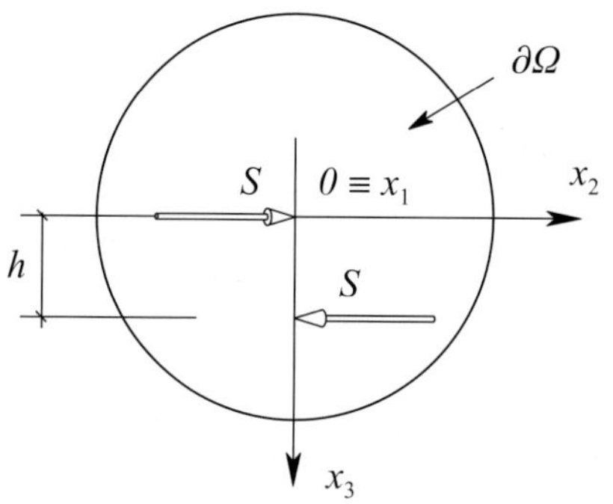

$$\lim_{h\to 0} Sh = M_0;$$

Fig. 1.3.

(6°) the twisting moment

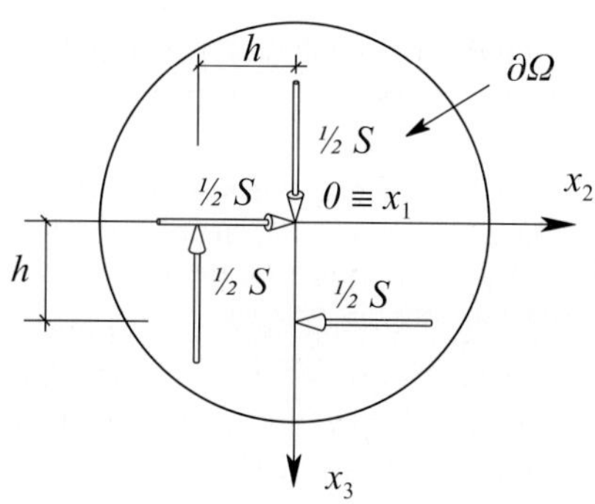

Fig. 1.4.

where P_0, S_0, M_μ and M_0 — constant values.

---

[23] The index μ in M_μ means that it refers to a twisting couple of couple-stress type. $[\mu_{ij}] = [\mathrm{N}\cdot\mathrm{m}^{-1}]$, $[\delta(x_2)\delta(x_3)] = [\mathrm{m}^{-2}]$, $[M_\mu] = [\mathrm{N}\cdot\mathrm{m}]$.

In boundary value problems denoted by (1°)–(6°), the singular parts of physical fields (displacements, rotations, force-stresses, couple-stresses) were extracted in explicit forms. The character and orders of those singularities were discussed in detail. In the references mentioned above we also gave solutions of boundary value problems with conditions (1°)–(6°) for the elastic half-space in (K-M), (HM) and (H) models. A number of observations and conclusions concerning physical fields were provided. In particular, we analysed the correspondence between the results obtained within the framework of micropolar theory and the results for the limiting theories (it concerns results in the form of integral representations as well as their singular parts given in an explicit form). As an example we shall give here (due to extensiveness of [79]–[82]) only the results referring to the orders of singularities of physical fields in the Cerruti problem (on the boundary $\partial\Omega$ of the half-space Ω a concentrated tangential force is prescribed); the results referring to the boundary value problem of Boussinesq type (1°), as well as to the problems of the half-space, both non-classically (3°) and classically (6°) twisted on the boundary, will be discussed together with the axially-symmetric problems.

The results are put together in Table 1.2.

Table 1.2.

Physical fields	(1)	(2)	(3)
u_i	$O(R^{-1})$	$O(R^{-1})$	$O(R^{-1})$
φ_α	$O(1)$	$O(1)$	$O(R^{-2})$
φ_3	$O(\ln(R+x_1))$	$O(\ln(R+x_1))$	$O(R^{-2})$
σ_{ij}	$O(R^{-2})$	$s_{ij}=O(R^{-2})$	$O(R^{-2})$
μ_{ij}	$O(R^{-1})$	$m_{ij}=O(R^{-1})$	–

Numbers (1), (2) and (3) denote[24], respectively, micropolar theory, couple-stress theory, and the classical theory of elasticity; $R = \sqrt{x_1^2+x_2^2+x_3^2}$, the symbols s_{ij}, m_{ij} denote, respectively, the symmetric part of the tensor σ_{ij} and the deviatoric part of the tensor μ_{ij}, and they refer to couple-stress theory

The symbols $O(R^{-n})$, $(n=1,\ 2)$, $O(\ln(R+x_1))$ mean that in a neighbourhood of the point where a tangential force is applied, a given physical quantity increases without bound as $1/R^n$ or logarithmically, respectively. By $O(1)$ we denote the regular part of a solution in the sense that it is bounded

[24] In this work, the convention concerning the numbering of theories is of local character (e.g., notations in Fig. 1.1 and Table 1.2 are different).

at the point $(0,0,0)$. It is interesting to discuss the limits referring to the results of micropolar theory and the limiting theories. The correspondence for the integral representations of the results (for displacements, rotations, force-stresses and couple-stresses) from theories (1), (2) and (3) depend, in general, on the elasticity constants α, $l^\star$.

The problem of taking the limits for the singular parts of solutions is connected with the coefficient standing by a singularity (called the coefficient of singularity intensity) and is of a more complex character. Let us note the following general conclusions:

1. The solutions for u_i, φ_i, σ_{ji} and μ_{ji} (by which we mean here the integral representation of results) in the problem of the micropolar half-space (E-N) with the boundary conditions (1°)–(6°) are transformed in the limit $(\alpha \to \infty)$ into the solutions obtained within the framework of the couple-stress theory (K-M); so that $\frac{1}{2}(\sigma_{ij}+\sigma_{ji}) \to s_{ij}$. A similar correspondence for singular parts of solutions is true only in the case of the boundary conditions (1°), (4°). In turn, in problems with the boundary conditions (2°), (3°), (5°), (6°) one should take the iterated limits $(\alpha \to \infty,\ \beta \to \infty)$; so that for σ_{ji} and μ_{ji} the following rule holds:

$$\tfrac{1}{2}(\overset{s}{\sigma}_{ij}+\overset{s}{\sigma}_{ji}) \xrightarrow{\alpha\to\infty,\ \beta\to\infty} \overset{s}{s}_{ij}, \qquad (\overset{s}{\mu}_{ij}-\hat{\mu}^s\delta_{ij}) \xrightarrow{\alpha\to\infty,\ \beta\to\infty} \overset{s}{m}_{ij},$$

where the index s denotes the singular part of a solution and $\hat{\mu}$ is an invariant: $\hat{\mu}=\frac{1}{3}\mu_{ii}$.

2. Consider the limiting case $\alpha \to 0^+$. In the problems of micropolar theory (E-N) with the boundary conditions (1°), (2°) (4°)–(6°) it holds $(\mu_{ji}, \overset{s}{\mu}_{ji}) \to (0,0)$ and we obtain for u_i, $\overset{s}{u}_i$ and σ_{ji}, $\overset{s}{\sigma}_{ji}$ the classical results (H). An exception is furnished by the rotations φ_i, $\overset{s}{\varphi}_i$, which do not lead to the classical results since they tend to zero. In the problem with the boundary conditions (3°), the limit $\alpha \to 0^+$ transforms the formulae for the pair $(\varphi_i, \overset{s}{\varphi}_i)$ from micropolar theory (E-N) into the formulae for a hypothetical medium (HM), whereas the solutions for u_i, σ_{ji} and their singular parts $\overset{s}{u}_i$, $\overset{s}{\sigma}_{ji}$ tend to zero.

3. The solutions for u_i, φ_i, s_{ij}, m_{ji} obtained in couple-stress theory (K-M) (referring to the boundary conditions (1°)–(6°)) go over into the classical results (H) $(m_{ji} \to 0)$ for $l^\star \to 0^+$. On the contrary, the singular parts of these solutions $\overset{s}{u}_i$, $\overset{s}{\varphi}_i$, $\overset{s}{s}_{ij}$, $\overset{s}{m}_{ij}$ do not correspond at all to those in the classical theory (H).

1.2.9 Stress equations. Fundamental solutions

Let us consider the following stress equations in the rectangular system of coordinates $Ox_1x_2x_3$:

— for force-stresses,

$$\begin{aligned}
&2\mu\nabla^2\nabla^2(\nu_1^2\nabla^2-1)(l^2\nabla^2-1)\sigma_{ji}\\
&\quad=(\nu_1^2\nabla^2-1)\Big\{(\mu+\alpha)[\nabla^2\epsilon_{ilk}Y_{k,l}-2\nabla^2(l^2\nabla^2-1)X_i\\
&\quad-\mu_0^{-1}\nabla^2\epsilon_{imp}\epsilon_{plk}X_{k,lm}]_{,j}\\
&\quad+(\mu-\alpha)[\nabla^2\epsilon_{ljk}Y_{k,l}-2\nabla^2(l^2\nabla^2-1)X_j\\
&\quad-\mu_0^{-1}\nabla^2\epsilon_{jmp}\epsilon_{plk}X_{k,lm}]_{,i}\\
&\quad+2\mu(1-\nu)^{-1}(l^2\nabla^2-1)X_{l,lij}\Big\}\\
&\quad+2\alpha\epsilon_{kji}\nabla^2\Big\{\mu(2\alpha)^{-1}(l^2\nabla^2-1)Y_{l,lk}\\
&\quad-(\nu_1^2\nabla^2-1)[\epsilon_{klm}X_{m,l}+\alpha_0^{-1}\epsilon_{kmp}\epsilon_{pln}Y_{n,lm}]\Big\}\\
&\quad-2\mu\lambda(2\mu+\lambda)^{-1}\nabla^2(\nu_1^2\nabla^2-1)(l^2\nabla^2-1)X_{l,l}\delta_{ij};
\end{aligned}\tag{1.265}$$

— for couple-stresses,

$$\begin{aligned}
&2\mu\nabla^2(\nu_1^2\nabla^2-1)(l^2\nabla^2-1)\mu_{ji}\\
&\quad=(\gamma+\varepsilon)\big[-(2\alpha)^{-1}\mu(l^2\nabla^2-1)Y_{l,li}+(\nu_1^2\nabla^2-1)\epsilon_{ilm}X_{m,l}\\
&\quad+\alpha_0^{-1}(\nu_1^2\nabla^2-1)\epsilon_{imp}\epsilon_{plk}Y_{k,lm}\big]_{,j}\\
&\quad+(\gamma-\varepsilon)\big[-(2\alpha)^{-1}\mu(l^2\nabla^2-1)Y_{l,lj}\\
&\quad+(\nu_1^2\nabla^2-1)\epsilon_{jlm}X_{m,l}+\alpha_0^{-1}(\nu_1^2\nabla^2-1)\epsilon_{jmp}\epsilon_{plk}Y_{k,lm}\big]_{,i}\\
&\quad-\beta\mu(2\alpha)^{-1}\nabla^2(l^2\nabla^2-1)Y_{k,k}\delta_{ij}.
\end{aligned}\tag{1.266}$$

Let us recall some notation:

$$\mu_0=\frac{2\mu}{\gamma+\varepsilon},\qquad \alpha_0=\frac{2\alpha}{\mu+\alpha},\qquad \nu_1^2=\frac{2\gamma+\beta}{4\alpha},\qquad l^2=\frac{(\gamma+\varepsilon)(\mu+\alpha)}{4\alpha\mu}.$$

Equations (1.265) and (1.266) can be a basis for determining fundamental solutions for stresses. By analogy to the body forces (1.38), one can make use of equations (1.265) and (1.266) to find fundamental solutions in $\mathbb{R}^3$ for σ_{ji} and μ_{ji} in the problem with the distortion field (1.199). As an example, consider the field of body loadings given by formula (1.182). Applying the Fourier integral transformation (0.50) (for $n=3$) to equations (1.265) and (1.266), one arrives at the following formulae:

— for the loading $(1.182)_1$ ($P_0 \neq 0$, $M_0 = 0$),

$$
\begin{aligned}
8\pi\mu P_0^{-1}\sigma_{ji}^{(s)} &= 2\mu\Bigg[\left(\frac{1}{R}\right)_{,j}\delta_{is} + \left(\frac{1}{R}\right)_{,i}\delta_{js} \\
&\quad + \nu(1-\nu)^{-1}\left(\frac{1}{R}\right)_{,s}\delta_{ij} - \tfrac{1}{2}(1-\nu)^{-1}R_{,ijs}\Bigg] \\
&\quad - (\gamma+\varepsilon)\left(\frac{1-e^{-R/l}}{R}\right)_{,ijs} - \frac{4\alpha\mu}{\mu+\alpha}\left(\frac{e^{-R/l}}{R}\right)_{,i}\delta_{js}, \qquad (1.267) \\
8\pi\mu P_0^{-1}\mu_{ji}^{(s)} &= (\gamma+\varepsilon)\epsilon_{ils}\left(\frac{1-e^{-R/l}}{R}\right)_{,jl} \\
&\quad + (\gamma-\varepsilon)\epsilon_{jls}\left(\frac{1-e^{-R/l}}{R}\right)_{,il};
\end{aligned}
$$

— for the loading $(1.182)_2$ ($P_0 = 0$, $M_0 \neq 0$),

$$
\begin{aligned}
4\pi M_0^{-1}\sigma_{ji}^{(s)} &= \tfrac{1}{2}\epsilon_{kji}\left(\frac{1-e^{-R/\nu_1}}{R}\right)_{,ks} + \epsilon_{sjl}\left(\frac{1-e^{-R/l}}{R}\right)_{,li}, \\
4\pi M_0^{-1}\mu_{ji}^{(s)} &= -\frac{\gamma}{2\alpha}\left(\frac{1-e^{-R/\nu_1}}{R}\right)_{,sij} + \frac{\beta}{2\gamma+\beta}\left(\frac{e^{-R/\nu_1}}{R}\right)_{,s}\delta_{ij} \qquad (1.268) \\
&\quad + \frac{2\gamma}{\gamma+\varepsilon}l^2\left(\frac{1-e^{-R/l}}{R}\right)_{,sij} + \left(\frac{e^{-R/l}}{R}\right)_{,j}\delta_{is} \\
&\quad + \frac{\gamma-\varepsilon}{\gamma+\varepsilon}\left(\frac{e^{-R/l}}{R}\right)_{,i}\delta_{js}.
\end{aligned}
$$

The constant ν_1 in formula (1.268) (in the expression $\exp(-R/\nu_1)$) is given by formula (1.189), whereas ν in (1.267) is the Poisson ratio. Formulae (1.267), (1.268) represent the fundamental solutions for force-stresses and couple-stresses with the loadings (1.182), since, having substituted $\sigma_{ji}^{(s)}$, $\mu_{ji}^{(s)}$ to the equations of equilibrium

$$
\begin{aligned}
&\sigma_{ji,j} + X_i = 0, \\
&\epsilon_{ijk}\sigma_{jk} + \mu_{ji,j} + Y_i = 0
\end{aligned} \qquad (1.269)
$$

we obtain identities for the loadings (1.182), by virtue of (1.206). The limiting cases of formulae (1.267) and (1.268) are interesting. As an example, let us analyse formulae (1.267). Having extracted from $(1.267)_1$ the symmetric part $\sigma_{ji}^{(s)}$, whereas from $(1.267)_2$ — the deviatoric part $\mu_{ji}^{(s)}$, then taking the limit $\alpha \to \infty$ and taking into account (1.202), we can see that formulae (1.267) are transformed into the respective formulae:

$$
\begin{aligned}
4\pi P_0^{-1} s_{ij}^{(s)} = & -\tfrac{1}{2}(1-\nu)^{-1} R_{,ijs} + \left(\frac{1}{R}\right)_{,j} \delta_{is} + \left(\frac{1}{R}\right)_{,i} \delta_{js} \\
& + \nu(1-\nu)^{-1} \left(\frac{1}{R}\right)_{,s} \delta_{ij} - 2l^{\star 2} \left(\frac{1-e^{-R/l^\star}}{R}\right)_{,ijs} \\
& - \left(\frac{e^{-R/l^\star}}{R}\right)_{,j} \delta_{is} - \left(\frac{e^{-R/l^\star}}{R}\right)_{,i} \delta_{js}, \qquad (1.270) \\
4\pi P_0^{-1} \mu_{ji}^{(s)} = & \; 2l^{\star 2} \epsilon_{ils} \left(\frac{1-e^{-R/l^\star}}{R}\right)_{,jl} \\
& + 2\eta l^{\star 2} \epsilon_{jls} \left(\frac{1-e^{-R/l^\star}}{R}\right)_{,il} .
\end{aligned}
$$

We will show that formulae (1.270) give a basic solution for the stresses s_{ij}, $m_{ji} = \mu_{ji}$ within the framework of couple-stress theory. From the equations of couple-stress theory in the form

$$
\begin{aligned}
\nabla^2\nabla^2(l^{\star 2}\nabla^2 - 1)s_{ij} = & (1-\nu)^{-1}(l^{\star 2}\nabla^2 - 1)X_{l,lij} - 2l^{\star 2}\nabla^2 F_{r,rij} \\
& + \nabla^2(F_{i,j} + F_{j,i}) - \nu(1-\nu)^{-1}\nabla^2(l^{\star 2}\nabla^2 - 1)X_{k,k}\delta_{ij}, \qquad (1.271) \\
\nabla^2(l^{\star 2}\nabla^2 - 1)m_{ji} = & \; 2l^{\star 2}(\epsilon_{ilk}F_{k,lj} + \eta\epsilon_{jlk}F_{k,li}),
\end{aligned}
$$

where $F_i = X_i + \frac{1}{2}\epsilon_{ijk}Y_{k,j}$, we obtain (1.270) for the loading $(1.182)_1$. Moreover, the reduced equations of equilibrium of couple-stress theory:

$$
s_{ji,j} - \tfrac{1}{2}\epsilon_{kji}(m_{lk,lj} + Y_{k,j}) + X_i = 0, \qquad (1.272)
$$

become identities when formulae (1.270) and $(1.182)_1$ are substituted and (1.206) is taken into account. Formulae (1.267) and (1.268) were derived in [48], [52] and [49], where the superposition method was applied to stress equations. Also, the limiting cases of these formulae were studied there in detail; these give the fundamental solutions for stresses in the limiting theories (4), (6) and (8), put together in Fig. 1.1.

The homogeneous equations (1.265) and (1.266) in the form

$$
\nabla^2(\nu_1^2\nabla^2 - 1)(l^2\nabla^2 - 1)(\nabla^2\sigma_{ji}, \mu_{ji}) = (0, 0) \qquad (1.273)
$$

may be used in solving the boundary value problems. We will obtain a solution in the form

$$
\sigma_{ji} = \sigma_{ji}^{(1)} + \sigma_{ji}^{(2)} + \sigma_{ji}^{(3)}, \qquad \mu_{ji} = \mu_{ji}^{(1)} + \mu_{ji}^{(2)} + \mu_{ji}^{(3)}, \qquad (1.274)
$$

where the integrals for stresses denoted by indices (1), (2) and (3) satisfy the following equations:

$$
\begin{aligned}
&\nabla^2\nabla^2\sigma_{ji}^{(1)} = 0, \qquad \nabla^2\mu_{ji}^{(1)} = 0,\\
&(\nu_1^2\nabla^2 - 1)\sigma_{ji}^{(2)} = 0, \qquad (\nu_1^2\nabla^2 - 1)\mu_{ji}^{(2)} = 0,\\
&(l^2\nabla^2 - 1)\sigma_{ji}^{(3)} = 0, \qquad (l^2\nabla^2 - 1)\mu_{ji}^{(3)} = 0.
\end{aligned} \tag{1.275}
$$

The above procedure is valid by virtue of the Boggio theorem [180, p. 506] and is used repeatedly in this book. Performing the Fourier transformation (0.50) (for $n = 2$) on equations (1.273), from the partial differential equations we obtain the ordinary differential equations

$$
(\mathbb{D}_1^2 - \xi^2)(\mathbb{D}_1^2 - \eta^2)(\mathbb{D}_1^2 - \rho^2)[(\mathbb{D}_1^2 - \xi^2)\tilde{\sigma}_{ji}, \tilde{\mu}_{ji}] = (0, 0) \tag{1.276}
$$

for the transforms $\tilde{\sigma}_{ji}$, $\tilde{\mu}_{ji}$, where

$$
\mathbb{D}_1 \stackrel{\text{def}}{=} d/dx_1, \qquad \rho^2 = \xi^2 + l^{-2}, \qquad \eta^2 = \xi^2 + \nu_1^{-2}, \qquad \xi^2 = \xi_L\xi_L,
$$

$$
l^2 = \frac{(\gamma+\varepsilon)(\mu+\alpha)}{4\alpha\mu}, \qquad \nu_1^2 = \frac{2\gamma+\beta}{4\alpha}.
$$

Equations (1.276) may be applied in the case of boundary value problems formulated in terms of stresses for a micropolar layer and an elastic half-space. As an example, let us discuss such a procedure of constructing a solution for the problem of the half-space Ω (1.209). We assume the solution for the transforms of stresses $\tilde{\sigma}_{ji}$, $\tilde{\mu}_{ji}$ in the form of general solutions of equations (1.276)

$$
\begin{aligned}
\tilde{\sigma}_{ji} &= (A_{ji} + B_{ji}\xi x_1)e^{-\xi x_1} + C_{ji}e^{-\eta x_1} + D_{ji}e^{-\rho x_1},\\
\tilde{\mu}_{ji} &= E_{ji}e^{-\xi x_1} + F_{ji}e^{-\eta x_1} + G_{ji}e^{-\rho x_1},
\end{aligned} \tag{1.277}
$$

for which we took into account the physical condition

$$
\sigma_{ji} \to 0, \qquad \mu_{ji} \to 0, \qquad \text{for } R = \sqrt{x_i x_i} \to \infty. \tag{1.278}
$$

The matrices of coefficients A_{ji}–G_{ji} should be determined. For that purpose we have at our disposal 6 equations of equilibrium

$$
\sigma_{ji,j} = 0, \qquad \epsilon_{ijk}\sigma_{jk} + \mu_{ji,j} = 0 \qquad \text{in } \Omega, \tag{1.279}
$$

the compatibility equations (0.2), which, after making use of the constitutive relations (0.4) (for $\theta = 0$), should be expressed in terms of stresses, and finally, 6 boundary conditions,

$$
\sigma_{ji}n_j = p_i, \qquad \mu_{ji}n_j = m_i \qquad \text{on } \partial\Omega. \tag{1.280}
$$

To determine uniquely the coefficients A_{ji}–G_{ji} in a simply-connected region, it is necessary and sufficient to satisfy the field equations and boundary conditions mentioned above. Such a stress formulation is an analogue of Beltrami-Michell's formulation known from the classical theory of elasticity. The author of this monograph has used such a procedure many times when solving boundary value problems of elastostatics and micropolar thermoelastostatics, formulated as two-dimensional problems (plane and axially-symmetric state of strain). Such problems are discussed in [43]. In two-dimensional problems, when setting the number of independent compatibility equations, we have found the following algorithm to be helpful: the sum of the sought (independent) components of stress is the same as the sum of the stress equations. This is explained by Table 1.3.

Table 1.3.

Problem	σ_{ji}	μ_{ji}	Equations of equilibrium	Strain compatibility equations
first plane	4	2	3	3
second plane	4	4	3	5
first axially-symmetric	5	3	3	5
second axially-symmetric	4	5	3	6

The above procedure is also valid for the three-dimensional problem. In that case, there are 6 equations of equilibrium, hence, out of 81 compatibility equations (0.2) one should select 12 independent ones (there are 18 independent components of the stress state). Such equations were derived in [49]. Here, however, we give the general result given by Kozák & Szeidl [134]. Consider the two following square matrices of rank three:

$$\begin{bmatrix} \kappa_{31,2}-\kappa_{21,3} & \kappa_{32,2}-\kappa_{22,3} & \kappa_{33,2}-\kappa_{23,3} \\ \kappa_{11,3}-\kappa_{31,1} & \kappa_{12,3}-\kappa_{32,1} & \kappa_{13,3}-\kappa_{33,1} \\ \kappa_{21,1}-\kappa_{11,2} & \kappa_{22,1}-\kappa_{12,2} & \kappa_{23,1}-\kappa_{13,2} \end{bmatrix},$$

$$\begin{bmatrix} \gamma_{31,2}-\gamma_{21,3}+\kappa_{22}+\kappa_{33} & \gamma_{32,2}-\gamma_{22,3}-\kappa_{21} & \gamma_{33,2}-\gamma_{23,3}-\kappa_{31} \\ \gamma_{11,3}-\gamma_{31,1}-\kappa_{12} & \gamma_{12,3}-\gamma_{32,1}+\kappa_{11}+\kappa_{33} & \gamma_{13,3}-\gamma_{33,1}-\kappa_{32} \\ \gamma_{21,3}-\gamma_{11,2}-\kappa_{13} & \gamma_{22,1}-\gamma_{12,2}-\kappa_{23} & \gamma_{23,1}-\gamma_{13,2}+\kappa_{11}+\kappa_{22} \end{bmatrix}.$$

We shall obtain the compatibility equations by crossing out one entry from each column of both matrices and equating the remaining entries to zero. In this way we obtain 12 compatibility equations, valid on the set Ω[25]. Finally,

[25] Sandru [206] deals with spacial stress equations of micropolar elastostatics.

let us add that, when determining the coefficients A_{ji}–G_{ji} from equations (1.277), it is convenient, due to operational reasons, to make use of the additional equations:

— for force-stresses,

$$\begin{aligned}
&\nabla^2\sigma_{ii} = 0,\\
&\nabla^2\nabla^2(l^2\nabla^2 - 1)\sigma_{kk} = 0 \qquad \text{(we do not sum over } k\text{)},\\
&\nabla^2\nabla^2(l^2\nabla^2 - 1)(\sigma_{mn} + \sigma_{nm}) = 0 \qquad (m \neq n),\\
&(\nu_1^2\nabla^2 - 1)(l^2\nabla^2 - 1)(\sigma_{lk} - \sigma_{kl}) = 0 \qquad (k \neq l);
\end{aligned} \tag{1.281}$$

— for couple-stresses,

$$\begin{aligned}
&(\nu_1^2\nabla^2 - 1)\mu_{ii} = 0,\\
&\nabla^2(l^2\nabla^2 - 1)(\mu_{nm} - \mu_{mn}) = 0 \qquad (n \neq m).
\end{aligned} \tag{1.282}$$

1.2.10 Generalized representation of Papkovich-Neuber. Micropolar theory and couple-stress theory

Let us introduce to equations (1.155) the following substitution (we set $\mathbf{X} = \mathbf{0}$, $\mathbf{Y} = \mathbf{0}$) in Ω:

$$\begin{aligned}
\mathbf{u} &= A\,\mathrm{grad}\,\Phi + B\,\mathrm{curl}\,\boldsymbol{\Psi}, \qquad \mathrm{div}\,\boldsymbol{\Psi} = 0,\\
\boldsymbol{\varphi} &= C\,\mathrm{grad}\,\Sigma + D\,\mathrm{curl}\,\mathbf{H}, \qquad \mathrm{div}\,\mathbf{H} = 0,
\end{aligned} \tag{1.283}$$

where A, B, C, D are constants, which will be determined later.

The system of equations (1.155) is transformed into a system of equations with potentials Φ, Σ, $\boldsymbol{\Psi}$ and $\mathbf{H}$

$$\begin{aligned}
&A(2\mu + \lambda)\,\mathrm{grad}\,\nabla^2\Phi + B(\mu + \alpha)\vec{\nabla}^2\,\mathrm{curl}\,\boldsymbol{\Psi} - 2\alpha D\vec{\nabla}^2\mathbf{H} = \mathbf{0},\\
&C[(2\gamma + \beta)\vec{\nabla}^2 - 4\alpha]\,\mathrm{grad}\,\Sigma\\
&\qquad + D[(\gamma + \varepsilon)\vec{\nabla}^2 - 4\alpha]\,\mathrm{curl}\,\mathbf{H} - 2\alpha B\vec{\nabla}^2\boldsymbol{\Psi} = \mathbf{0}.
\end{aligned} \tag{1.284}$$

We assume $A = -\dfrac{\mu + \alpha}{\mu + \lambda}$, $B = 2(1 - \nu)$ and transform $(1.284)_1$ into the form

$$\vec{\nabla}^2\left(\mathrm{grad}\,\Phi - \mathrm{curl}\,\boldsymbol{\Psi} + \frac{\alpha}{\mu + \alpha}\cdot\frac{1}{1 - \nu}D\mathbf{H}\right) = \mathbf{0}, \tag{1.285}$$

where $\nu = \dfrac{\lambda}{2(\mu + \lambda)}$.

Let us introduce the vector $\boldsymbol{\psi}$ as follows:

$$2\boldsymbol{\psi} = \operatorname{grad} \Phi - \operatorname{curl} \boldsymbol{\Psi} + \frac{\alpha}{\mu+\alpha} \cdot \frac{1}{1-\nu} D\mathbf{H}. \tag{1.286}$$

It satisfies the equation

$$\vec{\nabla}^2 \boldsymbol{\psi} = \mathbf{0} \qquad \text{in } \Omega. \tag{1.287}$$

In turn, the potential Φ satisfies

$$\nabla^2 \Phi = 2 \operatorname{div} \boldsymbol{\psi} \qquad \text{in } \Omega. \tag{1.288}$$

Since

$$\nabla^2(\varphi + \mathbf{R} \circ \boldsymbol{\psi}) = \nabla^2 \varphi + \mathbf{R} \circ \vec{\nabla}^2 \boldsymbol{\psi} + 2 \operatorname{div} \boldsymbol{\psi},$$

where $\mathbf{R}$ is the radius vector, we obtain from equation (1.288) that

$$\Phi = \varphi + \mathbf{R} \circ \boldsymbol{\psi}, \tag{1.289}$$

where φ satisfies the equation

$$\nabla^2 \varphi = 0 \qquad \text{in } \Omega. \tag{1.290}$$

Once the values of constants A, B and relations (1.286), (1.289) are taken into account, the representation given by $(1.283)_1$ for the vector $\mathbf{u}$ takes the form

$$\mathbf{u} = \left(\frac{\mu+\lambda-\alpha}{\mu+\lambda}\right) \operatorname{grad}(\varphi + \mathbf{R} \circ \boldsymbol{\psi}) - 4(1-\nu)\boldsymbol{\psi} + \frac{2\alpha}{\mu+\alpha} D\mathbf{H}. \tag{1.291}$$

Having performed the curl operation on (1.286), we obtain

$$\vec{\nabla}^2 \boldsymbol{\Psi} = 2 \operatorname{curl} \boldsymbol{\psi} - \frac{\alpha}{\mu+\alpha} \cdot \frac{1}{1-\nu} D \operatorname{curl} \mathbf{H}. \tag{1.292}$$

By use of (1.292) the equation $(1.284)_2$ is transformed to the form

$$C(\nu_1^2 \vec{\nabla}^2 - 1) \operatorname{grad} \Sigma = B \operatorname{curl} \boldsymbol{\psi} - \frac{\mu}{\mu+\alpha} D(l^2 \vec{\nabla}^2 - 1) \operatorname{curl} \mathbf{H}, \tag{1.293}$$

where $\nu_1^2 = \dfrac{2\gamma+\beta}{4\alpha}$.

The equation for the potential Σ has the form

$$\nabla^2(\nu_1^2 \nabla^2 - 1)\Sigma = 0 \qquad \text{in } \Omega. \tag{1.294}$$

Eliminating the expression $\operatorname{grad} \Sigma$ from the representation for the vector $\boldsymbol{\varphi}$ given by the equation $(1.283)_3$, and using (1.293), we shall write it as

$$\boldsymbol{\varphi} = C\nu_1^2 \vec{\nabla}^2 \operatorname{grad} \Sigma + D \frac{\mu}{\mu+\alpha} \left(l^2 \vec{\nabla}^2 + \frac{\alpha}{\mu}\right) \operatorname{curl} \mathbf{H} - B \operatorname{curl} \boldsymbol{\psi}. \tag{1.295}$$

It enables us to introduce a new scalar function $\sigma = \nabla^2 \Sigma$ in place of Σ, which satisfies the equation

$$(\nu_1^2 \nabla^2 - 1)\sigma = 0 \qquad \text{in } \Omega. \tag{1.296}$$

The representation (1.295) takes now the form

$$\boldsymbol{\varphi} = C\nu_1^2 \operatorname{grad}\sigma + D\frac{\mu}{\mu+\alpha}\left(l^2\overrightarrow{\nabla}^2 + \frac{\alpha}{\mu}\right)\operatorname{curl}\mathbf{H} - 2(1-\nu)\operatorname{curl}\boldsymbol{\psi}. \tag{1.297}$$

We shall yet give equations for the vector $\mathbf{H}$. Performing the curl operation on (1.293), we obtain

$$\overrightarrow{\nabla}^2(l^2\overrightarrow{\nabla}^2 - 1)\mathbf{H} = -\frac{\mu+\alpha}{\mu}\cdot\frac{B}{D}\operatorname{grad}\operatorname{div}\boldsymbol{\psi}. \tag{1.298}$$

In turn, applying the operator $\overrightarrow{\nabla}^2$ to equation (1.298), we get

$$\overrightarrow{\nabla}^2\overrightarrow{\nabla}^2(l^2\overrightarrow{\nabla}^2 - 1)\mathbf{H} = \mathbf{0}. \tag{1.299}$$

The representation given by (1.291) and (1.297), with the field equations (1.287), (1.290), (1.296), (1.298), (1.299) and the field condition $\operatorname{div}\mathbf{H} = 0$, can be viewed as the Papkovich-Neuber representation[26] generalized to micropolar theory. For, when $\alpha \to 0$, we obtain from $(1.155)_1$ (for $\mathbf{X} = \mathbf{0}$) the classical equation

$$\mu\overrightarrow{\nabla}^2\mathbf{u} + (\lambda+\mu)\operatorname{grad}\operatorname{div}\mathbf{u} = \mathbf{0} \qquad \text{in } \Omega \tag{1.300}$$

whereas from (1.291) and (1.297) (for instance, for $C = D = \alpha^2$)[27] we obtain the representation of Papkovich-Neuber in Ω:

$$\begin{aligned} \mathbf{u} &= \operatorname{grad}(\varphi + \mathbf{R}\circ\boldsymbol{\psi}) - 4(1-\nu)\boldsymbol{\psi}, \\ \boldsymbol{\varphi} &= -2(1-\nu)\operatorname{curl}\boldsymbol{\psi}, \end{aligned} \tag{1.301}$$

where φ, $\boldsymbol{\psi}$ satisfy the scalar and vector Laplace equations (1.287) and (1.290), respectively (see Nowacki [180, p. 163]). The problem of completness

[26] We can obtain an analogy by plugging arbitrary constants into the symbol of unknown functions, e.g., one can assume $C\sigma = \hat{\sigma}$, $D\mathbf{H} = \hat{\mathbf{H}}$ etc.

[27] Behaviour of the constants A, B, C, D in the representation (1.283) plays an auxiliary role. Setting $A = -\dfrac{\mu+\alpha}{\mu+\lambda}$, $B = 2(1-\nu)$, we obtain equation (1.285), which for $\alpha \to 0$ goes over into the classical equation [180, p. 163]. Assuming $C = D = \alpha^p$ (where $p \geq 2$), it enables us to obtain from equations (1.291) and (1.297) the representation of Papkovich-Neuber (1.301) for $\alpha \to 0$ (both for $\mathbf{u}$ and $\boldsymbol{\varphi}$). This step is not necessary, since for arbitrary $C \in \mathbb{R}$ and D such that $\dfrac{2\alpha}{\mu+\alpha}D \to 0$ for $\alpha \to 0$, the representation given by $(1.301)_1$ can be obtained from (1.291), and rotations can be determined from the formula $\boldsymbol{\varphi} = \frac{1}{2}\operatorname{curl}\mathbf{u}$.

of the Papkovich-Neuber representation for the classical theory was resolved by Slobodianski, Eubanks and Sternberg (see [180, p. 166]) and others[28]. The problem of completness of the representation (1.291) and (1.297) remains open. For $\alpha \to 0$ we obtain from $(1.155)_2$ (for $\mathbf{Y} = \mathbf{0}$) the equations for a hypothetical medium (HM)

$$(\gamma + \varepsilon)\vec{\nabla}^2\boldsymbol{\varphi} + (\beta + \gamma - \varepsilon)\,\mathrm{grad}\,\mathrm{div}\,\boldsymbol{\varphi} = \mathbf{0} \qquad \text{in } \Omega. \tag{1.302}$$

Taking for the vector $\boldsymbol{\varphi}$ the decomposition given by $(1.283)_3$ (for $C = -(\gamma+\varepsilon)$, $D = (2\gamma + \beta)$), we obtain from equation (1.302) the equation

$$\vec{\nabla}^2(\mathrm{grad}\,\Sigma - \mathrm{curl}\,\mathbf{H}) = \mathbf{0}. \tag{1.303}$$

Introducing the vector $\mathbf{h}$ by

$$2\mathbf{h} = \mathrm{grad}\,\Sigma - \mathrm{curl}\,\mathbf{H} \tag{1.304}$$

we obtain

$$\vec{\nabla}^2\mathbf{h} = \mathbf{0} \qquad \text{in } \Omega. \tag{1.305}$$

Since the potential Σ satisfies the equation $\nabla^2\Sigma = 2\,\mathrm{div}\,\mathbf{h}$, we obtain: $\Sigma = \sigma + \mathbf{R} \circ \mathbf{h}$, where it holds

$$\nabla^2\sigma = 0 \qquad \text{in } \Omega. \tag{1.306}$$

Finally, the representation of Papkovich-Neuber for a hypothetical medium takes in Ω the form

$$\boldsymbol{\varphi} = (\gamma + \beta - \varepsilon)\,\mathrm{grad}(\sigma + \mathbf{R} \circ \mathbf{h}) - 2(2\gamma + \beta)\mathbf{h}. \tag{1.307}$$

Let us go over to couple-stress theory. Within its framework, consider the displacement equations (1.233) (for $\mathbf{F} = \mathbf{0}$). Assuming the decomposition given by $(1.283)_1$ (for $A = -(1 - 2\nu)$, $B = 2(1 - \nu)$), we find that equation (1.233) takes the form

$$\vec{\nabla}^2[\mathrm{grad}\,\Phi - (1 - l^{\star 2}\vec{\nabla}^2)\,\mathrm{curl}\,\boldsymbol{\Psi}] = \mathbf{0}. \tag{1.308}$$

Introduce the vector $\boldsymbol{\psi}$:

$$2\boldsymbol{\psi} = \mathrm{grad}\,\Phi - (1 - l^{\star 2}\vec{\nabla}^2)\,\mathrm{curl}\,\boldsymbol{\Psi}, \tag{1.309}$$

which satisfies the equation

$$\vec{\nabla}^2\boldsymbol{\psi} = \mathbf{0} \qquad \text{in } \Omega. \tag{1.310}$$

Since the potential Φ satisfies $\nabla^2\Phi = 2\,\mathrm{div}\,\boldsymbol{\psi}$, we obtain that $\Phi = \varphi + \mathbf{R} \circ \boldsymbol{\psi}$, where φ satisfies the equation

[28] Gurtin [105, p. 139].

$$\nabla^2 \varphi = 0 \qquad \text{in } \Omega. \tag{1.311}$$

The displacements $\mathbf{u}$ can be now represented in the form

$$\mathbf{u} = \operatorname{grad}(\varphi + \mathbf{R} \circ \boldsymbol{\psi}) - 4(1-\nu)\boldsymbol{\psi} + 2(1-\nu) l^{\star 2} \overrightarrow{\nabla}^2 \operatorname{curl} \boldsymbol{\Psi}, \tag{1.312}$$

where we have at our disposal the following equation for $\operatorname{curl} \boldsymbol{\Psi}$:

$$\overrightarrow{\nabla}^2 (l^{\star 2} \overrightarrow{\nabla}^2 - 1) \operatorname{curl} \boldsymbol{\Psi} = -2 \operatorname{grad} \operatorname{div} \boldsymbol{\psi}. \tag{1.313}$$

Introduce a new vector $\mathbf{u}^\star = 2(1-\nu)\overrightarrow{\nabla}^2 \operatorname{curl} \boldsymbol{\Psi}$, for which $\operatorname{div} \mathbf{u}^\star = 0$. Moreover, we have the equations

$$(l^{\star 2} \overrightarrow{\nabla}^2 - 1)\mathbf{u}^\star = -4(1-\nu) \operatorname{grad} \operatorname{div} \boldsymbol{\psi}, \tag{1.314}$$

$$\overrightarrow{\nabla}^2 (l^{\star 2} \overrightarrow{\nabla}^2 - 1)\mathbf{u}^\star = \mathbf{0}. \tag{1.315}$$

The vector $\mathbf{u}$ can be now written in the form

$$\mathbf{u} = \mathbf{u}^0 + l^{\star 2} \mathbf{u}^\star, \tag{1.316}$$

where $\mathbf{u}^0 = \operatorname{grad} \operatorname{div}(\varphi + \mathbf{R} \circ \boldsymbol{\psi}) - 4(1-\nu)\boldsymbol{\psi}$.

By virtue of the Boggio theorem we can assume $\mathbf{u}^\star = \mathbf{u}_1^\star + \mathbf{u}_2^\star$, where the vectors $\mathbf{u}_\alpha^\star$ satisfy the equations

$$\overrightarrow{\nabla}^2 \mathbf{u}_1^\star = \mathbf{0}, \qquad (l^{\star 2} \overrightarrow{\nabla}^2 - 1)\mathbf{u}_2^\star = \mathbf{0}. \tag{1.317}$$

The vector $\mathbf{u}_1^\star$ can be eliminated from further considerations, since from equation (1.314) we obtain the relation

$$\mathbf{u}_1^\star = 4(1-\nu) \operatorname{grad} \operatorname{div} \boldsymbol{\psi} \qquad \text{in } \Omega. \tag{1.318}$$

Note that $\operatorname{div} \mathbf{u}_1^\star = 0$. Since $\operatorname{div} \mathbf{u}^\star = 0$, the following condition holds:

$$\operatorname{div} \mathbf{u}_2^\star = 0 \qquad \text{in } \Omega. \tag{1.319}$$

The vector $\mathbf{u}$ can be finally represented in the form

$$\begin{aligned} \mathbf{u} = {} & \operatorname{grad}(\varphi + \mathbf{R} \circ \boldsymbol{\psi}) \\ & - 4(1-\nu)[\boldsymbol{\psi} - l^{\star 2} \operatorname{grad} \operatorname{div} \boldsymbol{\psi}] + l^{\star 2} \mathbf{u}_2^\star \qquad \text{in } \Omega. \end{aligned} \tag{1.320}$$

The representation (1.320) with the field equations (1.310), (1.311), $(1.317)_2$, and the field condition (1.319) can be treated as the Papkovich-Neuber representation generalized to couple-stress theory[29]. For $l^\star \to 0$ we have $\mathbf{u}_2^\star = \mathbf{0}$ and the representation (1.320) is transformed into the representation given by $(1.301)_1$.

---

[29] Mindlin & Tiersten [158] generalized the representation of Papkovich-Neuber for elastostatics onto couple-stress theory and gave examples of its applications.

1.2.11 Applications of the generalized Papkovich-Neuber representation

The micropolar model of Eringen-Nowacki (E-N). In [61], we considered the micropolar elastic half-space $\overline{\Omega} = \Omega \cup \partial\Omega$ (formula (1.209)) with boundary conditions of Boussinesq type:

$$\sigma_{11} = -P_0 \delta(x_2)\delta(x_3), \qquad \sigma_{1K} = 0, \qquad \mu_{1i} = 0 \qquad \text{on } \partial\Omega, \tag{1.321}$$

where $\delta(.)$ is the Dirac delta distribution and P_0 is a real constant. Moreover, we postulate that the displacements, rotations, strains and stresses vanish for $R = \sqrt{x_i x_i} \to \infty$. This problem is illustrated by Fig. 1.5.

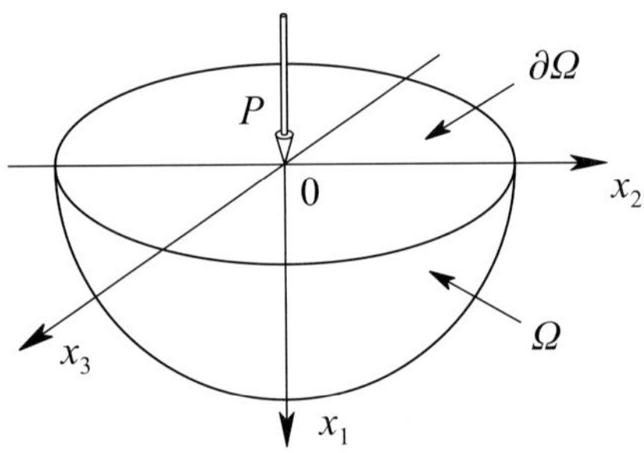

Fig. 1.5. The Boussinesq problem

In this boundary problem we expressed the fields of displacements, rotations, strains and stresses by means of the generalized representation of Papkovich-Neuber from Subsection 1.2.10, i.e. equations (1.291) and (1.297) with the field equations (1.287), (1.290), (1.296), (1.298), (1.299) and the field condition $\operatorname{div} \mathbf{H} = 0$. We will briefly discuss the procedure of finding the solution of this boundary value problem on the basis of [61]. From the Boggio theorem [180, p. 506] the vector $\mathbf{H}$ can be represented as a sum of two vectors, namely

$$H_i = H_i^1 + H_i^2, \tag{1.322}$$

where the vectors H_i^1 and H_i^2 satisfy in Ω the following equations:

$$\nabla^2\nabla^2 H_i^1 = 0, \qquad (l^2\nabla^2 - 1)H_i^2 = 0, \tag{1.323}$$

respectively, where we used equation (1.299). Now, using (1.322) and (1.323), we obtain equation (1.298) in the new form

$$D\nabla^2 H_i^1 = 2\left(1 + \frac{\alpha}{\mu}\right)(1-\nu)\psi_{j,ji} \qquad \text{in } \Omega. \tag{1.324}$$

Moreover, we obtain the following conditions in Ω:

$$H^1_{i,i} = 0, \qquad H^2_{i,i} = 0. \tag{1.325}$$

Applying the curl operator to equation (1.324), we obtain the equation

$$\nabla^2 \epsilon_{ijk} H^1_{k,j} = 0 \qquad \text{in } \Omega. \tag{1.326}$$

Now, taking into account (in this order) (1.322), (1.323) and (1.324), the representation for the rotation vector φ_i from formula (1.297) takes in Ω the form

$$\varphi_i = C\nu_1^2 \sigma_{,i} + D\epsilon_{ijk}\left(\tfrac{1}{2}\alpha_0 H^1_{k,j} + H^2_{k,j}\right) - 2(1-\nu)\epsilon_{ijk}\psi_{k,j}. \tag{1.327}$$

Substituting the representation (1.291), i.e.

$$u_i = \alpha_1(\varphi + x_j\psi_j)_{,i} - 4(1-\nu)\psi_i + \alpha_0 D H_i \qquad \text{in } \Omega, \tag{1.328}$$

and the representation (1.327) to the fundamental equations of the problem, expressed in terms of displacements and rotations as

$$\begin{aligned} &(\mu+\alpha)\nabla^2 u_i + (\lambda+\mu-\alpha)u_{j,ji} + 2\alpha\epsilon_{ijk}\varphi_{k,j} = 0, \\ &[(\gamma+\varepsilon)\nabla^2 - 4\alpha]\varphi_i + (\beta+\gamma-\varepsilon)\varphi_{j,ji} + 2\alpha\epsilon_{ijk}u_{k,j} = 0 \qquad \text{in } \Omega, \end{aligned} \tag{1.329}$$

we obtain equation (1.324) and equation (1.326), respectively. Let us finally collect the equations which comprise the generalized representation of Papkovich-Neuber, by means of which we will solve the stated boundary value problem of Boussinesq (1.321). These are equations (1.287), (1.290) and (1.296), which we rewrite here for convenience:

$$\nabla^2 \psi_i = 0, \tag{1.330}$$

$$\nabla^2 \varphi = 0, \tag{1.331}$$

$$(\nu_1^2 \nabla^2 - 1)\sigma = 0, \tag{1.332}$$

and equations (1.323)–(1.328) in Ω. We use the following notations:

$$\nu = \frac{\lambda}{2(\mu+\lambda)}, \qquad \nu_1^2 = \frac{2\gamma+\beta}{4\alpha},$$

$$\alpha_0 = \frac{2\alpha}{\mu+\alpha}, \qquad \alpha_1 = 1 - \frac{\alpha}{\mu+\alpha}, \qquad l^2 = \frac{(\gamma+\varepsilon)(\mu+\alpha)}{4\alpha\mu}.$$

The constants C and D are assumed in the form

$$C = D = \alpha^2. \tag{1.333}$$

Assuming now that

$$\varphi = (1-2\nu)\chi, \qquad \psi_1 = \chi_{,1}, \qquad \psi_N = 0 \qquad \text{in } \Omega, \tag{1.334}$$

we obtain the representation for the pair (u_i, φ_i) given by formulae (1.327) and (1.328) in the following form:

$$u_i = \alpha_1[(1-2\nu)\chi + x_1\chi_{,1}]_{,i} - 4(1-\nu)\chi_{,1}\delta_{i1} + \alpha_0\alpha^2 H_i, \tag{1.335}$$

$$\varphi_i = \alpha^2\left[\nu_1^2\sigma_{,i} + \epsilon_{ijk}\left(\tfrac{1}{2}\alpha_0 H^1_{k,j} + H^2_{k,j}\right) - 2(1-\nu)\epsilon_{ij1}\chi_{,1j}\right], \tag{1.336}$$

where the function χ satisfies the Laplace equation

$$\nabla^2\chi = 0 \qquad \text{in } \Omega. \tag{1.337}$$

In the sequel, we will use the method of Fourier transforms to find the solution of the problem. Thus, applying formula (0.50) for $n = 2$ to equations (1.323), (1.332) and (1.337), we arrive at ordinary differential equations in the transformed region of the form

$$\begin{aligned} (\mathbb{D}_1^2 - \eta^2)\tilde{\sigma} &= 0, \qquad (\mathbb{D}_1^2 - \xi^2)^2 \tilde{H}_i^1 = 0,\\ (\mathbb{D}_1^2 - \rho^2)\tilde{H}_i^2 &= 0, \qquad (\mathbb{D}_1^2 - \xi^2)\tilde{\chi}_i = 0, \end{aligned} \tag{1.338}$$

where

$$\mathbb{D}_1 \stackrel{\text{def}}{=} \frac{d}{dx_1}, \qquad \xi = \sqrt{\xi_N\xi_N}, \qquad \eta = \sqrt{\xi^2 + \nu_1^{-2}}, \qquad \rho = \sqrt{\xi^2 + l^{-2}}.$$

The general solutions of equations (1.338), for the transforms $\tilde{\sigma}$, $\tilde{H}_i^1$, $\tilde{H}_i^2$ and $\tilde{\chi}$, take in the case of the half-space Ω the form

$$\begin{aligned} \tilde{\sigma} &= b e^{-\eta x_1}, \qquad \tilde{H}_i^1 = (B_i + C_i\xi x_1)e^{-\xi x_1},\\ \tilde{H}_i^2 &= D_i e^{-\rho x_1}, \qquad \tilde{\chi} = a e^{-\xi x_1}, \end{aligned} \tag{1.339}$$

where a, b, B_i, C_i, D_i are arbitrary real-valued constants, which shall be determined. The number of those constants is reduced from 11 to 6 by satisfying the transformed relations (1.324)–(1.326). Taking a, b, B_N, D_N as the basic constants, we obtain

$$\begin{aligned} C_1 &= \xi\hat{\alpha}a, \qquad C_K = i\xi_K\hat{\alpha}a,\\ B_1 &= \xi\hat{\alpha}a - i\xi^{-1}\xi_N B_N, \qquad D_1 = -i\rho^{-1}\xi_N D_N, \end{aligned} \tag{1.340}$$

where $\hat{\alpha} = \dfrac{2(1-\nu)}{\alpha_0\alpha\mu}$.

The constants a, b, B_N, D_N are determined from the six boundary conditions (1.321). Taking into account that

$$\frac{1}{2\pi}\int_{-\infty}^{+\infty}\int_{-\infty}^{+\infty} P_0\delta(x_2)\delta(x_3)e^{i\xi_N x_N}\,dx_2\,dx_3 = \frac{P_0}{2\pi}$$

we obtain on $\partial\Omega$ (for $x_1 = 0$) the following conditions:

$$
\begin{aligned}
\tilde{\sigma}_{11} &= (2\mu + \lambda)\tilde{u}_{1,1} - \lambda i \xi_N \tilde{u}_N = -\frac{P_0}{2\pi}, \\
\tilde{\sigma}_{1K} &= (\mu + \alpha)\tilde{u}_{K,1} - i\xi_K(\mu - \alpha)\tilde{u}_1 - 2\alpha \epsilon_{1KL}\tilde{\varphi}_L = 0, \\
\tilde{\mu}_{11} &= (2\gamma + \beta)\tilde{\varphi}_{1,1} - \beta i \xi_N \tilde{\varphi}_N = 0, \\
\tilde{\mu}_{1K} &= (\gamma + \varepsilon)\tilde{\varphi}_{K,1} - i\xi_K(\gamma - \varepsilon)\tilde{\varphi}_1 = 0
\end{aligned}
\tag{1.341}
$$

from which we get

$$
\begin{aligned}
a &= \frac{A_0}{2l^2\hat{\alpha}}, \qquad b = 0, \qquad D_K = i\xi_K \xi^2 A_0, \\
B_K &= -i\xi_K \left(\frac{\xi^3}{\rho} - \frac{1 - 2\nu}{2l^2} \right) A_0,
\end{aligned}
\tag{1.342}
$$

where

$$
\begin{aligned}
A_0 &= \frac{P_0 a_0}{2\pi\mu\xi^2\alpha_0\alpha^2\Delta_0}, \qquad \Delta_0 = 1 + 2a_0\xi^2\left(1 - \frac{\xi}{\rho}\right), \\
a_0 &= \frac{(\gamma + \varepsilon)(2\mu + \lambda)}{4\mu(\lambda + \mu)}.
\end{aligned}
$$

Finally, the displacements $\tilde{u}_i$ take the form

$$
\begin{aligned}
\tilde{u}_1 &= \frac{P_0}{4\pi\mu} \cdot \frac{1}{\Delta_0} \left\{ \frac{1}{\xi}[2(1 - \nu) + \xi x_1]e^{-\xi x_1} \right. \\
&\quad \left. + \frac{2a_0\xi^2}{\rho} \left(e^{-\rho x_1} - e^{-\xi x_1} \right) \right\}, \\
\tilde{u}_K &= \frac{i\xi_K}{\xi} \cdot \frac{P_0}{4\pi\mu} \cdot \frac{1}{\Delta_0} \left\{ \frac{1}{\xi}[-(1 - 2\nu) + \xi x_1]e^{-\xi x_1} \right. \\
&\quad \left. + \frac{2a_0\xi^2}{\rho} \left(\frac{\rho}{\xi} e^{-\rho x_1} - e^{-\xi x_1} \right) \right\}.
\end{aligned}
\tag{1.343}
$$

For the rotations $\tilde{\varphi}_i$ we have

$$
\begin{aligned}
\tilde{\varphi}_1 &= 0, \\
\tilde{\varphi}_K &= -i\epsilon_{1KN} \cdot \frac{\xi_N}{\xi} \cdot \frac{P_0(1 - \nu)}{2\pi\mu} \cdot \frac{1}{\Delta_0} \left(e^{-\xi x_1} - \frac{\xi}{\rho} e^{-\rho x_1} \right).
\end{aligned}
\tag{1.344}
$$

Performing the inverse Fourier transformation, we obtain from formulae (1.343) and (1.344) the solution of the Boussinesq problem in the form of complex-valued integrals. Further, in [61], we analyse the singularities of physical fields in the half-space by means of the geometric relations (0.3) and the constitutive relations (0.4) (in Ω for $\theta = 0$). For example, using the asymptotic expansions

$$\rho = \xi + \frac{1}{2l^2\xi} - \frac{1}{8l^4\xi^3} + O(\xi^{-5}),$$
$$\frac{\xi}{\rho} = 1 - \frac{1}{2l^2\xi^2} + \frac{3}{8l^4\xi^4} + O(\xi^{-6}),$$
$$\Delta_0 = 1 + \frac{a_0}{l^2} - \frac{3a_0}{4l^4\xi^2} + O(\xi^{-4}), \tag{1.345}$$
$$\frac{1}{\Delta_0} = \frac{l^2}{l^2+a_0} + \frac{3a_0}{4(l^2+a_0)^2}\cdot\frac{1}{\xi^2} - \frac{a_0(10l^2+a_0)}{16l^2(l^2+a_0)^3}\cdot\frac{1}{\xi^4} + O(\xi^{-6}),$$
$$e^{-\rho x_1} = e^{-\xi x_1}\left[1 - \frac{x_1}{2l^2\xi} + \frac{x_1^2}{8l^4\xi^2} + \frac{6l^2x_1 - x_1^3}{48l^6\xi^3} + O(\xi^{-4})\right],$$

where $O(\xi^{-n})$ denotes the expression which behaves as the function $g(\xi) = \frac{1}{\xi^n}$ for $\xi \to \infty$, and the results of [101], we obtain

— the displacements u_i,

$$u_i = \frac{P_0}{2\pi\mu}\left[(1-\nu)\frac{l^2}{l^2+a_0}\cdot\frac{1}{R} + \frac{l^2-a_0}{2(l^2+a_0)}\cdot\frac{x_1^2}{R^3}\right] + O(1),$$
$$u_K = \frac{P_0}{2\pi\mu}\left[\frac{a_0-(1-2\nu)l^2}{2(l^2+a_0)}\cdot\frac{x_K}{R(R+x_1)} + \frac{l^2-a_0}{2(l^2+a_0)}\cdot\frac{x_1x_K}{R^3}\right] + O(1); \tag{1.346}$$

— the non-vanishing rotations φ_i,

$$\varphi_K = -\frac{P_0(1-\nu)}{4\pi\mu}\cdot\frac{1}{l^2+a_0}\epsilon_{1KN}\frac{x_N}{R} + O(1), \tag{1.347}$$

where $R = \sqrt{x_ix_i}$ and $O(1)$ denotes the regular part of u_i or φ_i ($O(1)$ has finite value at $(0,0,0)$). Nowacki and Dyszlewicz [192, p. 159] investigated the considered boundary value problem by a different method.

The couple-stress model of Koiter-Mindlin (K-M). Dyszlewicz & Żal have considered in [85] the problem of an elastic half-space within the framework of the elastostatics of couple-stress theory. Our aim was to determine and study the physical fields in the half-space twisted on the boundary by concentrated moments. We have considered two cases of twisting.

(a) The first case is the classical case of twisting, which refers to a concentrated torque M understood as a limiting case of the action of two pairs of tangent forces (for $h \to 0$). This is illustrated by Fig. 1.6.

The boundary conditions take the form

$$\begin{aligned}
\sigma_{11} &= 0,\\
\sigma_{12} &= \frac{S}{2}\delta(x_2)\delta(x_3-h) - \frac{S}{2}\delta(x_2)\delta(x_3),\\
\sigma_{13} &= \frac{S}{2}\delta(x_2+h)\delta(x_3) - \frac{S}{2}\delta(x_2)\delta(x_3),\\
\mu_{1i} &= 0 \quad \text{on } \partial\Omega.
\end{aligned} \tag{1.348}$$

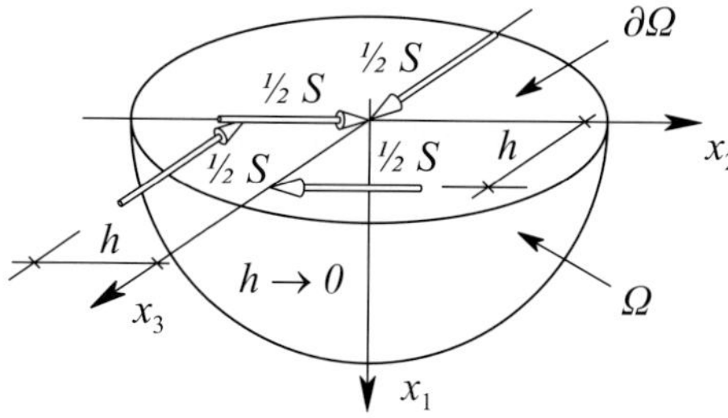

Fig. 1.6. Classically twisted half-space

We assume that

$$\lim_{h\to 0} \frac{S}{2} \cdot h = \frac{M}{2}, \tag{1.349}$$

where S, M are constant values.

(b) The second case of twisting is non-classical and refers to the problem in which a concentrated moment loading of torsion couple-stress type acts on the boundary of the half-space. This is illustrated by Fig. 1.7.

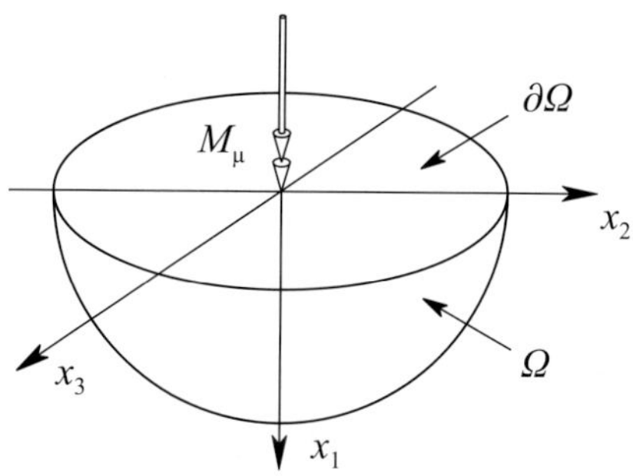

Fig. 1.7. Non-classically twisted half-space

The boundary conditions take the form

$$\sigma_{1i} = 0, \qquad \mu_{11} = M_\mu \delta(x_2)\delta(x_3), \qquad \mu_{1K} = 0 \qquad \text{on } \partial\Omega, \tag{1.350}$$

where M_μ is a constant.

To solve the stated problems of twisted half-space, we start from the displacement equations of equilibrium of couple-stress theory (0.28), written below in the vector form with disregarded temperature and body loadings

$$\mu\overrightarrow{\nabla}^2\mathbf{u} + (\lambda+\mu)\operatorname{grad}\operatorname{div}\mathbf{u} + 2\mu l^{\star 2}\overrightarrow{\nabla}^2\operatorname{curl}\boldsymbol{\varphi} = \mathbf{0} \qquad \text{in } \Omega, \tag{1.351}$$

where

$$\boldsymbol{\varphi} = \tfrac{1}{2}\operatorname{curl}\mathbf{u}, \tag{1.352}$$

which can be separated by means of the generalized Papkovich-Neuber representation for couple-stress theory, discussed in Subsection 1.2.10. The vector form of this representation is in Ω the following

$$\mathbf{u} = \operatorname{grad}(\varphi + \mathbf{R} \circ \boldsymbol{\psi}) - 4(1-\nu)(1 - l^{\star 2} \operatorname{grad}\operatorname{div})\boldsymbol{\psi} + l^{\star 2} \overset{*}{\mathbf{u}}_2 \tag{1.353}$$

with the additional field condition

$$\operatorname{div} \overset{*}{\mathbf{u}}_2 = 0, \tag{1.354}$$

where $\mathbf{R} \overset{\text{def}}{=} (x_1, x_2, x_3)$ is the radius vector. The function φ satisfies the scalar Laplace equation

$$\nabla^2 \varphi = 0. \tag{1.355}$$

In turn, the vector $\boldsymbol{\psi}$ satisfies the vector Laplace equation

$$\vec{\nabla}^2 \boldsymbol{\psi} = \mathbf{0} \tag{1.356}$$

and $\overset{*}{\mathbf{u}}_2$ is determined from

$$(l^{\star 2} \vec{\nabla}^2 - 1) \overset{*}{\mathbf{u}}_2 = \mathbf{0}, \tag{1.357}$$

where, $\nabla^2 \overset{\text{def}}{=} \operatorname{div}\operatorname{grad}$ and $\vec{\nabla}^2 \overset{\text{def}}{=} \operatorname{grad}\operatorname{div} - \operatorname{curl}\operatorname{curl}$.

Let us now find the general form of the Papkovich-Neuber function for the couple-stress elastic half-space $\overline{\Omega} = \Omega \cup \partial\Omega$, where

$$\Omega = \{(x_1, x_2, x_3) : x_1 > 0, |x_2| < \infty, |x_3| < \infty\}.$$

Applying the Fourier transformation (0.50) for $n = 2$, we transform equation (1.356) into the equation

$$(\mathbb{D}_1^2 - \xi^2)\tilde{\psi}_i = 0 \tag{1.358}$$

from which we obtain

$$\tilde{\psi}_i = K_i e^{-\xi x_1} \tag{1.359}$$

taking into account the condition

$$\mathbf{u} \to (0,0,0) \qquad \text{for } R = \sqrt{x_i x_i} \to \infty. \tag{1.360}$$

Next, we introduce the notation

$$K_1 = B(\xi_2, \xi_3), \qquad K_2 = C(\xi_2, \xi_3), \qquad K_3 = D(\xi_2, \xi_3). \tag{1.361}$$

In order to satisfy the boundary conditions, it is convenient to use in further considerations the substitution

$$\Phi = \varphi + \mathbf{R} \circ \boldsymbol{\psi} \qquad \text{in } \Omega. \tag{1.362}$$

Equation (1.355) takes now the form

$$(\Phi - x_j\psi_j)_{,ii} = 0. \tag{1.363}$$

Note that

$$(x_j\psi_j)_{,ii} = x_j\psi_{j,ii} + 2\psi_{j,j}. \tag{1.364}$$

Taking into account equation (1.356) and identity (1.364), we transform equation (1.363) into

$$\Phi_{,ii} - 2\psi_{j,j} = 0 \qquad \text{in } \Omega. \tag{1.365}$$

Applying the Fourier transformation, we arrive at

$$(\mathbb{D}_1^2 - \xi^2)\tilde{\Phi} = 2(\mathbb{D}_1\tilde{\psi}_1 - i\xi_N\tilde{\psi}_N). \tag{1.366}$$

In view of (1.359)–(1.361), the solution of equation (1.366) takes the form

$$\tilde{\Phi} = \left[A + \left(B + \frac{i\xi_2}{\xi}C + \frac{i\xi_3}{\xi}D\right)\right]e^{-\xi x_1}. \tag{1.367}$$

Let us go over to equation (1.357). After the application of the Fourier transformation, we obtain

$$(\mathbb{D}_1^2 - \rho^2)\tilde{\overset{*}{u}}_{2i} = 0 \tag{1.368}$$

from which, in view of (1.360), we obtain the solution in the form

$$\tilde{\overset{*}{u}}_{2i} = E_i e^{-\rho x_1}, \tag{1.369}$$

where we denote

$$E_1 = E(\xi_2, \xi_3), \qquad E_2 = F(\xi_2, \xi_3), \qquad E_3 = G(\xi_2, \xi_3). \tag{1.370}$$

The number of constants A–G, on which the Papkovich-Neuber functions φ, ψ_i, $\overset{*}{u}_{2i}$ depend, can be reduced to 6, by satisfying the transformed field condition (1.354)

$$\mathbb{D}_1\tilde{\overset{*}{u}}_{12} - i\xi_N\tilde{\overset{*}{u}}_{N2} = 0, \tag{1.371}$$

from which we obtain

$$E = -\frac{i}{\rho}(\xi_2 F + \xi_3 G). \tag{1.372}$$

In couple-stress theory (K-M model), we have to satisfy 5 boundary conditions, but there are 6 thus-far independent constants of integration, on which the Papkovich-Neuber functions depend (A, B, C, D, F and G). In [85], we have shown that in order solve the boundary value problem for the half-space it is enough to assume the vector $\boldsymbol{\psi}$ to be in the form

$$\psi_i \overset{\text{def}}{=} (\psi_1, \psi_2, 0) \tag{1.373}$$

which eliminates the constant D:

$$D(\xi_2, \xi_3) = 0. \tag{1.374}$$

Formulae (1.359), (1.367), (1.369) and (1.372)–(1.374) consitute the complete set of formulae in the transformed space, which will be used in solving the boundary value problem for the couple-stress half-space.

We shall briefly discuss the solution of problem (a), i.e. the classically twisted half-space, by means of the superposition of solutions. Since the torque is realized as a superposition of four tangent forces, appropriately applied to the boundary of the half-space, we shall first consider the basic case of the boundary value problem, namely the half-space Ω with a concentrated tangent force applied to the boundary $\partial\Omega$, which is illustrated in Fig. 1.8.

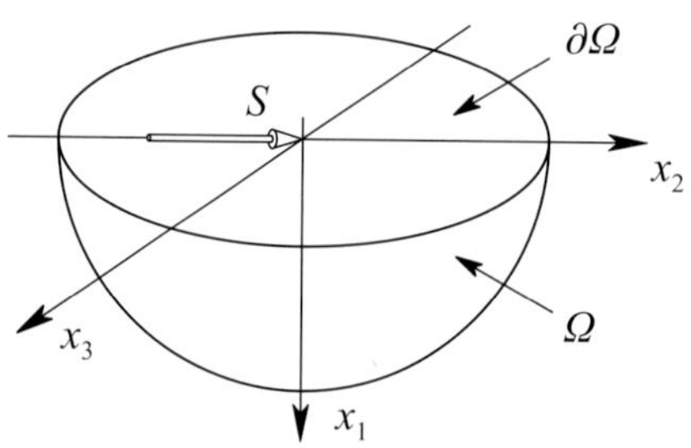

Fig. 1.8. The boundary value problem of Cerruti type. Half-space loaded by a tangent force

The boundary conditions take the form

$$\begin{aligned} \sigma_{11} &= 0, \\ \sigma_{12} &= -S\delta(x_2)\delta(x_3), \\ \sigma_{13} &= 0, \\ \mu_{1i} &= 0 \qquad \text{on } \partial\Omega. \end{aligned} \tag{1.375}$$

Transforming the conditions (1.375) for the K-M model, we obtain the following 5 boundary conditions:

$$\begin{aligned} &s_{11} = 0, \\ &s_{12} - \tfrac{1}{2}(m_{i3,i} - m_{11,3}) = -S\delta(x_2)\delta(x_3), \\ &s_{13} + \tfrac{1}{2}(m_{i2,i} - m_{11,2}) = 0, \\ &m_{1K} = 0 \qquad \text{on } \partial\Omega. \end{aligned} \tag{1.376}$$

From those 5 conditions, we determined in [85] five constants of integration A, B, C, F, G. The next step in problem (a) consists in finding the solution for the half-space loaded on the boundary by a twisting moment with a pair of tangent forces. This is shown in Fig. 1.9.

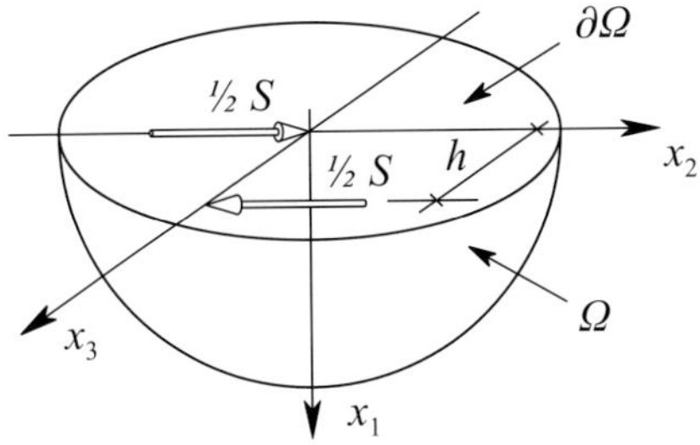

Fig. 1.9. Half-space loaded by a twisting moment with a pair of tangent forces

The formulae for displacements, rotations and stresses in the stated boundary value problem are obtained as follows. Let the functions

$$f^{\star}(x_1, x_2, x_3; 0, 0, 0)$$

and

$$f^{\star}(x_1, x_2, x_3; 0, 0, h)$$

denote the solutions for the fields of displacements, rotations and stresses in two cases: of a unit tangent force applied at the origin of the coordinate system in the direction of the Ox_2 axis, and of a unit tangent force applied at $(0, 0, h)$ in the direction opposite to Ox_2, respectively. Then the formulae for the physical fields f resulting from the loading by a single dipole will have the form

$$f = \frac{S}{2}\left[f^{\star}(x_1, x_2, x_3; 0, 0, 0) - f^{\star}(x_1, x_2, x_3; 0, 0, h)\right]. \tag{1.377}$$

Multiplying and dividing the right-hand side of (1.377) by h and assuming that

$$\lim_{h \to 0} S \cdot h = M \tag{1.378}$$

we obtain the solution of the problem in the form

$$f = \frac{M}{2} \cdot \frac{\partial}{\partial x_3} f^{\star}(x_1, x_2, x_3). \tag{1.379}$$

The expressions for $f^{\star}$, i.e. u_i, φ_i, s_{ij}, m_{ij}, ε_{ij} and κ_{ij}, have been obtained by solving the Cerruti problem with the half-space loaded by a tangent force (Fig. 1.8).

Note that the half-space loaded by a classical torque (Fig. 1.6) is a superposition of the problems shown in Fig. 1.10[30]:

[30] See R. Świtka, On the definition of a concentrated moment acting on an elastic medium, Eng. Trans., **14**, 3, 403, 1966 (in Polish).

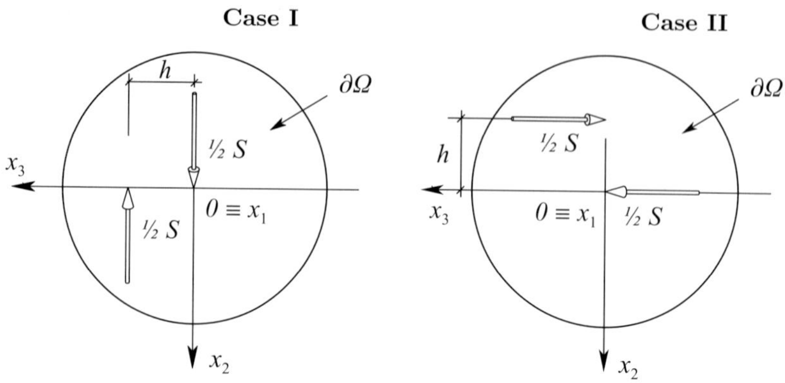

Fig. 1.10. Twisting moment with a pair of tangent forces

The fields of displacements, rotations and stresses for Case I have been determined when solving the problem described in Fig. 1.9. Note that the solution for Case II can be obtained from that for Case I if we make the following transformation of the position variables:

$$x_2 \mapsto x_3, \qquad x_3 \mapsto -x_2, \qquad x_1 \mapsto x_1. \tag{1.380}$$

The formulae for u_i, φ_i, s_{ij}, m_{ij}, ε_{ij} and κ_{ij} in Case II are obtained by appropriate transformation of the indices for the physical fields of Case I.

We end the discussion of problem (a) by giving the final formulae for the constants of integration in the transforms of the Papkovich-Neuber functions, which will be justified in the sequel:

$$\begin{aligned} A &= -\frac{M}{2\pi\mu}\alpha W\frac{\xi_2}{\xi_3\xi}, \\ B &= \frac{M}{2\pi\mu}\cdot\frac{\alpha}{4(1-\nu)}W\frac{\xi_2}{\xi_3}, \\ C &= \frac{M}{2\pi\mu}\cdot\frac{\alpha}{4(1-\nu)}W\frac{i\xi}{\xi_3}, \\ F &= \frac{M}{2\pi\mu}\cdot\frac{1}{l^{\star 2}}W\frac{i\xi_3}{\xi}, \\ G &= -\frac{M}{2\pi\mu}\cdot\frac{1}{l^{\star 2}}W\frac{i\xi_2}{\xi}, \end{aligned} \tag{1.381}$$

where

$$W = \frac{2(1+\eta)l^{\star 2}\left(\xi - \frac{\rho}{\alpha}\right)\xi}{-4\alpha\left[1 + (1+\eta)l^{\star 2}\xi\left(\xi - \frac{\rho}{\alpha}\right)\right]} + \frac{1}{2\alpha}, \qquad \alpha = \frac{\xi^2(1+\eta)l^{\star 2}+1}{\xi^2(1+\eta)l^{\star 2}}.$$

(b) Consider now the non-classical twisting of the half-space. Here, the boundary conditions (1.350) take the form

$$\begin{aligned} &s_{11} = 0, \qquad s_{12} - \tfrac{1}{2}(m_{j3,j} - m_{11,3}) = -\tfrac{1}{2}[M_\mu \delta(x_2)\delta(x_3)]_{,3}, \\ &s_{13} + \tfrac{1}{2}(m_{j2,j} - m_{11,2}) = \tfrac{1}{2}[M_\mu \delta(x_2)\delta(x_3)]_{,2}, \\ &m_{1K} = 0 \qquad \text{on } \partial\Omega. \end{aligned} \tag{1.382}$$

Applying the Fourier transformation and using relations (0.22) and (0.26) for $\theta = 0$, we obtain from the conditions (1.382) a system of 5 algebraic Cramer equations for the constants A, B, C, F and G, with the solution in the form identical to formulae (1.381).

In [85], we made a number of observations and conclusions. Let us recall here two interesting final conclusions.

Conclusion 1. The couple-stress half-space (K-M model) does not distinguish the classical twisting discussed in (a) from the non-classical one discussed in (b). The solutions are identical.

Conclusion 2. In the K-M model, we notice a visible influence of the couple-stresses on the stress concentration. The concentration of force-stresses in the neighbourhood of the moment M is in the couple-stress half-space (K-M model) two times smaller than in the classical elastic half-space (Hooke's model). This phenomenon is illlustrated in Table 1.4.

Table 1.4.

Physical fields	K-M model	H model
u_1	0	0
u_K	$O(1)$	$O(R^{-2})$
φ_i	$O(R^{-1})$	$O(R^{-3})$
σ_{11}, s_{11}	0	0
remaining stresses σ_{ij}, s_{ij}	$O(R^{-1})$	$O(R^{-3})$
m_{ij}	$O(R^{-2})$	–

Singularities of physical fields for K-M and H models; s_{ij} and m_{ij} refer to the K-M model

Below we shall present (Fig. 1.11–Fig. 1.17) numerical analysis of the phenomenon of twisting for the couple-stress half-space (K-M) and the classical half-space (H), based on the material constants (0.52).

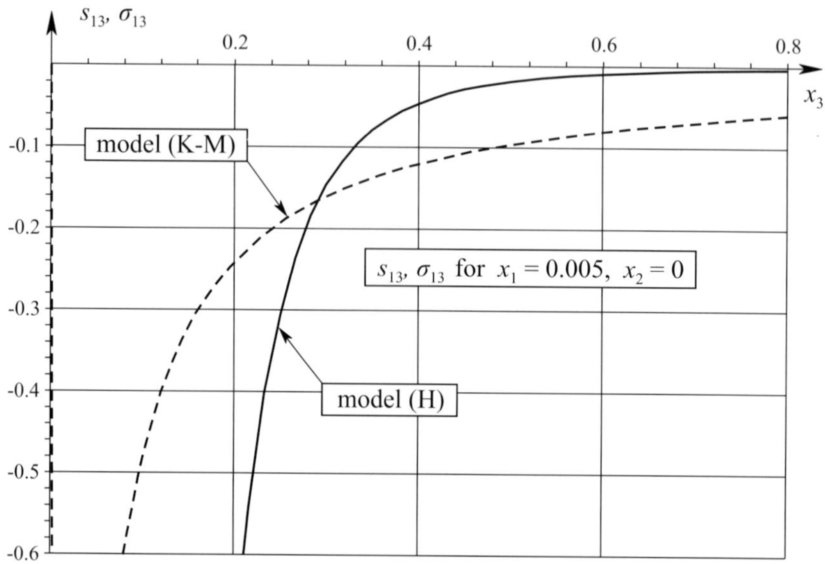

Fig. 1.11. Concentration of force-stresses, tangent at the same depth, in the vicinity of the point of action of the moment M

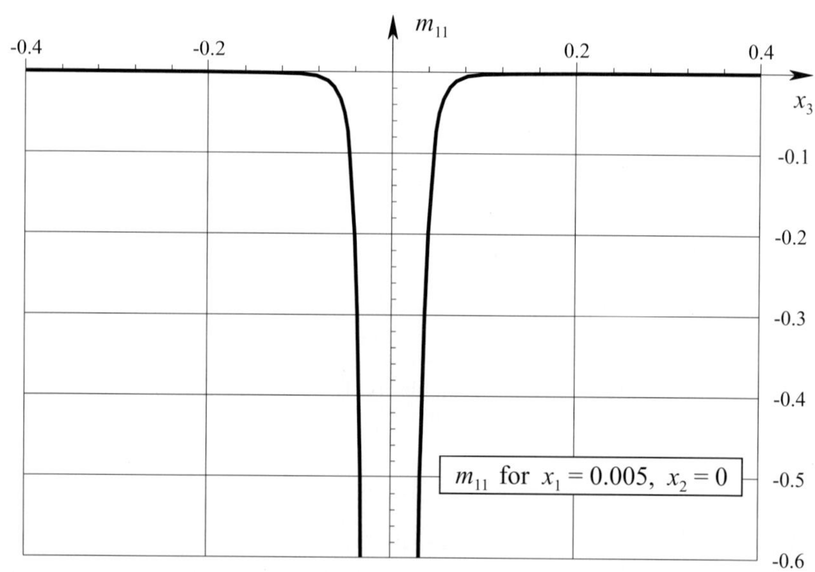

Fig. 1.12. The graph of the twisting couple-stresses m_{11} in the vicinity of the point where the moment M is applied

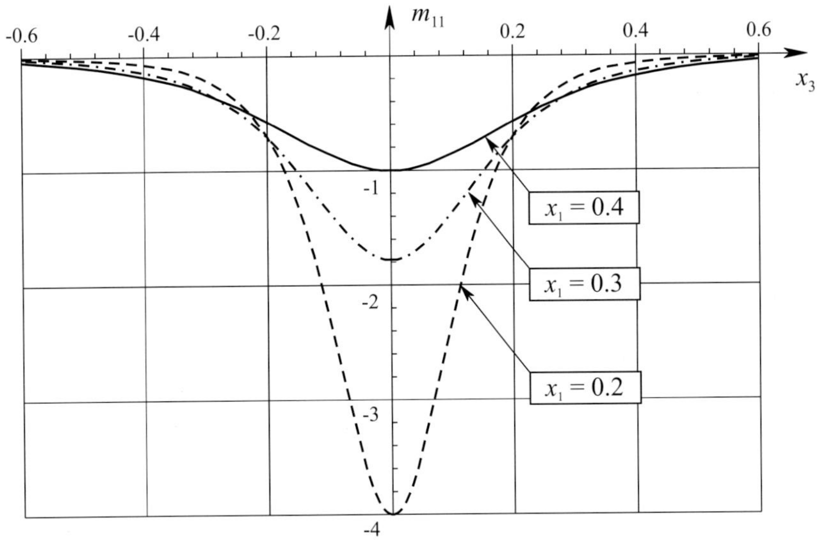

Fig. 1.13. The distribution of the couple-stresses m_{11} under the moment M, at various depths. From the graph follows the importance of the Saint-Venant's principle for couple-stresses

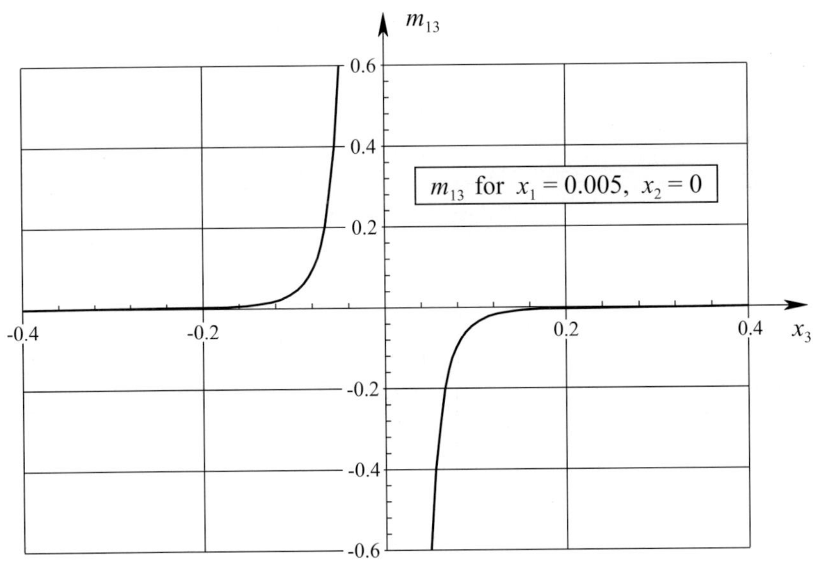

Fig. 1.14. The graph of the bending couple-stresses m_{13} in the vicinity of the point where the moment M is applied

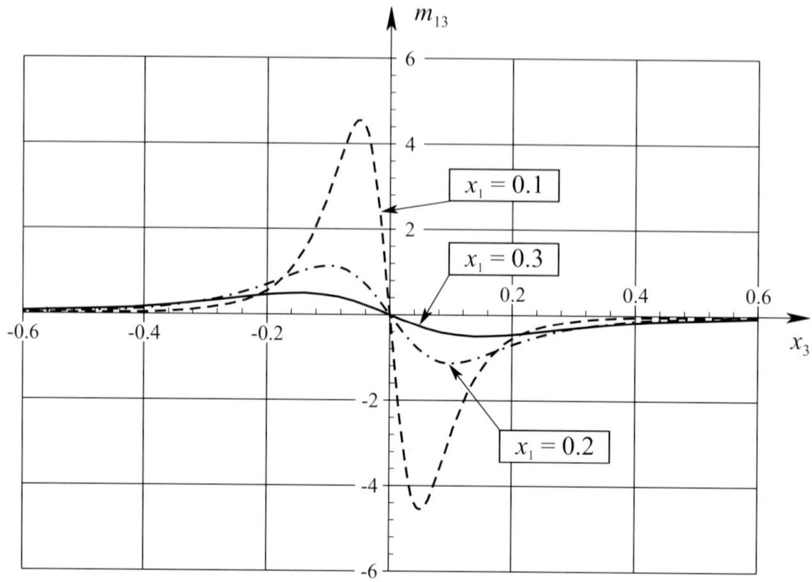

Fig. 1.15. The distribution of the couple-stresses m_{13} at various depths, under the moment M

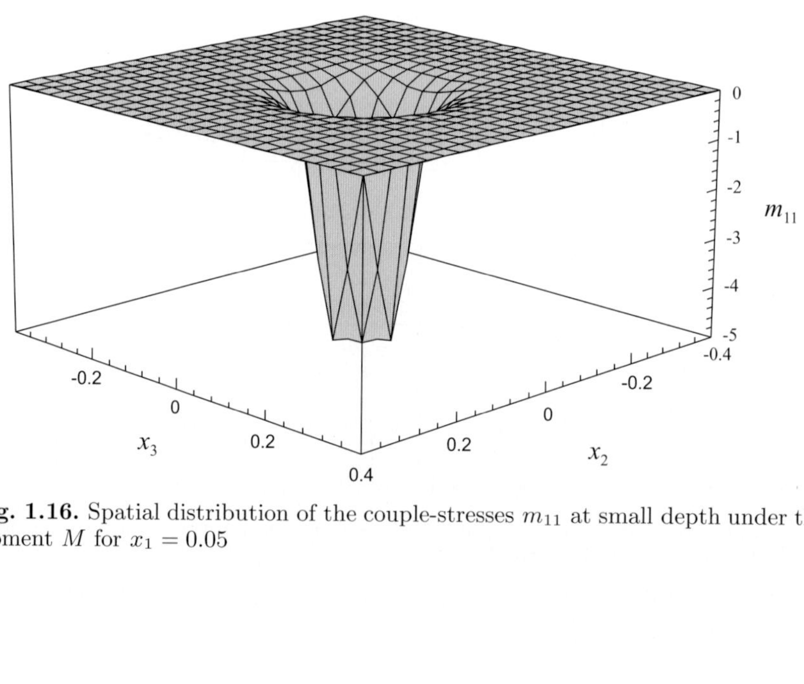

Fig. 1.16. Spatial distribution of the couple-stresses m_{11} at small depth under the moment M for $x_1 = 0.05$

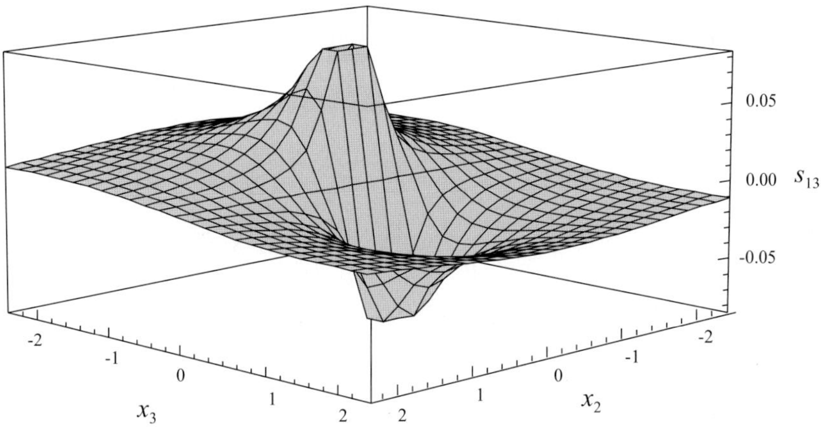

Fig. 1.17. Spatial distribution of the force-stresses s_{13} at small depth under the moment M for $x_1 = 0.05$

1.2.12 Generalized representation of Papkovich-Neuber. The case of nonhomogenous equations of micropolar elastostatics (E-N model)

Let us consider in Ω the nonhomogenous equations (1.155) in displacements and rotations of the miropolar model (E-N), written in the vector form with body loadings. We introduce the Helmholtz decomposition (1.283) to (1.155). Then the system of equations (1.155) is transformed into

$$A(2\mu+\lambda)\,\mathrm{grad}\,\nabla^2\Phi + B(\mu+\alpha)\vec{\nabla}^2\,\mathrm{curl}\,\boldsymbol{\Psi} - 2\alpha D\vec{\nabla}^2\mathbf{H} + \mathbf{X} = \mathbf{0}, \quad (1.383)$$

$$\begin{aligned} &C[(2\gamma+\beta)\vec{\nabla}^2 - 4\alpha]\,\mathrm{grad}\,\Sigma + D[(\gamma+\varepsilon)\vec{\nabla}^2 - 4\alpha]\,\mathrm{curl}\,\mathbf{H} \\ &\quad - 2\alpha B\vec{\nabla}^2\boldsymbol{\Psi} + \mathbf{Y} = \mathbf{0}. \end{aligned} \quad (1.384)$$

Let

$$A = -\frac{\mu+\alpha}{\mu+\lambda}, \qquad B = 2(1-\nu), \quad (1.385)$$

where $\nu = \dfrac{\lambda}{2(\mu+\lambda)}$. Equation (1.383) is transformed into the form

$$\vec{\nabla}^2\left(\mathrm{grad}\,\Phi - \mathrm{curl}\,\boldsymbol{\Psi} + \frac{\alpha}{\mu+\alpha}\cdot\frac{1}{1-\nu}D\mathbf{H}\right) = \frac{1}{\mu+\alpha}\cdot\frac{1}{2(1-\nu)}\mathbf{X}. \quad (1.386)$$

We introduce a new vector:

$$2\boldsymbol{\psi} = \operatorname{grad} \Phi - \operatorname{curl} \boldsymbol{\Psi} + \frac{\alpha}{\mu+\alpha} \cdot \frac{1}{1-\nu} D\mathbf{H}, \tag{1.387}$$

which satisfies

$$\vec{\nabla}^2 \boldsymbol{\psi} = \frac{1}{\mu+\alpha} \cdot \frac{1}{4(1-\nu)} \mathbf{X}. \tag{1.388}$$

From (1.387) and (1.388) we obtain

$$\nabla^2 \Phi = 2 \operatorname{div} \boldsymbol{\psi}, \tag{1.389}$$

$$\nabla^2 \nabla^2 \Phi = \frac{1}{\mu+\alpha} \cdot \frac{1}{2(1-\nu)} \operatorname{div} \mathbf{X}. \tag{1.390}$$

The following relation holds:

$$\nabla^2 (\varphi + \mathbf{R} \circ \boldsymbol{\psi}) = \nabla^2 \varphi + \mathbf{R} \circ \vec{\nabla}^2 \boldsymbol{\psi} + 2 \operatorname{div} \boldsymbol{\psi}, \tag{1.391}$$

where $\mathbf{R}$ is the radius vector for $\mathbf{x} \in \Omega$, the circle $\circ$ denotes the scalar product of vectors. Using (1.388) and (1.389), we obtain from (1.391) the equation

$$\Phi = \varphi + \mathbf{R} \circ \boldsymbol{\psi}, \tag{1.392}$$

where

$$\nabla^2 \varphi = -\frac{1}{\mu+\alpha} \cdot \frac{1}{4(1-\nu)} \cdot \mathbf{R} \circ \mathbf{x}. \tag{1.393}$$

The representation for the displacement fields takes the form

$$\mathbf{u} = \frac{\mu+\lambda-\alpha}{\mu+\lambda} \operatorname{grad}(\varphi + \mathbf{R} \circ \boldsymbol{\psi}) - 4(1-\nu)\boldsymbol{\psi} + \frac{2\alpha}{\mu+\alpha} D\mathbf{H}. \tag{1.394}$$

Let us go over to equation (1.384), taking into account that

$$\vec{\nabla}^2 \boldsymbol{\Psi} = 2 \operatorname{curl} \boldsymbol{\psi} - \frac{\alpha}{\mu+\alpha} \cdot \frac{1}{1-\nu} D \operatorname{curl} \mathbf{H}. \tag{1.395}$$

After some transformations, equation (1.384) takes the form

$$\begin{aligned} C(\nu_1^2 \vec{\nabla}^2 - 1) \operatorname{grad} \Sigma + \frac{\mu}{\mu+\alpha} D(l^2 \vec{\nabla}^2 - 1) \operatorname{curl} \mathbf{H} \\ - 2(1-\nu) \operatorname{curl} \boldsymbol{\psi} + \frac{1}{4\alpha} \mathbf{Y} = \mathbf{0}, \end{aligned} \tag{1.396}$$

where $\nu_1^2 = \dfrac{2\gamma+\beta}{4\alpha}$ and $l^2 = \dfrac{(\gamma+\varepsilon)(\mu+\alpha)}{4\alpha\mu}$. Applying the operator of divergence to equation (1.396), we obtain

$$\nabla^2 (\nu_1^2 \nabla^2 - 1) \Sigma = -\frac{1}{4\alpha C} \operatorname{div} \mathbf{Y}. \tag{1.397}$$

From $(1.283)_3$, using (1.396), we have

$$\begin{aligned}\boldsymbol{\varphi} &= C\nu_1^2 \overrightarrow{\nabla}^2 \operatorname{grad} \Sigma + D\frac{\mu}{\mu+\alpha}\left(l^2\overrightarrow{\nabla}^2 + \frac{\alpha}{\mu}\right) \operatorname{curl} \mathbf{H} \\ &\quad - 2(1-\nu) \operatorname{curl} \boldsymbol{\psi} + \frac{1}{4\alpha}\mathbf{Y}.\end{aligned} \tag{1.398}$$

Such a procedure enables us to eliminate the potential Σ by substituting

$$\sigma = \nabla^2 \Sigma, \tag{1.399}$$

where

$$(\nu_1^2 \nabla^2 - 1)\sigma = -\frac{1}{4\alpha C} \operatorname{div} \mathbf{Y}. \tag{1.400}$$

The representation for the rotation fields takes the form

$$\begin{aligned}\boldsymbol{\varphi} &= C\nu_1^2 \operatorname{grad} \sigma + D\frac{\mu}{\mu+\alpha}\left(l^2\overrightarrow{\nabla}^2 + \frac{\alpha}{\mu}\right) \operatorname{curl} \mathbf{H} \\ &\quad - 2(1-\nu) \operatorname{curl} \boldsymbol{\psi} + \frac{1}{4\alpha}\mathbf{Y},\end{aligned} \tag{1.401}$$

where the vector $\mathbf{H}$ is determined from the equation

$$\begin{aligned}D\overrightarrow{\nabla}^2\overrightarrow{\nabla}^2(l^2\overrightarrow{\nabla}^2 - 1)\mathbf{H} &= \overrightarrow{\nabla}^2\left(\frac{1}{2\mu}\mathbf{X} + \frac{\mu+\alpha}{4\alpha\mu} \operatorname{curl} \mathbf{Y}\right) \\ &\quad - \frac{1}{2\mu} \operatorname{grad} \operatorname{div} \mathbf{X}.\end{aligned} \tag{1.402}$$

The representations (1.394) and (1.401), with the field equations (1.388), (1.393), (1.400) and (1.402) as well as the field condition $(1.283)_4$ ($\operatorname{div} \mathbf{H} = 0$ in Ω) comprise the generalization of the Papkovich-Neuber representation of classical elastostatics to nonhomogenous equations for displacements and rotations of the micropolar model (E-N). Let us note that for $\alpha \to 0$, equation $(1.155)_1$ is transformed into the classical equation in Ω:

$$\mu\overrightarrow{\nabla}^2\mathbf{u} + (\lambda+\mu) \operatorname{grad} \operatorname{div} \mathbf{u} + \mathbf{X} = \mathbf{0}. \tag{1.403}$$

In turn, putting $\mathbf{Y} = \mathbf{0}$ and $C = D = \alpha^2$, we obtain from (1.388), (1.393), (1.394) and (1.401) the classical representation of Papkovich-Neuber with the field of body forces (cf. [180, p. 163]):

$$\mathbf{u} = \operatorname{grad}(\varphi + \mathbf{R} \circ \boldsymbol{\psi}) - 4(1-\nu)\boldsymbol{\psi}, \tag{1.404}$$

$$\boldsymbol{\varphi} = -2(1-\nu) \operatorname{curl} \boldsymbol{\psi}, \tag{1.405}$$

where

$$4(1-\nu)\overrightarrow{\nabla}^2\boldsymbol{\psi} - \frac{\mathbf{X}}{\mu} = \mathbf{0}, \tag{1.406}$$

$$4(1-\nu)\nabla^2\varphi + \frac{\mathbf{R} \circ \mathbf{x}}{\mu} = 0, \tag{1.407}$$

$$\boldsymbol{\varphi} = \frac{1}{2}\operatorname{curl}\mathbf{u}. \tag{1.408}$$

The result obtained for the micropolar model (E-N), i.e. $(1.283)_4$, (1.388), (1.393), (1.394) and (1.400)–(1.402), can be used to determine the fundamental solutions for the E-N model within the framework of elastostatics.

2. Axially-symmetric problems

2.1 The first axially-symmetric problem. Elastodynamics

2.1.1 Equations in displacements and rotations. Body loadings

From vector equations (1.1), for the cylindrical coordinate system (r, θ, z), and under the assumption of axial symmetry (causes and effects do not depend on the variable θ), we obtain equations in displacements and rotations for the first axially-symmetric problem represented by the vectors $\mathbf{u} \stackrel{\text{def}}{=} (u_r, 0, u_z)$ and $\boldsymbol{\varphi} \stackrel{\text{def}}{=} (0, \varphi_\theta, 0)$ in $\Omega \times \mathbf{T}_+$:

$$\begin{aligned}
&\Box_2^0 u_r + (\lambda + \mu - \alpha)\partial_r e - 2\alpha\partial_z\varphi_\theta + X_r = 0,\\
&\Box_2 u_z + (\lambda + \mu - \alpha)\partial_z e - 2\alpha r^{-1}\partial_r(r\varphi_\theta) + X_z = 0,\\
&\Box_4^0 \varphi_\theta + 2\alpha(\partial_z u_r - \partial_r u_z) + Y_\theta = 0.
\end{aligned} \tag{2.1}$$

Here, the operator $\vec{\nabla}^2$ has the form $\vec{\nabla}^2 \equiv (\nabla_0^2, \nabla_0^2, \nabla^2)$, where the scalar Laplace operator $\nabla^2 = \partial_r^2 + r^{-1}\partial_r + \partial_z^2$. Moreover, $\nabla_0^2 = \nabla^2 - r^{-2}$. The symbols ∂_r, ∂_z denote partial derivatives with respect to the position variables r, z, respectively, and the dilatation is given by $e = r^{-1}\partial_r(ru_r) + \partial_z u_z$. The symbols X_r, Y_θ, X_z denote body loadings. The operators $\Box_i^0$ $(i = 1, \ldots, 4)$ are defined through the operator ∇_0^2. In the Cartesian product $\Omega \times \mathbf{T}_+$, the role of the position variables $\mathbf{x}$ is played by the coordinates (r, z). Equations (2.1) have been given in the uncoupled form in [64] and [49] as follows:

— for the displacements u_r, u_z in $\Omega \times \mathbf{T}_+$,

$$\begin{aligned}
\Box_1^0(\Box_2^0\Box_4^0 + 4\alpha^2\nabla_0^2)u_r = {}& -2\alpha\partial_z\Box_1^0 Y_\theta - \Box_1^0\Box_4^0 X_r\\
&+ [(\lambda + \mu - \alpha)\Box_4^0 - 4\alpha^2]\partial_r X,\\
\Box_1(\Box_2\Box_4 + 4\alpha^2\nabla^2)u_z = {}& [(\lambda + \mu - \alpha)\Box_4 - 4\alpha^2]\partial_z X\\
&+ 2\alpha\Box_1[r^{-1}\partial_r(rY_\theta)] - \Box_1\Box_4 X_z,
\end{aligned} \tag{2.2}$$

where $X = r^{-1}\partial_r(rX_r) + \partial_z X_z$;

— for the rotation φ_θ in $\Omega \times \mathbf{T}_+$,

$$(\Box_2^0\Box_4^0 + 4\alpha^2\nabla_0^2)\varphi_\theta = 2\alpha(\partial_z X_r - \partial_r X_z) - \Box_2^0 Y_\theta. \tag{2.3}$$

On the basis of equations (2.2) and (2.3), the action in the infinite space of body loadings of the form

$$X_z = \frac{P_0}{2\pi r} e^{-i\omega t}\delta(r)\delta(z) \tag{2.4}$$

was studied. The remaining body loadings vanish ($X_r = Y_\theta = 0$). Applying to equations (2.2), (2.3) the method of integral Fourier transforms (0.50) for $n = 2$ (with respect to the time t and the position variable z), and Hankel transforms (0.51) with respect to the position variable r ($\nu = 0$ or $\nu = 1$), a solution in the closed form has been obtained:

$$\begin{aligned}
u_r &= \frac{P_0 e^{-i\omega t}}{4\pi\rho\omega^2}\partial_{rz}^2\left(A_1\frac{e^{i\lambda_1 R}}{R} + A_2\frac{e^{i\lambda_2 R}}{R} - \frac{e^{i\sigma_1 R}}{R}\right),\\
u_z &= \frac{P_0 e^{-i\omega t}}{4\pi\rho\omega^2}\left[A_1\lambda_1^2\frac{e^{i\lambda_1 R}}{R} + A_2\lambda_2^2\frac{e^{i\lambda_2 R}}{R}\right.\\
&\qquad \left. + \partial_z^2\left(A_1\frac{e^{i\lambda_1 R}}{R} + A_2\frac{e^{i\lambda_2 R}}{R} - \frac{e^{i\sigma_1 R}}{R}\right)\right],\\
\varphi_\theta &= -\frac{P_0 p e^{-i\omega t}}{4\pi\rho c_2^2(\lambda_1^2 - \lambda_2^2)}\partial_r\left(\frac{e^{i\lambda_1 R} - e^{i\lambda_2 R}}{R}\right),
\end{aligned} \tag{2.5}$$

where $R = \sqrt{r^2 + z^2}$. Here, the notation is identical to the one used in formulae (1.26), (1.27). Formulae (2.5) give the fundamental solution for the triple $(u_r, \varphi_\theta, u_z)$ when the body force field reduces to the action of a concentrated body force X_z varying in time harmonically. After substituting X_z and $(u_r, \varphi_\theta, u_z)$ from formulae (2.4), (2.5), equations (2.1) become satisfied identically. Here, the derivative is understood in the distributive sense:

$$\frac{1}{4\pi}\nabla^2\left(\frac{1}{R}\right) = -\frac{1}{2\pi r}\delta(r)\delta(z). \tag{2.6}$$

Moreover, the following condition is fulfilled:

$$(u_r, \varphi_\theta, u_z) \to (0,0,0) \qquad \text{for } R = \sqrt{r^2 + z^2} \to \infty. \tag{2.7}$$

Formulae (2.5) were derived by means of the method of potentials by Nowacki & Nowacki [197]. The limiting cases of formulae (2.5) were discussed in detail in [67]. Finding the limits may sometimes be fairly complex, but the results are interesting. Thus, it has been shown that using formulae (2.5) one can obtain the results for the limiting theories marked on the diagram (Fig. 1.1) by numbers (2)-(6). As an example, let us consider the limit $\alpha \to \infty$. Assuming the limits

$$J \to 0, \qquad \left(\frac{\gamma+\varepsilon}{4\mu}\right)^{\frac{1}{2}} \to l^\star, \tag{2.8}$$

formulae $(2.5)_{1,2}$ are transformed into the formulae that can be obtained after assuming in $(2.5)_{1,2}$ the correspondence

$$(A_1, A_2, \lambda_1, \lambda_2) \mapsto (A_1^\star, A_2^\star, \lambda_1^\star, \lambda_2^\star) \tag{2.9}$$

where it holds

$$\begin{gathered} A_1^\star = \frac{\lambda_2^{\star 2}}{\lambda_2^{\star 2} - \lambda_1^{\star 2}}, \qquad A_2^\star = \frac{\lambda_1^{\star 2}}{\lambda_1^{\star 2} - \lambda_2^{\star 2}}, \\ {\lambda_{1,2}^\star}^2 = \frac{1}{2l^{\star 2}}\left(-1 \pm \sqrt{1 + \frac{4\rho\omega^2}{\mu} l^{\star 2}}\right). \end{gathered} \tag{2.10}$$

Formula $(2.5)_3$ goes over into the formula

$$\varphi_\theta = -\frac{P_0 e^{-i\omega t}}{8\pi\mu l^{\star 2}(\lambda_1^{\star 2} - \lambda_2^{\star 2})}\,\partial_r\left(\frac{e^{i\lambda_1^\star R} - e^{i\lambda_2^\star R}}{R}\right). \tag{2.11}$$

For the triple $(u_r, \varphi_\theta, u_z)$ thus obtained formula (2.7) holds, as well as the following formula:

$$2\varphi_\theta = \partial_z u_r - \partial_r u_z. \tag{2.12}$$

Furthermore, if we separate the symmetric part from the force-stresses and the deviatoric part from the couple-stresses, and assume that for $\alpha \to \infty$

$$\frac{\gamma - \varepsilon}{\gamma + \varepsilon} \to \eta, \tag{2.13}$$

holds[1], where $l^\star$ and η are the elasticity constants of couple-stress theory, then the procedure leads to the results for displacements, rotations, force-stresses and couple-stresses in the framework of couple-stress theory.

2.1.2 Equations in displacements and rotations. Superposition method

In [49], the following substitution is introduced to the system of equations (2.1) in $\Omega \times \mathbf{T}_+$:

$$\zeta_\theta = \tfrac{1}{2}(\partial_z u_r - \partial_r u_z) - \varphi_\theta. \tag{2.14}$$

Then, we obtain in $\Omega \times \mathbf{T}_+$ the system of equations

$$\begin{gathered} \overset{*}{\Box}{}_2^0 u_r + (\lambda+\mu)\partial_r e + 2\alpha\partial_z\zeta_\theta + X_r = 0, \\ \overset{*}{\Box}{}_2 u_z + (\lambda+\mu)\partial_z e - 2\alpha r^{-1}\partial_r(r\zeta_\theta) + X_z = 0, \\ \overset{*}{\Box}{}_4^0(\partial_z u_r - \partial_r u_z) + 2Y_\theta = 2\Box_4^0\zeta_\theta. \end{gathered} \tag{2.15}$$

[1] Conditions (2.8) and (2.13) have physical motivation.

The operators $\overset{*}{\square}_2$, $\overset{*}{\square}{}_2^0$ and $\overset{*}{\square}{}_4^0$ are obtained from the operators $\square_2$, $\square_2^0$ and $\square_4^0$ by putting $\alpha = 0$. For further considerations it is essential that we apply the operator $\overset{*}{\square}{}_2^0$ to equation $(2.15)_3$. As a result, we then obtain the equation

$$\overset{*}{\square}{}_2^0\left[\overset{*}{\square}{}_4^0(\partial_z u_r - \partial_r u_z) + 2Y_\theta - 2\square_4^0\zeta_\theta\right] = 0. \tag{2.16}$$

Further, we take the triple $(u_r, \varphi_\theta, u_z)$ in the form

$$u_r = u_r^0 + u_r', \qquad \varphi_\theta = \varphi_\theta^0 + \varphi_\theta', \qquad u_z = u_z^0 + u_z', \tag{2.17}$$

where the triple $(u_r^0, \varphi_\theta^0, u_z^0)$ refers to the classical theory of elasticity. Moreover, we assume $\zeta_\theta = \zeta_\theta^0 + \zeta_\theta'$, $Y_\theta = Y_\theta^0 + Y_\theta'$. For $\zeta_\theta^0 = 0$ we obtain from equations $(2.15)_{1,2}$ the classical equations in $\Omega \times \mathbf{T}_+$:

$$\overset{*}{\square}{}_2^0 u_r^0 + (\lambda + \mu)\partial_r e^0 + X_r = 0, \qquad \overset{*}{\square}{}_2^0 u_z^0 + (\lambda + \mu)\partial_z e^0 + X_z = 0, \tag{2.18}$$

where $e^0 = r^{-1}\partial_r(ru_r^0) + \partial_z u_z^0$.

In the classical theory of elasticity the following relation holds:

$$2\varphi_\theta^0 = \partial_z u_r^0 - \partial_r u_z^0. \tag{2.19}$$

From equation (2.16) (for $\zeta_\theta^0 = 0$ and $Y_\theta^0 = 0$) we obtain the equation

$$\overset{*}{\square}{}_4^0\left[\overset{*}{\square}{}_2^0(\partial_z u_r^0 - \partial_r u_z^0) + (\partial_z X_r - \partial_r X_z)\right] = 0. \tag{2.20}$$

Equation (2.20) will become an identity, since from (2.18) we obtain the equation for φ_θ^0:

$$2\overset{*}{\square}{}_2^0\varphi_\theta^0 = \partial_r X_z - \partial_z X_r. \tag{2.21}$$

Let us also note separate equations for the pair (u_r^0, u_z^0) in $\Omega \times \mathbf{T}_+$:

$$\begin{aligned} \square_1^0\overset{*}{\square}{}_2^0 u_r^0 &= (\lambda + \mu)\partial_r\left[r^{-1}\partial_r(rX_r) + \partial_z X_z\right] - \square_1^0 X_r, \\ \square_1\overset{*}{\square}{}_2 u_z^0 &= (\lambda + \mu)\partial_z\left[r^{-1}\partial_r(rX_r) + \partial_z X_z\right] - \square_1 X_z. \end{aligned} \tag{2.22}$$

For the triple $(u_r', \zeta_\theta, u_z')$ we obtain from $(2.15)_{1,2}$ and (2.16) the equations

$$\begin{aligned} &\overset{*}{\square}{}_2^0 u_r' + (\lambda + \mu)\partial_r e' + 2\alpha\partial_z\zeta_\theta + \hat{X}_r = 0, \\ &\overset{*}{\square}{}_2^0 u_z' + (\lambda + \mu)\partial_z e' - 2\alpha r^{-1}\partial_r(r\zeta_\theta) + \hat{X}_z = 0, \\ &\overset{*}{\square}{}_2^0\left[\overset{*}{\square}{}_4^0(\partial_z u_r' - \partial_r u_z') - 2\square_4^0\zeta_\theta\right] = -2\hat{Y}_\theta, \end{aligned} \tag{2.23}$$

where it holds

$$\hat{X}_r = \hat{X}_z = 0, \qquad \hat{Y}_\theta = \tfrac{1}{2}\left[2\overset{*}{\square}{}_2^0 Y_\theta - \overset{*}{\square}{}_4^0(\partial_z X_r - \partial_r X_z)\right]. \tag{2.24}$$

From the system of equations (2.23) we derive the following uncoupled equations:

— for ζ_θ in $\Omega \times \mathbf{T}_+$,

$$2(\Box_2^0\Box_4^0 + 4\alpha^2\nabla_0^2)\zeta_\theta = 2\hat{Y}_\theta - \overset{*}{\Box}{}_4^0(\partial_z\hat{X}_r - \partial_r\hat{X}_z); \tag{2.25}$$

— for u'_r and u'_z in $\Omega \times \mathbf{T}_+$,

$$\begin{aligned}
&\overset{*}{\Box}{}_2^0\Box_1^0(\Box_2^0\Box_4^0 + 4\alpha^2\nabla_0^2)u'_r \\
&\quad = \overset{*}{\Box}{}_2^0\left\{\left[(\lambda+\mu-\alpha)\Box_4^0 - 4\alpha^2\right]\partial_r\hat{X} - \Box_1^0\Box_4^0\hat{X}_r\right\} \\
&\quad - 2\alpha\partial_z\Box_1^0\hat{Y}_\theta,
\end{aligned} \tag{2.26}$$

where $\hat{X} = r^{-1}\partial_r(r\hat{X}_r) + \partial_z\hat{X}_z$;

$$\begin{aligned}
&\overset{*}{\Box}{}_2\Box_1(\Box_2\Box_4 + 4\alpha^2\nabla^2)u'_z \\
&\quad = \overset{*}{\Box}{}_2\left\{\left[(\lambda+\mu-\alpha)\Box_4 - 4\alpha^2\right]\partial_z\hat{X} - \Box_1\Box_4\hat{X}_z\right\} \\
&\quad + 2\alpha\Box_1[r^{-1}\partial_r(r\hat{Y}_\theta)].
\end{aligned} \tag{2.27}$$

In further considerations we eliminate the function ζ_θ from the system of equations (2.23) and we obtain the system of equations for the triple $(u'_r, \varphi'_\theta, u'_z)$:

$$\begin{aligned}
&\Box_2^0 u'_r + (\lambda+\mu-\alpha)\partial_r e' - 2\alpha\partial_z\varphi'_\theta + \hat{X}_r = 0, \\
&\Box_2^0 u'_z + (\lambda+\mu-\alpha)\partial_z e' + 2\alpha r^{-1}\partial_r(r\varphi'_\theta) + \hat{X}_z = 0, \\
&\overset{*}{\Box}{}_2^0\left[\Box_4^0\varphi'_\theta + 2\alpha(\partial_z u'_r - \partial_r u'_z)\right] + \hat{Y}_\theta = 0.
\end{aligned} \tag{2.28}$$

From the system of equations (2.28) we derive the equation for φ'_θ:

$$\overset{*}{\Box}{}_2^0(\Box_2^0\Box_4^0 + 4\alpha^2\nabla_0^2)\varphi'_\theta = 2\alpha\overset{*}{\Box}{}_2^0(\partial_z\hat{X}_r - \partial_r\hat{X}_z) - \Box_2^0\hat{Y}_\theta. \tag{2.29}$$

The derived equations (2.18), (2.28) may be used to solve initial-boundary value problems by the superposition method. We add the following initial-boundary conditions to the system of equations (2.1):

— the boundary conditions on $\partial\Omega \times \mathbf{T}$,

$$\begin{aligned}
&n_r\sigma_{rr} + n_z\sigma_{zr} = p_r, \qquad n_r\sigma_{rz} + n_z\sigma_{zz} = p_z, \\
&n_r\mu_{r\theta} + n_z\mu_{z\theta} = m_\theta \qquad \text{on } \partial\Omega_\sigma \times \mathbf{T};
\end{aligned} \tag{2.30}$$

$$u_r = \hat{u}_r, \qquad u_z = \hat{u}_z, \qquad \varphi_\theta = \hat{\varphi}_\theta \qquad \text{on } \partial\Omega_u \times \mathbf{T}, \tag{2.31}$$

where the functions $p_r, p_z, m_\theta : \partial\Omega_\sigma \times \mathbf{T} \to \mathbb{R}$ and $\hat{u}_r, \hat{u}_z, \hat{\varphi}_\theta : \partial\Omega_u \times \mathbf{T} \to \mathbb{R}$ are given;

— the initial conditions in $\Omega \times \{0\}$,

$$\begin{aligned} u_r &= u_r^\star, & u_z &= u_z^\star, & \varphi_\theta &= \varphi_\theta^\star, \\ \dot{u}_r &= \dot{u}_r^\star, & \dot{u}_z &= \dot{u}_z^\star, & \dot{\varphi}_\theta &= \dot{\varphi}_\theta^\star, \end{aligned} \tag{2.32}$$

where the functions $u_r^\star$, $u_z^\star$, $\varphi_\theta^\star$, $\dot{u}_r^\star$, $\dot{u}_z^\star$, $\dot{\varphi}_\theta^\star : \Omega \to \mathbb{R}$ are given. The quantities n_r, n_z are the two components of a unit outward normal vector, expressed in the local basis related to the cylindrical coordinate system. The stresses σ_{rr}, σ_{zr}, σ_{rz}, σ_{zz}, $\mu_{r\theta}$, $\mu_{z\theta}$ will be expressed in terms of the triple $(u_r, \varphi_\theta, u_z)$ by using

— the constitutive relations in $\Omega \times \mathbf{T}$,

$$\begin{aligned} \sigma_{rr} &= 2\mu\gamma_{rr} + \lambda e, \qquad \sigma_{\theta\theta} = 2\mu\gamma_{\theta\theta} + \lambda e, \\ \sigma_{zz} &= 2\mu\gamma_{zz} + \lambda e, \\ \sigma_{rz} &= (\mu + \alpha)\gamma_{rz} + (\mu - \alpha)\gamma_{zr}, \\ \sigma_{zr} &= (\mu + \alpha)\gamma_{zr} + (\mu - \alpha)\gamma_{rz}, \\ \mu_{r\theta} &= (\gamma + \varepsilon)\kappa_{r\theta} + (\gamma - \varepsilon)\kappa_{\theta r}, \\ \mu_{\theta r} &= (\gamma + \varepsilon)\kappa_{\theta r} + (\gamma - \varepsilon)\kappa_{r\theta}, \\ \mu_{z\theta} &= (\gamma + \varepsilon)\kappa_{z\theta}, \qquad \mu_{\theta z} = \frac{\gamma - \varepsilon}{\gamma + \varepsilon}\mu_{z\theta}, \end{aligned} \tag{2.33}$$

where $e = \gamma_{rr} + \gamma_{\theta\theta} + \gamma_{zz}$, and

— the geometric relations in $\Omega \times \mathbf{T}_+$,

$$\begin{aligned} \gamma_{rr} &= \partial_r u_r, & \gamma_{\theta\theta} &= \frac{1}{r} u_r, & \gamma_{zz} &= \partial_z u_z, \\ \gamma_{rz} &= \partial_r u_z + \varphi_\theta, & \gamma_{zr} &= \partial_z u_r - \varphi_\theta, \\ \kappa_{r\theta} &= \partial_r \varphi_\theta, & \kappa_{\theta r} &= -\frac{1}{r}\varphi_\theta, & \kappa_{z\theta} &= \partial_z \varphi_\theta. \end{aligned} \tag{2.34}$$

In order to solve the initial-boundary value problem of micropolar theory described by equations (2.1) and (2.30)–(2.32) by means of the superposition method, we proceed as follows. We solve the classical equations (2.18) with classical initial-boundary conditions:

— the boundary conditions on $\partial\Omega \times \mathbf{T}$,

$$n_r \sigma_{rr}^0 + n_z \sigma_{zr}^0 = p_r, \qquad n_r \sigma_{rz}^0 + n_z \sigma_{zz}^0 = p_z \qquad \text{on } \partial\Omega_\sigma \times \mathbf{T}, \tag{2.35}$$

$$u_r^0 = \hat{u}_r, \qquad u_z^0 = \hat{u}_z \qquad \text{on } \partial\Omega_u \times \mathbf{T}, \tag{2.36}$$

where σ_{rr}^0, $\sigma_{zr}^0 = \sigma_{rz}^0$, σ_{zz}^0 are given by the classical constitutive relations in $\Omega \times \mathbf{T}$:

$$\begin{aligned}
&\sigma^0_{rr} = 2\mu\varepsilon^0_{rr} + \lambda e^0, \qquad \sigma^0_{\theta\theta} = 2\mu\varepsilon^0_{\theta\theta} + \lambda e^0,\\
&\sigma^0_{zz} = 2\mu\varepsilon^0_{zz} + \lambda e^0,\\
&\sigma^0_{rz} = 2\mu\varepsilon^0_{rz}, \qquad e^0 = \varepsilon^0_{rr} + \varepsilon^0_{\theta\theta} + \varepsilon^0_{zz}.
\end{aligned} \tag{2.37}$$

The quantities ε^0_{rr}, $\varepsilon^0_{\theta\theta}$, ε^0_{zz}, $\varepsilon^0_{rz} = \varepsilon^0_{zr}$ are connected with the pair (u^0_r, u^0_z) through the geometric relations in $\Omega \times \mathbf{T}_+$:

$$\begin{aligned}
&\varepsilon^0_{rr} = \partial_r u^0_r, \qquad \varepsilon^0_{\theta\theta} = \frac{1}{r} u^0_r, \qquad \varepsilon^0_{zz} = \partial_z u^0_z,\\
&\varepsilon^0_{rz} = \tfrac{1}{2}(\partial_z u^0_r + \partial_r u^0_z);
\end{aligned} \tag{2.38}$$

— the initial conditions in $\Omega \times \{0\}$,

$$u^0_r = u^\star_r, \qquad u^0_z = u^\star_z, \qquad \dot{u}^0_r = \dot{u}^\star_r, \qquad \dot{u}^0_z = \dot{u}^\star_z. \tag{2.39}$$

Further, we solve equations (2.28) for the triple $(u'_r, \varphi'_\theta, u'_z)$ with initial-boundary conditions (2.30)–(2.32) in the form:

— the boundary conditions on $\partial\Omega \times \mathbf{T}$,

$$\begin{aligned}
&n_r\sigma'_{rr} + n_z\sigma'_{zr} = 0, \qquad n_r\sigma'_{rz} + n_z\sigma'_{zz} = 0,\\
&n_r\mu'_{r\theta} + n_z\mu'_{z\theta} = m_\theta - m^0_\theta \qquad \text{on } \partial\Omega_\sigma \times \mathbf{T},
\end{aligned} \tag{2.40}$$

$$u'_r = 0, \qquad u'_z = 0, \qquad \varphi'_\theta = \hat{\varphi}_\theta - \varphi^0_\theta \qquad \text{on } \partial\Omega_u \times \mathbf{T}; \tag{2.41}$$

— the initial conditions in $\Omega \times \{0\}$,

$$\begin{aligned}
&u'_r = 0, \qquad u'_z = 0, \qquad \varphi'_\theta = \varphi^\star_\theta - \varphi^0_\theta,\\
&\dot{u}'_r = 0, \qquad \dot{u}'_z = 0, \qquad \dot{\varphi}'_\theta = \dot{\varphi}^\star_\theta - \dot{\varphi}^0_\theta,
\end{aligned} \tag{2.42}$$

where the quantities $m^0_\theta = n_r\mu^0_{r\theta} + n_z\mu^0_{z\theta}$, φ^0_θ, $\dot{\varphi}^0_\theta$ result from the solution of the classical initial-boundary value problem:

$$\begin{aligned}
&\varphi^0_\theta = \tfrac{1}{2}(\partial_z u^0_r - \partial_r u^0_z),\\
&\mu^0_{r\theta} = (\gamma+\varepsilon)\kappa^0_{r\theta} + (\gamma-\varepsilon)\kappa^0_{\theta r}, \qquad \mu^0_{z\theta} = (\gamma+\varepsilon)\kappa^0_{z\theta},\\
&\kappa^0_{r\theta} = \partial_r\varphi^0_\theta, \qquad \kappa^0_{\theta r} = -\frac{1}{r}\varphi^0_\theta, \qquad \kappa^0_{z\theta} = \partial_z\varphi^0_\theta.
\end{aligned} \tag{2.43}$$

We represent the stresses σ'_{rr}, σ'_{zr}, σ'_{rz}, σ'_{zz}, $\mu'_{r\theta}$, $\mu'_{z\theta}$ from the boundary conditions (2.40) in terms of u'_r, φ'_θ and u'_z, using relations (2.33) and (2.34), into which we put $\underset{\sim}{\sigma}'$, $\underset{\sim}{\mu}'$, $\underset{\sim}{\gamma}'$, $\underset{\sim}{\kappa}'$, $\mathbf{u}'$ and $\boldsymbol{\varphi}'$. The final result is obtained from formula (2.17). Similarly, we obtain: $\underset{\sim}{\sigma} = \underset{\sim}{\sigma}^0 + \underset{\sim}{\sigma}'$, $\underset{\sim}{\mu} = \underset{\sim}{\mu}^0 + \underset{\sim}{\mu}'$, $\underset{\sim}{\gamma} = \underset{\sim}{\varepsilon}^0 + \underset{\sim}{\gamma}'$, $\underset{\sim}{\kappa} = \underset{\sim}{\kappa}^0 + \underset{\sim}{\kappa}'$.

Equations (2.21), (2.22), (2.25)–(2.27) and (2.29) may be used to determine the fundamental solutions. In [67], formulae (2.5) were derived in this manner by the application of the Fourier-Hankel transforms. In the case of a loading given by formula (2.4) we obtain from (2.21) the following equation:

$$\varphi_\theta^0 = -\frac{P_0 \hat{\sigma}_2^2 e^{-i\omega t}}{8\pi\rho\omega^2}\,\partial_r\left(\frac{e^{i\hat{\sigma}_2 R}}{R}\right). \tag{2.44}$$

From equations (2.22) we obtain[2]

$$\begin{aligned} u_r^0 &= \frac{P_0 e^{-i\omega t}}{4\pi\rho\omega^2}\,\partial_{rz}^2\left(\frac{e^{i\hat{\sigma}_2 R} - e^{i\sigma_1 R}}{R}\right), \\ u_z^0 &= \frac{P_0 e^{-i\omega t}}{4\pi\rho\omega^2}\left[\partial_z^2\left(\frac{e^{i\hat{\sigma}_2 R} - e^{i\sigma_1 R}}{R}\right) + \hat{\sigma}_2^2\frac{e^{i\hat{\sigma}_2 R}}{R}\right], \end{aligned} \tag{2.45}$$

where $\hat{\sigma}_2 = \dfrac{\omega}{\hat{c}_2}$, $\hat{c}_2 = \sqrt{\frac{\mu}{\rho}}$, $R = \sqrt{r^2 + z^2}$.

For the function ζ_θ we obtain the formula

$$\zeta_\theta = -\frac{P_0 e^{-i\omega t}}{8\pi\rho\omega^2}\frac{\sigma_2^2}{\lambda_1^2 - \lambda_2^2}\partial_r\left[(\lambda_1^2 - \sigma_4^2)\frac{e^{i\lambda_1 R}}{R} - (\lambda_2^2 - \sigma_4^2)\frac{e^{i\lambda_2 R}}{R}\right]. \tag{2.46}$$

From equations (2.26), (2.27) and (2.29) we obtain

$$\begin{aligned} u_r' &= \frac{P_0 e^{-i\omega t}}{4\pi\rho\omega^2}\,\partial_{rz}^2\left(A_1\frac{e^{i\lambda_1 R}}{R} + A_2\frac{e^{i\lambda_2 R}}{R} - \frac{e^{i\hat{\sigma}_2 R}}{R}\right), \\ u_z' &= -\frac{P_0 e^{-i\omega t}}{4\pi\rho\omega^2}\left(r^{-1}\partial_r + \partial_r^2\right)\left(A_1\frac{e^{i\lambda_1 R}}{R} + A_2\frac{e^{i\lambda_2 R}}{R} - \frac{e^{i\hat{\sigma}_2 R}}{R}\right), \\ \varphi_\theta' &= -\varphi_\theta^0 - \frac{P_0 p e^{-i\omega t}}{4\pi\rho c_2^2(\lambda_1^2 - \lambda_2^2)}\partial_r\left(\frac{e^{i\lambda_1 R} - e^{i\lambda_2 R}}{R}\right). \end{aligned} \tag{2.47}$$

The notations appearing in formulae (2.44)–(2.47) are the same as in formulae (1.26)–(1.28). The result given by (2.5) in the form of (2.44)–(2.47) is very convenient for discussing the limiting cases.

Let us go back to the system of equations (2.28) and make the following observation. If we assume the formal correspondence

$$(u_r', \varphi_\theta', u_z') \mapsto (u_r, \varphi_\theta, u_z), \qquad (\hat{X}_r, \hat{Y}_\theta, \hat{X}_z) \mapsto (X_r, \overset{*}{\square}{}_2^0 Y_\theta, X_z), \tag{2.48}$$

then the system of equations (2.28) is transformed into the system (2.1), and equations (2.26), (2.27) and (2.29) are transformed into equations (2.2), (2.3), respectively (after removing the operator $\overset{*}{\square}_2$ or $\square_2^0$).

[2] Eason, Fulton & Sneddon [86]. The method consisting in applying integral transformations to the equations of the classical theory of elasticity is due to these authors and it has been commonly applied in many works.

2.1.3 Equations in displacements and rotations with a distortion field

In [49], uncoupled equations for displacements and rotations with a distortion field were introduced. Such equations will be discussed here in connection with the method of superposition presented in Subsection 2.1.2. We assume the following substitutions (which we call analogies of body loadings):

$$X_r = -\overset{*}{R}{}^0_1,\ X_z = -\overset{*}{R}{}^0_2,\ \hat{X}_r = -\overset{**}{R}{}^0_1,\ \hat{X}_z = -\overset{**}{R}{}^0_2,\ Y_\theta = -R^0_3 \tag{2.49}$$

and

$$\hat{Y}_\theta = -\hat{R}^0_3, \tag{2.50}$$

where we denote

$$\hat{R}^0_3 = \overset{*}{\Box}{}^0_2 R^0_3 - \tfrac{1}{2}\overset{*}{\Box}{}^0_4(\partial_z \overset{*}{R}{}^0_1 - \partial_r \overset{*}{R}{}^0_2),$$

$$R^0_3 = \sigma^0_{zr} - \sigma^0_{rz} + \partial_r \mu^0_{r\theta} + \partial_z \mu^0_{z\theta} + r^{-1}(\mu^0_{r\theta} + \mu^0_{\theta r}).$$

Moreover, we obtain the formula

$$R^0_\alpha = \overset{*}{R}{}^0_\alpha + \overset{**}{R}{}^0_\alpha \qquad (\alpha = 1,\ 2), \tag{2.51}$$

where we denote

$$R^0_1 = \partial_r \sigma^0_{rr} + \partial_z \sigma^0_{zr} + r^{-1}(\sigma^0_{rr} - \sigma^0_{\theta\theta}), \qquad R^0_2 = \partial_r \sigma^0_{rz} + \partial_z \sigma^0_{zz} + r^{-1}\sigma^0_{rz}.$$

The symbols σ^0_{rr}, $\sigma^0_{\theta\theta}$, σ^0_{zz}, σ^0_{rz}, σ^0_{zr}, and $\mu^0_{r\theta}$, $\mu^0_{\theta r}$, $\mu^0_{z\theta}$, $\mu^0_{\theta z}$, have the form of the right-hand sides of the constitutive relations (2.33) written for distortions γ^0_{rr}, $\gamma^0_{\theta\theta}$, γ^0_{zz}, γ^0_{rz}, γ^0_{zr} and $\kappa^0_{r\theta}$, $\kappa^0_{\theta r}$, $\kappa^0_{z\theta}$. For example,

$$\sigma^0_{rz} = (\mu+\alpha)\gamma^0_{rz} + (\mu-\alpha)\gamma^0_{zr}, \qquad \mu^0_{\theta r} = (\gamma+\varepsilon)\kappa^0_{\theta r} + (\gamma-\varepsilon)\kappa^0_{r\theta} \qquad \text{etc.}$$

Thus for the expressions $\overset{*}{R}{}^0_\alpha$, $\overset{**}{R}{}^0_\alpha$ we obtain the formulae:

$$\begin{aligned}
\overset{*}{R}{}^0_1 &= \partial_r(2\mu\gamma^0_{rr} + \lambda e_0) + 2\mu[r^{-1}(\gamma^0_{rr} - \gamma^0_{\theta\theta}) + \partial_z \gamma^0_{(zr)}],\\
\overset{*}{R}{}^0_2 &= 2\mu[(r^{-1} + \partial_r)\,\gamma^0_{(rz)} + \partial_z \gamma^0_{zz}] + \lambda\partial_z e_0,\\
\overset{**}{R}{}^0_1 &= 2\alpha\partial_z \gamma^0_{[zr]},\\
\overset{**}{R}{}^0_2 &= 2\alpha\,(r^{-1} + \partial_r)\,\gamma^0_{[rz]},
\end{aligned} \tag{2.52}$$

where $e_0 = \gamma^0_{rr} + \gamma^0_{\theta\theta} + \gamma^0_{zz}$.

Now, from equations (2.21) and (2.22), for case $(2.49)_{1,2}$, we obtain the classical equations with a distortion field in $\Omega \times \mathbf{T}_+$:

$$
\begin{aligned}
&\square_1^0 \overset{*}{\square}{}_2^0 u_r^0 = -(\lambda+\mu)\partial_r \left[r^{-1}\partial_r (r\overset{*}{R}{}_1^0) + \partial_z \overset{*}{R}{}_2^0 \right] + \square_1^0 \overset{*}{R}{}_1^0, \\
&\square_1 \overset{*}{\square}{}_2 u_z^0 = -(\lambda+\mu)\partial_z \left[r^{-1}\partial_r (r\overset{*}{R}{}_1^0) + \partial_z \overset{*}{R}{}_2^0 \right] + \square_1^0 \overset{*}{R}{}_2^0, \\
&\overset{*}{\square}{}_2^0 \varphi_\theta^0 = \tfrac{1}{2}(\partial_z \overset{*}{R}{}_1^0 - \partial_r \overset{*}{R}{}_2^0).
\end{aligned}
\tag{2.53}
$$

Further, from equations (2.26), (2.27) and (2.29) for case $(2.49)_{3\text{–}4}$–(2.50) we obtain the equations

— for the displacements u'_r, u'_z,

$$
\begin{aligned}
&\overset{*}{\square}{}_2^0 \square_1^0 (\square_2^0 \square_4^0 + 4\alpha^2 \nabla_0^2) u'_r \\
&\qquad = -\overset{*}{\square}{}_2^0 \Big\{ [(\lambda+\mu-\alpha)\square_4^0 - 4\alpha^2] \partial_r [r^{-1}\partial_r (r\overset{**}{R}{}_1^0) + \partial_z \overset{**}{R}{}_2^0] \\
&\qquad\quad - \square_1^0 \square_4^0 \overset{**}{R}{}_1^0 \Big\} + 2\alpha \partial_z \square_1^0 \hat{R}_3^0, \\
&\overset{*}{\square}{}_2 \square_1 (\square_2 \square_4 + 4\alpha^2 \nabla^2) u'_z \\
&\qquad = -\overset{*}{\square}{}_2 \Big\{ [(\lambda+\mu-\alpha)\square_4 - 4\alpha^2] \partial_z [r^{-1}\partial_r (r\overset{**}{R}{}_1^0) + \partial_z \overset{**}{R}{}_2^0] \\
&\qquad\quad - \square_1 \square_4 \overset{**}{R}{}_2^0 \Big\} + 2\alpha \square_1 [r^{-1}\partial_r (r\hat{R}_3^0)];
\end{aligned}
\tag{2.54}
$$

— for the rotation φ'_θ,

$$
\overset{*}{\square}{}_2^0 (\square_2^0 \square_4^0 + 4\alpha^2 \nabla_0^2) \varphi'_\theta = -2\alpha \overset{*}{\square}{}_2^0 (\partial_z \overset{**}{R}{}_1^0 - \partial_r \overset{**}{R}{}_2^0) + \square_2^0 \hat{R}_3^0. \tag{2.55}
$$

If we now assume $(2.48)_1$ and the correspondence

$$
(\overset{**}{R}{}_1^0, \hat{R}_3^0, \overset{**}{R}{}_2^0) \mapsto (R_1^0, \overset{*}{\square}{}_2^0 R_3^0, R_2^0) \tag{2.56}
$$

then, after removing the operator $\overset{*}{\square}{}_2$ or $\overset{*}{\square}{}_2^0$, equations (2.54) and (2.55) go over into the equations

$$
\begin{aligned}
\square_1^0 (\square_2^0 \square_4^0 + 4\alpha^2 \nabla_0^2) u_r &= 2\alpha \partial_z \square_1^0 R_3^0 + \square_1^0 \square_4^0 R_1^0 \\
&\quad - [(\lambda+\mu-\alpha)\square_4^0 - 4\alpha^2] \partial_r R^0, \\
\square_1 (\square_2 \square_4 + 4\alpha^2 \nabla^2) u_z &= -[(\lambda+\mu-\alpha)\square_4 - 4\alpha^2] \partial_z R^0 \\
&\quad - 2\alpha \square_1 \left[r^{-1}\partial_r (r R_3^0) \right] + \square_1 \square_4 R_2^0, \\
(\square_2^0 \square_4^0 + 4\alpha^2 \nabla_0^2) \varphi_\theta &= 2\alpha (\partial_r R_2^0 - \partial_z R_1^0) + \square_2^0 R_3^0,
\end{aligned}
\tag{2.57}
$$

respectively, where $R^0 = r^{-1}\partial_r (r R_1^0) + \partial_z R_2^0$.

These are uncoupled equations for the triple $(u_r, \varphi_\theta, u_z)$ with a distortion field. Let us note that we obtain equations (2.57) if in equations (2.1)–(2.3) we make the substitution

$$X_r = -R_1^0, \qquad Y_\theta = -R_3^0, \qquad X_z = -R_2^0. \tag{2.58}$$

In [24], action of the distortion field in the infinite space of the following form was considered:

$$\gamma_{zz}^0 = \frac{\gamma_0^{zz}}{2\pi r} e^{-i\omega t} \delta(r)\delta(z). \tag{2.59}$$

The remaining distortions are equal to zero. Making use of equations (2.53)–(2.55) as well as (2.57), and using the integral Fourier (0.50) and Hankel (0.51) transforms, the following fundamental result was obtained:

$$\begin{aligned}
u_r =& -\frac{\gamma_0 e^{-i\omega t}}{4\pi\rho\omega^2}\partial_r \left\{ 2\mu\partial_z^2 \left[A_1 \frac{e^{i\lambda_1 R}}{R} + A_2 \frac{e^{i\lambda_2 R}}{R} - \frac{e^{i\sigma_1 R}}{R} \right] \right. \\
& \left. + \lambda\sigma_1^2 \frac{e^{i\sigma_1 R}}{R} \right\}, \\
u_z =& \frac{\gamma_0 e^{-i\omega t}}{4\pi\rho\omega^2}\partial_z \left\{ 2\mu\left(\partial_r^2 + r^{-1}\partial_r\right) \left[A_1 \frac{e^{i\lambda_1 R}}{R} + A_2 \frac{e^{i\lambda_2 R}}{R} - \frac{e^{i\sigma_1 R}}{R} \right] \right. \\
& \left. - (\lambda + 2\mu)\sigma_1^2 \frac{e^{i\sigma_1 R}}{R} \right\}, \\
\varphi_\theta =& \frac{\gamma_0 e^{-i\omega t}}{4\pi l^2} \cdot \frac{1}{\lambda_1^2 - \lambda_2^2} \partial_{rz}^2 \left(\frac{e^{i\lambda_1 R} - e^{i\lambda_2 R}}{R} \right).
\end{aligned} \tag{2.60}$$

The notation in formulae (2.60) is the same as in formulae (1.26)–(1.28), $R = \sqrt{r^2 + z^2}$, $\gamma_0^{zz} \mapsto \gamma_0$. Equations(2.1) become identities after substituting (2.58)–(2.60) and taking into account (2.6). Also, condition (2.7) is satisfied. Many special cases related to the limiting theories (1)–(6) (Fig. 1.1) can be derived from formulae (2.60). Such analysis is carried out in [24]. For example, by use of (2.8)–(2.10) for $\alpha \to \infty$, formulae (2.60), as it has been shown, go over into the formulae of elastodynamics in couple-stress theory.

2.1.4 Stress functions

In [49], we introduced the generalized Galerkin vector of the form

$$\mathbf{F} \stackrel{\text{def}}{=} (F_r, 0, F_z), \qquad \mathbf{M} \stackrel{\text{def}}{=} (0, M_\theta, 0) \tag{2.61}$$

together with a representation for the triple $(u_r, \varphi_\theta, u_z)$ and equations for the stress functions[3] F_r, M_θ and F_z. Let us consider vectors of body loadings of the form

[3] In the axially-symmetric problem with the triple $(u_r, \varphi_\theta, u_z)$, body loadings are represented, in general, by vectors: $\mathbf{X} \stackrel{\text{def}}{=} (X_r, 0, X_z)$, $\mathbf{Y} \stackrel{\text{def}}{=} (0, Y_\theta, 0)$. Functions F_r, M_θ and F_z satisfy the following equations on $\Omega \times \mathbf{T}_+$:

$$(\Box_2^0\Box_4^0 + 4\alpha^2\nabla_0^2)M_\theta + Y_\theta = 0, \qquad \Box_1^0(\Box_2^0\Box_4^0 + 4\alpha^2\nabla_0^2)F_r + X_r = 0,$$

$$\Box_1(\Box_2\Box_4 + 4\alpha^2\nabla^2)F_z + X_z = 0.$$

$$\mathbf{X} \overset{\text{def}}{=} (0, 0, X_z), \qquad \mathbf{Y} \overset{\text{def}}{=} (0, 0, Y_z) \tag{2.62}$$

If we assume that in (2.61) only the component F_z does not vanish identically, we obtain for u_r, φ_θ, u_z the following representation in $\Omega \times \mathbf{T}_+$:

$$\begin{aligned} u_r &= -[(\lambda + \mu - \alpha)\square_4^0 - 4\alpha^2]\partial_{rz}^2 F_z, \\ u_z &= \left\{\square_1\square_4 - [(\lambda + \mu - \alpha)\square_4 - 4\alpha^2]\partial_z^2\right\} F_z, \\ \varphi_\theta &= 2\alpha\square_1^0\partial_r F_z. \end{aligned} \tag{2.63}$$

The field equation for the stress function F_z is obtained from equations (2.2) and (2.3) by applying (2.63)[4]:

$$\square_1(\square_2\square_4 + 4\alpha^2\nabla^2)F_z + X_z = 0 \qquad \text{in } \Omega \times \mathbf{T}_+. \tag{2.64}$$

Introducing another version of the stress function of that kind is connected with body loadings of the form

$$\mathbf{X} \overset{\text{def}}{=} (X_r, 0, 0), \qquad \mathbf{Y} \overset{\text{def}}{=} (Y_r, 0, 0). \tag{2.65}$$

Having assumed that in (2.61) only F_r does not vanish, we obtain the following representation in $\Omega \times \mathbf{T}_+$:

$$\begin{aligned} u_r &= \left\{\square_1^0\square_4^0 - [(\lambda + \mu - \alpha)\square_4^0 - 4\alpha^2]\partial_r(r^{-1} + \partial_r)\right\} F_r, \\ u_z &= -[(\lambda + \mu - \alpha)\square_4 - 4\alpha^2]\partial_z(r^{-1} + \partial_r)F_r, \\ \varphi_\theta &= -2\alpha\square_1^0\partial_z F_r, \end{aligned} \tag{2.66}$$

where the stress function F_r satisfies the field equation[5]

$$\square_1^0(\square_2^0\square_4^0 + 4\alpha^2\nabla_0^2)F_r + X_r = 0. \tag{2.67}$$

For $X_z = X_r = 0$, the relationship between the stress functions F_r, F_z takes the form

$$\square_1^0(\partial_r F_z - \partial_z F_r) = 0. \tag{2.68}$$

[4] Body moments Y_z are connected with the second axially-symmetric problem represented by the triples: $(\varphi_r, u_\theta, \varphi_z)$, (M_r, F_θ, M_z) and (Y_r, X_θ, Y_z). The appropriate equation containing Y_z is given in Subsection 2.3.8.

[5] The body moments Y_r are connected with the second axially-symmetric problem. If we take into account, apart from X_r, X_z, the body moments Y_θ, then, apart from the representations (2.63), (2.66) and equations (2.64), (2.67), we should include the representation in $\Omega \times \mathbf{T}_+$

$$u_r = 2\alpha\partial_z M_\theta, \qquad \varphi_r = \square_2^0 M_\theta, \qquad u_z = -2\alpha\left(r^{-1} + \partial_r\right) M_\theta$$

$$(\square_2^0\square_4^0 + 4\alpha^2\nabla_0^2)M_\theta + Y_\theta = 0.$$

For the initial-boundary value problem formulated in the framework of the first axially-symmetric problem without body loadings, it is sufficient to consider one of the two solving functions (F_r or F_z).

2.1.5 Method of potentials

In [49], we studied the vector equations (1.147) and (1.151)–(1.153) in relation to the axially-symmetric problem with the triple $(u_r, \varphi_\theta, u_z)$, which has already been discussed. In the problem, there appear a scalar potential Φ and vector potentials $\boldsymbol{\Psi} \stackrel{\text{def}}{=} (0, \Psi_\theta, 0)$, $\mathbf{H} \stackrel{\text{def}}{=} (H_r, 0, H_z)$.

From (1.147) we obtain in $\Omega \times \mathbf{T}_+$ the relations

$$\begin{aligned} u_r &= \partial_r \Phi - \partial_z \Psi_\theta, \\ u_z &= \partial_z \Phi + (r^{-1} + \partial_r)\Psi_\theta, \\ \varphi_\theta &= \partial_z H_r - \partial_r H_z. \end{aligned} \tag{2.69}$$

The equations for Φ, Ψ_θ, H_r and H_z are obtained from equations (1.152), (1.153). For Φ we obtain the equation

$$\nabla^2 \Box_1 \Phi + (r^{-1} + \partial_r) X_r + \partial_z X_z = 0. \tag{2.70}$$

In turn, for H_r and H_z we get

$$\begin{aligned} \nabla_0^2(\Box_2^0 \Box_4^0 + 4\alpha^2 \nabla_0^2) H_r &= -\Box_2^0 \partial_z Y_\theta + 2\alpha \partial_z (\partial_z X_r - \partial_r X_z), \\ \nabla^2(\Box_2 \Box_4 + 4\alpha^2 \nabla^2) H_z &= \Box_2 (r^{-1} + \partial_r) Y_\theta \\ &\quad - 2\alpha (r^{-1} + \partial_r)(\partial_z X_r - \partial_r X_z). \end{aligned} \tag{2.71}$$

The functions Φ, Ψ_θ, H_r and H_z must satisfy equations following from equations (1.151), which will enable us to simplify the representation (2.69). The condition $\operatorname{div} \boldsymbol{\Psi} = 0$ is an identity. From the condition $\operatorname{div} \mathbf{H} = 0$ we obtain the relation

$$\left(r^{-1} + \partial_r\right) H_r + \partial_z H_z = 0 \qquad \text{in } \Omega \times \mathbf{T}_+. \tag{2.72}$$

Let us introduce a new function H, assuming that

$$H_r = \partial_z H, \qquad H_z = -\left(r^{-1} + \partial_r\right) H. \tag{2.73}$$

Relation (2.72) is an identity, whereas from $(2.69)_3$ we obtain

$$\varphi_\theta = \nabla_0^2 H \qquad \text{in } \Omega \times \mathbf{T}_+. \tag{2.74}$$

Let us make the decomposition (1.148) for the vectors of body loadings, $\mathbf{X} \stackrel{\text{def}}{=} (X_r, 0, X_z)$ and $\mathbf{Y} \stackrel{\text{def}}{=} (0, Y_\theta, 0)$. In the problem, a scalar function ϑ and vectors $\boldsymbol{\chi} \stackrel{\text{def}}{=} (0, \chi_\theta, 0)$, $\boldsymbol{\eta} \stackrel{\text{def}}{=} (\eta_r, 0, \eta_z)$ will appear.

We obtain in $\Omega \times \mathbf{T}$:

$$\begin{aligned} X_r &= \partial_r \vartheta - \partial_z \chi_\theta, \\ X_z &= \partial_z \vartheta + \left(r^{-1} + \partial_r\right) \chi_\theta, \\ Y_\theta &= \partial_z \eta_r - \partial_r \eta_z. \end{aligned} \tag{2.75}$$

The condition $\operatorname{div}\boldsymbol{\chi} = 0$ is an identity, whereas the condition $\operatorname{div}\boldsymbol{\eta} = 0$ can be transformed into an identity by introducing a new function η:

$$\eta_r = \partial_z \eta, \qquad \eta_z = -\left(r^{-1} + \partial_r\right)\eta.$$

Formula $(2.75)_3$ takes now in $\Omega \times \mathbf{T}$ the form

$$Y_\theta = \nabla_0^2 \eta. \tag{2.76}$$

Let us go back to the vector equations (1.151), from which we obtain the following equations in $\Omega \times \mathbf{T}_+$:

$$\begin{aligned} &\Box_1^0 \partial_r \Phi - \Box_2^0 \partial_z \Psi_\theta - 2\alpha \nabla_0^2 H_r + \partial_r \vartheta - \partial_z \chi_\theta = 0, \\ &\Box_1 \partial_z \Phi + \Box_2 \left(r^{-1} + \partial_r\right) \Psi_\theta - 2\alpha \nabla^2 H_z + \partial_z \vartheta + \left(r^{-1} + \partial_r\right) \chi_\theta = 0, \\ &\Box_4^0 (\partial_z H_r - \partial_r H_z) - 2\alpha \nabla_0^2 \Psi_\theta + \nabla_0^2 \eta = 0. \end{aligned} \tag{2.77}$$

We eliminate the function Ψ_θ from further considerations in such a way that, using equations $(2.77)_3$ and (2.73), we assume

$$\Psi_\theta = \frac{1}{2\alpha}(\Box_4^0 H + \eta). \tag{2.78}$$

The representation (2.69) takes the form (2.74) and the form

$$\begin{aligned} u_r &= \partial_r \Phi - \frac{1}{2\alpha} \partial_z (\Box_4^0 H + \eta), \\ u_z &= \partial_z \Phi + \frac{1}{2\alpha} \left(r^{-1} + \partial_r\right) (\Box_4^0 H + \eta). \end{aligned} \tag{2.79}$$

The field equations for the functions Φ, H in $\Omega \times \mathbf{T}_+$ are obtained from equations $(2.77)_{1,2}$ by means of (2.73) and (2.78):

$$\begin{aligned} &2\alpha \partial_r (\Box_1 \Phi + \vartheta) - \partial_z \left[(\Box_2^0 \Box_4^0 + 4\alpha^2 \nabla_0^2) H + \Box_2^0 \eta + 2\alpha \chi_\theta\right] = 0, \\ &2\alpha \partial_z (\Box_1 \Phi + \vartheta) \\ &\qquad + \left(r^{-1} + \partial_r\right) \left[(\Box_2^0 \Box_4^0 + 4\alpha^2 \nabla_0^2) H + \Box_2^0 \eta + 2\alpha \chi_\theta\right] = 0. \end{aligned} \tag{2.80}$$

Equations (2.80) lead to the following wave equations in $\Omega \times \mathbf{T}_+$:

$$\begin{aligned} &\Box_1 \Phi + \vartheta = 0, \\ &(\Box_2^0 \Box_4^0 + 4\alpha^2 \nabla_0^2) H + \Box_2^0 \eta + 2\alpha \chi_\theta = 0. \end{aligned} \tag{2.81}$$

Finally, from equations (2.80), preserving body forces X_r, Y_θ, X_z, we obtain equation (2.70) and the equation

$$\nabla_0^2 (\Box_2^0 \Box_4^0 + 4\alpha^2 \nabla_0^2) H + \Box_2^0 Y_\theta + 2\alpha (\partial_r X_z - \partial_z X_r) = 0. \tag{2.82}$$

Equation (2.82) also follows from (2.71) and (2.73).

2.1.6 Stress equations of motion of Ignaczak type

In Subsection 1.1.5 we mentioned that Ignaczak [116] proved that solving the stress initial-boundary value problem (SIBVP) in linear micropolar elastodynamics is equivalent to solving the system of two coupled tensor stress equations with appropriately formulated initial-boundary conditions in terms of stresses. In this Subsection, we will consider the SIBVP for the plane and axially-symmetric problems of micropolar elastodynamics[6].

Let us consider the first axially-symmetric problem represented by the vector in $\Omega \times \mathbf{T}_+$:

$$\mathbf{u} \stackrel{\text{def}}{=} (u_r, 0, u_z), \qquad \boldsymbol{\varphi} \stackrel{\text{def}}{=} (0, \varphi_\theta, 0). \tag{2.83}$$

The system od field equations will be arranged in the following groups:

— the equations of motion in $\Omega \times \mathbf{T}_+$,

$$\begin{aligned} &\partial_r \sigma_{rr} + \partial_z \sigma_{zr} + r^{-1}(\sigma_{rr} - \sigma_{\theta\theta}) + X_r = \rho \ddot{u}_r, \\ &\left(r^{-1} + \partial_r\right) \sigma_{rz} + \partial_z \sigma_{zz} + X_z = \rho \ddot{u}_z, \\ &2\sigma_{[zr]} + \partial_r \mu_{r\theta} + \partial_z \mu_{z\theta} + 2r^{-1} \mu_{(r\theta)} + Y_\theta = J \ddot{\varphi}_\theta, \end{aligned} \tag{2.84}$$

where

$$\sigma_{[zr]} = \tfrac{1}{2}(\sigma_{zr} - \sigma_{rz}), \qquad \mu_{(r\theta)} = \tfrac{1}{2}(\mu_{r\theta} + \mu_{\theta r}).$$

The stress fields are represented by the matrices in $\Omega \times \mathbf{T}_+$,

$$\underset{\sim}{\sigma} \equiv \begin{bmatrix} \sigma_{rr} & 0 & \sigma_{rz} \\ 0 & \sigma_{\theta\theta} & 0 \\ \sigma_{zr} & 0 & \sigma_{zz} \end{bmatrix}, \qquad \underset{\sim}{\mu} \equiv \begin{bmatrix} 0 & \mu_{r\theta} & 0 \\ \mu_{\theta r} & 0 & \mu_{\theta z} \\ 0 & \mu_{z\theta} & 0 \end{bmatrix}; \tag{2.85}$$

— the geometric relations in $\Omega \times \mathbf{T}_+$,

$$\begin{aligned} &\gamma_{rr} = \partial_r u_r, \qquad \gamma_{\theta\theta} = r^{-1} u_r, \qquad \gamma_{zz} = \partial_z u_z, \\ &\gamma_{rz} = \partial_r u_z + \varphi_\theta, \qquad \gamma_{zr} = \partial_z u_r - \varphi_\theta, \\ &\kappa_{r\theta} = \partial_r \varphi_\theta, \qquad \kappa_{\theta r} = -r^{-1} \varphi_\theta, \qquad \kappa_{z\theta} = \partial_z \varphi_\theta, \end{aligned} \tag{2.86}$$

where the matrices

[6] The STIVBP's have been discussed in the following papers: Al-Hasan & Dyszlewicz "The coupled dynamical micropolar problem of thermoelasticity. Stress equations of motion for the axisymmetric problem" and "Plane state of deformation in the coupled dynamic micropolar problem of thermoelasticity. Stress equations of Ignaczak type", Reports No. 93 and 94 (1998), Institute of Mathematics, Wrocław University of Technology, Poland.

$$\underset{\sim}{\gamma} \equiv \begin{bmatrix} \gamma_{rr} & 0 & \gamma_{rz} \\ 0 & \gamma_{\theta\theta} & 0 \\ \gamma_{zr} & 0 & \gamma_{zz} \end{bmatrix}, \qquad \underset{\sim}{\kappa} \equiv \begin{bmatrix} 0 & \kappa_{r\theta} & 0 \\ \kappa_{\theta r} & 0 & 0 \\ 0 & \kappa_{z\theta} & 0 \end{bmatrix} \tag{2.87}$$

describe the strain fields in $\Omega \times \mathbf{T}_+$;

— the constitutive relations in $\Omega \times \mathbf{T}$,

$$\begin{aligned} &\sigma_{rr} = 2\mu\gamma_{rr} + \lambda e, \qquad \sigma_{\theta\theta} = 2\mu\gamma_{\theta\theta} + \lambda e, \qquad \sigma_{zz} = 2\mu\gamma_{zz} + \lambda e, \\ &(\sigma_{rz}, \sigma_{zr}) = (\mu + \alpha)(\gamma_{rz}, \gamma_{zr}) + (\mu - \alpha)(\gamma_{zr}, \gamma_{rz}), \\ &(\mu_{r\theta}, \mu_{\theta r}) = (\gamma + \varepsilon)(\kappa_{r\theta}, \kappa_{\theta r}) + (\gamma - \varepsilon)(\kappa_{\theta r}, \kappa_{r\theta}), \\ &\mu_{z\theta} = (\gamma + \varepsilon)\kappa_{z\theta}, \qquad \mu_{\theta z} = \frac{\gamma - \varepsilon}{\gamma + \varepsilon}\mu_{z\theta}, \end{aligned} \tag{2.88}$$

where $e = \gamma_{rr} + \gamma_{\theta\theta} + \gamma_{zz}$.

To the system of equations (2.84), (2.86) and (2.88) we add initial-boundary conditions in $\overline{\Omega} \times \mathbf{T}$:

— the boundary conditions on $\partial\Omega \times \mathbf{T}$,

$$\begin{aligned} &n_r\sigma_{rr} + n_z\sigma_{zr} = p_r, \qquad n_r\sigma_{rz} + n_z\sigma_{zz} = p_z, \\ &n_r\mu_{r\theta} + n_z\mu_{z\theta} = m_\theta, \end{aligned} \tag{2.89}$$

where p_r, p_z, $m_\theta : \partial\Omega \times \mathbf{T} \to \mathbb{R}$ are given functions, n_r, n_z are the two components of a unit outward normal vector $\partial\Omega$, expressed in terms of the local basis connected with the cylindrical system of coordinates;

— the initial conditions in $\Omega \times \{0\}$,

$$(u_r, \varphi_\theta, u_z) = (f_r, f_\theta, f_z), \qquad (\dot{u}_r, \dot{\varphi}_\theta, \dot{u}_z) = (g_r, g_\theta, g_z), \tag{2.90}$$

where the functions f_r, f_z, f_θ, g_r, g_z, $g_\theta : \Omega \to \mathbb{R}$ are given.

Let us now define what should be understood by a solution of the stress problem. By stress fields corresponding to a solution of the stress problem one should understand a pair $(\underset{\sim}{\sigma}, \underset{\sim}{\mu})$ described by matrices (2.85), which has such properties that there exists a quadruple $(\mathbf{u}, \boldsymbol{\varphi}, \underset{\sim}{\gamma}, \underset{\sim}{\kappa})$ described by matrices (2.83) and (2.87), respectively, and such that the set of functions $\{\mathbf{u}, \boldsymbol{\varphi}, \underset{\sim}{\gamma}, \underset{\sim}{\kappa}, \underset{\sim}{\sigma}, \underset{\sim}{\mu}\}$ is a solution of equations (2.84), (2.86) and (2.88)–(2.90).

Let us suppose now that we know the deformations $\underset{\sim}{\gamma}$, $\underset{\sim}{\kappa}$ from (2.87) and we want to find in $\overline{\Omega} \times \mathbf{T}$ the triple $(u_r, \varphi_\theta, u_z)$. This can be done by integrating the geometric relations (2.86) with respect to the position variables r, z, from which we obtain:

$$u_r = r\gamma_{\theta\theta}, \qquad \varphi_\theta = -r\kappa_{\theta r}, \tag{2.91}$$

$$\partial_z u_z = \gamma_{zz}, \qquad \partial_r u_z = \gamma_{rz} + r\kappa_{\theta r}. \tag{2.92}$$

From (2.92) we have

$$u_z = \int_{z_0}^{z} \gamma_{zz}\, dz + C_1(r,t) = \int_{r_0}^{r} (\gamma_{rz} + r\kappa_{\theta r})\, dr + C_2(z,t), \tag{2.93}$$

where $(r_0, z_0) \in \Omega$ — a fixed point, $C_1(r,t)$, $C_2(z,t)$ — arbitrary functions of the variables (r,t) and (z,t), respectively, which have to be determined. Here, it should be taken into account that when we substitute the geometric relations (2.86) into the compatibility equations[7], we must ensure that these equations hold in $\Omega \times \mathbf{T}$, here in the form given in [43]:

$$\begin{aligned}
&\kappa_{z\theta} + r\partial_z \kappa_{\theta r} = 0,\\
&\kappa_{r\theta} + \partial_r (r\kappa_{\theta r}) = 0,\\
&\partial_r (r\gamma_{\theta\theta}) - \gamma_{rr} = 0,\\
&\kappa_{z\theta} + \partial_r \gamma_{zz} - \partial_z \gamma_{rz} = 0,\\
&\partial_z \gamma_{\theta\theta} - r^{-1}\gamma_{zr} + \kappa_{\theta r} = 0.
\end{aligned} \tag{2.94}$$

In the stress equations of motion problem, the triple $(u_r, \varphi_\theta, u_z)$ is determined in such a way that we integrate the equations of motion (2.84) over the time t and take into account the initial conditions (2.90). Then we obtain in $\Omega \times \mathbf{T}$:

$$\begin{aligned}
u_r &= g_r t + f_r + \rho^{-1}(t * R_r),\\
\varphi_\theta &= g_\theta t + f_\theta + J^{-1}(t * R_\theta),\\
u_z &= g_z t + f_z + \rho^{-1}(t * R_z),
\end{aligned} \tag{2.95}$$

where

$$\begin{aligned}
R_r &= \partial_r \sigma_{rr} + \partial_z \sigma_{zr} + r^{-1}(\sigma_{rr} - \sigma_{\theta\theta}) + X_r = \hat{R}_r + X_r,\\
R_\theta &= 2\sigma_{[zr]} + \partial_r \mu_{r\theta} + \partial_z \mu_{z\theta} + 2r^{-1}\mu_{(r\theta)} + Y_\theta = \hat{R}_\theta + Y_\theta,\\
R_z &= \left(r^{-1} + \partial_r\right)\sigma_{rz} + \partial_z \sigma_{zz} + X_z = \hat{R}_z + X_z,
\end{aligned}$$

the asterisk stands for the convolution product on the t-axis (see Mikusiński [156]),

$$t * f(r,z,t) = \int_0^t (t-\tau) f(r,z,\tau)\, d\tau.$$

Obviously, the triple $(u_r, \varphi_\theta, u_z)$ given by formulae (2.91) and (2.93) and by formula (2.95) must be the same triple.

Let us turn to the derivation of the stress equations of motion problem. From equations (2.84), by use of (2.86), we obtain the following equations:

[7] See Sandru [206]; Günther [104]; Kozák & Szeidl [134].

$$\begin{aligned}
&\rho^{-1}\partial_r R_r = \ddot{\gamma}_{rr}, \qquad \rho^{-1}r^{-1}R_r = \ddot{\gamma}_{\theta\theta}, \qquad \rho^{-1}\partial_z R_z = \ddot{\gamma}_{zz},\\
&J^{-1}\partial_r R_\theta = \ddot{\kappa}_{r\theta}, \qquad -J^{-1}r^{-1}R_\theta = \ddot{\kappa}_{\theta r}, \qquad J^{-1}\partial_z R_\theta = \ddot{\kappa}_{z\theta}, \qquad (2.96)\\
&\rho^{-1}\partial_r R_z + J^{-1}R_\theta = \ddot{\gamma}_{rz}, \qquad \rho^{-1}\partial_z R_r - J^{-1}R_\theta = \ddot{\gamma}_{zr}.
\end{aligned}$$

Let us write down the constitutive relations (2.88) in an equivalent form:

$$\begin{aligned}
&2\mu\gamma_{rr} = \sigma_{rr} - \lambda e, \qquad 2\mu\gamma_{\theta\theta} = \sigma_{\theta\theta} - \lambda e, \qquad 2\mu\gamma_{zz} = \sigma_{zz} - \lambda e,\\
&\gamma_{rz} = \frac{1}{2\mu}\sigma_{(rz)} + \frac{1}{2\alpha}\sigma_{[rz]}, \qquad \gamma_{zr} = \frac{1}{2\mu}\sigma_{(rz)} - \frac{1}{2\alpha}\sigma_{[rz]},\\
&\kappa_{r\theta} = \frac{1}{2\gamma}\mu_{(r\theta)} + \frac{1}{2\varepsilon}\mu_{[r\theta]}, \qquad \kappa_{\theta r} = \frac{1}{2\gamma}\mu_{(r\theta)} - \frac{1}{2\varepsilon}\mu_{[r\theta]}, \qquad (2.97)\\
&\kappa_{z\theta} = \frac{1}{\gamma+\varepsilon}\mu_{z\theta},
\end{aligned}$$

where $e = \dfrac{1}{2\mu+3\lambda}(\sigma_{rr} + \sigma_{\theta\theta} + \sigma_{zz})$.

Equations (2.97) are used to transform (2.96) into

$$\begin{aligned}
&2\hat{c}_2^2\partial_r R_r - \ddot{\sigma}_{rr} + \lambda\ddot{e} = 0,\\
&2\hat{c}_2^2 r^{-1} R_r - \ddot{\sigma}_{\theta\theta} + \lambda\ddot{e} = 0,\\
&2\hat{c}_2^2\partial_z R_z - \ddot{\sigma}_{zz} + \lambda\ddot{e} = 0,\\
&J^{-1}\partial_r R_\theta - \frac{1}{2\gamma}\ddot{\mu}_{(r\theta)} - \frac{1}{2\varepsilon}\ddot{\mu}_{[r\theta]} = 0,\\
&J^{-1}r^{-1}R_\theta + \frac{1}{2\gamma}\ddot{\mu}_{(r\theta)} - \frac{1}{2\varepsilon}\ddot{\mu}_{[r\theta]} = 0, \qquad (2.98)\\
&c_4^2\partial_z R_\theta - \ddot{\mu}_{z\theta} = 0,\\
&\rho^{-1}\partial_r R_z + J^{-1}R_\theta - \frac{1}{2\mu}\ddot{\sigma}_{(rz)} - \frac{1}{2\alpha}\ddot{\sigma}_{[rz]} = 0,\\
&\rho^{-1}\partial_z R_r - J^{-1}R_\theta - \frac{1}{2\mu}\ddot{\sigma}_{(rz)} + \frac{1}{2\alpha}\ddot{\sigma}_{[rz]} = 0,
\end{aligned}$$

where

$$\hat{c}_2 = \sqrt{\frac{\mu}{\rho}}, \qquad c_4 = \sqrt{\frac{\gamma+\varepsilon}{J}},$$

$$\ddot{e} = \frac{1}{2\mu+3\lambda}(\ddot{\sigma}_{rr} + \ddot{\sigma}_{\theta\theta} + \ddot{\sigma}_{zz}).$$

The ultimate form of the stress equations of motion problem is the following: the field equations (2.98), the boundary conditions (2.89) and the initial conditions in $\Omega \times \{0\}$:

$$\begin{aligned} \underset{\sim}{\sigma} = \underset{\sim}{\sigma}^0 &\equiv \begin{bmatrix} \sigma^0_{rr} & 0 & \sigma^0_{rz} \\ 0 & \sigma^0_{\theta\theta} & 0 \\ \sigma^0_{zr} & 0 & \sigma^0_{zz} \end{bmatrix}, \qquad \underset{\sim}{\mu} = \underset{\sim}{\mu}^0 \equiv \begin{bmatrix} 0 & \mu^0_{r\theta} & 0 \\ \mu^0_{\theta r} & 0 & \mu^0_{\theta z} \\ 0 & \mu^0_{z\theta} & 0 \end{bmatrix}, \\ \dot{\underset{\sim}{\sigma}} = \dot{\underset{\sim}{\sigma}}^0 &\equiv \begin{bmatrix} \dot\sigma^0_{rr} & 0 & \dot\sigma^0_{rz} \\ 0 & \dot\sigma^0_{\theta\theta} & 0 \\ \dot\sigma^0_{zr} & 0 & \dot\sigma^0_{zz} \end{bmatrix}, \qquad \dot{\underset{\sim}{\mu}} = \dot{\underset{\sim}{\mu}}^0 \equiv \begin{bmatrix} 0 & \dot\mu^0_{r\theta} & 0 \\ \dot\mu^0_{\theta r} & 0 & \dot\mu_{\theta z} \\ 0 & \dot\mu^0_{z\theta} & 0 \end{bmatrix}, \end{aligned} \tag{2.99}$$

where $\underset{\sim}{\sigma}$, $\underset{\sim}{\mu}$, $\dot{\underset{\sim}{\sigma}}$, $\dot{\underset{\sim}{\mu}}$ are obtained from matrices (2.85), whereas $\underset{\sim}{\sigma}^0$, $\underset{\sim}{\mu}^0$, and $\dot{\underset{\sim}{\sigma}}^0$, $\dot{\underset{\sim}{\mu}}^0$ are implied by the initial conditions (2.90), the geometric relations (2.86) and the constitutive relations (2.88), i.e.

$$\sigma^0_{rr} = 2\mu\partial_r f_r + \lambda e^0, \qquad \sigma^0_{\theta\theta} = 2\mu r^{-1} f_r + \lambda e^0, \qquad \sigma^0_{zz} = 2\mu\partial_z f_z + \lambda e^0,$$

$$\sigma^0_{rz} = (\mu+\alpha)\gamma^0_{rz} + (\mu-\alpha)\gamma^0_{zr}, \qquad \sigma^0_{zr} = (\mu+\alpha)\gamma^0_{zr} + (\mu-\alpha)\gamma^0_{rz},$$

$$\mu^0_{r\theta} = (\gamma+\varepsilon)\partial_r f_\theta - (\gamma-\varepsilon)r^{-1} f_\theta, \qquad \mu^0_{\theta r} = -(\gamma+\varepsilon)r^{-1} f_\theta + (\gamma-\varepsilon)\partial_r f_\theta,$$

$$\mu^0_{z\theta} = (\gamma+\varepsilon)\partial_z f_\theta, \qquad \mu^0_{\theta z} = (\gamma-\varepsilon)\partial_z f_\theta,$$

where

$$\gamma^0_{rz} = \partial_r f_z + f_\theta, \qquad \gamma^0_{zr} = \partial_z f_r - f_\theta, \qquad e^0 = \left(r^{-1} + \partial_r\right) f_r + \partial_z f_z$$

and

$$\dot\sigma^0_{rr} = 2\mu\partial_r g_r + \lambda\dot e^0, \qquad \dot\sigma^0_{\theta\theta} = 2\mu r^{-1} g_r + \lambda\dot e^0, \qquad \dot\sigma^0_{zz} = 2\mu\partial_z g_z + \lambda\dot e^0,$$

$$\dot\sigma^0_{rz} = (\mu+\alpha)\dot\gamma^0_{rz} + (\mu-\alpha)\dot\gamma^0_{zr},$$

$$\dot\sigma^0_{zr} = (\mu+\alpha)\dot\gamma^0_{zr} + (\mu-\alpha)\dot\gamma^0_{rz},$$

$$\dot\mu^0_{r\theta} = (\gamma+\varepsilon)\partial_r g_\theta - (\gamma-\varepsilon)r^{-1} g_\theta,$$

$$\dot\mu^0_{\theta r} = -(\gamma+\varepsilon)r^{-1} g_\theta + (\gamma-\varepsilon)\partial_r g_\theta,$$

$$\dot\mu^0_{z\theta} = (\gamma+\varepsilon)\partial_z g_\theta, \qquad \dot\mu^0_{\theta z} = (\gamma-\varepsilon)\partial_z g_\theta,$$

where

$$\dot\gamma^0_{rz} = \partial_r g_z + g_\theta, \qquad \dot\gamma^0_{zr} = \partial_z g_r - g_\theta, \qquad \dot e^0 = \left(r^{-1} + \partial_r\right) g_r + \partial_z g_z.$$

The formulations given by relations (2.84), (2.86) and (2.88)–(2.90), and by relations (2.98), (2.89) and (2.99), are equivalent[8].

Theorem 2.1.1. *(Characterization of the traction problem in terms of stress).*

Let the system of functions $\mathbf{u}$, $\boldsymbol{\varphi}$, $\underset{\sim}{\sigma}$, $\underset{\sim}{\mu}$, $\underset{\sim}{\gamma}$ *and* $\underset{\sim}{\kappa}$ *be sufficiently smooth in* $\overline{\Omega} \times \mathbf{T}$. *More precisely, assume that:*

(a) p_r, p_z *and* m_θ *are continuous in time and piecewise regular on* $\partial\Omega \times \mathbf{T}$,

(b) f_r, f_θ, f_z, g_r, g_θ *and* g_z *are continuous in* $\overline{\Omega}$ *and of class* C^2 *in* Ω,

(c) X_r, X_z *and* Y_θ *are continuous in* $\overline{\Omega} \times \mathbf{T}$ *and of class* $C^{2,0}$ *in* $\Omega \times \mathbf{T}$,

(d) u_r, u_z *and* φ_θ *are of class* C^2 *in* $\Omega \times \mathbf{T}$, *and of class* C^1 *in* $\overline{\Omega} \times \mathbf{T}$,

(e) $\underset{\sim}{\sigma}$ *and* $\underset{\sim}{\mu}$ *are of class* $C^{3,2}$ *in* $\Omega \times \mathbf{T}$,

(f) $\underset{\sim}{\sigma}$, $\underset{\sim}{\mu}$, $\dot{\underset{\sim}{\sigma}}$, $\dot{\underset{\sim}{\mu}}$, $\hat{R}_r$, $\hat{R}_\theta$ *and* $\hat{R}_z$ *are continuous in* $\overline{\Omega} \times \mathbf{T}$.

Then the fields $\underset{\sim}{\sigma}$ *and* $\underset{\sim}{\mu}$ *correspond to a solution of the traction problem if and only if the equations* (2.98), *the initial conditions* (2.99) *as well as the traction boundary conditions* (2.89) *are satisfied.*

Proof. Necessity. Follows from the fact that equations (2.84), (2.86) and (2.88)–(2.90) imply the field equations (2.98), the traction boundary conditions (2.89) as well as the initial conditions (2.99).

Sufficiency. Assume that the stress fields $\underset{\sim}{\sigma}$ and $\underset{\sim}{\mu}$ correspond to a solution of the traction problem. Then the strain fields $\underset{\sim}{\gamma}$ and $\underset{\sim}{\kappa}$ (see (2.87)) are determined from (2.97). Next, determine the displacement field $\mathbf{u}$ and the rotation field $\boldsymbol{\varphi}$ from (2.95), and observe that (2.95), (2.97) and (2.98) imply relations (2.86). Since the system of functions $\mathbf{u}$, $\boldsymbol{\varphi}$, $\underset{\sim}{\sigma}$ and $\underset{\sim}{\mu}$ satisfies the equations of motion (2.84) as well as the initial conditions (2.90) if and only if (2.95) holds[9], we conclude that the set of functions $\mathbf{u}$, $\boldsymbol{\varphi}$, $\underset{\sim}{\gamma}$, $\underset{\sim}{\kappa}$, $\underset{\sim}{\sigma}$ and $\underset{\sim}{\mu}$ is a solution of the problem given by relations (2.84), (2.86) and (2.88)–(2.90). The proof is complete.

To complete this subsection, note that by integrating the stress equations of motion (2.98) twice with respect to time over the interval $[0, t]$, and taking into account the initial conditions (2.99), we arrive at the alternative form of stress equations of motion in $\Omega \times \mathbf{T}$:

---

[8] In the sequel we will sometimes use the abbreviation SEMP (Stress Equations of Motion Problem) for the stress equations of motion problem of Ignaczak type. SEMP may also be considered with mixed boundary conditions (see Gurtin [105, p. 220]). Mixed boundary conditions have the form:

(a) (2.89) on $\partial\Omega_\sigma \times \mathbf{T}$ and

(b) $\mathbf{u} = \hat{\mathbf{u}}$, $\boldsymbol{\varphi} = \hat{\boldsymbol{\varphi}}$ on $\partial\Omega_u \times \mathbf{T}$, where the vectors $\hat{\mathbf{u}}$, $\hat{\boldsymbol{\varphi}} : \partial\Omega_u \times \mathbf{T} \to \mathbb{R}^3$ are given, $\partial\Omega = \partial\Omega_\sigma \cup \partial\Omega_u$, and $\partial\Omega_\sigma \cap \partial\Omega_u = \emptyset$. Moreover, $\mathbf{u}$ and $\boldsymbol{\varphi}$ are expressed in terms of $\underset{\sim}{\sigma}$ and $\underset{\sim}{\mu}$ by means of formulae (2.95). When $\partial\Omega_\sigma = \partial\Omega$ ($\partial\Omega_u = \emptyset$) in equations (**a**) and (**b**), we refer to a traction problem. When $\partial\Omega_u = \partial\Omega$ ($\partial\Omega_\sigma = \emptyset$) in equations (**a**) and (**b**), we refer to a displacement-rotation problem.

[9] See Iesan [111].

$$
\begin{aligned}
&t * \rho^{-1}\partial_r R_r - \frac{1}{2\mu}(\sigma_{rr} - \lambda e) + \partial_r(g_r t + f_r) = 0,\\
&t * \rho^{-1} r^{-1} R_r - \frac{1}{2\mu}(\sigma_{\theta\theta} - \lambda e) + r^{-1}(g_r t + f_r) = 0,\\
&t * \rho^{-1}\partial_z R_z - \frac{1}{2\mu}(\sigma_{zz} - \lambda e) + \partial_z(g_z t + f_z) = 0,\\
&t * J^{-1}\partial_r R_\theta - \frac{1}{2\gamma}\mu_{(r\theta)} - \frac{1}{2\varepsilon}\mu_{[r\theta]} + \partial_r(g_\theta t + f_\theta) = 0,\\
&t * J^{-1} r^{-1} R_\theta + \frac{1}{2\gamma}\mu_{(r\theta)} - \frac{1}{2\varepsilon}\mu_{[r\theta]} + r^{-1}(g_\theta t + f_\theta) = 0,\\
&t * J^{-1}\partial_z R_\theta - \frac{1}{\gamma+\varepsilon}\mu_{z\theta} + \partial_z(g_\theta t + f_\theta) = 0,\\
&t * \rho^{-1}\partial_r R_z + t * J^{-1} R_\theta - \frac{1}{2\mu}\sigma_{(zr)} - \frac{1}{2\alpha}\sigma_{[rz]}\\
&\qquad + \partial_r(g_z t + f_z) + g_\theta t + f_\theta = 0,\\
&t * \rho^{-1}\partial_z R_r - t * J^{-1} R_\theta - \frac{1}{2\mu}\sigma_{(rz)} + \frac{1}{2\alpha}\sigma_{[rz]}\\
&\qquad + \partial_z(g_r t + f_r) - g_\theta t - f_\theta = 0.
\end{aligned}
\tag{2.100}
$$

The system (2.100) incorporates the initial conditions (2.99) explicitly in the field equations; it is a system of integro-differential equations.

Theorem 2.1.2. *Let $\underset{\sim}{\sigma}$ and $\underset{\sim}{\mu}$ (see* (2.85)*) be sufficiently smooth fields in $\overline{\Omega} \times \mathbf{T}$. Then $\underset{\sim}{\sigma}$ and $\underset{\sim}{\mu}$ are stress fields corresponding to a solution of the traction problem if and only if the integro-differential equations* (2.100) *as well as the traction boundary conditions* (2.89) *hold true.*

Proof. By taking into account Theorem 2.1.1 it is sufficient to prove equivalence of the formulation described by (2.98), (2.99) and (2.89) with the formulation described by (2.100) and (2.89). First suppose $\underset{\sim}{\sigma}$ and $\underset{\sim}{\mu}$ meet (2.100). Define $\underset{\sim}{\gamma}$ and $\underset{\sim}{\kappa}$ through (2.97) as well as $\mathbf{u}$ and $\boldsymbol{\varphi}$ through (2.95). Thus, and by Theorem 2.1.1, $\mathbf{u}$, $\boldsymbol{\varphi}$, $\underset{\sim}{\gamma}$, $\underset{\sim}{\kappa}$, $\underset{\sim}{\sigma}$ and $\underset{\sim}{\mu}$ are solution of the traction problem.

Necessity. The relations (2.98), (2.99) and (2.89), after integrating (2.98) twice with respect to time and taking into account the initial conditions (2.99), imply the field equations (2.100) and the traction boundary conditions (2.89).

Sufficiency. The equations (2.100) differentiated twice with respect to time imply the initial conditions (2.99) as well as the field equations (2.98), and the proof is complete.

Theorem 2.1.1 can be extended to a mixed problem.

Theorem 2.1.3. *(Characterization of the mixed problem in terms of stress). Let the system of functions* $\mathbf{u}$, $\boldsymbol{\varphi}$, $\underset{\sim}{\sigma}$, $\underset{\sim}{\mu}$, $\underset{\sim}{\gamma}$ *and* $\underset{\sim}{\kappa}$ *be sufficiently smooth in* $\overline{\Omega} \times \mathbf{T}$. *Then the stress fields* $\underset{\sim}{\sigma}$ *and* $\underset{\sim}{\mu}$ *correspond to a solution of the mixed problem if and only if the equations* (2.98), *the initial conditions* (2.99) *as well as the mixed boundary conditions:*

$$n_r\sigma_{rr} + n_z\sigma_{zr} = p_r, \qquad n_r\sigma_{rz} + n_z\sigma_{zz} = p_z,$$

$$n_r\mu_{r\theta} + n_z\mu_{z\theta} = m_\theta \qquad \text{on } \partial\Omega_\sigma \times \mathbf{T}$$

and

$$g_r t + f_r + \rho^{-1}(t * R_r) = \hat{u}_r, \qquad g_\theta t + f_\theta + J^{-1}(t * R_\theta) = \hat{\varphi}_\theta,$$

$$g_z t + f_z + \rho^{-1}(t * R_z) = \hat{u}_z \qquad \text{on } \partial\Omega_u \times \mathbf{T}$$

hold[10].

At the end of the discussion of the stress equations of motion problem we shall make several remarks.

Remark 1. In the SEMP, one does not need to care about the compatibility equations (2.94) since equations (2.100) imply equations (2.94).

Remark 2. In the SEMP, the number N of field equations (2.98) is the same as the number of components of the strain state or the number of independent components of the stress state, or the number of equations of motion and compatibility equations taken together. In the problem considered, $N = 8$. This result (parallel to the result from Subsection 1.2.9) is correct for plane, axially-symmetric and three-dimensional problems (one tensor equation corresponds to the number 9).

Remark 3. Some initial-boundary value problems of micropolar coupled thermoelasticity may also be formulated in terms of the system $(\underset{\sim}{\sigma}, \underset{\sim}{\mu}, \theta)$ (where θ is the temperature field) instead of the system $(\mathbf{u}, \boldsymbol{\varphi}, \theta)$. In other words, the thermoelastic process $(\mathbf{u}, \boldsymbol{\varphi}, \underset{\sim}{\gamma}, \underset{\sim}{\kappa}, \underset{\sim}{\sigma}, \underset{\sim}{\mu}, \theta)$ can be described in a natural way by means of the triple $(\underset{\sim}{\sigma}, \underset{\sim}{\mu}, \theta)$, similar to the case of the pair $(\underset{\sim}{\sigma}, \underset{\sim}{\mu})$ in the isothermical state (cf. Subsection 1.1.5, footnote 6 on p. 123 and [3]).

Remark 4. The SEMP is meaningful for arbitrary initial stress fields and stress velocities. However, at the initial moment $t = 0$, the compatibility equations will not be satisfied. The SEMP so formulated goes beyond linear micropolar elastodynamics.

---

[10] The proofs of the given theorems follow from the work of Ignaczak [119]. See also Dyszlewicz [57], [63].

2.2 The first axially-symmetric problem. Elastostatics and thermoelastostatics

2.2.1 Fields of body loadings. Equations for displacements and rotations. Direct method and superposition method

In [49], we considered a system of equations in displacements and rotations for the triple $(u_r, \varphi_\theta, u_z)$, where r, $z \in \Omega$ for each θ, namely

$$\begin{aligned}
&(\mu+\alpha)\nabla_0^2 u_r + (\lambda+\mu-\alpha)\partial_r e - 2\alpha\partial_z\varphi_\theta + X_r = 0,\\
&(\mu+\alpha)\nabla^2 u_z + (\lambda+\mu-\alpha)\partial_z e + 2\alpha r^{-1}\partial_r(r\varphi_\theta) + X_z = 0,\\
&[(\gamma+\varepsilon)\nabla_0^2 - 4\alpha]\varphi_\theta + 2\alpha(\partial_z u_r - \partial_r u_z) + Y_\theta = 0,
\end{aligned} \tag{2.101}$$

with the boundary conditions on $\partial\Omega$:

$$u_r = \hat{u}_r, \qquad u_z = \hat{u}_z, \qquad \varphi_\theta = \hat{\varphi}_\theta, \tag{2.102}$$

where the functions $\hat{u}_r$, $\hat{u}_z$, $\hat{\varphi}_\theta : \partial\Omega \to \mathbb{R}$ are given. Let us introduce the substitution (2.14) for ζ_θ to the system of equations (2.101). Then we obtain:

$$\begin{aligned}
&\mu\nabla_0^2 u_r + (\lambda+\mu)\partial_r e + 2\alpha\partial_z\zeta_\theta + X_r = 0,\\
&\mu\nabla^2 u_z + (\lambda+\mu)\partial_z e - 2\alpha r^{-1}\partial_r(r\zeta_\theta) + X_z = 0,\\
&(\gamma+\varepsilon)\nabla_0^2(\partial_z u_r - \partial_r u_z) + 2Y_\theta = 2[(\gamma+\varepsilon)\nabla_0^2 - 4\alpha]\zeta_\theta.
\end{aligned} \tag{2.103}$$

In the sequel we adopt a solution of the form (2.17), where the triple $(u_r^0, \varphi_\theta^0, u_z^0)$ refers to the classical theory of elasticity. Such effect is obtained by taking $\zeta_\theta = \zeta_\theta^0 + \zeta_\theta'$, $Y_\theta = Y_\theta^0 + Y_\theta'$. For $\zeta_\theta^0 = 0$ $(Y_\theta^0 = 0)$, from equations $(2.103)_{1,2}$ we obtain the classical equations in Ω:

$$\begin{aligned}
&\mu\nabla_0^2 u_r^0 + (\lambda+\mu)\partial_r e^0 + X_r = 0,\\
&\mu\nabla_0^2 u_z^0 + (\lambda+\mu)\partial_z e^0 + X_z = 0,
\end{aligned} \tag{2.104}$$

where e^0 is the dilatation. From (2.102) we obtain the boundary conditions on $\partial\Omega$,

$$u_r^0 = \hat{u}_r, \qquad u_z^0 = \hat{u}_z. \tag{2.105}$$

From equation $(2.103)_3$ we obtain

$$2\mu\nabla_0^2\varphi_\theta^0 + (\partial_z X_r - \partial_r X_z) = 0, \tag{2.106}$$

which also follows from equations (2.104)[11]. The rotation φ_θ^0 is expressed by formula (2.19). Let us write down further equations for the pair (u_r^0, u_z^0),

[11] Extracting equation (2.106) from equation $(2.103)_3$ is important, and the fact that the body loadings X_r, X_z do not appear in equation $(2.103)_3$ is no obstacle for doing that. Such a procedure (characteristic for the method of superposition in the entire book) enables us to obtain the classical formulation (2.104), (2.105) (equation (2.106) will be satisfied identically) and provides information on the form of field equations and boundary conditions for the complementary solution.

$$
\begin{aligned}
\mu\nabla_0^2\nabla_0^2 u_r^0 &= -\nabla_0^2 X_r + \frac{\lambda+\mu}{2\mu+\lambda}\partial_r X, \\
\mu\nabla^2\nabla^2 u_z^0 &= -\nabla^2 X_z + \frac{\lambda+\mu}{2\mu+\lambda}\partial_z X,
\end{aligned}
\tag{2.107}
$$

where $X = r^{-1}\partial_r(rX_r) + \partial_z X_z$.

For the triple $(u_r', \zeta_\theta, u_z')$ we obtain the system of equations,

$$
\begin{aligned}
&\mu\nabla_0^2 u_r' + (\lambda+\mu)\partial_r e' + 2\alpha\partial_z\zeta_\theta + \hat{X}_r = 0, \\
&\mu\nabla_0^2 u_z' + (\lambda+\mu)\partial_z e' - 2\alpha r^{-1}\partial_r(r\zeta_\theta) + \hat{X}_z = 0, \\
&(\gamma+\varepsilon)\nabla_0^2(\partial_z u_r' - \partial_r u_z') - 2[(\gamma+\varepsilon)\nabla_0^2 - 4\alpha]\zeta_\theta + 2\overset{*}{Y}_\theta = 0,
\end{aligned}
\tag{2.108}
$$

where it holds

$$
\hat{X}_r = \hat{X}_z = 0, \qquad \overset{*}{Y}_\theta = Y_\theta - \frac{\gamma+\varepsilon}{2\mu}(\partial_z X_r - \partial_r X_z).
\tag{2.109}
$$

From the system of equations (2.108) we derive the following uncoupled equations in Ω:

— for the displacements u_r', u_z',

$$
\begin{aligned}
2\mu\nabla_0^2\nabla_0^2(l^2\nabla_0^2 - 1)u_r' &= -\partial_z\nabla_0^2\overset{*}{Y}_\theta - 2\nabla_0^2(l^2\nabla_0^2 - 1)\hat{X}_r \\
&\quad + \frac{\gamma+\varepsilon}{2\mu}\nabla_0^2\partial_z(\partial_z\hat{X}_r - \partial_r\hat{X}_z) + \frac{2(\lambda+\mu)}{\lambda+2\mu}(l^2\nabla_0^2 - 1)\partial_r\hat{X}, \\
2\mu\nabla_0^2\nabla_0^2(l^2\nabla_0^2 - 1)u_z' &= \nabla^2[r^{-1}\partial_r(r\overset{*}{Y}_\theta)] - 2\nabla^2(l^2\nabla^2 - 1)\hat{X}_z \\
&\quad - \frac{\gamma+\varepsilon}{2\mu}\nabla^2\{r^{-1}\partial_r[r(\partial_z\hat{X}_r - \partial_r\hat{X}_z)]\} \\
&\quad + \frac{2(\lambda+\mu)}{\lambda+2\mu}(l^2\nabla^2 - 1)\partial_z\hat{X},
\end{aligned}
\tag{2.110}
$$

where $\hat{X} = r^{-1}\partial_r(r\hat{X}_r) + \partial_z\hat{X}_z$;

— for the function ζ_θ,

$$
4\alpha(l^2\nabla_0^2 - 1)\zeta_\theta = \overset{*}{Y}_\theta - \frac{\gamma+\varepsilon}{2\mu}(\partial_z\hat{X}_r - \partial_r\hat{X}_z).
\tag{2.111}
$$

In further considerations we obtain a system of equations for the triple $(u_r', \varphi_\theta', u_z')$ by eliminating the function ζ_θ from the system of equations (2.108), which we transform into the following form:

$$
\begin{aligned}
&(\mu+\alpha)\nabla_0^2 u_r' + (\lambda+\mu-\alpha)\partial_r e' - 2\alpha\partial_z\varphi_\theta' + \hat{X}_r = 0, \\
&(\mu+\alpha)\nabla^2 u_z' + (\lambda+\mu-\alpha)\partial_z e' + 2\alpha r^{-1}\partial_r(r\varphi_\theta') + \hat{X}_z = 0, \\
&\left[(\gamma+\varepsilon)\nabla_0^2 - 4\alpha\right]\varphi_\theta' + 2\alpha(\partial_z u_r' - \partial_r u_z') + \overset{*}{Y}_\theta = 0.
\end{aligned}
\tag{2.112}
$$

The rotations φ'_θ satisfy the equation

$$\nabla_0^2(l^2\nabla_0^2 - 1)\varphi'_\theta = \frac{1}{2\mu}(\partial_z \hat{X}_r - \partial_r \hat{X}_z) - \frac{\mu+\alpha}{4\alpha\mu}\nabla_0^2 \overset{*}{Y}_\theta. \tag{2.113}$$

To the system of equations (2.112) we add the boundary conditions, which we obtain from the original boundary conditions for a micropolar body (2.102), in which the boundary effect has been taken into account:

$$u'_r = 0, \qquad u'_z = 0, \qquad \varphi'_\theta = \hat{\varphi}_\theta - \varphi^0_\theta, \tag{2.114}$$

where φ^0_θ results from the solution of the classical boundary value problem (2.104), (2.105). This method of solving boundary value problems is used in [43].

In [52], we considered the action of body loadings in an infinite space, represented only by the component X_z ($X_r = Y_\theta = 0$) of the form

$$X_z = \frac{P_0}{2\pi r}\delta(r)\delta(z). \tag{2.115}$$

Using equations (2.106), (2.107), (2.110) and (2.113) and applying the integral Fourier transformations (0.50) for $n = 1$ with respect to the variable z, and the integral Hankel transformations (0.51) with respect to the variable r ($\nu = 0$ or $\nu = 1$), we arrive at the following formulae:

$$\begin{aligned} u^0_r &= \frac{P_0}{16\pi\mu(1-\nu)}\cdot\frac{zr}{R^3}, \qquad \varphi^0_\theta = \frac{P_0}{8\pi\mu}\cdot\frac{r}{R^3}, \\ u^0_z &= \frac{P_0}{16\pi\mu}\left(\frac{3-4\nu}{1-\nu}r^2 + 4z^2\right)\frac{1}{R^3}, \end{aligned} \tag{2.116}$$

and the formulae

$$\begin{aligned} u'_r &= -\frac{P_0}{8\pi\mu\mu_0}\partial^2_{rz}\left(\frac{1-e^{-R/l}}{R}\right), \\ u'_z &= \frac{P_0}{8\pi\mu\mu_0}(r^{-1}\partial_r + \partial^2_r)\left(\frac{1-e^{-R/l}}{R}\right), \\ \varphi'_\theta &= -\varphi^0_\theta - \frac{P_0}{8\pi\mu}\partial_r\left(\frac{1-e^{-R/l}}{R}\right), \end{aligned} \tag{2.117}$$

where

$$R = \sqrt{r^2 + z^2}, \qquad \mu_0 = \frac{2\mu}{\gamma+\varepsilon}, \qquad l = \sqrt{\frac{(\gamma+\varepsilon)(\mu+\alpha)}{4\alpha\mu}},$$

and $\nu = \dfrac{\lambda}{2(\mu+\lambda)}$ is the Poisson ratio.

The limiting cases of the formulae for u_r, φ_θ, u_z referring to the theories (2), (4) and (6) (Fig. 1.1) have been discussed. We emphasize that in the method of direct integration of the system of equations (2.101) we make use of the uncoupled equations for the triple $(u_r, \varphi_\theta, u_z)$. Let us note that such uncoupled equations are obtained from equations (2.112), (2.110) and (2.113), in which it suffices to adopt the correspondence

$$(u'_r, \varphi'_\theta, u'_z) \mapsto (u_r, \varphi_\theta, u_z), \qquad (\hat{X}_r, \overset{*}{\hat{Y}}_\theta, \hat{X}_z) \mapsto (X_r, Y_\theta, X_z). \tag{2.118}$$

For that reason, in the system of equations (2.108), the terms $\hat{X}_r$, $\hat{X}_z$ are preserved.

2.2.2 Half-space. The problem of Boussinesq-Mindlin type. Limiting cases

In [51] and [49], we considered the micropolar half-space

$$\Omega = \{(r, \theta, z) : r \geq 0, 0 \leq \theta < 2\pi, z > 0\} \tag{2.119}$$

loaded in the interior by a body force of the form

$$X_z = \frac{P_0}{2\pi r}\delta(r)\delta(z - \zeta), \qquad \zeta > 0, \tag{2.120}$$

with a free boundary $\partial\Omega$ $(z = 0)$ and the condition (2.7). This is illustrated in Fig. 2.1.

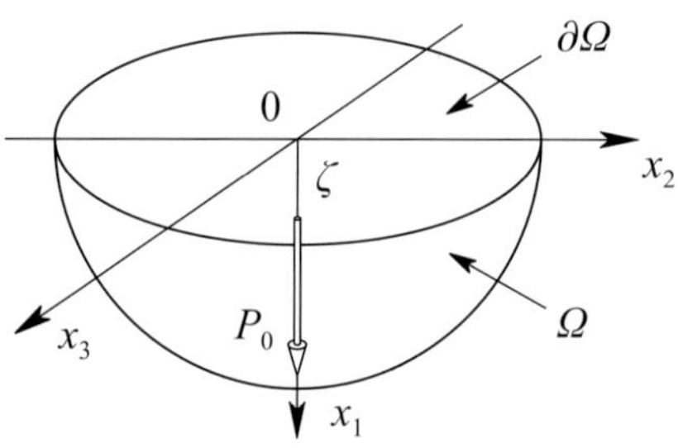

Fig. 2.1.

To solve the problem, the fundamental solution given by (2.116) and (2.117) was used. We obtain results in the form of the formula

$$u_r = u_r^1 + \bar{u}_r^2 + \bar{\bar{u}}_r^2, \qquad u_z = u_z^1 + \bar{u}_z^2 + \bar{\bar{u}}_z^2, \qquad \varphi_\theta = \varphi_\theta^1 + \bar{\varphi}_\theta^2 + \bar{\bar{\varphi}}_\theta^2. \tag{2.121}$$

The quantities u_r^1, φ_θ^1, u_z^1 are obtained by superposing the action of opposite loadings X_z in the infinite space at the points $(0, \zeta)$ and $(0, -\zeta)$:

$$
\begin{aligned}
u_r^1 &= (u_{r1}^0 + u_{r1}') - (u_{r2}^0 + u_{r2}'),\\
u_z^1 &= (u_{z1}^0 + u_{z1}') - (u_{z2}^0 + u_{z2}'),\\
\varphi_\theta^1 &= (\varphi_{\theta 1}^0 + \varphi_{\theta 1}') - (\varphi_{\theta 2}^0 + \varphi_{\theta 2}'),
\end{aligned}
\tag{2.122}
$$

where the triples $(u_{r1}^0, \varphi_{\theta 1}^0, u_{z1}^0)$, $(u_{r1}', \varphi_{\theta 1}', u_{z1}')$ are obtained from formulae (2.116) and (2.117) for $(z - \zeta)$, whereas the triples $(u_{r2}^0, \varphi_{\theta 2}^0, u_{z2}^0)$, $(u_{r2}', \varphi_{\theta 2}', u_{z2}')$ refer to the point $(z + \zeta)$. Taking into account the boundary effect is also analysed by the superposition method. The assumption of free boundary $\partial\Omega$ implies the conditions

$$
\sigma_{zz} = 0, \qquad \sigma_{zr} = 0, \qquad \mu_{z\theta} = 0 \qquad \text{on } \partial\Omega, \tag{2.123}
$$

which will not be satisfied by stresses corresponding to the triple $(u_r^1, \varphi_\theta^1, u_z^1)$ given by formulae (2.122). For, taking $\sigma_{zz} = \sigma_{zz}^1 + \sigma_{zz}^2$, $\sigma_{zr} = \sigma_{zr}^1 + \sigma_{zr}^2$, $\mu_{z\theta} = \mu_{z\theta}^1 + \mu_{z\theta}^2$, where the triple $(\sigma_{zz}^1, \sigma_{zr}^1, \mu_{z\theta}^1)$ corresponds to the triple (2.122), while the triple $(\sigma_{zz}^2, \sigma_{zr}^2, \mu_{z\theta}^2)$ is complementary, we find that for $z = 0$ only $\sigma_{zr}^1 = 0$, whereas $\sigma_{zz}^1 \neq 0$ and $\mu_{z\theta}^1 \neq 0$. Hence conditions (2.123) read

$$
\sigma_{zz}^2 = -\sigma_{zz}^1, \qquad \sigma_{zr}^2 = 0, \qquad \mu_{z\theta}^2 = -\mu_{z\theta}^1 \qquad \text{on } \partial\Omega \tag{2.124}
$$

and define a boundary value problem for the micropolar half-space (2.119), for the solution of which, as we have already noted, the superposition method discussed in Subsection 2.2.1 will be used. To the solution of the problem (2.124) there corresponds the triple $(u_r^2, \varphi_\theta^2, u_z^2)$, which, according to formula (2.121), has the form

$$
u_r^2 = \bar{u}_r^2 + \bar{\bar{u}}_r^2, \qquad u_z^2 = \bar{u}_z^2 + \bar{\bar{u}}_z^2, \qquad \varphi_\theta^2 = \bar{\varphi}_\theta^2 + \bar{\bar{\varphi}}_\theta^2.
$$

Let us also write the stresses σ_{zz}^2, σ_{zr}^2, $\mu_{z\theta}^2$ as

$$
\sigma_{zz}^2 = \bar{\sigma}_{zz}^2 + \bar{\bar{\sigma}}_{zz}^2, \qquad \sigma_{zr}^2 = \bar{\sigma}_{zr}^2 + \bar{\bar{\sigma}}_{zr}^2, \qquad \mu_{z\theta}^2 = \bar{\mu}_{z\theta}^2 + \bar{\bar{\mu}}_{z\theta}^2.
$$

The triple $(\bar{u}_r^2, \bar{\varphi}_\theta^2, \bar{u}_z^2)$ constitutes a solution of the classical boundary value problem for the half-space Ω with a vertical loading on the boundary, of the form

$$
\bar{\sigma}_{zz}^2 = -\sigma_{zz}^1, \qquad \bar{\sigma}_{zr}^2 = 0 \qquad \text{on } \partial\Omega, \tag{2.125}
$$

for which (by means of equations $\nabla_0^2 \nabla_0^2 \bar{u}_r^2 = 0$, $\nabla^2 \nabla^2 \bar{u}_z^2 = 0$, $\nabla_0^2 \bar{\varphi}_\theta^2 = 0$) we obtain the following formulae[12] in Ω:

[12] In the method of superposition classical solutions are formally classical since the Lamé constants μ, λ and the Poisson's ratio $\nu = \lambda/2(\mu + \lambda)$ refer to micropolar medium. If, however, we assume that $\bar{u}_r^2$, $\bar{\varphi}_\theta^2$ and $\bar{u}_z^2$ are numerically equal to the analogous classical solution, then in the sequel we must take into account not only the boundary effect, but also the effect of numerical change in elasticity constants.

$$
\begin{aligned}
2\mu\bar{u}_r^2 &= \mathcal{H}_1\left[\!\left[\xi^{-1}\tilde{p}_0[\xi z-(1-2\nu)]e^{-\xi z};\xi\to r\right]\!\right],\\
2\mu\bar{u}_z^2 &= \mathcal{H}_0\left[\!\left[\xi^{-1}\tilde{p}_0[\xi z+2(1-\nu)]e^{-\xi z};\xi\to r\right]\!\right],\\
2\mu\bar{\varphi}_\theta^2 &= 2(1-\nu)\mathcal{H}_1\left[\!\left[\tilde{p}_0 e^{-\xi z};\xi\to r\right]\!\right],
\end{aligned}
\tag{2.126}
$$

where $\tilde{p}_0$ is the transform of the loading $\sigma_{zz}^1(r,0)$:

$$
\tilde{p}_0(\xi)=\frac{P_0}{4\pi\mu_0 a_0}\left\{[2(1-\nu)+\zeta\xi]e^{-\zeta\xi}+2a_0\xi^2(e^{-\zeta\rho}-e^{-\zeta\xi})\right\}. \tag{2.127}
$$

The notations in formulae (2.126) and (2.127) are the following: the symbols $\mathcal{H}_0\left[\!\left[\cdot\right]\!\right]$, $\mathcal{H}_1\left[\!\left[\cdot\right]\!\right]$ refer to the inverse Hankel transform $(0.51)_2$, $\zeta>0$ is a fixed depth at which the body force acts, P_0 has the unit value, ν is the Poisson ratio,

$$
\mu_0=\frac{2\mu}{\gamma+\varepsilon},\ a_0=\frac{(\gamma+\varepsilon)(2\mu+\lambda)}{4\mu(\mu+\lambda)},\ \rho=\sqrt{\xi^2+\frac{1}{l^2}},\ l^2=\frac{(\gamma+\varepsilon)(\mu+\alpha)}{4\alpha\mu}.
$$

The triple $(\bar{u}_r^2,\bar{\varphi}_\theta^2,\bar{u}_z^2)$ takes into account the boundary effect in the micropolar half-space; on the boundary of the half-space only a loading of stress of type $\mu_{z\theta}$ will appear. Since the constitutive relations (2.33) and the geometric relations (2.34) imply that the rotations $\bar{\varphi}_\theta^2$ given by formula $(2.126)_3$ produce a stress field for $z\geq 0$: $\bar{\mu}_{z\theta}^2=(\gamma+\varepsilon)\partial_z\bar{\varphi}_\theta^2$, therefore the boundary conditions (2.124), after taking into account conditions (2.125), take the form

$$
\bar{\bar{\sigma}}_{zz}^2=0,\qquad \bar{\bar{\sigma}}_{zr}^2=0,\qquad \bar{\bar{\mu}}_{z\theta}^2=-\mu_{z\theta}^1-\bar{\mu}_{z\theta}^2\qquad \text{on }\partial\Omega. \tag{2.128}
$$

Now we should solve the system of equations for the triple $(\bar{\bar{u}}_r^2,\bar{\bar{\varphi}}_\theta^2,\bar{\bar{u}}_z^2)$ in Ω:

$$
\begin{aligned}
&(\mu+\alpha)\nabla_0^2\bar{\bar{u}}_r^2+(\lambda+\mu-\alpha)\partial_r\bar{\bar{e}}^2-2\alpha\partial_z\bar{\bar{\varphi}}_\theta^2=0,\\
&(\mu+\alpha)\nabla^2\bar{\bar{u}}_z^2+(\lambda+\mu-\alpha)\partial_z\bar{\bar{e}}^2+2\alpha r^{-1}\partial_r(r\bar{\bar{\varphi}}_\theta^2)=0,\\
&\left[(\gamma+\varepsilon)\nabla_0^2-4\alpha\right]\bar{\bar{\varphi}}_\theta^2+2\alpha(\partial_z\bar{\bar{u}}_r^2-\partial_r\bar{\bar{u}}_z^2)=0,
\end{aligned}
\tag{2.129}
$$

where $\bar{\bar{e}}^2=\left(r^{-1}+\partial_r\right)\bar{\bar{u}}_r^2+\partial_z\bar{\bar{u}}_z^2$, with the boundary conditions (2.128). To this end, we use the following equations:

$$
\begin{aligned}
&\nabla_0^2\nabla_0^2(l^2\nabla_0^2-1)\bar{\bar{u}}_r^2=0,\\
&\nabla^2\nabla^2(l^2\nabla^2-1)\bar{\bar{u}}_z^2=0,\\
&\nabla_0^2(l^2\nabla_0^2-1)\bar{\bar{\varphi}}_\theta^2=0.
\end{aligned}
\tag{2.130}
$$

Following [43], we give the final result for such a boundary value problem for the triple $(\bar{\bar{u}}_r^2,\bar{\bar{\varphi}}_\theta^2,\bar{\bar{u}}_z^2)$:

$$2\mu\bar{\bar{u}}_r^2 = \mathcal{H}_1\left[\!\left[\xi^{-1}\tilde{\bar{p}}_0\{(1-\Delta_0^{-1})[(1-2\nu)-\xi z]e^{-\xi z} + 2a_0\xi^2\Delta_0^{-1}(e^{-\rho z}-\xi\rho^{-1}e^{-\xi z})\};\xi\to r\right]\!\right],$$

$$2\mu\bar{\bar{u}}_z^2 = \mathcal{H}_0\left[\!\left[\xi^{-1}\tilde{\bar{p}}_0\{(1-\Delta_0^{-1})[2(1-\nu)+\xi z]e^{-\xi z} - 2a_0\xi^2\Delta_0^{-1}\xi\rho^{-1}(e^{-\rho z}-e^{-\xi z})\};\xi\to r\right]\!\right], \tag{2.131}$$

$$2\mu\bar{\bar{\varphi}}_\theta^2 = -2(1-\nu)\mathcal{H}_1\left[\!\left[\tilde{\bar{p}}_0[(1-\Delta_0^{-1})e^{-\xi z}+\xi\rho^{-1}\Delta_0^{-1}e^{-\rho z}];\xi\to r\right]\!\right],$$

where $\Delta_0 = 1 + 2a_0\xi^2(1-\xi\rho^{-1})$. In formulae (2.131), one should substitute

$$\tilde{\bar{p}}_0(\xi) = \tilde{p}_0 + \frac{P_0}{4\pi\mu_0 a_0}(e^{-\zeta\rho} - e^{-\zeta\xi}), \tag{2.132}$$

where $\tilde{p}_0$ is given by formula (2.127).

We have discussed the limiting cases of formulae (2.121) for u_r, φ_θ, u_z, which are illustrated by the diagram below:

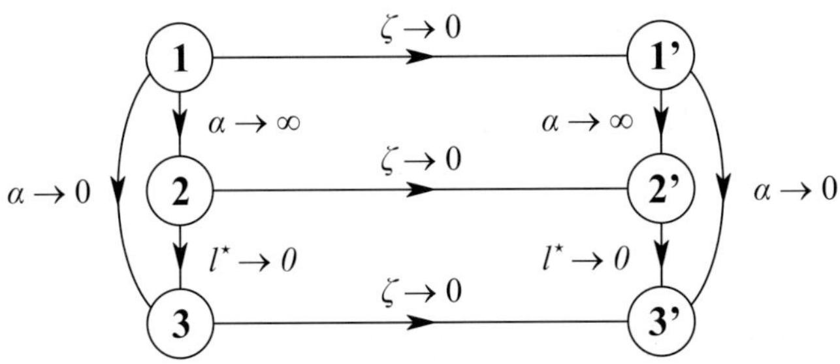

Fig. 2.2.

The numbers (1), (1'); (2), (2'); and (3), (3') in the diagram denote: micropolar theory (E-N), couple-stress theory (K-M), and the classical theory of elasticity (H), respectively. Moreover, the results concerning theories (1), (2) and (3) correspond to the loading (2.120), while in the results for (1'), (2') and (3'), the following condition holds on $\partial\Omega$:

$$\sigma_{zz} = -\frac{P_0}{2\pi r}\delta(r). \tag{2.133}$$

Bieńkowski & Dyszlewicz [14] have solved the problem of Boussinesq-Mindlin type (2.119), (2.120), without using equations (2.130). To solve the complementary problem for the triple $(\bar{\bar{u}}_r^2, \bar{\bar{\varphi}}_\theta^2, \bar{\bar{u}}_z^2)$, represented by the boundary conditions (2.128), we used the generalized Love representation in the form of equations (2.166) and (2.167) for $X_z = 0$, and obtained, by means of the Hankel transformations, results identical to those in [51] and [49]. Let us dwell a little longer on the results of [14] and the Boussinesq problem (Fig. 1.5) described by the boundary conditions (2.133) ($\sigma_{zr} = \mu_{z\theta} = 0$ on $\partial\Omega$). Using the superposition method, we assumed the solution of this problem in the form

$$u_r = u_r^0 + u_r', \qquad u_z = u_z^0 + u_z', \qquad \varphi_\theta = \varphi_\theta^0 + \varphi_\theta'. \tag{2.134}$$

Similar decompositions as for the displacements and rotations hold for the remaining physical fields. The classical solution (for model (H)) was derived in [14] from the system of equations in Ω given by

$$\begin{aligned} &\mu\nabla_0^2 u_r^0 + (\lambda+\mu)\partial_r e^0 = 0,\\ &\mu\nabla^2 u_z^0 + (\lambda+\mu)\partial_z e^0 = 0, \end{aligned} \tag{2.135}$$

with boundary conditions on $\partial\Omega$ of the form

$$\sigma_{zz}^0 = -\frac{P_0}{2\pi r}\delta(r), \qquad \sigma_{zr}^0 = 0 \tag{2.136}$$

and the condition $(u_r^0, \varphi_\theta^0, u_z^0) \to (0,0,0)$ for $R = \sqrt{r^2+z^2} \to \infty$.

To this end, we used the following separated equations in Ω:

$$\nabla^2 e^0 = 0, \qquad \nabla_0^2\nabla_0^2 u_r^0 = 0, \qquad \nabla^2\nabla^2 u_z^0 = 0, \tag{2.137}$$

where

$$e^0 = r^{-1}\partial_r(r u_r^0) + \partial_z u_z^0, \qquad \nabla^2 = \partial_r^2 + r^{-1}\partial_r + \partial_z^2$$

and

$$\nabla_0^2 = \nabla^2 - r^{-2}.$$

Applying the Hankel transformations, we obtained the classical result (for model (H)) in a closed form (see [180, p. 199]):

— displacements and rotations,

$$\begin{aligned} u_r^0 &= \frac{P_0}{4\pi\mu}\left(\frac{zr}{R^3} - \frac{\mu}{\lambda+\mu}\cdot\frac{r}{R(R+z)}\right),\\ u_z^0 &= \frac{P_0}{4\pi\mu}\left(\frac{z^2}{R^3} - \frac{\mu}{\lambda+\mu}\cdot\frac{1}{R}\right),\\ \varphi_\theta^0 &= \frac{P_0}{4\pi\mu}\cdot\frac{2\mu+\lambda}{\lambda+\mu}\cdot\frac{r}{R^3}, \end{aligned} \tag{2.138}$$

where $R = \sqrt{r^2+z^2}$. Let us list here the Hankel integrals with Bessel functions (see [101]), which will be used here and in the sequel:

$$\begin{aligned}
I_1 &= \int_0^\infty e^{-\xi z} \mathcal{J}_0(\xi r)\, d\xi = \frac{1}{R}, \\
I_2 &= \int_0^\infty e^{-\xi z} \mathcal{J}_1(\xi r)\, d\xi = \frac{r}{R(R+z)}, \\
I_3 &= \int_0^\infty \xi e^{-\xi z} \mathcal{J}_0(\xi r)\, d\xi = \frac{z}{R^3}, \\
I_4 &= \int_0^\infty \xi e^{-\xi z} \mathcal{J}_1(\xi r)\, d\xi = \frac{r}{R^3}, \\
I_5 &= \int_0^\infty \xi^{-1} e^{-\xi z} \mathcal{J}_1(\xi r)\, d\xi = \frac{r}{r+z}, \\
I_6 &= \int_0^\infty \xi^2 e^{-\xi z} \mathcal{J}_0(\xi r)\, d\xi = \frac{3z^2}{R^5} - \frac{1}{R^3}, \\
I_7 &= \int_0^\infty \xi^2 e^{-\xi z} \mathcal{J}_1(\xi r)\, d\xi = \frac{3rz}{R^5}.
\end{aligned} \tag{2.139}$$

Further, we have

— force-stresses (model (H)):

$$\begin{aligned}
\sigma^0_{zz} &= -\frac{P_0}{2\pi} \cdot \frac{3z^3}{R^5}, \\
\sigma^0_{zr} &= \sigma^0_{rz} = -\frac{P_0}{2\pi} \cdot \frac{3rz^2}{R^5}, \\
\sigma^0_{rr} &= \frac{P_0}{2\pi} \left[-\frac{3rz^2}{R^5} + (1-2\nu) \frac{1}{R(R+z)} \right], \\
\sigma^0_{\theta\theta} &= \frac{P_0(1-2\nu)}{2\pi} \left[\frac{z}{R^3} - \frac{1}{R(R+z)} \right].
\end{aligned} \tag{2.140}$$

The complementary solution $(u'_r, \varphi'_\theta, u'_z)$ will be obtained from the system of equations

$$\begin{aligned}
&(\mu+\alpha)\nabla_0^2 u'_r + (\lambda+\mu-\alpha)\partial_r e' - 2\alpha\partial_z \varphi'_\theta = 0, \\
&(\mu+\alpha)\nabla^2 u'_z + (\lambda+\mu-\alpha)\partial_z e' + 2\alpha r^{-1}\partial_r (r\varphi'_\theta) = 0, \\
&[(\gamma+\varepsilon)\nabla_0^2 - 4\alpha]\varphi'_\theta + 2\alpha(\partial_z u'_r - \partial_r u'_z) = 0,
\end{aligned} \tag{2.141}$$

where $e' = r^{-1}\partial_r(ru'_r) + \partial_z u'_z$. This is the fundamental system of equations in diplacements and rotations (without body forces and temperature) for the first axially-symmetric (E-N) model (cf. [43, p. 49]). However, for the complementary triple $(u'_r, \varphi'_\theta, u'_z)$ modified and transformed boundary conditions on $\partial\Omega$ hold, namely:

$$\tilde{\sigma}'_{zz0} = 0, \qquad \tilde{\sigma}'_{zr1} = 0, \qquad \tilde{\mu}'_{z\theta 1} = \frac{P_0}{2\pi} 2a_0\xi. \tag{2.142}$$

To solve the system of equations (2.141), we used in [14] a special type of the generalized Love function, equations (2.166) and (2.167). The Love function F_z satisfies in Ω a homogeneous equation of the form

$$\nabla^2\nabla^2(l^2\nabla^2 - 1)F_z = 0. \tag{2.143}$$

Multiplying both sides of equation (2.143) by $r\mathcal{J}_0(\xi r)$, integrating it with respect to r over the interval $[0, +\infty)$, and using the following properties of the Hankel transformation:

$$\int_0^\infty r\left(\frac{\partial^2}{\partial r^2} + \frac{1}{r}\frac{\partial}{\partial r}\right) f(r,z)\mathcal{J}_0(\xi r)\, dr = -\xi^2 \tilde{f}_0(\xi, z),$$

$$\int_0^\infty \frac{\partial}{\partial z} r f(r,z)\mathcal{J}_0(\xi r)\, dr = \frac{d}{dz}\tilde{f}_0(\xi, z),$$

we obtain an ordinary differential equation, namely

$$(\mathbb{D}_z^2 - \xi^2)^2(\mathbb{D}_z^2 - \rho^2)\tilde{F}_{z0} = 0, \tag{2.144}$$

where $\mathbb{D}_z \overset{\text{def}}{=} d/dz$ and

$$\xi^2 = \xi_N\xi_N, \qquad \rho^2 = \xi^2 + l^{-2}, \qquad l^2 = \frac{(\gamma+\varepsilon)(\mu+\alpha)}{4\alpha\mu}.$$

The transform of the Love function $\tilde{F}_{z0}$ is assumed in the form

$$\tilde{F}_{z0} = (A_1 + B_1\xi z)e^{-\xi z} + C_1 e^{-\rho z}. \tag{2.145}$$

Now, changing notation for the integration constants, we obtain for the triple $(u'_r, \varphi'_\theta, u'_z)$ the following expressions:

$$\begin{aligned} \tilde{u}'_{r1} &= (A + B\xi z)e^{-\xi z} + Ce^{-\rho z}, \\ \tilde{u}'_{z0} &= [A + (3-4\nu)B + B\xi z]e^{-\xi z} + \xi\rho^{-1}Ce^{-\rho z}, \\ \tilde{\varphi}'_{\theta 1} &= 2(1-\nu)\xi Be^{-\xi z} - \mu_0\rho^{-1}Ce^{-\rho z}, \end{aligned} \tag{2.146}$$

in view of (2.166), where $\mu_0 = \dfrac{2\mu}{\gamma+\varepsilon}$. From the boundary conditions (2.142) we obtain A, B and C:

$$\begin{aligned} A &= \frac{P_0}{4\pi\mu}\cdot\frac{1}{\xi}\left[\left(1 - \frac{1}{\Delta_0}\right)(1-2\nu) + \frac{2a_0\xi^2}{\Delta_0}\cdot\frac{\xi}{\rho}\right], \\ B &= -\frac{P_0}{4\pi\mu}\cdot\frac{1}{\xi}\left(1 - \frac{1}{\Delta_0}\right), \\ C &= \frac{P_0}{4\pi\mu}\cdot\frac{2a_0\xi}{\Delta_0}, \end{aligned} \tag{2.147}$$

where

$$\Delta_0 = 1 + 2a_0\xi^2\left(1 - \frac{\xi}{\rho}\right), \qquad a_0 = \frac{(\gamma+\varepsilon)(2\mu+\lambda)}{4\mu(\mu+\lambda)}.$$

Thus, the complementary solution has been determined. Further, in [14], we derive an approximate complementary solution. The Hankel integrals which determine u_r', u_z', φ_θ', as well as the complementary stresses $\underset{\sim}{\sigma}'$ and $\underset{\sim}{\mu}'$, are convergent in $\overline{\Omega} = \Omega \cup \partial\Omega$, with, perhaps, the exception of the point where the force P_0 is applied, i.e. for $z = 0$ and $r = 0$. The singularities of the integrals depend on integrands at the point $(0, 0; \xi)$ for $\xi \to \infty$ (cf. Muki & Sternberg [161]). In order to determine the character of the singularities of the integrals which appear in the complementary solution, one makes use of the asymptotic expansions of the integrands (cf. formulae (1.345) with formula $(1.345)_5$ for $e^{-\rho x_1}$ in the axially-symmetric problem considered here; we replace x_1 by z).

Let us give approximate formulae for u_z', u_r', φ_θ':

$$\begin{aligned}
u_z' &= -\frac{P_0}{2\pi\mu}\cdot\frac{a_0}{l^2+a_0}\left[\frac{2\mu+\lambda}{2(\mu+\lambda)}\cdot\frac{1}{R} + \frac{z^2}{R^3}\right] + O(1),\\
u_r' &= \frac{P_0}{2\pi\mu}\cdot\frac{a_0}{l^2+a_0}\left[\frac{2\mu+\lambda}{2(\mu+\lambda)}\cdot\frac{r}{R(R+z)} + \frac{zr}{R^3}\right] + O(1),\\
\varphi_\theta' &= \frac{P_0}{2\pi\mu}\cdot\frac{2\mu+\lambda}{\mu+\lambda}\left[-\frac{r}{R^3} + \frac{1}{2(l^2+a_0)}\cdot\frac{r}{R}\right] + O(1)
\end{aligned} \tag{2.148}$$

and for the complementary force-stresses:

$$\begin{aligned}
\sigma_{zz}' &= \frac{P_0}{\pi}\cdot\frac{a_0}{l^2+a_0}\left(\frac{3z^3}{R^5} - \frac{z}{R^3}\right) + O(1),\\
\sigma_{zr}' &= \frac{P_0}{\pi}\cdot\frac{a_0}{l^2+a_0}\cdot\frac{3rz^2}{R^5} + O(1),\\
\sigma_{rz}' &= \frac{P_0}{\pi}\cdot\frac{a_0}{l^2+a_0}\left(\frac{3rz^2}{R^5} - \frac{r}{R^3}\right) + O(1),\\
\sigma_{rr}' - \sigma_{\theta\theta}' &= \frac{P_0}{\pi}\cdot\frac{a_0}{l^2+a_0}\left[\frac{3r^2z}{R^5} + (1-\nu)\frac{z}{R^3} - 2(1-\nu)\frac{1}{R(R+z)}\right]\\
&\quad + O(1).
\end{aligned} \tag{2.149}$$

Let us present now a few graphs which illustrate the results of [14].

In [14], we also made a number of observations and conclusions concerning both, the decompositions of physical fields in the half-space, and the problem of the limiting solutions for the micropolar (E-N) model and their analysis. Thus, let us end with two such conclusions.

Conclusion 1. On the basis of Fig. 2.3 and Fig. 2.4, we can conclude that the solutions for the (H) model and for the (K-M) model are the envelopes of the solution for the (E-N) model. In Fig. 2.3 the graph for $\sigma_{rr} - \sigma_{\theta\theta}$ lies between the graphs for $s_{rr} - s_{\theta\theta}$ and $\sigma^0_{rr} - \sigma^0_{\theta\theta}$, whereas in Fig. 2.4 the graph for the force-stress σ_{zz} lies between the graphs for s_{zz} and σ^0_{zz} (cf. similar conlcusions in [58, pp. 225, 227]).

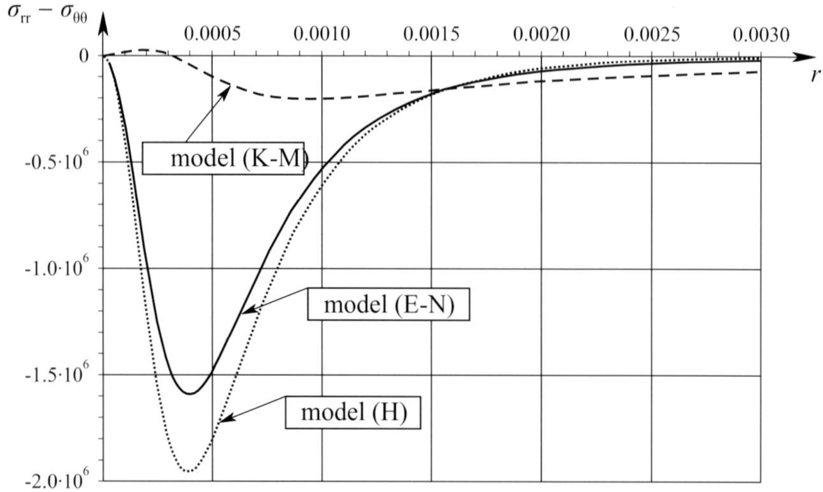

Fig. 2.3. The distribution of the difference of stresses $\sigma_{rr} - \sigma_{\theta\theta}$, due to concentrated loading for $z = 0.0005$. The influence of couple-stresses on stress concentration

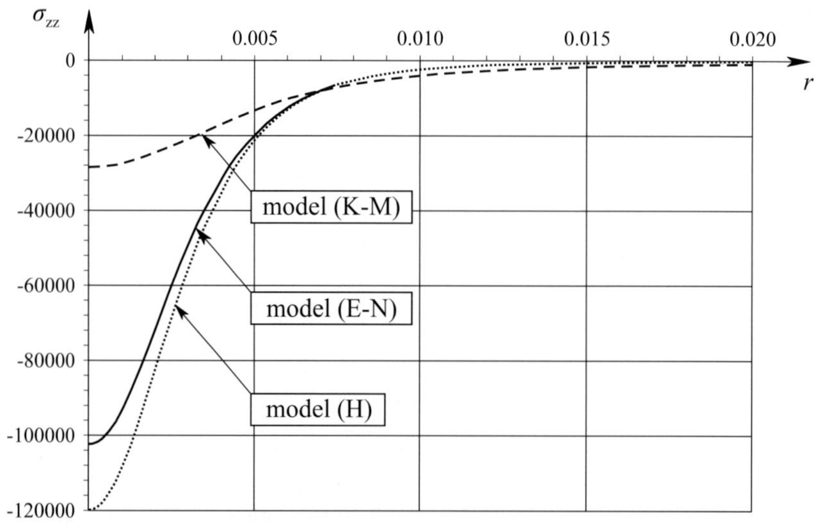

Fig. 2.4. Distribution of the stresses σ_{zz} right under the point of force application for $z = 0.005$

Conclusion 2. A number of graphs for couple-stresses given in [14] show that the Saint-Venant principle holds for couple-stresses since their distribution becomes more even as depth increases.

Let us also present the distributions of stresses σ_{zr} Fig. 2.5 and displacements u_z Fig. 2.6.

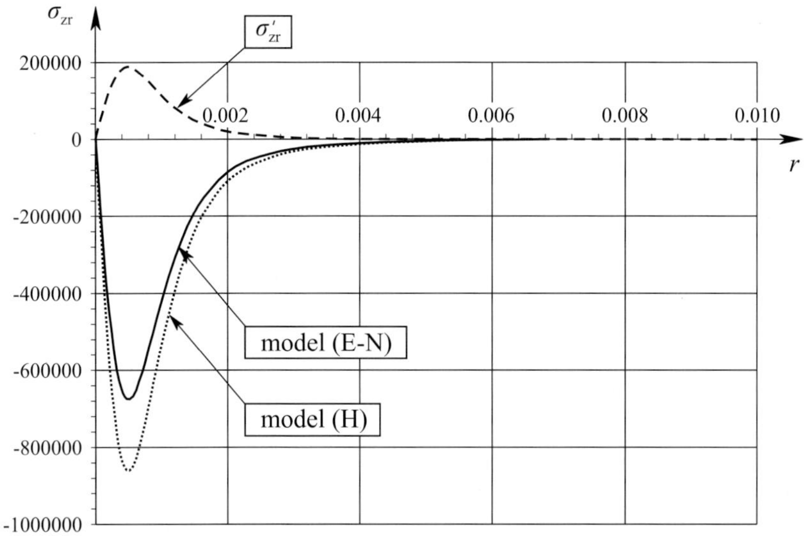

Fig. 2.5. Distribution of the stresses σ_{zr} under the point of force application for $z = 0.001$. The superposition of classical solution and the complementary solution

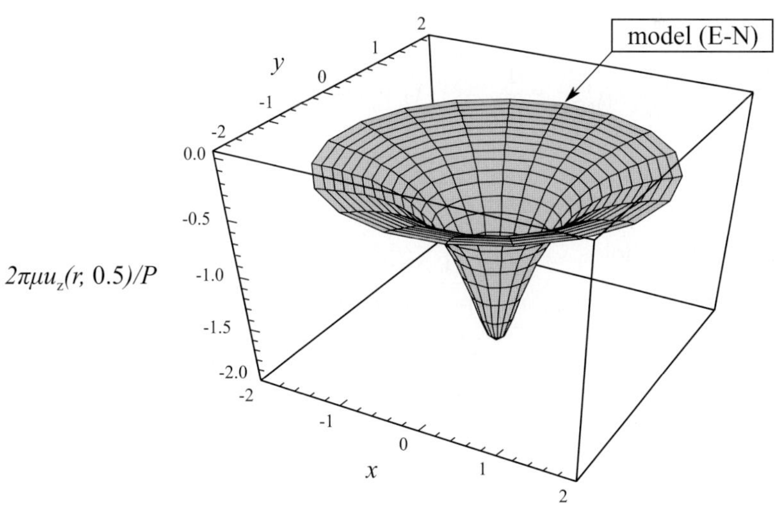

Fig. 2.6. The distribution of the displacements u_z at the depth $z = 0.5$

2.2.3 Elastic half-space. Problem of singularities of physical fields in elastostatics and thermoelastostatics

In [50] and [49], we discussed the problem of singularities of the displacements u_r, u_z, the rotations φ_θ, the force-stresses $\sigma_{rr}-\sigma_{\theta\theta}$, σ_{zz}, σ_{rz}, σ_{zr} and the couple-stresses $\mu_{r\theta}$, $\mu_{\theta r}$, $\mu_{z\theta}$, $\mu_{\theta z}$ in the half-space Ω (2.119), loaded and heated pointwise on the boundary. The starting point for further considerations is given by the formulae for displacements, rotations, force-stresses and couple-stresses in the form of Hankel integrals for the following boundary value problems:

$$
\begin{array}{llll}
1^0 & \sigma_{zz}=-\dfrac{P_0}{2\pi r}\delta(r), & \sigma_{zr}=\mu_{z\theta}=0 & \text{on } \partial\Omega, \\
2^0 & \theta=\dfrac{Q_0}{2\pi r}\delta(r), & \sigma_{zz}=\sigma_{zr}=\mu_{z\theta}=0 & \text{on } \partial\Omega.
\end{array}
$$

The improper integrals appearing in the integral representation of solutions of boundary value problems 1^0 and 2^0 converge in the set

$$\overline{\Omega}=\Omega\cup\partial\Omega=\{(r,z): r\geq 0\wedge z\geq 0\},$$

except, possibly, at the point $(r=0\wedge z=0)$. In the author's opinion, representing these integrals in a closed form is probably not possible. In this work, however, we extracted in a closed form those parts of the solution which are a basis for the determination of the nature of singularities of physical fields in the neighbourhood of the point where a loading or heat is applied. To this end, in the integrands, we make use of asymptotic expansions for expressions: ρ, $\dfrac{\xi}{\rho}$, $e^{-\rho z}$, $\dfrac{1}{\Delta_0}$, where

$$\rho=\sqrt{\xi^2+\frac{1}{l^2}}, \qquad \Delta_0=1+2a_0\xi^2\left(1-\frac{\xi}{\rho}\right),$$

$$a_0=\frac{(\gamma+\varepsilon)(2\mu+\lambda)}{4\mu(\mu+\lambda)}, \qquad l^2=\frac{(\gamma+\varepsilon)(\mu+\alpha)}{4\alpha\mu}.$$

For example, it holds:

$$\frac{1}{\Delta_0}=\frac{l^2}{l^2+a_0}+\frac{3a_0}{4(l^2+a_0)^2}\cdot\frac{1}{\xi^2}-\frac{a_0(10l^2+a_0)}{16l^2(l^2+a_0)^3}\cdot\frac{1}{\xi^4}+O(\xi^{-6}) \quad \text{for } \xi\to\infty,$$

(cf. formulae (1.345)).

By the symbol $O(\xi^{-n})$, $(n\in\mathbb{N})$ we denote an expression, which for $\xi\to\infty$ behaves like $g(\xi)=1/\xi^n$. Once the asymptotic expansions are made, the range at which the formulae remain valid is limited to a neighbourhood of the point of force (heat) application. We shall write now the physical fields in the form $f=f^s+O(1)$, where f^s denotes the singular or leading part of a solution, whereas $O(1)$ stands for the regular part of f in the sense that it assumes finite values at the point $r = 0 \wedge z = 0$. This procedure requires careful analysis as to which elements of the integrand contribute to the singular part of a solution. As an example, we give the result for $\mu_{z\theta}$ and φ_θ in the problem with a concentrated force:

$$\mu_{z\theta} = -\frac{P_0}{2\pi} \cdot \frac{a_0}{l^2 + a_0} \cdot \frac{rz}{R^3} + O(1), \qquad \varphi_\theta = O(1).$$

The singular part of the stress $\mu_{z\theta}$ is equal to

$$\mu^s_{z\theta} = A\frac{rz}{R^3},$$

where $A = -\dfrac{P_0}{2\pi} \cdot \dfrac{a_0}{l^2 + a_0}$ is a coefficient standing by the singularity $O(R^{-1})$ (called the coefficient of intensity of the singularity). The rotations φ_θ at the point of application of the force P_0 are bounded, but they should be presented in the form[13]:

$$\varphi_\theta = \frac{P_0}{4\pi\mu} \cdot \frac{2\mu + \lambda}{\lambda + \mu} \cdot \frac{1}{2(l^2 + a_0)} \cdot \frac{r}{R} + O(1), \tag{2.150}$$

whereas $\varphi^s_\theta = B\dfrac{r}{R}$, where

$$B = \frac{P_0}{4\pi\mu} \cdot \frac{2\mu + \lambda}{\lambda + \mu} \cdot \frac{1}{2(l^2 + a_0)},$$

is the leading part of the field φ_θ. Notice that $\varphi^s_\theta = O(1)$, but the leading part is not unique at the point $(0, 0)$: the double limit $\lim\limits_{(r,z)\to(0,0)} \dfrac{r}{\sqrt{r^2 + z^2}}$ does not exist:

$$\lim_{z\to 0}\left(\lim_{r\to 0} \frac{r}{\sqrt{r^2 + z^2}}\right) = 0, \qquad \lim_{r\to 0}\left(\lim_{z\to 0} \frac{r}{\sqrt{r^2 + z^2}}\right) = 1.$$

However, extracting the leading part φ^s_θ is of crucial importance for correct determination of the singular part $\mu_{z\theta}$ of the field, since

$$\mu^s_{z\theta} = (\gamma + \varepsilon)\partial_z \varphi^s_\theta = A\frac{rz}{R^3}.$$

Extracting the leading part of a solution is also relevant for the analysis of limits. For $\alpha \to \infty$, $\dfrac{1}{a_0 + l^2} \to \dfrac{1}{3 - 2\nu} \cdot \dfrac{1}{l^{\star 2}}$ we obtain from formula (2.150) the result for the couple-stress theory:

[13] Cf. Dhaliwal [27], where, in the problem with a concentrated force and point-heating on $\partial\Omega$, a formula of the following type is obtained: $\varphi^s_\theta = C\dfrac{r}{R + z}$, where C denotes a constant which does not lead to the correct formulae for $\underset{\sim}{\mu}^s$ in micropolar theory as well as φ^s_θ and $\underset{\sim}{\mu}^s$ in couple-stress theory.

$$\varphi_\theta = \frac{P_0}{8\pi\mu} \cdot \frac{1}{l^{\star 2}} \cdot \frac{\beta^2}{2\beta^2 - 1} \cdot \frac{r}{R} + O(1),$$

where $\beta^2 = \dfrac{2(1-\nu)}{1-2\nu}$, from which we obtain the correct result:

$$\mu^s_{z\theta} = -\frac{P_0}{2\pi} \cdot \frac{\beta^2}{2\beta^2 - 1} \cdot \frac{rz}{R^3}.$$

The question of extracting singular parts from solutions of the problem of a point-heated half-space, loaded by concentrated forces and moments and by a continuous loading with a finite jump (plane, axially-symmetric and three-dimensional state of strain) was analysed in detail in [79]–[82], [50], [49] and in earlier works [72]–[75], [43]. Let us go back to [50], where we obtained the results for the displacements u_r and u_z, which were given in a different way by Puri [202] and Dhaliwal [27]. Apart from displacements and rotations, singular parts in the explicit form were extracted for force-stresses and couple-stresses. The nature and order of those singularities were discussed, and solutions for the limiting (classical- and couple-stress-) theories were found. Let us note some conclusions:

1. The integral representations of solutions and the singular parts of those solutions for displacements, rotations, force-stresses and couple-stresses of micropolar theory, referring to the boundary value problems 1^0 and 2^0, go over in the limit ($\alpha \to \infty$) into the integral representations of the results and their singular parts, respectively, of the couple-stress theory. In that theory, one uses the symmetric part of the force-stress tensor and the deviatoric part of the couple-stress tensor. For $\alpha \to \infty$, we obtain:

$$\begin{aligned}
&(\sigma_{rr}, \sigma_{\theta\theta}, \sigma_{zz}) \to (s_{rr}, s_{\theta\theta}, s_{zz}),\\
&(\mu_{r\theta}, \mu_{\theta r}, \mu_{z\theta}, \mu_{\theta z}) \to (m_{r\theta}, m_{\theta r}, m_{z\theta}, m_{\theta z}),\\
&\tfrac{1}{2}(\sigma_{rz} + \sigma_{zr}) \to s_{zr} = s_{rz},
\end{aligned} \tag{2.151}$$

respectively.

2. For $\alpha \to 0^+$, the integral results and their singular parts for couple-stresses and rotations in micropolar theory (for the boundary value problems 1^0 and 2^0) tend to zero, while the integral form of displacements and stresses and their singular parts go over into closed classical formulae for displacements and stresses.

3. Within the framework of couple-stress theory, the integral representations of solutions for displacements, rotations and force-stresses for the boundary value problems 1^0 and 2^0 go over in the limit $l^\star \to 0^+$ into closed classical formulae for displacements, rotations and stresses, respectively; whereas couple-stresses in the integral form tend to zero. However, singular parts of solutions for displacements, rotations, force- and couple-stresses in couple-stress theory, for the boundary value problems 1^0 and 2^0, do not correspond at all with the classical results.

4. Results concerning the orders of singularities of physical fields are arranged in Table 2.1.

Table 2.1.

<table>
<tr><td rowspan="2">Physical fields</td><td colspan="2">(1)</td><td colspan="2">(2)</td><td colspan="2">(3)</td></tr>
<tr><td>1^0</td><td>2^0</td><td>1^0</td><td>2^0</td><td>1^0</td><td>2^0</td></tr>
<tr><td>u_r, u_z</td><td colspan="2">$O(R^{-1})$</td><td colspan="2">$O(R^{-1})$</td><td colspan="2">$O(R^{-1})$</td></tr>
<tr><td>φ_θ</td><td colspan="2">$O(1)$</td><td colspan="2">$O(1)$</td><td colspan="2">$O(R^{-2})$</td></tr>
<tr><td>σ_{rr}-$\sigma_{\theta\theta}$</td><td colspan="2">$O(R^{-2})$</td><td colspan="2">$O(R^{-2})$</td><td colspan="2">$O(R^{-2})$</td></tr>
<tr><td>$\sigma_{rz}, \sigma_{zr}, \sigma_{zz}$</td><td colspan="2">$O(R^{-2})$</td><td colspan="2">$O(R^{-2})$</td><td>$O(R^{-2})$</td><td>0</td></tr>
<tr><td>$\mu_{r\theta}, \mu_{\theta r}, \mu_{z\theta}, \mu_{\theta z}$</td><td colspan="2">$O(R^{-1})$</td><td colspan="2">$O(R^{-1})$</td><td colspan="2">–</td></tr>
<tr><td>θ</td><td>–</td><td>$O(R^{-2})$</td><td>–</td><td>$O(R^{-2})$</td><td>–</td><td>$O(R^{-2})$</td></tr>
</table>

Numbers (1), (2), (3) denote, respectively: micropolar theory (E-N), couple-stress theory (K-M), the classical theory of elasticity (H). In couple-stress theory (2), the stress symbols should be read in the way consistent with formula (2.151)

2.2.4 Displacement-rotation equations with a distortion field

In [49], the equations for displacements and rotations with a distortion field were uncoupled for the triple $(u_r, \varphi_\theta, u_z)$. In the superposition method we shall adopt the analogy (2.49) and

$$\overset{*}{Y}_\theta = -\overset{*}{R}{}_3^0, \tag{2.152}$$

where

$$\overset{*}{R}{}_3^0 = R_3^0 - \frac{\gamma+\varepsilon}{2\mu}(\partial_z \overset{*}{R}{}_1^0 - \partial_r \overset{*}{R}{}_2^0).$$

It suffices to obtain from equations (2.107), (2.110), (2.113) uncoupled classical and complementary equations with a distortion field, respectively. Assuming the correspondence $(2.118)_1$ and

$$(\overset{**}{R}{}_1^0, \overset{*}{R}{}_3^0, \overset{**}{R}{}_2^0) \mapsto (R_1^0, R_3^0, R_2^0), \tag{2.153}$$

we transform equations (2.110) and (2.113) (expressed in terms of distortions) into the following equations:

— for the displacements u_r, u_z,

$$2\mu\nabla_0^2\nabla_0^2(l^2\nabla_0^2-1)u_r = \partial_z\nabla_0^2 R_3^0 + 2\nabla_0^2(l^2\nabla_0^2-1)R_1^0 + \frac{\gamma+\varepsilon}{2\mu}\nabla_0^2\partial_z(\partial_r R_2^0 - \partial_z R_1^0) - \frac{2(\lambda+\mu)}{\lambda+2\mu}(l^2\nabla_0^2-1)\partial_r R^0,$$

$$2\mu\nabla^2\nabla^2(l^2\nabla^2-1)u_z = -\nabla^2[r^{-1}\partial_r(rR_3^0)] + 2\nabla^2(l^2\nabla^2-1)R_2^0 - \frac{\gamma+\varepsilon}{2\mu}\nabla^2\left\{r^{-1}\partial_r[r(\partial_r R_2^0 - \partial_z R_1^0)]\right\} - \frac{2(\lambda+\mu)}{\lambda+2\mu}(l^2\nabla^2-1)\partial_z R^0; \tag{2.154}$$

— for the rotations φ_θ,

$$\nabla_0^2(l^2\nabla_0^2-1)\varphi_\theta = \frac{1}{2\mu}(\partial_r R_2^0 - \partial_z R_1^0) + \frac{\mu+\alpha}{4\alpha\mu}\nabla_0^2 R_3^0, \tag{2.155}$$

where $R^0 = r^{-1}\partial_r(rR_1^0) + \partial_z R_2^0$.

The notations for R_i^0, $\overset{*}{R}{}_\alpha^0$ and $\overset{**}{R}{}_\alpha^0$ are the same as in Subsection 2.1.3. Equations (2.154) and (2.155) are, in fact, the separated equations (2.101), in which the analogy (2.58) has been assumed. In [24], equations (2.154) and (2.155) were used to determine displacements and rotations in the case when the distortion

$$\gamma_{zz}^0 = \frac{\gamma_0^{zz}}{2\pi r}\delta(r)\delta(z) \tag{2.156}$$

in the infinite space is given.

The remaining distortions are equal to zero. Using the Fourier-Hankel transforms (0.50), (0.51), the following equations were obtained:

$$\begin{aligned} u_r &= -\frac{\gamma_0}{8\pi}\left\{\partial_z\left[\frac{1}{1-\nu}\cdot\frac{zr}{R^3} - 2\mu_0^{-1}\partial_{rz}^2\left(\frac{1-e^{-R/l}}{R}\right)\right] + \frac{2\nu}{1-\nu}\partial_r\left(\frac{1}{R}\right)\right\}, \\ u_z &= -\frac{\gamma_0}{8\pi}\partial_z\left[\left(\frac{3-4\nu}{1-\nu}r^2+4z^2\right)\frac{1}{R^3} + 2\mu_0^{-1}(r^{-1}\partial_r+\partial_r^2)\left(\frac{1-e^{-R/l}}{R}\right) + \frac{2\nu}{1-\nu}\left(\frac{1}{R}\right)\right], \\ \varphi_\theta &= \frac{\gamma_0}{4\pi}\partial_{rz}^2\left(\frac{1-e^{-R/l}}{R}\right), \end{aligned} \tag{2.157}$$

where $\gamma_0^{zz}\mapsto\gamma_0$ and

$$R=\sqrt{r^2+z^2},\qquad \nu=\frac{\lambda}{2(\mu+\lambda)},\qquad \mu_0=\frac{2\mu}{\gamma+\varepsilon},\qquad l=\sqrt{\frac{(\gamma+\varepsilon)(\mu+\alpha)}{4\alpha\mu}}.$$

Let us emphasize that once we have calculated the complex limiting transition $\omega \to 0$ in formulae (2.60), we obtain formulae (2.157), which, in turn, are a basis for obtaining the results for theories (4) and (6) (Fig. 1.1).

2.2.5 The generalized Love function

In [49], we consider the case of the representation (1.137) related to the triple $(u_r, \varphi_\theta, u_z)$. Then, the generalized Galerkin vector has the form (2.61). The equations for the stress functions F_r, M_θ and F_z with body loadings X_r, Y_θ and X_z are derived and a representation for u_r, φ_θ, u_z is given. We will study certain interesting special cases of this representation, associated with the body loadings (2.62). The body moments Y_z do not appear in this axially-symmetric problem. Let us apply the superposition method. With the classical problem, we associate the stress vectors

$$\mathbf{F} \stackrel{\text{def}}{=} (0, 0, F_z^0), \qquad \mathbf{M} \stackrel{\text{def}}{=} (0, 0, M_z^0) \tag{2.158}$$

in the following manner in Ω:

$$\begin{aligned} \mu u_r^0 &= -\tfrac{1}{2}(1-\nu)^{-1}\partial_{rz}^2 F_z^0, \\ \mu u_z^0 &= \nabla_0^2 F_z^0 - \tfrac{1}{2}(1-\nu)^{-1}\partial_z^2 F_z^0, \\ 2\mu\varphi_\theta^0 &= -\partial_r M_z^0. \end{aligned} \tag{2.159}$$

From equations (2.107) and (2.159) we obtain the equations

$$\nabla^2\nabla^2 F_z^0 + X_z = 0, \qquad \nabla^2 M_z^0 + X_z = 0. \tag{2.160}$$

Between the stress functions F_z^0, M_z^0 we have the relation $M_z^0 = \nabla^2 F_z^0$. With the complementary solution we associate the stress vectors

$$\mathbf{F}' \stackrel{\text{def}}{=} (0, 0, F_z'), \qquad \mathbf{M}' \stackrel{\text{def}}{=} (0, M_\theta', 0) \tag{2.161}$$

as follows:

$$\begin{aligned} 2\mu u_r' &= [-(1-\nu)^{-1}(l^2\nabla_0^2 - 1) + \mu_0^{-1}\nabla_0^2]\partial_{rz}^2 F_z' + \partial_z M_\theta', \\ 2\mu u_z' &= [-(1-\nu)^{-1}(l^2\nabla^2 - 1)\partial_z^2 - \mu_0^{-1}\nabla^2\left(r^{-1} + \partial_r\right)\partial_r \\ &\quad + 2\nabla^2(l^2\nabla^2 - 1)]F_z' - \left(r^{-1} + \partial_r\right)M_\theta', \\ 2\mu\varphi_\theta' &= \alpha_0^{-1}\nabla_0^2 M_\theta' + \nabla_0^2\partial_r F_z'. \end{aligned} \tag{2.162}$$

Equations for the functions F_z', M_θ' are obtained from equations (2.110), (2.113) and (2.162) in Ω:

$$\nabla^2\nabla^2(l^2\nabla^2 - 1)F_z' + \hat{X}_z = 0, \qquad \nabla_0^2(l^2\nabla_0^2 - 1)M_\theta' + \overset{*}{Y}_\theta = 0, \tag{2.163}$$

where

$$\hat{X}_z = 0, \qquad \overset{*}{Y}_\theta = \mu_0^{-1}\partial_r X_z, \qquad \mu_0 = \frac{2\mu}{\gamma+\varepsilon}, \qquad \alpha_0 = \frac{2\alpha}{\mu+\alpha}.$$

In the infinite space $F_z' \equiv 0$ and it remains to solve equation $(2.163)_2$. It should be noted that after performing subsequent operations ∂_z, $-(r^{-1}+\partial_r)$, α_0^{-1}, -1 on equation $(2.163)_2$, we obtain the equations:

$$\begin{aligned} &2\mu\nabla_0^2(l^2\nabla^2-1)u_r' + \partial_z\overset{*}{Y}_\theta = 0,\\ &2\mu\nabla^2(l^2\nabla^2-1)u_z' - \left(r^{-1}+\partial_r\right)\overset{*}{Y}_\theta = 0,\\ &2\mu(l^2\nabla_0^2-1)\varphi_\theta' + \alpha_0^{-1}\overset{*}{Y}_\theta = 0,\\ &4\alpha(l^2\nabla_0^2-1)\zeta_\theta - \overset{*}{Y}_\theta = 0, \end{aligned} \tag{2.164}$$

which have been derived in [43] and then applied to determine the fundamental solutions. Let us go back to the representation (2.162). For the stated boundary value problem (without body loadings) we put $M_\theta' \overset{\text{def}}{=} 0$ and it is sufficient to find only the vector $\mathbf{F}$ with the component $F_z^0 + F_z'$. We obtain the solution of the boundary value problem in the form: $u_r = u_r^0 + u_r'$, $u_z = u_z^0 + u_z'$, $\varphi_\theta = \varphi_\theta^0 + \varphi_\theta'$, where the triple $(u_r^0, \varphi_\theta^0, u_z^0)$ is expressed in terms of F_z^0 by relations (2.159) and the triple $(u_r', \varphi_\theta', u_z')$ is expressed in terms of F_z' by relations (2.162), taken for $M_\theta' = 0$. We determine the function F_z^0 for the classical boundary conditions, whereas the function F_z' is determined for the micropolar boundary conditions, in which the boundary effect is taken into account. Further considerations lead to the observation that once we have assumed the correspondence

$$\hat{X}_z \mapsto X_z, \qquad F_z' \mapsto F_z, \tag{2.165}$$

we obtain the following representation in Ω:

$$\begin{aligned} 2\mu u_r &= \left[-(1-\nu)^{-1}(l^2\nabla_0^2-1) + \mu_0^{-1}\nabla_0^2\right]\partial_{rz}^2 F_z,\\ 2\mu u_z &= \left[-(1-\nu)^{-1}(l^2\nabla^2-1)\partial_z^2 - \mu_0^{-1}\nabla^2\left(r^{-1}+\partial_r\right)\partial_r\right.\\ &\qquad \left. + 2\nabla^2(l^2\nabla^2-1)\right]F_z,\\ 2\mu\varphi_\theta &= \nabla_0^2\partial_r F_z, \end{aligned} \tag{2.166}$$

where the function F_z satisfies

$$\nabla^2\nabla^2(l^2\nabla^2-1)F_z + X_z = 0. \tag{2.167}$$

Substituting the formula

$$F_z = 4\alpha\mu(\lambda+2\mu)\chi \tag{2.168}$$

into (2.166) and (2.167), we obtain a representation for the generalized Love function, which was introduced to micropolar theory by Nowacki[14] [178].

[14] See [186, p. 185], [192, p. 196].

Introducing another version of the generalized Love function is connected with the body loading (2.65). The vector $\mathbf{F}$ has only one component F_r. The final form of such a representation is given by

$$
\begin{aligned}
2\mu u_r &= [-(1-\nu)^{-1}(l^2\nabla_0^2 - 1)\partial_r\left(r^{-1}+\partial_r\right) \\
&\quad - \mu_0^{-1}\nabla_0^2\partial_z^2 + 2\nabla_0^2(l^2\nabla_0^2 - 1)]F_r, \\
2\mu u_z &= [-(1-\nu)^{-1}(l^2\nabla^2 - 1) + \mu_0^{-1}\nabla^2]\partial_z\left(r^{-1}+\partial_r\right)F_r, \\
2\mu\varphi_\theta &= -\nabla_0^2\partial_z F_r,
\end{aligned}
\tag{2.169}
$$

where F_r satisfies

$$\nabla_0^2\nabla_0^2(l^2\nabla_0^2 - 1)F_r + X_r = 0. \tag{2.170}$$

Between the functions F_z, F_r the following relation holds:

$$\nabla_0^2(\partial_r F_z - \partial_z F_r) = 0. \tag{2.171}$$

2.2.6 Half-space. Application of the generalized Love functions

In [49], the functions F_z and F_r were applied to solve the boundary value problem for the half-space Ω (2.119). It is assumed that

$$u_z = f, \qquad u_r = g, \qquad \varphi_\theta = \varphi \qquad \text{on } \partial\Omega, \tag{2.172}$$

where f, g, $\varphi : \partial\Omega \to \mathbb{R}$ are given functions and we assume conditions (2.7).

In the problem we leave out body loadings. We will discuss the solution by the superposition method in two ways, by means of the function F_z and the function F_r. We will determine the functions F_z^0 and F_r^0 from the equations

$$\nabla^2\nabla^2 F_z^0 = 0, \qquad \nabla_0^2\nabla_0^2 F_r^0 = 0, \qquad \text{in } \Omega, \tag{2.173}$$

respectively, with the boundary conditions

$$u_z^0 = f, \qquad u_r^0 = g, \qquad \text{on } \partial\Omega. \tag{2.174}$$

Moreover, $(u_r^0, u_z^0) \to (0,0)$ for $R = \sqrt{r^2+z^2} \to \infty$.

After applying to equations (2.173) the Hankel transformation (0.51), the representation (2.159) and the representation

$$
\begin{aligned}
\mu u_r^0 &= \nabla_0^2 F_r^0 - \tfrac{1}{2}(1-\nu)^{-1}\partial_r\left(r^{-1}+\partial_r\right)F_r^0, \\
\mu u_z^0 &= -\tfrac{1}{2}(1-\nu)^{-1}\partial_z\left(r^{-1}+\partial_r\right)F_r^0, \\
2\mu\varphi_\theta^0 &= \nabla_0^2\partial_z F_r^0,
\end{aligned}
\tag{2.175}
$$

we arrive at the classical result (for the transforms $\tilde{u}_{z0}^0$, $\tilde{u}_{r1}^0$, $\tilde{\varphi}_{\theta1}^0$):

$$\begin{aligned}\tilde{u}_{z0}^{0} &= \left[\tilde{f}_0 + (\tilde{f}_0 - \tilde{g}_1)\frac{\xi z}{3-4\nu}\right]e^{-\xi z},\\ \tilde{u}_{r1}^{0} &= \left[\tilde{g}_1 + (\tilde{f}_0 - \tilde{g}_1)\frac{\xi z}{3-4\nu}\right]e^{-\xi z},\\ \tilde{\varphi}_{\theta 1}^{0} &= (\tilde{f}_0 - \tilde{g}_1)\frac{2\xi(1-\nu)}{3-4\nu}e^{-\xi z},\end{aligned} \tag{2.176}$$

where $\tilde{f}_0$, $\tilde{g}_1$ are the transforms of the functions f, g, respectively.

The functions F'_z, F'_r are found from the following equations in Ω:

$$\begin{aligned}\nabla^2\nabla^2(l^2\nabla^2 - 1)F'_z &= 0,\\ \nabla_0^2\nabla_0^2(l^2\nabla_0^2 - 1)F'_r &= 0.\end{aligned} \tag{2.177}$$

We take the transforms of these functions in the form

$$\begin{aligned}\tilde{F}'_{z0} &= (C + D\xi z)e^{-\xi z} + Ee^{-\rho z},\\ \tilde{F}'_{r1} &= (\bar{C} + \bar{D}\xi z)e^{-\xi z} + \bar{E}e^{-\rho z},\end{aligned} \tag{2.178}$$

where

$$\rho = \sqrt{\xi^2 + \frac{1}{l^2}}, \qquad l^2 = \frac{(\gamma+\varepsilon)(\mu+\alpha)}{4\alpha\mu}.$$

The three constants: C, D, E and $\bar{C}$ $\bar{D}$, $\bar{E}$, are determined from conditions (2.172) after taking into account the boundary effect for a micropolar body:

$$u'_z = 0, \qquad u'_r = 0, \qquad \varphi'_\theta = \varphi - \varphi_\theta^0 \qquad \text{on } \partial\Omega, \tag{2.179}$$

where $\varphi_\theta^0 = \frac{1}{2}(\partial_z u_r^0 - \partial_r u_z^0)\big|_{z=0}$, by applying the representation (2.162) (for $M'_\theta = 0$), and the representation in Ω:

$$\begin{aligned}2\mu u'_r &= [-(1-\nu)^{-1}(l^2\nabla_0^2 - 1)\partial_r\left(r^{-1} + \partial_r\right)\\ &\quad - \mu_0^{-1}\nabla_0^2\partial_z^2 + 2\nabla_0^2(l^2\nabla_0^2 - 1)]F'_r,\\ 2\mu u'_z &= [-(1-\nu)^{-1}(l^2\nabla^2 - 1) + \mu_0^{-1}\nabla^2]\partial_z\left(r^{-1} + \partial_r\right)F'_r,\\ 2\mu\varphi'_\theta &= -\nabla_0^2\partial_z F_r,\end{aligned} \tag{2.180}$$

respectively.

Let us make the following observation. If, in conditions (2.179), we assume that on $\partial\Omega$ it holds $\varphi(r) - \varphi_\theta^0(0, r) = 0$, then we will obtain that $u'_z \equiv 0$, $u'_r \equiv 0$, $\varphi'_\theta \equiv 0$. However, a micropolar effect in the form of a couple-stress field will appear in the half-space (there exist rotations $\varphi_\theta^0(r, z) \neq 0$ and material constants γ, ε):

$$\mu_{r\theta}^0 = (\gamma+\varepsilon)\partial_r\varphi_\theta^0 - (\gamma-\varepsilon)r^{-1}\varphi_\theta^0, \qquad \mu_{\theta r}^0 = (\gamma-\varepsilon)\partial_r\varphi_\theta^0 - (\gamma+\varepsilon)r^{-1}\varphi_\theta^0,$$

$$\mu_{z\theta}^0 = (\gamma+\varepsilon)\partial_z\varphi_\theta^0, \qquad \mu_{\theta z}^0 = (\gamma-\varepsilon)\partial_z\varphi_\theta^0.$$

Suppose that $\varphi(r) \stackrel{\text{def}}{=} 0$. For the pair (u_r^0, u_z^0) we may model the boundary conditions in such a way that we obtain $\varphi_\theta^0 \equiv 0$ with a simultaneous appearance of a displacement field; e.g., it is sufficient to take $f(r) = (\mathcal{J}_1(\xi r)/\mathcal{J}_0(\xi r)) \cdot g(r)$ where $\mathcal{J}_0(\xi r)$ and $\mathcal{J}_1(\xi r)$ are the Bessel functions. We obtain: $u_z' \equiv 0$, $u_r' \equiv 0$, $\varphi_\theta \equiv 0$ and $\underset{\sim}{\mu} \equiv 0$. In a micropolar half-space, a classical effect will take place: the displacement field u_r^0, u_z^0 and the stress field σ_{rr}^0, σ_{zz}^0, $\sigma_{\theta\theta}^0$, $\sigma_{zr}^0 = \sigma_{rz}^0$. This conclusion may be generalized: modelling classical conditions on $\partial\Omega$ for a simply-connected micropolar body $\overline{\Omega} = \Omega \cup \partial\Omega$ in such a way that $\operatorname{curl}\mathbf{u}^0 = \mathbf{0}$ in $\overline{\Omega}$, and assuming homogenous non-classical conditions: $\boldsymbol{\varphi} = \mathbf{0}$ on $\partial\Omega$ (or $\underset{\sim}{\mu} = 0$ on $\partial\Omega$), we will see that a micropolar body does not exhibit the effect of microstructure.

2.2.7 Method of potentials

The discussion on the application of the method of potentials described by vector equations (1.147) and (1.237)–(1.239) to two-dimensional problems of elastostatics (plane and axially-symmetric deformation states) has an interesting practical aspect and is presented in detail in [49]. Here, we give some observations and results referring to the axial problem discussed, where $\mathbf{u} \stackrel{\text{def}}{=} (u_r, 0, u_z)$, $\boldsymbol{\varphi} \stackrel{\text{def}}{=} (0, \varphi_\theta, 0)$, $\mathbf{X} \stackrel{\text{def}}{=} (X_r, 0, X_z)$, $\mathbf{Y} \stackrel{\text{def}}{=} (0, Y_\theta, 0)$. A formal procedure leads to the conclusion that a scalar potential Φ and vector potentials $\boldsymbol{\Psi} \equiv (0, \Psi_\theta, 0)$ and $\mathbf{H} \equiv (H_r, 0, H_z)$ will appear in the problem. We represent body loadings X_r, Y_θ and X_z in the form (2.75) and (2.76). The representation for u_r, φ_θ, u_z has the form (2.69) in Ω, with the condition $\operatorname{div}\mathbf{H} = 0$. The functions Φ, Ψ_θ, H_r, H_z are related through equations which follow from equations (1.237)–(1.239). We shall demonstrate that by taking into account the distinct features of the first axially-symmetric problem, we will get rid of the vector $\mathbf{H}$. From equation $(1.237)_2$ we obtain the following equation in Ω:

$$[(\gamma + \varepsilon)\nabla_0^2 - 4\alpha](\partial_z H_r - \partial_r H_z) - 2\alpha \nabla_0^2 \Psi_\theta + \nabla_0^2 \eta = 0. \tag{2.181}$$

The condition $\operatorname{div}\mathbf{H} = 0$ requires that we introduce a function H; then we obtain

$$H_r = \partial_z H, \qquad H_z = -\left(r^{-1} + \partial_r\right) H, \qquad \varphi_\theta = \nabla_0^2 H. \tag{2.182}$$

Equation (2.181) takes now the form

$$\nabla_0^2 \Psi_\theta = 2(k^2 \nabla_0^2 - 1)\varphi_\theta + \frac{1}{2\alpha}\nabla_0^2 \eta, \tag{2.183}$$

where $k^2 = \dfrac{\gamma + \varepsilon}{4\alpha}$, and relates the functions Ψ_θ and φ_θ.

From equation $(1.237)_1$ we obtain two equations

$$\begin{aligned}
&(2\mu+\lambda)\nabla_0^2\partial_r\Phi-(\mu+\alpha)\nabla_0^2\partial_z\Psi_\theta-2\alpha\nabla_0^2H_r+\partial_r\vartheta-\partial_z\chi_\theta=0,\\
&(2\mu+\lambda)\nabla^2\partial_z\Phi-(\mu+\alpha)\nabla^2\left(r^{-1}+\partial_r\right)\Psi_\theta-2\alpha\nabla^2H_z\\
&\qquad+\partial_z\vartheta+\left(r^{-1}+\partial_r\right)\chi_\theta=0,
\end{aligned}\tag{2.184}$$

from which we eliminate $\nabla_0^2\Psi_\theta$, $\nabla_0^2 H$ by means of (2.182) and (2.183).

We obtain the following equations in Ω:

$$\begin{aligned}
&(2\mu+\lambda)\nabla_0^2\partial_r\Phi-2\mu(l^2\nabla_0^2-1)\partial_z\varphi_\theta\\
&\qquad-\frac{\mu+\alpha}{2\alpha}\nabla_0^2\partial_z\eta+\partial_r\vartheta-\partial_z\chi_\theta=0,\\
&(2\mu+\lambda)\nabla^2\partial_z\Phi+2\mu(l^2\nabla^2-1)\left(r^{-1}+\partial_r\right)\varphi_\theta\\
&\qquad+\frac{\mu+\alpha}{2\alpha}\nabla^2\left(r^{-1}+\partial_r\right)\eta+\partial_z\vartheta+\left(r^{-1}+\partial_r\right)\chi_\theta=0,
\end{aligned}\tag{2.185}$$

where the functions Φ, φ_θ satisfy

$$\begin{aligned}
&(2\mu+\lambda)\nabla^2\nabla^2\Phi+\nabla^2\vartheta=0,\\
&2\mu\nabla_0^2(l^2\nabla_0^2-1)\varphi_\theta+\frac{\mu+\alpha}{2\alpha}\nabla_0^2\nabla_0^2\eta+\nabla_0^2\chi_\theta=0.
\end{aligned}\tag{2.186}$$

The displacements are found from the formulae

$$u_r=\partial_r\Phi-\partial_z\Psi_\theta,\qquad u_z=\partial_z\Phi+\left(r^{-1}+\partial_r\right)\Psi_\theta,\tag{2.187}$$

where Ψ_θ satisfies equation (2.183).

Let us stress the following observation. Constructing representations on the basis of vector equations (1.147) and (1.237) allows for still another approach to the problem, different than that in the case of a representation associated with equations (2.185). It consists in eliminating not the vector $\mathbf{H}$, but the vector $\boldsymbol{\Psi}$ (as we did in the dynamical problem, Subsection 2.1.5).

Using (2.182) we obtain from equation (2.181)

$$\Psi_\theta=2(k^2\nabla_0^2-1)H+\frac{1}{2\alpha}\eta.$$

Now we obtain the following representation in Ω (we disregard the body loadings):

$$\begin{aligned}
&u_r=\partial_r\Phi-2(k^2\nabla_0^2-1)\partial_zH,\\
&u_z=\partial_z\Phi+2(k^2\nabla_0^2-1)\left(r^{-1}+\partial_r\right)H,\\
&\varphi_\theta=\nabla_0^2H,
\end{aligned}\tag{2.188}$$

where Φ, H satisfy

$$\nabla_0^2\nabla_0^2\Phi = 0, \qquad \nabla_0^2\nabla_0^2(l^2\nabla_0^2 - 1)H = 0 \tag{2.189}$$

and the differential relations

$$\begin{aligned} &\partial_r\Phi - \frac{2\mu}{2\mu+\lambda}(l^2\nabla_0^2 - 1)\partial_z H = 0, \\ &\partial_z\Phi + \frac{2\mu}{2\mu+\lambda}(l^2\nabla^2 - 1)\left(r^{-1} + \partial_r\right)H = 0. \end{aligned} \tag{2.190}$$

Equations (2.189) and (2.190) may be written in a different manner. The functions Φ, φ_θ are determined from

$$\nabla^2\nabla^2\Phi = 0, \qquad \nabla_0^2(l^2\nabla_0^2 - 1)\varphi_\theta = 0 \tag{2.191}$$

and the following relations in Ω must hold:

$$\begin{aligned} &\nabla_0^2\partial_r\Phi - \frac{2\mu}{2\mu+\lambda}(l^2\nabla_0^2 - 1)\partial_z\varphi_\theta = 0, \\ &\nabla^2\partial_z\Phi + \frac{2\mu}{2\mu+\lambda}(l^2\nabla^2 - 1)\left(r^{-1} + \partial_r\right)\varphi_\theta = 0, \end{aligned} \tag{2.192}$$

with the representation (2.188) remaining valid.

Let us make another important observation. Disregarding the body loadings and assuming the correspondence

$$\Psi_\theta \mapsto -\partial_r\psi, \qquad \varphi_\theta \mapsto \partial_r\vartheta \tag{2.193}$$

we obtain from (2.183) and (2.185)–(2.187) the representation

$$u_r = \partial_r\Phi + \partial_{rz}^2\psi, \qquad u_z = \partial_z\Phi - (\nabla^2 - \partial_z^2)\psi, \qquad \varphi_\theta = \partial_r\vartheta, \tag{2.194}$$

the equations

$$\begin{aligned} &\nabla^2\Phi - \frac{2\mu}{2\mu+\lambda}(l^2\nabla^2 - 1)\partial_z\vartheta = 0, \\ &\nabla^2\partial_z\Phi + \frac{2\mu}{2\mu+\lambda}(l^2\nabla^2 - 1)\left(r^{-1} + \partial_r\right)\partial_r\vartheta = 0 \end{aligned} \tag{2.195}$$

as well as

$$\begin{aligned} &\nabla^2\nabla^2\Phi = 0, \qquad \nabla^2(l^2\nabla^2 - 1)\vartheta = 0, \\ &\nabla^2\psi = -\frac{1}{2\alpha}[(\gamma+\varepsilon)\nabla^2 - 4\alpha]\vartheta. \end{aligned} \tag{2.196}$$

This is the representation of Nowacki [183], which may be directly derived from the system of equations (2.101).

2.2.8 Half-space. Application of the method of potentials

Let us consider the boundary value problem (2.172) for the half-space Ω (2.119), using representation (2.187) or (2.188) and the loading problem

$$\sigma_{zz} = -p(r), \qquad \sigma_{zr} = -s(r), \qquad \mu_{z\theta} = -m(r) \qquad \text{on } \partial\Omega, \tag{2.197}$$

where p, s, $m : \partial\Omega \to \mathbb{R}$ are given functions. We postulate now that physical fields vanish at infinity. From equations (2.189) we take the potentials Φ and H in the form of the following transforms:

$$\tilde{\Phi}_0 = (A + B\xi z)e^{-\xi z}, \qquad \tilde{H}_1 = (C + D\xi z)e^{-\xi z} + Ee^{-\rho z}, \tag{2.198}$$

where

$$\rho = \sqrt{\xi^2 + \frac{1}{l^2}}, \qquad l^2 = \frac{(\gamma+\varepsilon)(\mu+\alpha)}{4\alpha\mu}.$$

Out of the five constants $A(\xi)$–$E(\xi)$, we eliminate two, C, D, satisfying the transformed differential relations (2.190), and find the remaining three E, A, B from the three transformed boundary conditions, (2.172) or (2.197). The stresses σ_{zz}, σ_{zr}, $\mu_{z\theta}$ are expressed in terms of the potentials Φ, H by means of the constitutive relations (2.33) and the geometric relations (2.34). We may also assume the functions Φ, φ_θ in the form

$$\tilde{\Phi}_0 = (A + B\xi z)e^{-\xi z}, \qquad \tilde{\varphi}_{\theta 1} = Fe^{-\xi z} + Ge^{-\rho z} \tag{2.199}$$

from equations (2.191).

Out of the four constants, A, B, F, G, we eliminate one by fulfilling relations (2.192):

$$B = \frac{2\mu}{2\mu+\lambda} \cdot \frac{1}{2\xi} F.$$

Since boundary conditions are expressed in terms of the function H, the equation $\nabla_0^2 H = \varphi_\theta$ must be solved to yield

$$\tilde{H}_1 = -\frac{1}{2\xi} Fze^{-\xi z} + l^2 Ge^{-\rho z}. \tag{2.200}$$

Three constants A, F, G are determined from the boundary conditions (2.172) or (2.197).

2.2.9 Half-space (E-N) with an inside heat source. Thermoelastostatics

Dyszlewicz & Sikora [77] have given a solution of the half-space (E-N) problem with an inside heat source and free boundary, slightly different than the solution derived in [43].

On the basis of [77] we shall briefly discuss the method of solving the half-space (E-N) problem with heat sources inside and free boundary. Basic relations and differential equations for the axially-symmetric problem of thermoelastostatics without body forces and moments take the form of (2.83), (2.84) (for $X_r = X_z = Y_\theta = 0$; we disregard the dynamical terms $\rho\ddot{u}_r$, $J\ddot{\varphi}_\theta$, $\rho\ddot{u}_z$), (2.85)–(2.87) and (2.89). In addition, we have in Ω:

— the compatibility equations:

$$\begin{aligned}
&\kappa_{z\theta} + r\partial_z\kappa_{\theta r} = 0,\\
&\kappa_{r\theta} + \partial_r(r\kappa_{\theta r}) = 0,\\
&\kappa_{z\theta} + \partial_r\gamma_{zz} - \partial_z\gamma_{rz} = 0,\\
&\kappa_{\theta r} + \partial_z\gamma_{\theta\theta} - r^{-1}\gamma_{zr} = 0,\\
&\partial_z(r\gamma_{\theta\theta}) - \gamma_{rr} = 0;
\end{aligned} \tag{2.201}$$

— the constitutive relations:

$$\begin{aligned}
\sigma_{rr} &= 2\mu\gamma_{rr} + \lambda e - \nu_T T,\\
\sigma_{\theta\theta} &= 2\mu\gamma_{\theta\theta} + \lambda e - \nu_T T,\\
\sigma_{zz} &= 2\mu\gamma_{zz} + \lambda e - \nu_T T,\\
\sigma_{rz} &= (\mu+\alpha)\gamma_{rz} + (\mu-\alpha)\gamma_{zr},\\
\sigma_{zr} &= (\mu+\alpha)\gamma_{zr} + (\mu-\alpha)\gamma_{rz},\\
\mu_{r\theta} &= (\gamma+\varepsilon)\kappa_{r\theta} + (\gamma-\varepsilon)\kappa_{\theta r},\\
\mu_{\theta r} &= (\gamma+\varepsilon)\kappa_{\theta r} + (\gamma-\varepsilon)\kappa_{r\theta},\\
\mu_{z\theta} &= (\gamma+\varepsilon)\kappa_{z\theta},\\
\mu_{\theta z} &= (\gamma-\varepsilon)\kappa_{z\theta}.
\end{aligned} \tag{2.202}$$

The temperature[15] T satisfies the equation of heat conduction

$$\nabla^2 T = -\frac{W}{\lambda_0} \qquad \text{in } \Omega, \tag{2.203}$$

where the function W determines the intensity of the heat sources inside the body, whereas λ_0 is the coefficient of heat conduction, $\nu_T = (3\lambda + 2\mu)\alpha_t$, α_t being the thermal coefficient of linear expansion of the medium.

Substituting the constitutive relations (2.202) to equations (2.84) (for statics and without body loadings) and using the geometric relations (2.86), we obtain the equations of equilibrium in Ω expressed in terms of displacements, rotations and temperature in the form (the compatibility equations (2.201) become identities):

---

[15] Since θ stands for the position variable, we denote (only in this subsection and Subsection 3.2.3) the temperature by T.

$$\begin{aligned}
&(\mu+\alpha)\nabla_0^2 u_r + (\lambda+\mu-\alpha)\partial_r e - 2\alpha\partial_z\varphi_\theta = \nu_T\partial_r T,\\
&(\mu+\alpha)\nabla^2 u_z + (\lambda+\mu-\alpha)\partial_z e + 2\alpha r^{-1}\partial_r(r\varphi_\theta) = \nu_T\partial_z T, \qquad (2.204)\\
&[(\gamma+\varepsilon)\nabla_0^2 - 4\alpha]\varphi_\theta + 2\alpha(\partial_z u_r - \partial_r u_z) = 0,
\end{aligned}$$

where

$$\nabla^2 = \partial_r^2 + r^{-1}\partial_r + \partial_z^2,\ \nabla_0^2 = \nabla^2 - r^{-2} \qquad \text{and} \qquad e = r^{-1}\partial_r(ru_r) + \partial_z u_z.$$

We consider the half-space $\overline{\Omega} = \Omega \cup \partial\Omega$, where

$$\Omega = \{(r,\theta,z) : z > 0,\ 0 \le \theta < 2\pi,\ 0 \le r < \infty\}.$$

In the interior of $\overline{\Omega}$ we have a pointwise heat source (temperature kernel) of the form

$$W(r,z) = W_0(2\pi r)^{-1}\delta(r)\delta(z-\zeta), \qquad (2.205)$$

where $\zeta > 0$ and W_0 is a positive constant. On the boundary $\partial\Omega$ of the half-space we assume homogenous thermo-mechanical conditions

$$\sigma_{zz} = 0, \qquad \sigma_{zr} = 0, \qquad \mu_{z\theta} = 0, \qquad (2.206)$$

$$T = 0. \qquad (2.207)$$

As we have noticed earlier, the problem is characterized by axial symmetry, i.e. causes and effects do not depend on the angle θ. This is illustrated in Fig. 2.7.

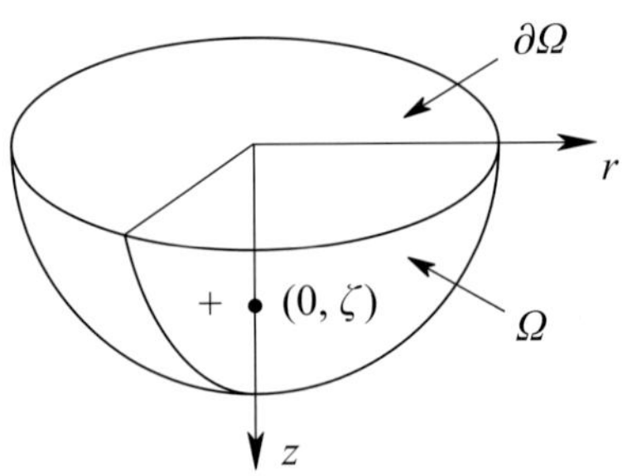

Fig. 2.7.

In addition, we postulate that the physical fields vanish at infinity, i.e.

$$\begin{aligned}
&\mathbf{u}\to\mathbf{0},\ \boldsymbol{\varphi}\to\mathbf{0},\ \underset{\sim}{\gamma}\to\underset{\sim}{0},\ \underset{\sim}{\kappa}\to\underset{\sim}{0},\ \underset{\sim}{\sigma}\to\underset{\sim}{0},\ \underset{\sim}{\mu}\to\underset{\sim}{0}\\
&\text{and } T\to 0 \qquad \text{for } R=\sqrt{r^2+z^2}\to\infty.
\end{aligned} \qquad (2.208)$$

The solution of the problem can be obtained by the superposition method, as in [43]. The procedure is illustrated by Fig. 2.8.

	I stage	**II stage**	**III stage**
$\sigma_{zz} = 0,\ \sigma_{zr} = 0,$ $\mu_{z\theta} = 0,\ T = 0.$	$\bar{\sigma}^0_{zr} \neq 0,\ \bar{\sigma}^0_{zz} = 0,$ $\bar{\mu}^0_{z\theta} = 0,\ T = 0.$	$\bar{\bar{\sigma}}^0_{zr} = -\bar{\sigma}^0_{zr},$ $\bar{\bar{\sigma}}^0_{zz} = 0,\ \bar{\bar{\mu}}^0_{z\theta} \neq 0.$	$\sigma'_{zr} = 0,\ \sigma'_{zz} = 0,$ $\mu'_{z\theta} = -\bar{\bar{\mu}}^0_{z\theta}.$

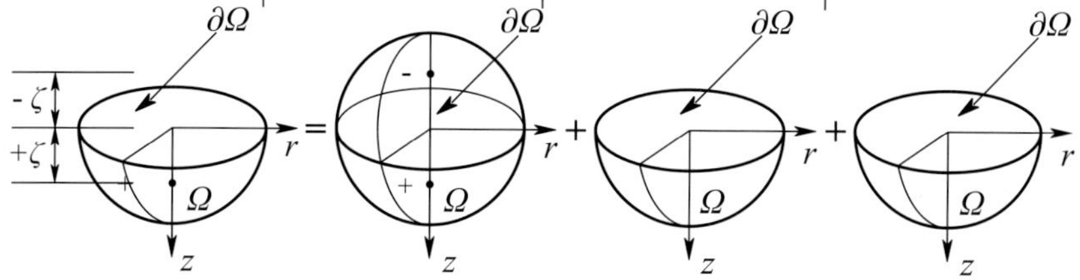

Fig. 2.8.

$$f(r, z) = \bar{f}^0(r, z) + \bar{\bar{f}}^0(r, z) + f'(r, z).$$

The physical quantity sought for consists of the classical solution in the form $\bar{f}^0 + \bar{\bar{f}}^0$ and the complementary solution f'. Such an effect is obtained if we introduce to (2.204) the Schaefer vector

$$\zeta_\theta = \tfrac{1}{2}(\partial_z u_r - \partial_r u_z) - \varphi_\theta, \tag{2.209}$$

(see [43, p. 49]). Substituting (2.209) to (2.204) and assuming that $\zeta_\theta = \zeta^0_\theta + \zeta'_\theta$, where $\zeta^0_\theta = 0$, we obtain two systems of equations in Ω equivalent to (2.204), namely:

— the equations of classical thermoelastostatics:

$$\begin{aligned} \mu\nabla_0^2 u_r^0 + (\lambda+\mu)\partial_r e^0 &= -\nu_T \partial_r T, \\ \mu\nabla_0^2 u_z^0 + (\lambda+\mu)\partial_z e^0 &= -\nu_T \partial_z T, \end{aligned} \tag{2.210}$$

where $e^0 = r^{-1}\partial_r(ru_r^0) + \partial_z u_z^0$;

— the complementary equations (without temperature):

$$\begin{aligned} &\mu\nabla_0^2 u'_r + (\lambda+\mu)\partial_r e' = -2\alpha\partial_z\zeta'_\theta, \\ &\mu\nabla^2 u'_z + (\lambda+\mu)\partial_z e' = 2\alpha r^{-1}\partial_r(r\zeta'_\theta), \\ &k^2\nabla_0^2(\partial_z u'_r - \partial_r u'_z) = 2(k^2\nabla_0^2 - 1)\zeta'_\theta, \end{aligned} \tag{2.211}$$

where $e' = r^{-1}\partial_r(ru'_r) + \partial_z u'_z$ and $k^2 = \dfrac{\gamma+\varepsilon}{4\alpha}$.

Now we shall be concerned with solving the equations of classical thermoelastostatics (2.210). Let

$$u_r^0 = \bar{u}_r^0 + \bar{\bar{u}}_r^0, \qquad u_z^0 = \bar{u}_z^0 + \bar{\bar{u}}_z^0, \qquad \varphi_\theta^0 = \bar{\varphi}_\theta^0 + \bar{\bar{\varphi}}_\theta^0. \tag{2.212}$$

To equations (2.210) we introduce the thermoelastic potential $\Phi(r, z)$:

$$\bar{u}_r^0 = \partial_r \Phi, \qquad \bar{u}_z^0 = \partial_z \Phi. \tag{2.213}$$

Thus we obtain the following equation in Ω:

$$\nabla^2 \Phi = mT, \tag{2.214}$$

where $m = \dfrac{\nu_T}{2\mu + \lambda}$. The fundamental solution of the heat conduction equation (2.203) for the source $W(r,z) = W_0(2\pi r)^{-1}\delta(r)\delta(z)$ has the form

$$T = \frac{W_0}{4\pi\lambda_0} \cdot \frac{1}{R}, \tag{2.215}$$

where $R = \sqrt{r^2 + z^2}$.

Using (2.215), we obtain from equation (2.214) the thermoelastic potential Φ in the form

$$\Phi = \frac{mW_0}{8\pi\lambda_0} R. \tag{2.216}$$

The triple $(\bar{u}_r^0, \bar{u}_z^0, \bar{\varphi}_\theta^0)$ can be determined by means of (2.213), (2.216) and the relation

$$\bar{\varphi}_\theta^0 = \tfrac{1}{2}(\partial_z \bar{u}_r^0 - \partial_r \bar{u}_z^0). \tag{2.217}$$

Now, in the first step (see Fig. 2.8), we proceed as follows. In the infinite space we place the source $W(r,z)$ at the point $(0,\zeta)$ and the source $-W(r,z)$ — at the point $(0,-\zeta)$. Using (2.215), we obtain the distribution of the temperature field T in the region Ω, which satisfies the boundary condition (2.207):

$$T(r,z) = W_0(4\pi\lambda_0)^{-1}(R_1^{-1} - R_2^{-1}), \tag{2.218}$$

where

$$R_1 = \sqrt{r^2 + (z-\zeta)^2}, \qquad R_2 = \sqrt{r^2 + (z+\zeta)^2}.$$

Here, the thermoelastic potential has the form

$$\Phi(r,z) = mW_0(8\pi\lambda_0)^{-1}(R_1 - R_2) = A_0(R_1 - R_2), \tag{2.219}$$

where $A_0 = \dfrac{mW_0}{8\pi\lambda_0}$.

On the basis of formulae (2.213) we determine

$$\begin{aligned} \bar{u}_r^0 &= A_0 r(R_1^{-1} - R_2^{-1}), \\ \bar{u}_z^0 &= A_0[(z-\zeta)R_1^{-1} - (z+\zeta)R_2^{-1}]. \end{aligned} \tag{2.220}$$

From (2.217) and (2.220) we obtain

$$\bar{\varphi}_\theta^0 \equiv 0 \tag{2.221}$$

and thus, in view of the constitutive relations $(2.202)_{6-9}$ and the geometric relations (2.86), we conclude that

$$\bar{\mu}^0_{z\theta} \equiv \bar{\mu}^0_{\theta z} \equiv \bar{\mu}^0_{r\theta} \equiv \bar{\mu}^0_{\theta r} \equiv 0. \tag{2.222}$$

From formulae (2.220), (2.221) and relations (2.86) and $(2.202)_{1-5}$ we determine the classical force-stresses with one bar:

$$\begin{aligned}
\bar{\sigma}^0_{rr} &= -2\mu A_0[(1+r^2R_1^{-2})R_1^{-1} - (1+r^2R_2^{-2})R_2^{-1}],\\
\bar{\sigma}^0_{\theta\theta} &= -2\mu A_0(R_1^{-1} - R_2^{-1}),\\
\bar{\sigma}^0_{zz} &= -2\mu A_0[(2-r^2R_1^{-2})R_1^{-1} - (2-r^2R_2^{-2})R_2^{-1}],\\
\bar{\sigma}^0_{rz} &= \bar{\sigma}^0_{zr} = -2\mu A_0 r[(z-\zeta)R_1^{-3} - (z+\zeta)R_2^{-3}].
\end{aligned} \tag{2.223}$$

The stresses (2.223) do not satisfy the mechanical boundary conditions (2.206), since on $\partial\Omega$ we have

$$\bar{\sigma}^0_{zz} = 0, \qquad \bar{\sigma}^0_{zr} = 4\mu A_0\zeta r R_0^{-3}, \qquad \bar{\mu}^0_{z\theta} = 0, \tag{2.224}$$

where $R_0 = \sqrt{r^2+\zeta^2}$. Hence, the above choice of the particular solution leads (in the second step — see Fig. 2.8) to the problem of the half-space Ω with the conditions on $\partial\Omega$ of the form

$$\bar{\bar{\sigma}}^0_{zz} + \sigma'_{zz} = 0, \qquad \bar{\bar{\sigma}}^0_{zr} + \sigma'_{zr} = -4\mu A_0\zeta r R_0^{-3}, \qquad \bar{\bar{\mu}}^0_{z\theta} + \mu'_{z\theta} = 0. \tag{2.225}$$

The displacements $\bar{\bar{u}}^0_r$ and $\bar{\bar{u}}^0_z$ satisfy equations (2.210) without the temperature field. Their general forms for the half-space can be obtained from the equations

$$\nabla_0^2\nabla_0^2\bar{\bar{u}}^0_r = 0, \qquad \nabla^2\nabla^2\bar{\bar{u}}^0_z = 0, \qquad \text{in } \Omega \tag{2.226}$$

with the boundary conditions

$$\bar{\bar{\bar{\sigma}}}^0_{zz0} = 0, \qquad \bar{\bar{\bar{\sigma}}}^0_{zr1} = -4\mu A_0\zeta e^{-\zeta\xi} \qquad \text{on } \partial\Omega. \tag{2.227}$$

The displacements $\bar{\bar{u}}^0_r$ and $\bar{\bar{u}}^0_z$ are given by

$$\begin{aligned}
\bar{\bar{u}}^0_r &= 2A_0\zeta r R_2^{-1}\left\{2(1-\nu)[R_2+(z+\zeta)]^{-1} + zR_2^{-2}\right\},\\
\bar{\bar{u}}^0_z &= -2A_0\zeta R_2^{-1}[(1-2\nu) + z(z+\zeta)]R_2^{-2}].
\end{aligned} \tag{2.228}$$

The formula for the rotation $\bar{\bar{\varphi}}^0_\theta$ takes the form

$$\bar{\bar{\varphi}}^0_\theta = -4A_0\zeta(1-\nu)rR_2^{-3}. \tag{2.229}$$

Next, we determine the classical force-stresses with two bars:

$$\begin{aligned}
\bar{\bar{\sigma}}^0_{zr} &= \bar{\bar{\sigma}}^0_{rz} = -4\mu A_0\zeta r R_2^{-3}[1 - 3z(z+\zeta)R_2^{-2}],\\
\bar{\bar{\sigma}}^0_{zz} &= -4\mu A_0\zeta z R_2^{-3}[1 - 3(z+\zeta)^2R_2^{-2}],\\
\bar{\bar{\sigma}}^0_{rr} &= 4\mu A_0\zeta\left\{2(z+\zeta)R_2^{-3} + 2(1-\nu)r^{-2}[(z+\zeta)R_2^{-1} - 1]\right.\\
&\quad \left. + zR_2^{-3}(3r^2R_2^{-2} - 1)\right\},\\
\bar{\bar{\sigma}}^0_{\theta\theta} &= 4\mu A_0\zeta\left\{2\nu(z+\zeta)R_2^{-3}\right.\\
&\quad \left. - 2(1-\nu)r^{-2}[(z+\zeta)R_2^{-1} - 1] - zR_2^{-3}\right\}
\end{aligned} \tag{2.230}$$

and the generated non-classical couple-stresses of the form

$$\begin{aligned}
\bar{\bar{\mu}}^0_{z\theta} &= 24\mu A_0\zeta a_0 r(z+\zeta)R_2^{-5},\\
\bar{\bar{\mu}}^0_{r\theta} &= -8\mu A_0\zeta a_0 R_2^{-3}[2\varepsilon(\gamma+\varepsilon)^{-1} - 3r^2R_2^{-2}],\\
\bar{\bar{\mu}}^0_{\theta r} &= 8\mu A_0\zeta a_0 R_2^{-3}[2\varepsilon(\gamma+\varepsilon)^{-1} + 3r^2R_2^{-2}].
\end{aligned} \tag{2.231}$$

In particular, formula $(2.231)_1$ for $\bar{\bar{\mu}}^0_{z\theta}$ on $\partial\Omega$ takes the form

$$\bar{\bar{\mu}}^0_{z\theta} = 8\mu A_0\zeta a_0 \mathcal{H}_1\left[\!\left[\xi e^{-\zeta\xi}; \xi\to r\right]\!\right], \tag{2.232}$$

where $\mathcal{H}_1\,[\![.]\!]$ denotes the Hankel transform (0.51) of the Bessel function $\mathcal{J}_1(\xi r)$.

We still have to determine the complementary triple $(u'_z, \varphi'_\theta, u'_r)$, the third superposition step (see Fig. 2.8). Eliminating the vector ζ'_θ from the system of equations (2.211), we obtain a system of equations for the triple $(u'_r, \varphi'_\theta, u'_z)$ which is identical to (2.204) without the temperature T. This is the system of equations (2.141), for which the following boundary conditions must be satisfied on $\partial\Omega$:

$$\sigma'_{zz} = 0, \qquad \sigma'_{zr} = 0, \qquad \mu'_{z\theta} = -8\mu A_0\zeta a_0\mathcal{H}_1\left[\!\left[\xi e^{-\zeta\xi}; \xi\to r\right]\!\right]. \tag{2.233}$$

In [43, pp. 54–59], we gave the solution to the complementary boundary value problem, i.e. to the system of equations (2.141) with the boundary conditions (2.233), by using the analogy between boundary conditions, which can be noticed in the process of solving the first axially-symmetric problem with the boundary conditions of Boussinesq type, i.e.

$$\sigma_{zz} = -p(r), \qquad \sigma_{zr} = 0, \qquad \mu_{z\theta} = 0 \qquad \text{on } \partial\Omega \tag{2.234}$$

and the thermoelastic axial problem of the (E-N) model with the conditions (2.206) and (2.207) on $\partial\Omega$. To see that, use the relation

$$\tilde{\mu}'_{z\theta 1} = 2a_0\xi\tilde{p}_0(\xi) \tag{2.235}$$

in the boundary conditions (2.142), where $\tilde{p}_0(\xi) = \mathcal{H}_0\,[\![p(r); r\to\xi]\!]$ is the transform of the given loading normal to $\partial\Omega$ and satisfying the condition

$$\int_0^\infty \sqrt{r}\,|p(r)|\,dr < +\infty,$$

and take into account that, in view of (2.233), we have

$$\tilde{\mu}'_{z\theta 1} = -8\mu A_0\zeta a_0\xi e^{-\zeta\xi}. \tag{2.236}$$

Now, it is enough to assume in the solution of Boussinesq type (2.234) the substitution

$$\tilde{p}_0(\xi) = -4\mu A_0 \zeta e^{-\zeta\xi} \tag{2.237}$$

to obtain the complementary solution $(u'_r, \varphi'_\theta, u'_z)$, i.e. the solution to the boundary value problem (2.233) for the system of equations (2.141).

In [77], on the other hand, to solve the system of equations (2.141) and the boundary conditions (2.233), we used (also for instructive purposes) the displacement-rotation potentials in Ω of the form (see [58, p. 140]):

$$\begin{aligned} u'_r &= \partial_r \Phi - \partial_z \Psi_\theta, \\ u'_z &= \partial_z \Phi + \left(r^{-1} + \partial_r\right) \Psi_\theta \end{aligned} \tag{2.238}$$

and

$$\nabla_0^2 \Psi_\theta = 2(k^2 \nabla_0^2 - 1)\varphi'_\theta \tag{2.239}$$

where $k^2 = \dfrac{\gamma + \varepsilon}{4\alpha}$.

Substituting (2.238) to (2.141) and using (2.239), we obtain in Ω a system of equations for Φ and φ'_θ in the form

$$\begin{aligned} &(\lambda + 2\mu)\nabla_0^2 \partial_r \Phi - 2\mu(l^2 \nabla_0^2 - 1)\partial_z \varphi'_\theta = 0, \\ &(\lambda + 2\mu)\nabla^2 \partial_z \Phi + 2\mu(l^2 \nabla^2 - 1)\left(r^{-1} + \partial_r\right)\varphi'_\theta = 0. \end{aligned} \tag{2.240}$$

Separating the system of equations (2.240), we arrive at

$$\nabla_0^2 (l^2 \nabla_0^2 - 1)\varphi'_\theta = 0, \qquad \nabla^2 \nabla^2 \Phi = 0 \qquad \text{in } \Omega. \tag{2.241}$$

Applying the integral Hankel transformation (0.51) and satisfying the differential relations (2.240) and the condition (2.208), we obtain

$$\begin{aligned} \tilde{\Phi}_0 &= (A' + B'\xi z)e^{-\xi z}, \\ \tilde{\varphi}'_{\theta 1} &= C' e^{-\xi z} + D' e^{-\rho z}, \end{aligned} \tag{2.242}$$

where

$$B' = \frac{\mu}{2\mu + \lambda} \cdot \frac{1}{\xi^2} C'.$$

The potential Ψ_θ can be determined from equation (2.239) by use of $(2.242)_2$. Thus we have

$$\tilde{\Psi}_{\theta 1} = 2D'(k^2 - l^2)e^{-\rho z} + \xi^{-1} C' z e^{-\xi z}. \tag{2.243}$$

Using formulae (2.238), (2.242) and (2.243), we obtain the complementary triple $(u'_z, \varphi'_\theta, u'_r)$ exactly in the form of (2.146) if we take

$$A = -(\xi^{-1} C' + \xi A'), \qquad B = \frac{1}{2(1-\nu)} \cdot \frac{1}{\xi} C', \qquad C = -\frac{\rho}{\mu_0} D'. \tag{2.244}$$

The constants A, B and C are determined from the boundary conditions (2.233). Finally, we arrive at

$$
\begin{aligned}
u'_r = & \frac{1}{2\mu}\mathcal{H}_1\left[\!\!\left[\frac{1}{\xi}\tilde{p}_0(\xi)\left\{\left(1-\frac{1}{\Delta_0}\right)[(1-2\nu)-\xi z]e^{-\xi z}\right.\right.\right.\\
& \left.\left.\left.+\frac{2a_0\xi^2}{\Delta_0}(e^{-\rho z}-\xi\rho^{-1}e^{-\xi z})\right\};\xi\to r\right]\!\!\right],\\
u'_z = & \frac{1}{2\mu}\mathcal{H}_0\left[\!\!\left[\frac{1}{\xi}\tilde{p}_0(\xi)\left\{\left(1-\frac{1}{\Delta_0}\right)[2(1-\nu)+\xi z]e^{-\xi z}\right.\right.\right.\\
& \left.\left.\left.-\frac{2a_0\xi^2}{\Delta_0}\frac{\xi}{\rho}(e^{-\rho z}-e^{-\xi z})\right\};\xi\to r\right]\!\!\right],\\
\varphi'_\theta = & -2(1-\nu)\frac{1}{2\mu}\mathcal{H}_1\left[\!\!\left[\tilde{p}_0(\xi)\left\{\left(1-\frac{1}{\Delta_0}\right)e^{-\xi z}\right.\right.\right.\\
& \left.\left.\left.+\frac{1}{\Delta_0}\frac{\xi}{\rho}e^{-\rho z}\right\};\xi\to r\right]\!\!\right],
\end{aligned}
\tag{2.245}
$$

where

$$
\tilde{p}_0(\xi) = -4\mu A_0\zeta e^{-\zeta\xi},\quad \Delta_0 = 1+2a_0\xi^2\left(1-\frac{\xi}{\rho}\right),\quad a_0 = \frac{(\gamma+\varepsilon)(2\mu+\lambda)}{4\mu(\mu+\lambda)}.
$$

Next, let us write the formulae for the stresses which enter the boundary conditions:

$$
\begin{aligned}
\sigma'_{zz} = & \mathcal{H}_0\left[\!\!\left[\tilde{p}_0(\xi)\left\{\left(1-\frac{1}{\Delta_0}\right)(1+\xi z)e^{-\xi z}\right.\right.\right.\\
& \left.\left.\left.+2a_0\xi^2\frac{1}{\Delta_0}\left(\frac{\xi}{\rho}e^{-\xi z}-e^{-\rho z}\right)\right\};\xi\to r\right]\!\!\right],\\
\sigma'_{zr} = & \mathcal{H}_1\left[\!\!\left[\tilde{p}_0(\xi)\left\{\left(1-\frac{1}{\Delta_0}\right)\xi z e^{-\xi z}\right.\right.\right.\\
& \left.\left.\left.+2a_0\xi^2\frac{1}{\Delta_0}\frac{\xi}{\rho}(e^{-\xi z}-e^{-\rho z})\right\};\xi\to r\right]\!\!\right],\\
\sigma'_{z\theta} = & 2a_0\mathcal{H}_1\left[\!\!\left[\xi\tilde{p}_0(\xi)\left[\left(1-\frac{1}{\Delta_0}\right)e^{-\xi z}+\frac{1}{\Delta_0}e^{-\rho z}\right];\xi\to r\right]\!\!\right].
\end{aligned}
\tag{2.246}
$$

In [77], we presented a number of interesting remarks. Let us give a few of them at the end of this subsection.

Remark 1. We have found out that the influence of the complementary terms on the solution of the problem increases with decreasing ζ (the depth at which there is a heat source).

This observation has enabled us determine the leading part of the complementary solution. Assuming small depth ζ and using the asymptotic expansions (1.345), we have obtained approximate expressions for the Hankel integrals of types (2.245) and (2.246), valid in the neighbourhood of the point $(0,0)$ for each θ. The approximate formulae have been obtained in the form

$$u_r = A_0\zeta\left\{\frac{r}{\zeta}\left(\frac{1}{R_1}-\frac{1}{R_2}\right)\Bigg|^{I} + \left(\frac{4(1-\nu)r}{R_2[R_2+(z+\zeta)]}-\frac{2rz}{R_2^3}\right)\Bigg|^{II}\right.$$
$$\left. -\frac{4a_0}{l^2+a_0}\left[(1-\nu)\frac{r}{R_2[R_2+(z+\zeta)]}-\frac{rz}{R_2^3}\right]\Bigg|^{III}\right\}.$$

Remark 2. If we want to obtain the solution of the problem with the following conditions on $\partial\Omega$:

$$\sigma_{zz}=0, \qquad \sigma_{zr}=0, \qquad \mu_{z\theta}=0, \qquad T=\frac{T_0}{2\pi r}\delta(r), \tag{2.247}$$

where T_0 is a positive constant, from the results for the half-space with a heat source inside, it is sufficient (see [77]) to assume

$$A_0(\zeta)\stackrel{\text{def}}{=}\frac{mT_0}{8\pi}\cdot\frac{1}{\zeta}, \qquad \frac{W_0}{\lambda_0}\stackrel{\text{def}}{=}\frac{T_0}{\zeta}$$

and then evaluate the limit $\zeta\to 0^+$.

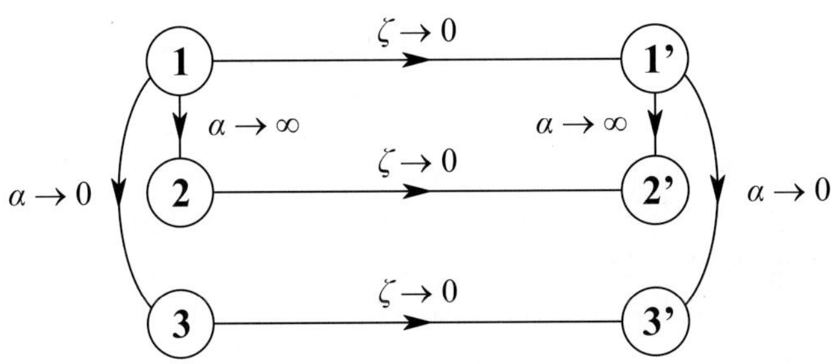

Fig. 2.9.

In [77], we studied the limiting transitions which are illustrated in Fig. 2.9, where numbers (1) and (1′) refer to the (E-N) model, (2) and (2′) — to the (K-M) model, finally, (3) and (3′) — to the classical model (H). In all cases, the obtained results were in agreement with the results of other authors (see [58, p. 131]).

Remark 3. We have found out [77] that the orders of singularities of physical fields in the half-space Ω with the boundary conditions (2.247) are higher (grow as $1/R$) than in the problem with a heat source inside the body (Fig. 2.7, the boundary conditions (2.206) and (2.207)). Only the rotations φ_θ behave differently. This is illustrated by Tables 2.2 and 2.3.

Table 2.2. Refers to the problem of the half-space with a heat source inside

Physical fields	(E-N) model	(K-M) model	(H) model
$u_r,\ u_z$	$O(1)$	$O(1)$	$O(1)$
φ_θ	$O(1)$	$O(1)$	$O(1)$
$\sigma_{zz},\ \sigma_{zr},\ \sigma_{rz}$	$O(R^{-1})$	$O(R^{-1})$	$O(R^{-1})$
$\mu_{z\theta}$	$O(1)$	$O(1)$	–
T	$O(R^{-1})$	$O(R^{-1})$	$O(R^{-1})$

Table 2.3. Refers to the boundary value problem (2.247)

Physical fields	(E-N) model	(K-M) model	(H) model
$u_r,\ u_z$	$O(R^{-1})$	$O(R^{-1})$	$O(R^{-1})$
φ_θ	$O(1)$	$O(1)$	$O(R^{-2})$
$\sigma_{zz},\ \sigma_{zr},\ \sigma_{rz}$	$O(R^{-2})$	$O(R^{-2})$	$O(R^{-2})$
$\mu_{z\theta}$	$O(R^{-1})$	$O(R^{-1})$	–
T	$O(R^{-2})$	$O(R^{-2})$	$O(R^{-2})$

Remark 4. From the analysis of the graphs we conclude that the respective graphs for the (H) model and the (K-M) model are the envelopes for the graphs for the (E-N) model.

Remark 5. If we assume the vertical Oz axis to be the direction of displacement, then we observe (for three models: (E-N), (K-M) and (H)) the phenomenon of decreasing concentration of stresses as the distance from the heat source increases, which confirms the Saint Venant principle.

Remark 6. The fields of couple-stresses $\mu_{z\theta}$ and rotations φ_θ do not exhibit singularities in the neighbourhood of the heat source.

Moreover, in [77] we carried out an extensive numerical analysis of the obtained results, based on the material constants (0.52). Below we shall present a number of graphs illustrating the analytical results Fig. 2.10–2.25.

Graphs of temperature T

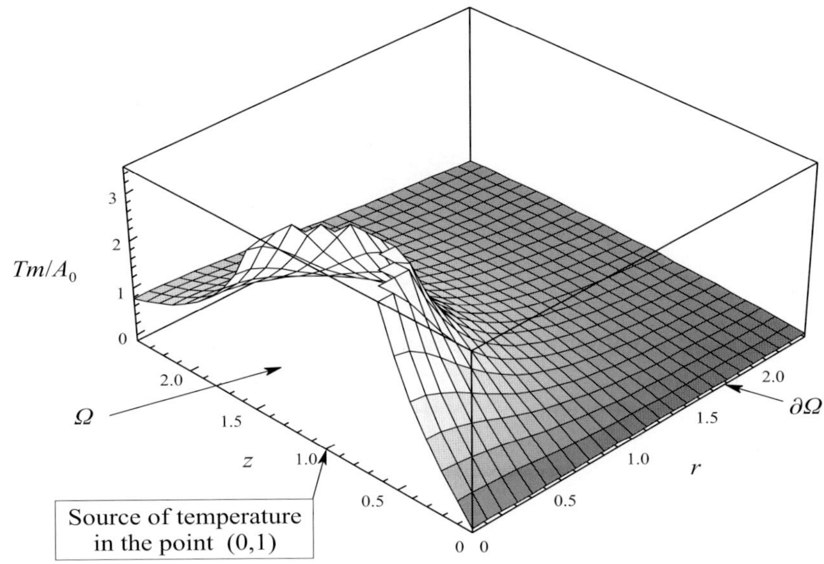

Fig. 2.10. 3D graph of temperature $T(r, z)$

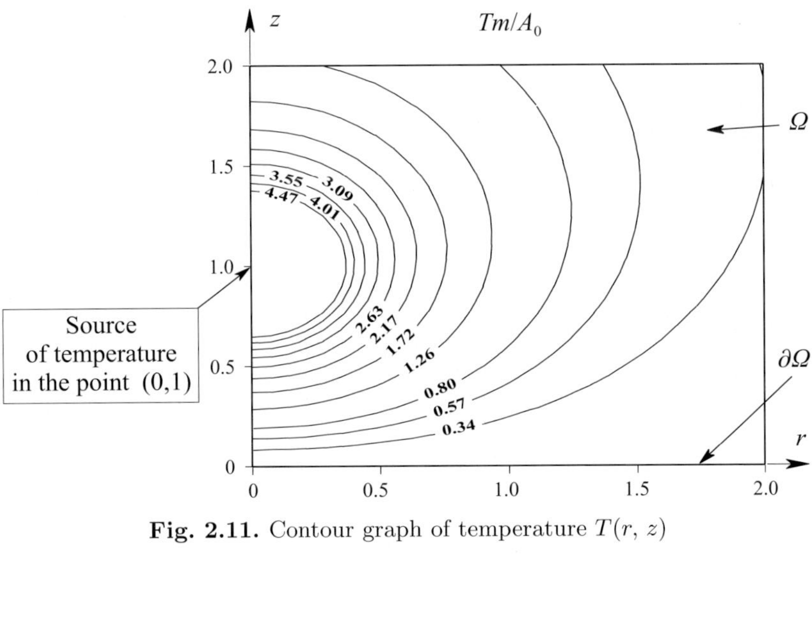

Fig. 2.11. Contour graph of temperature $T(r, z)$

Graphs of stresses σ_{zz}

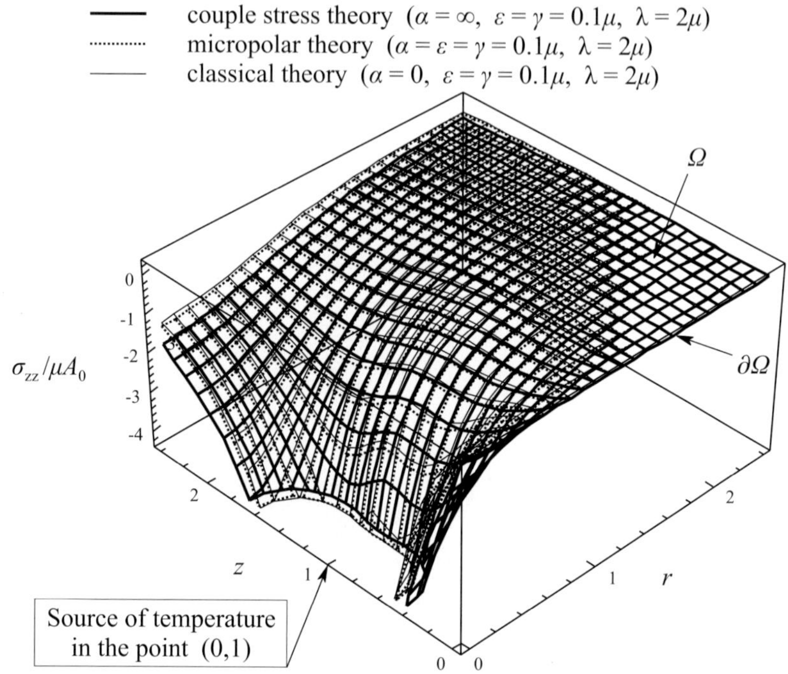

Fig. 2.12. 3D graph of stresses $\sigma_{zz}(r, z)$

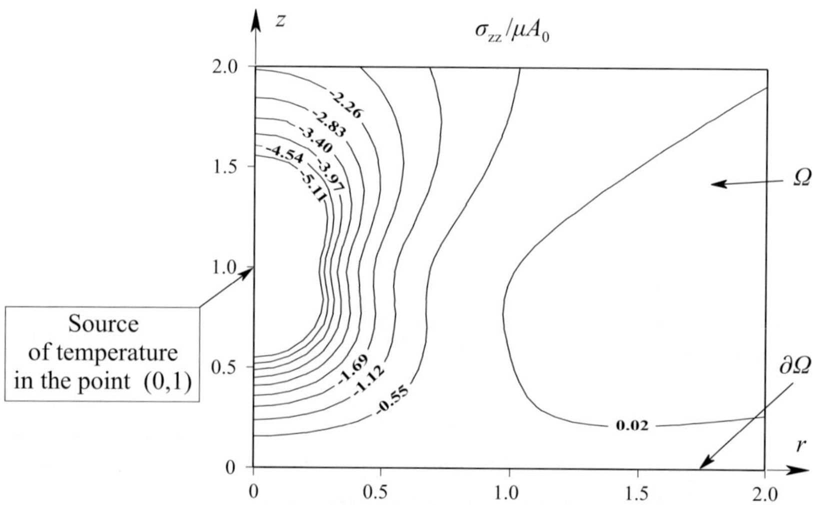

Fig. 2.13. Contour graph of stresses $\sigma_{zz}(r, z)$ for micropolar theory

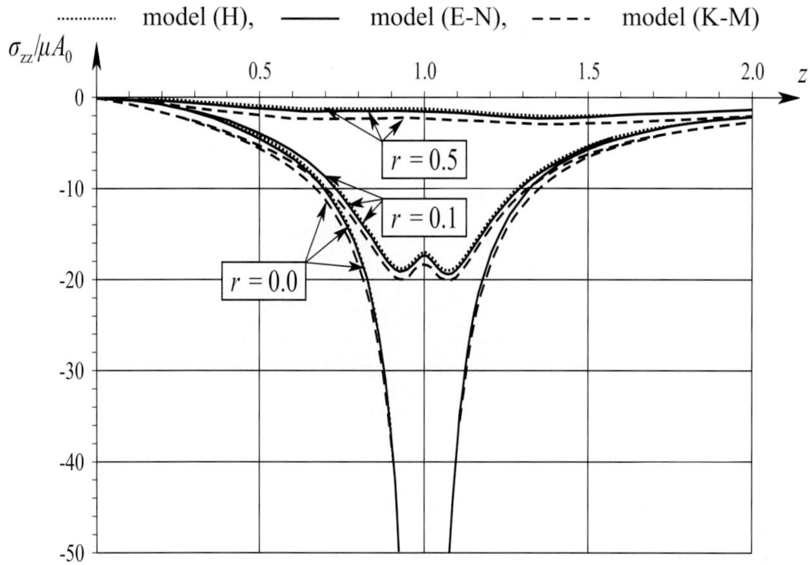

Fig. 2.14. The graph of σ_{zz} versus depth z for the temperature kernel, for $\zeta = 1$

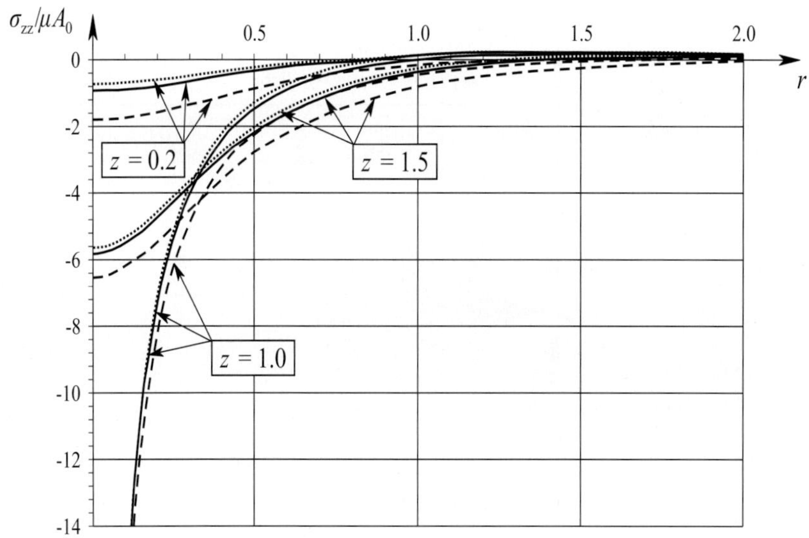

Fig. 2.15. The graph of σ_{zz} versus radius r for the temperature kernel, for $\zeta = 1$

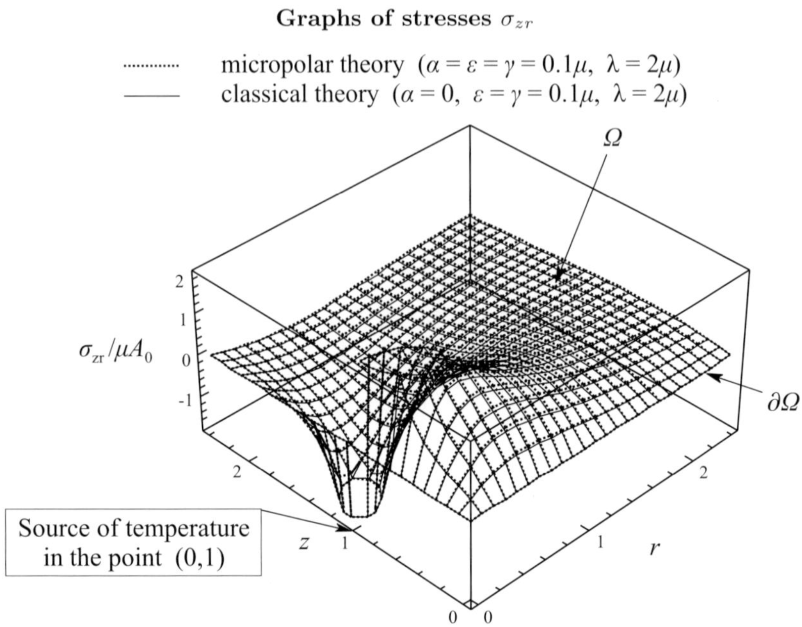

Fig. 2.16. 3D graph of stresses $\sigma_{zr}(r, z)$

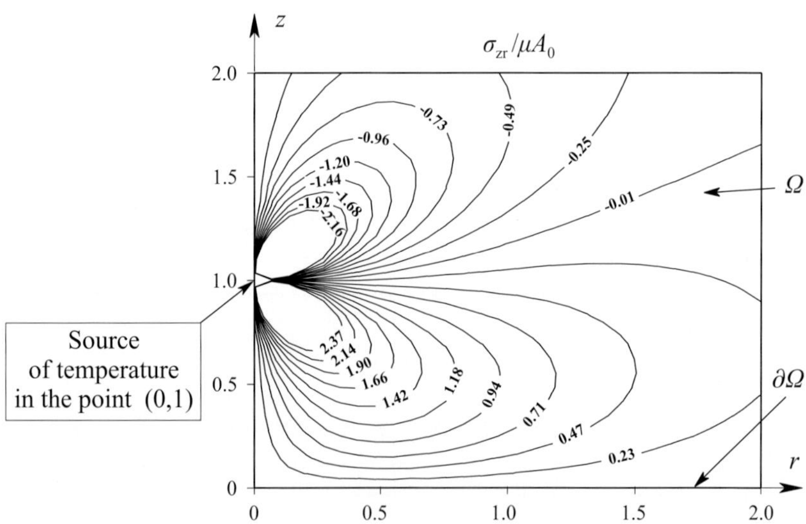

Fig. 2.17. Contour graph of stresses $\sigma_{zr}(r, z)$ for micropolar theory

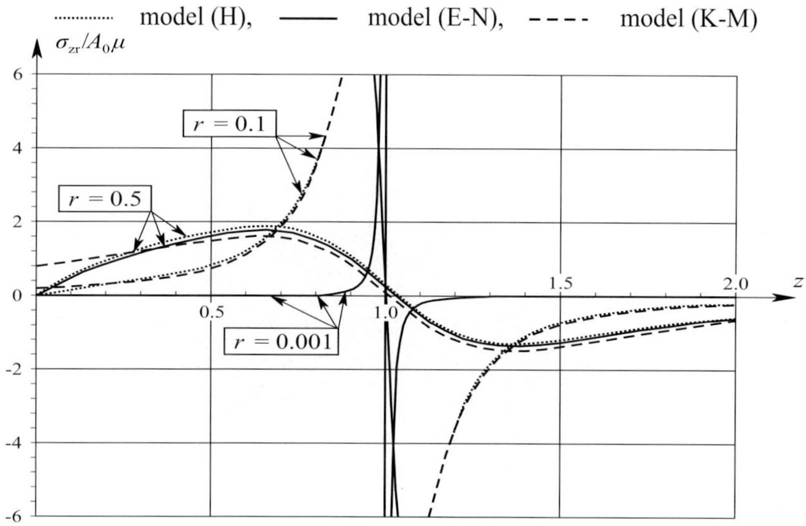

Fig. 2.18. The graph of σ_{zr} versus depth z for the temperature kernel, for $\zeta = 1$

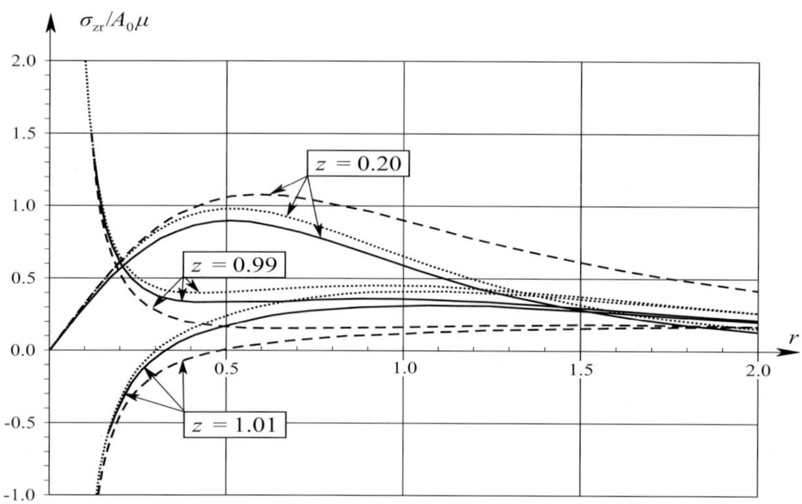

Fig. 2.19. The graph of σ_{zr} versus radius r for the temperature kernel, for $\zeta = 1$

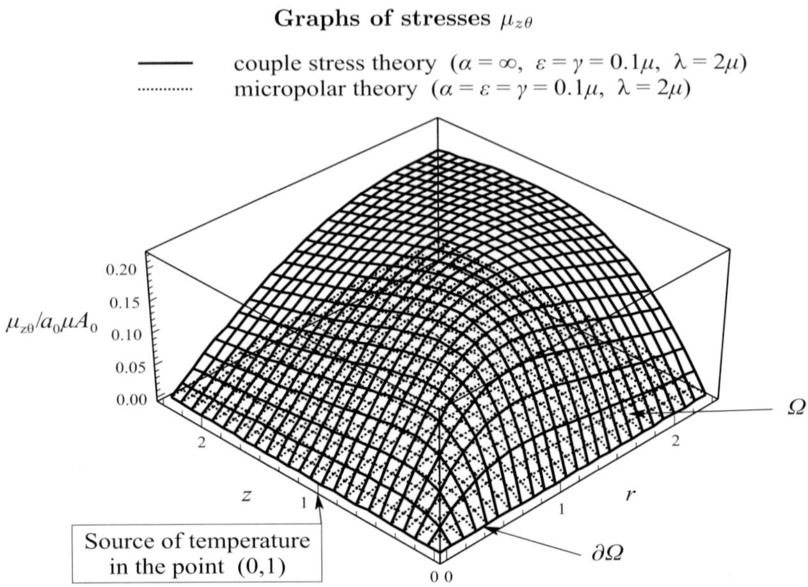

Fig. 2.20. 3D graph of couple stresses $\mu_{z\theta}(r, z)$

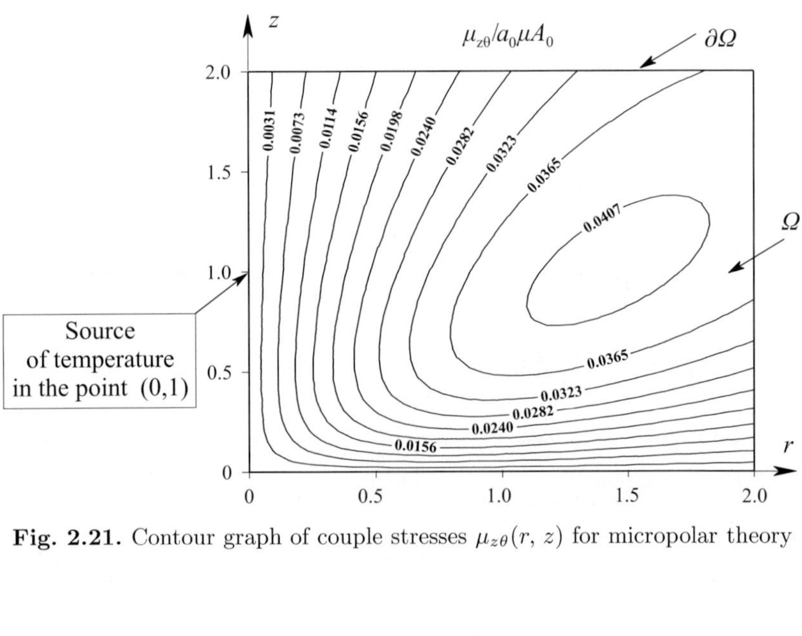

Fig. 2.21. Contour graph of couple stresses $\mu_{z\theta}(r, z)$ for micropolar theory

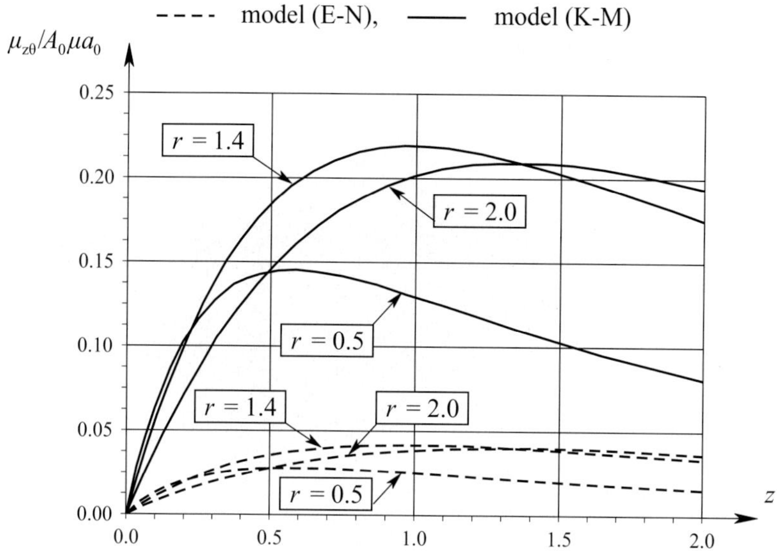

Fig. 2.22. The graph of $\mu_{z\theta}$ versus depth z for the temperature kernel, for $\zeta = 1$

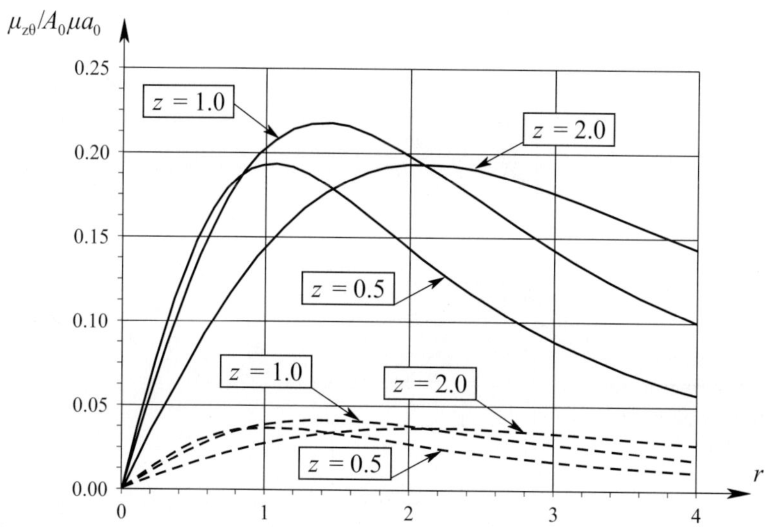

Fig. 2.23. The graph of $\mu_{z\theta}$ versus radius r for the temperature kernel, for $\zeta = 1$

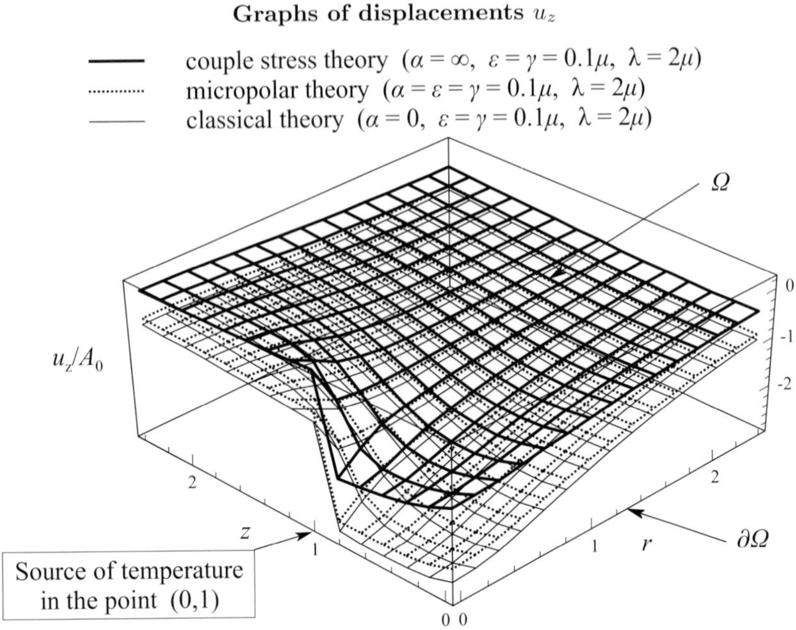

Fig. 2.24. 3D graph of displacements $u_z(r, z)$

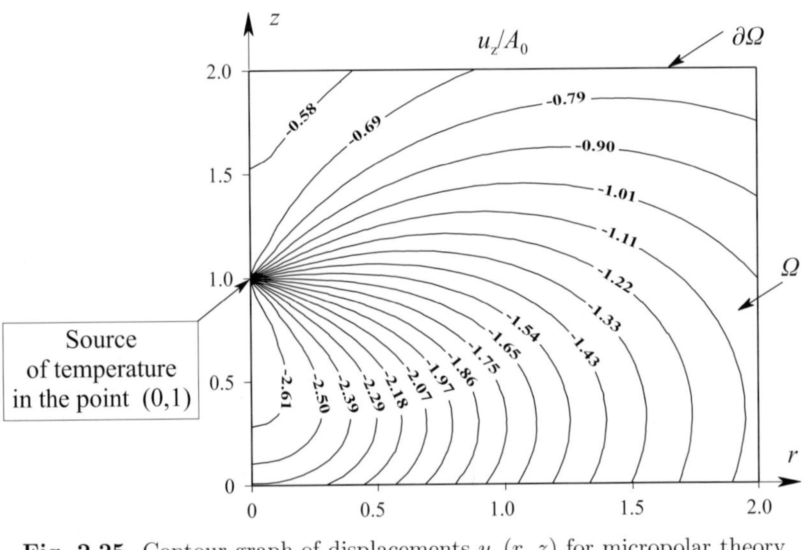

Fig. 2.25. Contour graph of displacements $u_z(r, z)$ for micropolar theory

2.3 The second axially-symmetric problem. Elastodynamics

2.3.1 Equations in displacements and rotations. Body loadings

From vector equations (1.1) we obtain for the triple $(\varphi_r, u_\theta, \varphi_z)$ the following equations in $\Omega \times \mathbf{T}_+$:

$$
\begin{aligned}
&\Box_4^0 \varphi_r + (\beta+\gamma-\varepsilon)\partial_r \kappa - 2\alpha \partial_z u_\theta + Y_r = 0, \\
&\Box_4 \varphi_z + (\beta+\gamma-\varepsilon)\partial_z \kappa + 2\alpha r^{-1}\partial_r(r u_\theta) + Y_z = 0, \\
&\Box_2^0 u_\theta + 2\alpha(\partial_z \varphi_r - \partial_r \varphi_z) + X_\theta = 0,
\end{aligned}
\tag{2.248}
$$

where $\kappa = r^{-1}\partial_r(r\varphi_r) + \partial_z \varphi_z$, $\quad Y_r, X_\theta, Y_z$ — body loadings.

In [64] and [49] uncoupled equations for the triple $(\varphi_r, u_\theta, \varphi_z)$ in $\Omega \times \mathbf{T}_+$ were given:

$$
\begin{aligned}
\Box_3^0(\Box_2^0\Box_4^0 + 4\alpha^2\nabla_0^2)\varphi_r = &- 2\alpha\partial_z \Box_3^0 X_\theta - \Box_2^0\Box_3^0 Y_r \\
&+ [(\beta+\gamma-\varepsilon)\Box_2^0 - 4\alpha^2]\partial_r Y, \\
\Box_3(\Box_2\Box_4 + 4\alpha^2\nabla^2)\varphi_z = &\,[(\beta+\gamma-\varepsilon)\Box_2 - 4\alpha^2]\partial_z Y \\
&+ 2\alpha\Box_3[r^{-1}\partial_r(rX_\theta)] - \Box_2\Box_3 Y_z, \\
(\Box_2^0\Box_4^0 + 4\alpha^2\nabla_0^2)u_\theta = &\,2\alpha(\partial_z Y_r - \partial_r Y_z) - \Box_4^0 X_\theta,
\end{aligned}
\tag{2.249}
$$

where $Y = r^{-1}\partial_r(rY_r) + \partial_z Y_z$.

Recall the notation for operators:

$$
\begin{aligned}
\Box_2 &= (\mu+\alpha)\nabla^2 - \rho\partial_t^2, \\
\Box_3 &= (\beta+2\gamma)\nabla^2 - 4\alpha - J\partial_t^2, \\
\Box_4 &= (\gamma+\varepsilon)\nabla^2 - 4\alpha - J\partial_t^2,
\end{aligned}
$$

where $\nabla^2 = \partial_r^2 + r^{-1}\partial_r + \partial_z^2$.

The operators $\Box_2^0$, $\Box_3^0$, $\Box_4^0$ are determined by $\nabla_0^2 = \nabla^2 - r^{-2}$. We obtain starred operators from the corresponding unstarred operators by putting $\alpha = 0$; e.g., $\overset{*}{\Box}{}_2^0 = \mu\nabla_0^2 - \rho\partial_t^2$ etc.

By using equations of type (2.249), the fundamental solution for the triple $(\varphi_r, u_\theta, \varphi_z)$ has been found in the case of the infinite space loaded by a body moment Y_z of the form

$$
Y_z = \frac{M_0}{2\pi r} e^{-i\omega t}\delta(r)\delta(z), \qquad Y_r = X_\theta = 0. \tag{2.250}
$$

Having applied the Fourier-Hankel integral transformations, we obtain the result in the closed form:

$$
\begin{aligned}
\varphi_r &= \frac{M_0 e^{-i\omega t}}{4\pi J\omega^2}\partial^2_{rz}\left(\frac{\sigma_4^2 A_2}{\lambda_1^2}\cdot\frac{e^{i\lambda_1 R}}{R}+\frac{\sigma_4^2 A_1}{\lambda_2^2}\cdot\frac{e^{i\lambda_2 R}}{R}-\frac{\sigma_3^2}{\lambda_3^2}\cdot\frac{e^{i\lambda_3 R}}{R}\right),\\
\varphi_z &= \frac{M_0 e^{-i\omega t}}{4\pi J\omega^2}\left[\sigma_4^2 A_2\frac{e^{i\lambda_1 R}}{R}+\sigma_4^2 A_1\frac{e^{i\lambda_2 R}}{R}\right.\\
&\quad\left.+\,\partial_z^2\left(\frac{\sigma_4^2 A_2}{\lambda_1^2}\cdot\frac{e^{i\lambda_1 R}}{R}+\frac{\sigma_4^2 A_1}{\lambda_2^2}\cdot\frac{e^{i\lambda_2 R}}{R}-\frac{\sigma_3^2}{\lambda_3^2}\cdot\frac{e^{i\lambda_3 R}}{R}\right)\right],\\
u_\theta &= -\frac{M_0 s e^{-i\omega t}}{4\pi J c_4^2(\lambda_1^2-\lambda_2^2)}\partial_r\left(\frac{e^{i\lambda_1 R}-e^{i\lambda_2 R}}{R}\right).
\end{aligned}
\tag{2.251}
$$

The notations in formulae (2.251) are the same as in formulae (1.26)–(1.28), $R=\sqrt{r^2+z^2}$, and M_0 is of unit value. Formulae (2.251) were derived by the method of potentials by Nowacki & Nowacki [197]. In [68] and [69] the limiting cases were derived from formulae (2.251), giving fundamental solutions in the framework of the limiting theories (2)–(8) (Fig. 1.1). For example, by applying the complex limiting procedure $\omega\to 0$, formulae (2.251) are transformed into the formulae:

$$
\begin{aligned}
\varphi_r &= \frac{M_0}{16\pi\alpha}\partial^2_{rz}\left[2\mu_1^2\left(\frac{1-e^{-R/l}}{R}\right)\right.\\
&\quad\left.-\left(\frac{1-e^{-R/\nu_1}}{R}\right)\right],\\
\varphi_z &= -\frac{M_0}{16\pi\alpha}\left[2\mu_1^2(r^{-1}\partial_r+\partial_r^2)\left(\frac{1-e^{-R/l}}{R}\right)\right.\\
&\quad\left.+\,\partial_z^2\left(\frac{1-e^{-R/\nu_1}}{R}\right)\right],\\
u_\theta &= -\frac{M_0}{8\pi\mu}\partial_r\left(\frac{1-e^{-R/l}}{R}\right),
\end{aligned}
\tag{2.252}
$$

where

$$
\mu_1=\sqrt{\frac{\mu+\alpha}{2\mu}},\qquad \nu_1=\sqrt{\frac{2\gamma+\beta}{4\alpha}},\qquad l=\sqrt{\frac{(\gamma+\varepsilon)(\mu+\alpha)}{4\alpha\mu}}.
$$

For $\alpha\to\infty$, for instance, limits (2.8) should be taken and symbols (2.10) and formulae (2.251) are transformed into the formulae which are valid within the framework of couple-stress elastodynamics:

$$
\begin{aligned}
u_\theta &= -\frac{M_0 e^{-i\omega t}}{8\pi\mu l^{\star 2}(\lambda_1^{\star 2}-\lambda_2^{\star 2})}\partial_r\left(\frac{e^{i\lambda_1^\star R}-e^{i\lambda_2^\star R}}{R}\right),\\
\varphi_r &= -\tfrac{1}{2}\partial_z u_\theta,\qquad \varphi_z=\tfrac{1}{2}\left(r^{-1}+\partial_r\right)u_\theta.
\end{aligned}
\tag{2.253}
$$

2.3.2 The generalized Lamb problem

The classical Lamb problem plays the same fundamental role in elastokinetics as the Boussinesq problem in elastostatics. Nowacki & Nowacki [192, pp. 102 & 117] generalized the Lamb problem to micropolar elastokinetics. The authors of the method of potentials provided formulae for displacements and rotations for the half-space in plane and axially-symmetric state of strain. One of these generalizations is the problem of the half-space (2.119) with axially-symmetric loadings on the boundary of tangential stress $\sigma_{z\theta}$ type and couple-stresses μ_{zz} and μ_{zr} type, aperiodically dependent on time t. Dyszlewicz & Kołodziej solved and discussed this problem in [71]. Namely, the method of direct integration of basic equations of the problem was chosen. The starting point is given by the uncoupled equations (2.249) without body loadings for the triple $(\varphi_r, u_\theta, \varphi_z)$, in $\Omega \times \mathbf{T}_+$:

$$\begin{aligned} &\Box_3^0(\Box_2^0\Box_4^0 + 4\alpha^2\nabla_0^2)\varphi_r = 0,\\ &\Box_3(\Box_2\Box_4 + 4\alpha^2\nabla^2)\varphi_z = 0,\\ &(\Box_2^0\Box_4^0 + 4\alpha^2\nabla_0^2)u_\theta = 0 \end{aligned} \tag{2.254}$$

with the conditions:

$$\sigma_{z\theta} = -h_1, \qquad \mu_{zz} = -h_2, \qquad \mu_{zr} = -h_3 \qquad \text{on } \partial\Omega \times \mathbf{T}, \tag{2.255}$$

$$(\varphi_r, u_\theta, \varphi_z) \to (0,0,0) \qquad \text{for } R = \sqrt{r^2+z^2} \to \infty. \tag{2.256}$$

The functions $h_i : \partial\Omega \times \mathbf{T} \to \mathbb{R}$ are given. When solving the stated Lamb problem, the procedure is the following. We apply integral transformations: Fourier's — over the time t and Hankel's — over the position variable r to equations (2.254). In order to conform with the notation of the original works [69] and [71], note that

$$\tilde{f}_\nu(\eta,\tau) = \mathcal{F}\left\{H_\nu\left[\!\left[f(r,t); r \to \eta\right]\!\right]; t \to \tau\right\},$$

$$f(r,t) = \mathcal{F}^\star\left\{H_\nu\left[\!\left[\tilde{f}_\nu(\eta,\tau); \eta \to r\right]\!\right]; \tau \to t\right\},$$

where symbols $\mathcal{F}$, $\mathcal{F}^\star$ and H_ν ($\nu = 0$ or $\nu = 1$) refer to the Fourier and Hankel transformations, respectively. The partial differential equations (2.254) are transformed into ordinary differential equations for the transforms $\tilde{\varphi}_{r1}$, $\tilde{u}_{\theta 1}$, $\tilde{\varphi}_{z0}$:

$$\begin{aligned} &(D_z^2 - \bar{\lambda}_1^2)(D_z^2 - \bar{\lambda}_2^2)\tilde{u}_{\theta 1} = 0,\\ &(D_z^2 - \bar{\lambda}_1^2)(D_z^2 - \bar{\lambda}_2^2)(D_z^2 - \bar{\lambda}_3^2)(\tilde{\varphi}_{r1}, \tilde{\varphi}_{z0}) = (0,0), \end{aligned} \tag{2.257}$$

where $D_z \overset{\text{def}}{=} d/dz$, $\bar{\lambda}_i^2 = \eta^2 - \lambda_i^2$, the expressions λ_i^2 are defined in Subsection 1.1.2.

From such equations we obtain general solutions, and, applying condition (2.256), get them in the form

$$\begin{aligned}\tilde{u}_{\theta 1} &= Ae^{-\bar{\lambda}_1 z} + Be^{-\bar{\lambda}_2 z},\\ \tilde{\varphi}_{r1} &= Ce^{-\bar{\lambda}_1 z} + De^{-\bar{\lambda}_2 z} + Ee^{-\bar{\lambda}_3 z},\\ \tilde{\varphi}_{z0} &= Fe^{-\bar{\lambda}_1 z} + Ge^{-\bar{\lambda}_2 z} + He^{-\bar{\lambda}_3 z}.\end{aligned} \tag{2.258}$$

Coefficients $A, \ldots, H$ are constants of integration depending on the parameters η, τ, which need to be determined. To this end, we transform equations (2.248) (for $Y_r = X_\theta = Y_z = 0$):

$$\begin{aligned}&(\gamma+\varepsilon)(D_z^2 - \eta^2 - \nu_0^2 + \sigma_4^2)\tilde{\varphi}_{r1}\\ &\qquad - (\beta+\gamma-\varepsilon)\eta(\eta\tilde{\varphi}_{r1} + D_z\tilde{\varphi}_{z0}) - 2\alpha D_z\tilde{u}_{\theta 1} = 0,\\ &(\gamma+\varepsilon)(D_z^2 - \eta^2 - \nu_0^2 + \sigma_4^2)\tilde{\varphi}_{z0}\\ &\qquad + (\beta+\gamma-\varepsilon)D_z(D_z\tilde{\varphi}_{z0} + \eta\tilde{\varphi}_{r1}) + 2\alpha\eta\tilde{u}_{\theta 1} = 0,\\ &(\mu+\alpha)(D_z^2 - \eta^2 + \sigma_2^2)\tilde{u}_{\theta 1} + 2\alpha(D_z\tilde{\varphi}_{r1} + \eta\tilde{\varphi}_{z0}) = 0,\end{aligned} \tag{2.259}$$

where

$$\nu_0^2 = \frac{4\alpha}{\gamma+\varepsilon}, \qquad \sigma_2^2 = \frac{\omega}{c_2}, \qquad \sigma_4^2 = \frac{\omega}{c_4}, \qquad c_2 = \sqrt{\frac{\mu+\alpha}{\rho}}, \qquad c_4 = \sqrt{\frac{\gamma+\varepsilon}{J}}.$$

Equations (2.259), transformed in such a way, must be satisfied for every $z > 0$ by the triple $(\tilde{u}_{\theta 1}, \tilde{\varphi}_{r1}, \tilde{\varphi}_{z0})$ from (2.258). Equating the coefficients standing by the exponential functions $\exp(-\bar{\lambda}_j z)$, $(j = 1, 2, 3)$ to zero, one obtains 9 algebraic homogeneous linear equations linking constants $A, \ldots, H$. The largest non-vanishing minor of the main determinant of this system of equations is of rank five[16]. We obtain five linearly independent equations for eight constants $A, \ldots, H$, which enables us to make the solution of the Lamb problem depend on three basic constants C, D, E. The basic constants are determined from the three conditions (2.255) written for $z = 0$ in the form:

$$\begin{aligned}(\mu+\alpha)D_z\tilde{u}_{\theta 1} + 2\alpha\tilde{\varphi}_{r1} &= -\tilde{h}_{11},\\ (\beta+2\gamma)D_z\tilde{\varphi}_{z0} + \eta\beta\tilde{\varphi}_{r1} &= -\tilde{h}_{20},\\ (\gamma+\varepsilon)D_z\tilde{\varphi}_{r1} - \eta(\gamma-\varepsilon)\tilde{\varphi}_{z0} &= -\tilde{h}_{31}.\end{aligned} \tag{2.260}$$

The final result for the triple $(\varphi_r, u_\theta, \varphi_z)$ is obtained in the form of improper Fourier-Hankel integrals in $\overline{\Omega} \times \mathbf{T}$:

---

[16] When analysing this system of equations, it is more convenient, due to operational reasons, to use equations in $\Omega \times \mathbf{T}_+$

$$\square_3[r^{-1}\partial_r(r\varphi_r) + \partial_z\varphi_z] = 0, \qquad (\square_2^0\square_4^0 + 4\alpha^2\nabla_0^2)(\partial_z\varphi_r - \partial_r\varphi_z) = 0.$$

$$
\begin{aligned}
u_\theta &= \frac{1}{\sqrt{2\pi}} \int_{-\infty}^{+\infty} e^{-i\tau t}\, d\tau \times \\
&\quad \int_0^\infty \eta(\kappa_1 \bar{\lambda}_1^{-1} \beta_{j1} \tilde{h}_j e^{-\bar{\lambda}_1 z} + \kappa_2 \bar{\lambda}_2^{-1} \beta_{j2} \tilde{h}_j e^{-\bar{\lambda}_2 z}) \mathcal{J}_1(\eta r)\, d\eta, \\
\varphi_r &= \frac{1}{\sqrt{2\pi}} \int_{-\infty}^{+\infty} e^{-i\tau t}\, d\tau \times \\
&\quad \int_0^\infty \eta(\beta_{j1} \tilde{h}_j e^{-\bar{\lambda}_1 z} + \beta_{j2} \tilde{h}_j e^{-\bar{\lambda}_2 z} + \beta_{j3} \tilde{h}_j e^{-\bar{\lambda}_3 z}) \mathcal{J}_1(\eta r)\, d\eta, \\
\varphi_z &= \frac{1}{\sqrt{2\pi}} \int_{-\infty}^{+\infty} e^{-i\tau t}\, d\tau \times \\
&\quad \int_0^\infty \eta(\eta \bar{\lambda}_1^{-1} \beta_{j1} \tilde{h}_j e^{-\bar{\lambda}_1 z} + \eta \bar{\lambda}_2^{-1} \beta_{j2} \tilde{h}_j e^{-\bar{\lambda}_2 z} \\
&\quad + \eta^{-1} \bar{\lambda}_3 \beta_{j3} \tilde{h}_j e^{-\bar{\lambda}_3 z}) \mathcal{J}_0(\eta r)\, d\eta.
\end{aligned}
\tag{2.261}
$$

The symbols $\beta_{jl}\tilde{h}_j$ have the form ($l = 1,\ 2,\ 3$):

$$
\begin{aligned}
\beta_{j1}\tilde{h}_j &= \frac{1}{\Delta}[(4\gamma^2\eta^2\bar{\lambda}_1\bar{\lambda}_2\bar{\lambda}_3 - \xi q_2 \bar{\lambda}_1)\tilde{h}_{11} - 2\bar{\lambda}_1 \eta(\alpha q_2 + \gamma r_2 \bar{\lambda}_3)\tilde{h}_{20} \\
&\quad + \bar{\lambda}_1(\xi r_2 + 4\alpha\gamma\eta^2\bar{\lambda}_2)\tilde{h}_{31}], \\
\beta_{j2}\tilde{h}_j &= \frac{1}{\Delta}[-\bar{\lambda}_2(4\gamma^2\eta^2\bar{\lambda}_1\bar{\lambda}_3 - \xi q_1)\tilde{h}_{11} + 2\eta\bar{\lambda}_2(\gamma r_1 \bar{\lambda}_3 + \alpha q_1)\tilde{h}_{20} \\
&\quad - \bar{\lambda}_2(\xi r_1 + 4\alpha\gamma\eta^2\bar{\lambda}_1)\tilde{h}_{31}], \\
\beta_{j3}\tilde{h}_j &= \frac{1}{\Delta}[2\gamma\eta^2(q_2\bar{\lambda}_1 - q_1\bar{\lambda}_2)\tilde{h}_{11} - \eta(r_1 q_2 - r_2 q_1)\tilde{h}_{20} \\
&\quad + 2\gamma\eta^2(r_1\bar{\lambda}_2 - r_2\bar{\lambda}_1)\tilde{h}_{31}],
\end{aligned}
\tag{2.262}
$$

where

$$
\begin{aligned}
&\xi = (\beta + 2\gamma)\bar{\lambda}_3^2 - \beta\eta^2, \qquad r_\delta = \bar{\lambda}_\delta[(\mu + \alpha)\kappa_\delta - 2\alpha], \\
&q_\delta = (\gamma + \varepsilon)\bar{\lambda}_\delta^2 + (\gamma - \varepsilon)\eta^2, \qquad \kappa_\delta = \frac{1}{p}(\eta^2 - \bar{\lambda}_\delta^2 + \nu_0^2 - \sigma_4^2), \\
&\Delta = [4\alpha\gamma\eta^2(q_1\bar{\lambda}_2 - q_2\bar{\lambda}_1) - r_1(\xi q_2 - 4\gamma^2\eta^2\bar{\lambda}_2\bar{\lambda}_3) \\
&\qquad + r_2(\xi q_1 - 4\gamma^2\eta^2\bar{\lambda}_1\bar{\lambda}_3)], \\
&\delta = 1,\ 2.
\end{aligned}
\tag{2.263}
$$

Limiting cases referring to theories (3), (5) and (7) (Fig. 1.1) for the result given by (2.261) have been given. Moreover, we also took into account loadings which harmonically vary in time:

$$
h_j(r, t) = f_j(r) \exp(-i\omega t).
$$

Another axial Lamb problem is connected with the triple $(u_r, \varphi_\theta, u_z)$ (the first axially-symmetric problem). In an extensive work [71], results for both axial Lamb problems were derived. The limiting cases for these results referring to limiting theories (3), (5) and (7) (Fig. 1.1) can also be obtained from the fundamental equations of respective theories. Such discussion of limiting cases of the results for Lamb problems leads to many observations and conclusions concerning the correspondence between the results of the theories (1), (3), (5) and (7) (Fig. 1.1). The solution (2.261) represents a formal solution to the Lamb problem for the triple $(\varphi_r, u_\theta, \varphi_z)$. Nevertheless, such results can be of theoretical and practical importance[17] (they may be helpful in studying surface waves in the half-space and in experimental verification of the micropolar theory).

2.3.3 Stress equations of motion problem (SEMP)

Let us consider the initial-boundary value problem for the second axially-symmetric problem represented by the vectors in $\Omega \times \mathbf{T}_+$:

$$\mathbf{u} \stackrel{\text{def}}{=} (0, u_\theta, 0), \qquad \boldsymbol{\varphi} \stackrel{\text{def}}{=} (\varphi_r, 0, \varphi_z). \tag{2.264}$$

We shall arrange field equations in the following groups:

— the equations of motion in $\Omega \times \mathbf{T}_+$,

$$\begin{aligned}
&2\sigma_{[\theta z]} + \partial_r \mu_{rr} + \partial_z \mu_{zr} + r^{-1}(\mu_{rr} - \mu_{\theta\theta}) + Y_r = J\ddot{\varphi}_r, \\
&2\sigma_{[r\theta]} + \left(r^{-1} + \partial_r\right) \mu_{rz} + \partial_z \mu_{zz} + Y_z = J\ddot{\varphi}_z, \\
&\partial_r \sigma_{r\theta} + \partial_z \sigma_{z\theta} + 2r^{-1}\sigma_{(r\theta)} + X_\theta = \rho \ddot{u}_\theta,
\end{aligned} \tag{2.265}$$

where

$$\sigma_{[\theta z]} = \tfrac{1}{2}(\sigma_{\theta z} - \sigma_{z\theta}), \qquad \sigma_{[r\theta]} = \tfrac{1}{2}(\sigma_{r\theta} - \sigma_{\theta r}), \qquad \sigma_{(r\theta)} = \tfrac{1}{2}(\sigma_{r\theta} + \sigma_{\theta r}),$$

and the matrices

$$\underset{\sim}{\sigma} \equiv \begin{bmatrix} 0 & \sigma_{r\theta} & 0 \\ \sigma_{\theta r} & 0 & \sigma_{\theta z} \\ 0 & \sigma_{z\theta} & 0 \end{bmatrix}, \qquad \underset{\sim}{\mu} \equiv \begin{bmatrix} \mu_{rr} & 0 & \mu_{rz} \\ 0 & \mu_{\theta\theta} & 0 \\ \mu_{zr} & 0 & \mu_{zz} \end{bmatrix}, \tag{2.266}$$

define stress fields in $\Omega \times \mathbf{T}_+$;

— the geometric relations in $\Omega \times \mathbf{T}_+$,

$$\begin{aligned}
&(\kappa_{rr}, \kappa_{\theta\theta}, \kappa_{zz}) = (\partial_r \varphi_r, r^{-1}\varphi_r, \partial_z \varphi_z), \\
&(\kappa_{rz}, \kappa_{zr}) = (\partial_r \varphi_z, \partial_z \varphi_r), \\
&\gamma_{r\theta} = \partial_r u_\theta - \varphi_z, \qquad \gamma_{\theta r} = \varphi_z - r^{-1} u_\theta, \\
&\gamma_{z\theta} = \partial_z u_\theta + \varphi_r, \qquad \gamma_{\theta z} = -\varphi_r,
\end{aligned} \tag{2.267}$$

[17] See Gauthier & Jahsman [98] and Nowacki & Nowacki [195].

where the matrices

$$\underset{\sim}{\gamma} \equiv \begin{bmatrix} 0 & \gamma_{r\theta} & 0 \\ \gamma_{\theta r} & 0 & \gamma_{\theta z} \\ 0 & \gamma_{z\theta} & 0 \end{bmatrix}, \qquad \underset{\sim}{\kappa} \equiv \begin{bmatrix} \kappa_{rr} & 0 & \kappa_{rz} \\ 0 & \kappa_{\theta\theta} & 0 \\ \kappa_{zr} & 0 & \kappa_{zz} \end{bmatrix}, \tag{2.268}$$

define strain fields in $\Omega \times \mathbf{T}_+$;

— the constitutive relations in $\Omega \times \mathbf{T}$,

$$\begin{aligned} &\mu_{rr} = 2\gamma\kappa_{rr} + \beta\kappa, \qquad \mu_{\theta\theta} = 2\gamma\kappa_{\theta\theta} + \beta\kappa, \\ &\mu_{zz} = 2\gamma\kappa_{zz} + \beta\kappa, \\ &(\mu_{rz}, \mu_{zr}) = (\gamma + \varepsilon)(\kappa_{rz}, \kappa_{zr}) + (\gamma - \varepsilon)(\kappa_{zr}, \kappa_{rz}), \\ &(\sigma_{r\theta}, \sigma_{\theta r}) = (\mu + \alpha)(\gamma_{r\theta}, \gamma_{\theta r}) + (\mu - \alpha)(\gamma_{\theta r}, \gamma_{r\theta}), \\ &(\sigma_{\theta z}, \sigma_{z\theta}) = (\mu + \alpha)(\gamma_{\theta z}, \gamma_{z\theta}) + (\mu - \alpha)(\gamma_{z\theta}, \gamma_{\theta z}), \end{aligned} \tag{2.269}$$

where $\kappa = \kappa_{rr} + \kappa_{\theta\theta} + \kappa_{zz}$.

Using relations (2.269) and (2.267), we find that the equations of motion (2.265) are transformed into equations (2.248) for the triple $(\varphi_r, u_\theta, \varphi_z)$. Let us formulate the following initial-boundary conditions in $\overline{\Omega} \times \mathbf{T}$:

— the boundary conditions on $\partial\Omega \times \mathbf{T}$,

$$\begin{aligned} &n_r\mu_{rr} + n_z\mu_{zr} = m_r, \qquad n_r\mu_{rz} + n_z\mu_{zz} = m_z, \\ &n_r\sigma_{r\theta} + n_z\sigma_{z\theta} = p_\theta \qquad \text{on } \partial\Omega_\sigma \times \mathbf{T}; \end{aligned} \tag{2.270}$$

$$(\varphi_r, u_\theta, \varphi_z) = (\hat{\varphi}_r, \hat{u}_\theta, \hat{\varphi}_z) \qquad \text{on } \partial\Omega_u \times \mathbf{T}, \tag{2.271}$$

where $m_r,\ m_z,\ p_\theta : \partial\Omega_\sigma \times \mathbf{T} \to \mathbb{R}$, $\hat{\varphi}_r,\ \hat{u}_\theta,\ \hat{\varphi}_z : \partial\Omega_u \times \mathbf{T} \to \mathbb{R}$ are given functions;

— the initial conditions in $\Omega \times \{0\}$,

$$(\varphi_r, u_\theta, \varphi_z) = (h_r, h_\theta, h_z), \qquad (\dot{\varphi}_r, \dot{u}_\theta, \dot{\varphi}_z) = (l_r, l_\theta, l_z), \tag{2.272}$$

where functions $h_r,\ h_z,\ h_\theta,\ l_r,\ l_z,\ l_\theta : \Omega \to \mathbb{R}$ are given.

By means of (2.267) we obtain from equations of motion (2.265) the following equations:

$$\begin{aligned} &J^{-1}(-1, \partial_r, r^{-1}, \partial_z)R_r = (\ddot{\gamma}_{\theta z}, \ddot{\kappa}_{rr}, \ddot{\kappa}_{\theta\theta}, \ddot{\kappa}_{zr}), \\ &J^{-1}(\partial_z, \partial_r)R_z = (\ddot{\kappa}_{zz}, \ddot{\kappa}_{rz}), \\ &\rho^{-1}\partial_r R_\theta - J^{-1}R_z = \ddot{\gamma}_{r\theta}, \\ &J^{-1}R_z - \rho^{-1}r^{-1}R_\theta = \ddot{\gamma}_{\theta r}, \\ &\rho^{-1}\partial_z R_\theta + J^{-1}R_r = \ddot{\gamma}_{z\theta}, \end{aligned} \tag{2.273}$$

where the symbols R_r, R_z, R_θ denote the left-hand sides of equations $(2.265)_{1,2,3}$, respectively. Solving relations (2.269) with respect to strains, we obtain:

$$
\begin{aligned}
&2\gamma\kappa_{rr} = \mu_{rr} - \beta\kappa, \qquad 2\gamma\kappa_{\theta\theta} = \mu_{\theta\theta} - \beta\kappa, \qquad 2\gamma\kappa_{zz} = \mu_{zz} - \beta\kappa,\\
&\kappa_{rz} = \frac{1}{2\gamma}\mu_{(rz)} + \frac{1}{2\varepsilon}\mu_{[rz]}, \qquad \kappa_{zr} = \frac{1}{2\gamma}\mu_{(rz)} - \frac{1}{2\varepsilon}\mu_{[rz]},\\
&\gamma_{r\theta} = \frac{1}{2\mu}\sigma_{(r\theta)} + \frac{1}{2\alpha}\sigma_{[r\theta]}, \qquad \gamma_{\theta r} = \frac{1}{2\mu}\sigma_{(r\theta)} - \frac{1}{2\alpha}\sigma_{[r\theta]},\\
&\gamma_{\theta z} = \frac{1}{2\mu}\sigma_{(\theta z)} + \frac{1}{2\alpha}\sigma_{[\theta z]}, \qquad \gamma_{z\theta} = \frac{1}{2\mu}\sigma_{(\theta z)} - \frac{1}{2\alpha}\sigma_{[\theta z]},
\end{aligned} \tag{2.274}
$$

where

$$\kappa = \frac{1}{2\gamma + 3\beta}(\mu_{rr} + \mu_{\theta\theta} + \mu_{zz}).$$

Now, using (2.274), we see that equations (2.273) are transformed into the equations:

$$
\begin{aligned}
&2\gamma J^{-1}\partial_r R_r - \ddot{\mu}_{rr} + \beta\ddot{\kappa} = 0,\\
&2\gamma J^{-1} r^{-1} R_r - \ddot{\mu}_{\theta\theta} + \beta\ddot{\kappa} = 0,\\
&2\gamma J^{-1}\partial_z R_z - \ddot{\mu}_{zz} + \beta\ddot{\kappa} = 0,\\
&J^{-1}\partial_z R_r - \frac{1}{2\gamma}\ddot{\mu}_{(rz)} + \frac{1}{2\varepsilon}\ddot{\mu}_{[rz]} = 0,\\
&J^{-1}\partial_r R_z - \frac{1}{2\gamma}\ddot{\mu}_{(rz)} - \frac{1}{2\varepsilon}\ddot{\mu}_{[rz]} = 0,\\
&J^{-1}R_r + \frac{1}{2\mu}\ddot{\sigma}_{(\theta z)} + \frac{1}{2\alpha}\ddot{\sigma}_{[\theta z]} = 0,\\
&\rho^{-1}\partial_r R_\theta - J^{-1}R_z - \frac{1}{2\mu}\ddot{\sigma}_{(r\theta)} - \frac{1}{2\alpha}\ddot{\sigma}_{[r\theta]} = 0,\\
&J^{-1}R_z - \rho^{-1}r^{-1}R_\theta - \frac{1}{2\mu}\ddot{\sigma}_{(r\theta)} + \frac{1}{2\alpha}\ddot{\sigma}_{[r\theta]} = 0,\\
&\rho^{-1}\partial_z R_\theta + J^{-1}R_r - \frac{1}{2\mu}\ddot{\sigma}_{(\theta z)} + \frac{1}{2\alpha}\ddot{\sigma}_{[\theta z]} = 0,
\end{aligned} \tag{2.275}
$$

where

$$\ddot{\kappa} = \frac{1}{2\gamma + 3\beta}(\ddot{\mu}_{rr} + \ddot{\mu}_{\theta\theta} + \ddot{\mu}_{zz}).$$

The stress equations of motion problem (SEMP) for the triple $(\varphi_r, u_\theta, \varphi_z)$ consists of: the field equations (2.275), the boundary conditions (2.270) and the initial conditions in $\Omega \times \{0\}$ in the form[18]:

[18] For the mixed problem boundary we add conditions (2.271) on $\partial\Omega_u \times \mathbf{T}$ in the form

$$
\begin{aligned}
&l_r t + h_r + J^{-1}(t * R_r) = \hat{\varphi}_r\\
&l_\theta t + h_\theta + \rho^{-1}(t * R_\theta) = \hat{u}_\theta,\\
&l_z t + h_z + J^{-1}(t * R_z) = \hat{\varphi}_z.
\end{aligned}
$$

$$\underset{\sim}{\sigma} = \underset{\sim}{\sigma}^0, \qquad \underset{\sim}{\mu} = \underset{\sim}{\mu}^0, \qquad \dot{\underset{\sim}{\sigma}} = \dot{\underset{\sim}{\sigma}}^0, \qquad \dot{\underset{\sim}{\mu}} = \dot{\underset{\sim}{\mu}}^0, \tag{2.276}$$

where $\underset{\sim}{\sigma}$, $\underset{\sim}{\mu}$ are given by matrices (2.266), while the matrices of initial values $\underset{\sim}{\sigma}^0$, $\underset{\sim}{\mu}^0$, $\dot{\underset{\sim}{\sigma}}^0$, $\dot{\underset{\sim}{\mu}}^0$ for stresses and their velocities result from relations (2.267), (2.269) and (2.272). We obtain:

$$\begin{aligned}
&\mu^0_{rr} = 2\gamma\partial_r h_r + \beta\kappa^0, \qquad \mu^0_{\theta\theta} = 2\gamma r^{-1} h_r + \beta\kappa^0,\\
&\mu^0_{zz} = 2\gamma\partial_z h_z + \beta\kappa^0,\\
&(\mu^0_{rz}, \mu^0_{zr}) = (\gamma+\varepsilon)(\partial_r h_z, \partial_z h_r) + (\gamma-\varepsilon)(\partial_z h_r, \partial_r h_z),\\
&(\sigma^0_{r\theta}, \sigma^0_{\theta r}) = (\mu+\alpha)(\gamma^0_{r\theta}, \gamma^0_{\theta r}) + (\mu-\alpha)(\gamma^0_{\theta r}, \gamma^0_{r\theta}),\\
&(\sigma^0_{\theta z}, \sigma^0_{z\theta}) = (\mu+\alpha)(\gamma^0_{\theta z}, \gamma^0_{z\theta}) + (\mu-\alpha)(\gamma^0_{z\theta}, \gamma^0_{\theta z}),
\end{aligned} \tag{2.277}$$

where

$$\gamma^0_{r\theta} = \partial_r h_\theta - h_z, \qquad \gamma^0_{\theta r} = -r^{-1} h_\theta + h_z, \qquad \gamma^0_{z\theta} = \partial_z h_\theta + h_r,$$

$$\gamma^0_{\theta z} = -h_r, \qquad \kappa^0 = \left(r^{-1} + \partial_r\right) h_r + \partial_z h_z;$$

and

$$\begin{aligned}
&\dot{\mu}^0_{rr} = 2\gamma\partial_r l_r + \beta\dot{\kappa}^0, \qquad \dot{\mu}^0_{\theta\theta} = 2\gamma r^{-1} l_r + \beta\dot{\kappa}^0,\\
&\dot{\mu}^0_{zz} = 2\gamma\partial_z l_z + \beta\dot{\kappa}^0,\\
&(\dot{\mu}^0_{rz}, \dot{\mu}^0_{zr}) = (\gamma+\varepsilon)(\partial_r l_z, \partial_z l_r) + (\gamma-\varepsilon)(\partial_z l_r, \partial_r l_z),\\
&(\dot{\sigma}^0_{r\theta}, \dot{\sigma}^0_{\theta r}) = (\mu+\alpha)(\dot{\gamma}^0_{r\theta}, \dot{\gamma}^0_{\theta r}) + (\mu-\alpha)(\dot{\gamma}^0_{\theta r}, \dot{\gamma}^0_{r\theta}),\\
&(\dot{\sigma}^0_{\theta z}, \dot{\sigma}^0_{z\theta}) = (\mu+\alpha)(\dot{\gamma}^0_{\theta z}, \dot{\gamma}^0_{z\theta}) + (\mu-\alpha)(\dot{\gamma}^0_{z\theta}, \dot{\gamma}^0_{\theta z}),
\end{aligned} \tag{2.278}$$

where

$$\dot{\gamma}^0_{r\theta} = \partial_r l_\theta - l_z, \qquad \dot{\gamma}^0_{\theta r} = -r^{-1} l_\theta + l_z, \qquad \dot{\gamma}^0_{z\theta} = \partial_z l_\theta + l_r,$$

$$\dot{\gamma}^0_{\theta z} = -l_r, \qquad \dot{\kappa}^0 = \left(r^{-1} + \partial_r\right)_r + \partial_z l_z.$$

The triple $(\varphi_r, u_\theta, \varphi_z)$ is obtained from the following formulae in $\Omega \times \mathbf{T}$:

$$\begin{aligned}
\varphi_r &= l_r t + h_r + J^{-1}(t * R_r),\\
\varphi_z &= l_z t + h_z + J^{-1}(t * R_z),\\
u_\theta &= l_\theta t + h_\theta + \rho^{-1}(t * R_\theta),
\end{aligned} \tag{2.279}$$

where

$$t * R(r,z,t) = \int_0^t (t-\tau) R(r,z,\tau)\, d\tau.$$

For sufficiently smooth physical fields and initial-boundary conditions, the initial-boundary value problems defined by relations (2.265), (2.267) and (2.269)–(2.272) [or relations (2.248), (2.267) and (2.269)–(2.272)] and by relations (2.270), (2.271), (2.274)–(2.276) and (2.279) are equivalent. The proof of equivalence follows from the paper of Ignaczak [116].

2.3.4 Fundamental solutions for stresses

In [64], [69] and [49] we derived the fundamental solution for the force-stresses $\sigma_{\theta z}$, $\sigma_{z\theta}$, $\sigma_{r\theta}$, $\sigma_{\theta r}$ and the couple-stresses μ_{zz}, μ_{rz}, μ_{zr}, μ_{rr}, $\mu_{\theta\theta}$ in the case of the infinite space with the body loading (2.250) ($Y_r = X_\theta = 0$). The fundamental solution was obtained by means of differential equations for individual stress components or their combination in the form:

— force-stresses,

$$
\begin{aligned}
&\Box_3^0(\Box_2^0\Box_4^0 + 4\alpha^2\nabla_0^2)\sigma_{\theta z} = -2\alpha[(\mu-\alpha)\Box_3^0 + L_1^0]\partial_{rz}^2 Y_z,\\
&\Box_3^0(\Box_2^0\Box_4^0 + 4\alpha^2\nabla_0^2)\sigma_{z\theta} = -2\alpha[(\mu+\alpha)\Box_3^0 - L_1^0]\partial_{rz}^2 Y_z,\\
&\Box_3(\Box_2\Box_4 + 4\alpha^2\nabla^2)(\sigma_{r\theta} - \sigma_{\theta r})\\
&\qquad = -4\alpha[\alpha\Box_3(r^{-1}\partial_r + \partial_r^2) + L_1\partial_z^2 - \Box_2\Box_3]Y_z,
\end{aligned}
\tag{2.280}
$$

where

$$
L_1 = (\beta+\gamma-\varepsilon)\Box_2 - 4\alpha^2, \qquad L_1^0 = (\beta+\gamma-\varepsilon)\Box_2^0 - 4\alpha^2;
$$

— couple-stresses,

$$
\begin{aligned}
&\Box_3(\Box_2\Box_4 + 4\alpha^2\nabla^2)\mu_{zz}\\
&\qquad = [2\gamma L_1\partial_z^2 - 2\gamma\Box_2\Box_3 - \beta(\Box_2\Box_4 + 4\alpha^2\nabla^2)]\partial_z Y_z,\\
&\Box_3^0(\Box_2^0\Box_4^0 + 4\alpha^2\nabla_0^2)\mu_{rz} = [2\gamma L_1^0\partial_z^2 - (\gamma+\varepsilon)\Box_2^0\Box_3^0]\partial_r Y_z,\\
&\Box_3^0(\Box_2^0\Box_4^0 + 4\alpha^2\nabla_0^2)\mu_{zr} = [2\gamma L_1^0\partial_z^2 - (\gamma-\varepsilon)\Box_2^0\Box_3^0]\partial_r Y_z,\\
&\Box_3(\Box_2\Box_4 + 4\alpha^2\nabla^2)(\mu_{rr} + \mu_{\theta\theta})\\
&\qquad = [2\gamma L_1(r^{-1}\partial_r + \partial_r^2) - 2\beta(\Box_2\Box_4 + 4\alpha^2)]\partial_z Y_z.
\end{aligned}
\tag{2.281}
$$

The separation of expressions $\sigma_{r\theta} - \sigma_{\theta r}$, $\mu_{rr} + \mu_{\theta\theta}$ into components was obtained by using appropriately transformed equations of motion (2.265):

$$
\begin{aligned}
\partial_r(r^2\mu_{rr}) &= r(\mu_{rr} + \mu_{\theta\theta}) - r^2\partial_z\mu_{zr} - 2r^2\sigma_{[\theta z]}\\
&\quad - Jr^2\left(\frac{1}{2\mu}\ddot{\sigma}_{(z\theta)} - \frac{1}{2\alpha}\ddot{\sigma}_{[z\theta]}\right),\\
\partial_{rz}^2(r^2\sigma_{r\theta}) &= 2r\partial_z\sigma_{[r\theta]} - r^2\partial_z^2\sigma_{z\theta} + \frac{\rho}{\mu}r^2\ddot{\sigma}_{(z\theta)}.
\end{aligned}
\tag{2.282}
$$

Having applied Fourier transformations (0.50) for $n = 2$ (over z and t) and Hankel transformations (0.51) over r ($\nu = 0$ or $\nu = 1$) to equations (2.280) and (2.281), with Y_r, X_θ being taken into account, we have derived integral representations for the stress fields. Subsequently, having taken the loading (2.250) into account, we obtained the results for stresses in a closed form:

— force-stresses,

$$\begin{aligned}
\sigma_{\theta z} &= -\frac{M_0 e^{-i\omega t}}{4\pi}\partial^2_{rz}\left(\frac{\mu-\alpha}{\mu+\alpha}\cdot\frac{p}{\lambda_1^2-\lambda_2^2}F_1+\frac{2\alpha}{J\omega^2}G\right),\\
\sigma_{z\theta} &= -\frac{M_0 e^{-i\omega t}}{4\pi}\partial^2_{rz}\left(\frac{p}{\lambda_1^2-\lambda_2^2}F_1-\frac{2\alpha}{J\omega^2}G\right),\\
\sigma_{r\theta} &= -\frac{M_0 \alpha e^{-i\omega t}}{2\pi J\omega^2}\left[\frac{\sigma_4^2}{\lambda_1^2-\lambda_2^2}\left(\partial_r^2-\frac{\mu-\alpha}{\mu+\alpha}r^{-1}\partial_r\right)F_1\right.\\
&\quad\left.+\sigma_4^2F_2+\partial_z^2G\right],\\
\sigma_{\theta r} &= \frac{M_0 \alpha e^{-i\omega t}}{2\pi J\omega^2}\left[\frac{\sigma_4^2}{\lambda_1^2-\lambda_2^2}\left(r^{-1}\partial_r-\frac{\mu-\alpha}{\mu+\alpha}\partial_r^2\right)F_1\right.\\
&\quad\left.+\sigma_4^2F_2+\partial_z^2G\right],
\end{aligned}\tag{2.283}$$

where

$$F_1=\frac{1}{R}(e^{i\lambda_1R}-e^{i\lambda_2R}),\qquad F_2=\frac{1}{R}(A_2e^{i\lambda_1R}+A_1e^{i\lambda_2R}),$$

$$G=\left[\sigma_4^2\left(\frac{A_2}{\lambda_1^2}e^{i\lambda_1R}+\frac{A_1}{\lambda_2^2}e^{i\lambda_2R}\right)-\frac{\sigma_3^2}{\lambda_3^2}e^{i\lambda_3R}\right]\frac{1}{R},$$

(the remaining notation is the same as in formulae (2.251));

— couple-stresses,

$$\begin{aligned}
\mu_{zz} &= \frac{M_0 e^{-i\omega t}}{4\pi J\omega^2}\partial_z\left(2\gamma\partial_z^2G+2\gamma\sigma_4^2F_2+\beta\sigma_3^2\frac{e^{i\lambda_3R}}{R}\right),\\
\mu_{rz} &= \frac{M_0 e^{-i\omega t}}{4\pi J\omega^2}\partial_r\left[2\gamma\partial_z^2G+(\gamma+\varepsilon)\sigma_4^2F_2\right],\\
\mu_{zr} &= \frac{M_0 e^{-i\omega t}}{4\pi J\omega^2}\partial_r\left[2\gamma\partial_z^2G+(\gamma-\varepsilon)\sigma_4^2F_2\right],\\
\mu_{rr} &= \frac{M_0 e^{-i\omega t}}{4\pi J\omega^2}\partial_z\left(2\gamma\partial_r^2G+\beta\sigma_3^2\frac{e^{i\lambda_3R}}{R}\right),\\
\mu_{\theta\theta} &= \frac{M_0 e^{-i\omega t}}{4\pi J\omega^2}\partial_z\left(2\gamma r^{-1}\partial_rG+\beta\sigma_3^2\frac{e^{i\lambda_3R}}{R}\right).
\end{aligned}\tag{2.284}$$

Having determined the stress fields (2.283) and (2.284), we shall find displacement and rotation fields through integrating the geometric relations (2.267) and using relations (2.274), (2.283) and (2.284). We obtain the expressions:

$$\begin{aligned}\varphi_r &= -\frac{1}{2\mu}\sigma_{(z\theta)} + \frac{1}{2\alpha}\sigma_{[\theta z]},\\ u_\theta &= r\left(\varphi_z - \frac{1}{2\mu}\sigma_{(r\theta)} + \frac{1}{2\alpha}\sigma_{[r\theta]}\right),\\ \partial_r\varphi_z &= \frac{1}{2\gamma}\mu_{(zr)} + \frac{1}{2\varepsilon}\mu_{[rz]},\\ \partial_z\varphi_z &= \frac{1}{2\gamma}(\mu_{zz} - \mu_{\theta\theta}) + r^{-1}\varphi_r.\end{aligned} \tag{2.285}$$

From $(2.285)_1$ we obtain

$$\varphi_r = A(t)\partial^2_{rz}G, \tag{2.286}$$

where

$$A(t) = A_0 e^{-i\omega t}, \qquad A_0 = \frac{M_0}{4\pi J\omega^2}.$$

We transform the expressions $(2.285)_{3,4}$ into the form

$$\partial_r[\varphi_z - A(\partial^2_z G + \sigma_4^2 F_2)] = 0, \qquad \partial_z[\varphi_z - A(\partial^2_z G + \sigma_4^2 F_2)] = 0,$$

from which we obtain $\varphi_z = A(\partial_z^2 G + \sigma_4^2 F_2) + C$, where $C \in \mathbb{R}$. Assuming $C = 0$, we get

$$\varphi_z = A(\partial^2_z G + \sigma_4^2 F_2). \tag{2.287}$$

In turn, from the expression $(2.285)_2$ we obtain

$$u_\theta = -\frac{2\alpha}{\mu+\alpha}\cdot\frac{\sigma_4^2}{\lambda_1^2 - \lambda_2^2} A\partial_r F_1. \tag{2.288}$$

The triple $(\varphi_r, u_\theta, \varphi_z)$ given by formulae (2.286)–(2.288) can also be obtained by the method discussed in Subsection 2.3.3. To this end, one needs to set initial values for displacement and rotation fields as well as their velocities. By use of (2.283) and (2.284), the analysis of the systems of equations (2.277) and (2.278) leads to the observation that

$$(l_r, l_z, l_\theta) = -i\omega(h_r, h_z, h_\theta), \tag{2.289}$$

where

$$h_r = A_0\partial^2_{rz}G, \qquad h_z = A_0(\partial^2_z G + \sigma_4^2 F_2), \qquad h_\theta = -\frac{2\alpha}{\mu+\alpha}\cdot\frac{\sigma_4^2}{\lambda_1^2-\lambda_2^2}A_0\partial_r F_1.$$

Now, from formulae (2.279) we obtain formulae (2.286)–(2.288) by using equations (2.283), (2.284) and (2.289), and taking into account that

$$\int_0^t (t-\tau)e^{-i\omega\tau}\, d\tau = \omega^{-2}(1 - i\omega t - e^{-i\omega t}),$$

$$\nabla^2\left(\frac{e^{i\lambda_1 R}}{R}\right) = \nabla^2\left(\frac{1}{R}\right) - \lambda_1^2\frac{e^{i\lambda_1 R}}{R},$$

$$A_1 + A_2 = 1.$$

When $R \to \infty$, then the stress, displacement and rotation fields tend to zero. Moreover, substituting (2.250), (2.283), (2.284) and (2.286)–(2.288) to the equations of motion (2.265), and using

$$\frac{1}{4\pi}\nabla^2\left(\frac{1}{R}\right) = -\frac{1}{2\pi r}\delta(r)\delta(z)$$

we see that equations (2.265) become identities.

2.3.5 Equations in displacements and rotations. Superposition method

In [49], we analysed the initial-boundary value problem defined by: the field equations (2.248), (2.267) and (2.269), the boundary conditions (2.270) and (2.271), and the initial conditions (2.272), by means of the superposition method. Let us introduce the following substitutions to the system of equations (2.248):

$$\zeta_r = -\tfrac{1}{2}\partial_z u_\theta - \varphi_r, \qquad \zeta_z = \tfrac{1}{2}\left(r^{-1} + \partial_r\right) u_\theta - \varphi_z. \tag{2.290}$$

We assume a solution in the form

$$\varphi_r = \varphi_r^0 + \varphi_r', \qquad u_\theta = u_\theta^0 + u_\theta', \qquad \varphi_z = \varphi_z^0 + \varphi_z' \tag{2.291}$$

where the triple $(\varphi_r^0, u_\theta^0, \varphi_z^0)$ refers to the classical theory of elasticity. We obtain such an effect by assuming $\zeta_r = \zeta_r^0 + \zeta_r'$, $\zeta_z = \zeta_z^0 + \zeta_z'$, $Y_r = Y_r^0 + Y_r'$, $Y_z = Y_z^0 + Y_z'$. Here, we shall give the final result of such investigations. When $\zeta_r^0 = \zeta_z^0 = 0$ ($Y_r^0 = Y_z^0 = 0$), then from the system of equations (2.248) and (2.290) we obtain the classical equation in $\Omega \times \mathbf{T}_+$:

$$\overset{*}{\Box}{}_2^0 u_\theta^0 + X_\theta = 0 \tag{2.292}$$

with classical initial-boudary conditions:

— the boundary conditions on $\partial\Omega \times \mathbf{T}$,

$$n_r \sigma_{r\theta}^0 + n_z \sigma_{z\theta}^0 = p_\theta \qquad \text{on } \partial\Omega_\sigma \times \mathbf{T}, \tag{2.293}$$

$$u_\theta^0 = \hat{u}_\theta^0 \qquad \text{on } \partial\Omega_u \times \mathbf{T}; \tag{2.294}$$

— the initial conditions in $\Omega \times \{0\}$,

$$u_\theta^0 = h_\theta, \qquad \dot{u}_\theta^0 = l_\theta. \tag{2.295}$$

The classical stresses $\sigma_{r\theta}^0 = \sigma_{\theta r}^0$, $\sigma_{z\theta}^0 = \sigma_{\theta z}^0$ are determined from the constitutive relations in $\Omega \times \mathbf{T}$:

$$\sigma_{r\theta}^0 = 2\mu\varepsilon_{r\theta}^0, \qquad \sigma_{z\theta}^0 = 2\mu\varepsilon_{z\theta}^0. \tag{2.296}$$

The quantities $\varepsilon^0_{r\theta}=\varepsilon^0_{\theta r}$, $\varepsilon^0_{z\theta}=\varepsilon^0_{\theta z}$ are related to the displacement u^0_θ through the geometric relations in $\Omega\times\mathbf{T}_+$:

$$\varepsilon^0_{r\theta}=\tfrac{1}{2}(\partial_r-r^{-1})u^0_\theta, \qquad \varepsilon^0_{z\theta}=\tfrac{1}{2}\partial_z u^0_\theta. \tag{2.297}$$

Equations for the triple $(\varphi'_r,u'_\theta,\varphi'_z)$ are obtained in the following form in $\Omega\times\mathbf{T}_+$:

$$\begin{aligned}
&\Box^0_2 u'_\theta+2\alpha(\partial_z\varphi'_r-\partial_r\varphi'_z)+\hat{X}_\theta=0,\\
&\overset{*}{\Box}{}^0_2[\Box^0_4\varphi'_r+(\beta+\gamma-\varepsilon)\partial_r\kappa'-2\alpha\partial_z u'_\theta]+\hat{Y}_r=0,\\
&\overset{*}{\Box}_2[\Box_4\varphi'_z+(\beta+\gamma-\varepsilon)\partial_z\kappa'+2\alpha\left(r^{-1}+\partial_r\right)u'_\theta]+\hat{Y}_z=0,
\end{aligned}\tag{2.298}$$

where $\kappa'=r^{-1}\partial_r(r\varphi'_r)+\partial_z\varphi'_z$. Moreover, it holds:

$$\begin{aligned}
&\hat{X}_\theta=0, \qquad \hat{Y}_r=\overset{*}{\Box}{}^0_2 Y_r+\tfrac{1}{2}\overset{*}{\Box}{}^0_4\partial z X_\theta,\\
&\hat{Y}_z=\overset{*}{\Box}_2 Y_z-\tfrac{1}{2}\overset{*}{\Box}_4\left(r^{-1}+\partial_r\right)X_\theta.
\end{aligned}\tag{2.299}$$

We obtain initial-boundary conditions for the system of equations (2.298) from conditions (2.270)–(2.272), in which we take into account the initial-boundary effect for a micropolar body:

— the boundary conditions on $\partial\Omega\times\mathbf{T}$,

$$\begin{aligned}
&n_r\sigma'_{r\theta}+n_z\sigma'_{z\theta}=0, \qquad n_r\mu'_{rr}+n_z\mu'_{zr}=m_r-m^0_r,\\
&n_r\mu'_{rz}+n_z\mu'_{zz}=m_z-m^0_z \qquad \text{on } \partial\Omega_\sigma\times\mathbf{T},
\end{aligned}\tag{2.300}$$

$$\varphi'_r=\hat{\varphi}_r-\varphi^0_r, \qquad \varphi'_z=\hat{\varphi}_z-\varphi^0_z, \qquad u'_\theta=0 \qquad \text{on } \partial\Omega_u\times\mathbf{T}; \tag{2.301}$$

— the initial conditions in $\Omega\times\{0\}$,

$$\begin{aligned}
&(\varphi'_r,u'_\theta,\varphi'_z)=(h_r-\varphi^0_r,0,h_z-\varphi^0_z),\\
&(\dot{\varphi}'_r,\dot{u}'_\theta,\dot{\varphi}'_z)=(l_r-\dot{\varphi}^0_r,0,l_z-\dot{\varphi}^0_z).
\end{aligned}\tag{2.302}$$

The quantities $m^0_r=n_r\mu^0_{rr}+n_z\mu^0_{zr}$, $m^0_z=n_r\mu^0_{rz}+n_z\mu^0_{zz}$, φ^0_r, φ^0_z, $\dot{\varphi}^0_r$, $\dot{\varphi}^0_z$ (under sufficiently strong assumptions concerning the class of functions in $\overline{\Omega}\times\mathbf{T}$) follow from the classical solution of the problem. It holds:

$$\begin{aligned}
&\varphi^0_r=-\tfrac{1}{2}\partial_z u^0_\theta, \qquad \varphi^0_z=\tfrac{1}{2}\left(r^{-1}+\partial_r\right)u^0_\theta,\\
&\mu^0_{rr}=2\gamma\kappa^0_{rr}+\beta\kappa^0, \qquad \mu^0_{zz}=2\gamma\kappa^0_{zz}+\beta\kappa^0,\\
&\mu^0_{zr}=(\gamma+\varepsilon)\kappa^0_{zr}+(\gamma-\varepsilon)\kappa^0_{rz},\\
&\mu^0_{rz}=(\gamma+\varepsilon)\kappa^0_{rz}+(\gamma-\varepsilon)\kappa^0_{zr},\\
&\kappa^0=\kappa^0_{rr}+\kappa^0_{\theta\theta}+\kappa^0_{zz}, \qquad \kappa^0_{rr}=\partial_r\varphi^0_r, \qquad \kappa^0_{\theta\theta}=r^{-1}\varphi^0_r,\\
&\kappa^0_{zz}=\partial_z\varphi^0_z, \qquad \kappa^0_{rz}=\partial_r\varphi^0_z, \qquad \kappa^0_{zr}=\partial_z\varphi^0_r,\\
&\dot{\varphi}^0_i=\lim_{t\to 0}\frac{\partial\varphi^0_i(r,z,t)}{\partial t}, \qquad i=r,\ z.
\end{aligned}\tag{2.303}$$

The stresses $\sigma'_{r\theta}$, $\sigma'_{z\theta}$ and μ'_{rr}, μ'_{zr}, μ'_{rz}, μ'_{zz}, appearing in the boundary conditions (2.300), are expressed in terms of the triple $(\varphi'_r, u'_\theta, \varphi'_z)$ by means of relations (2.267) and (2.269), into which we substitute $\underset{\sim}{\sigma}'$, $\underset{\sim}{\mu}'$, $\underset{\sim}{\gamma}'$ $\underset{\sim}{\kappa}'$, $\mathbf{u}'$, $\boldsymbol{\varphi}'$. We obtain a solution in the form of formula (2.291) and in the form

$$\underset{\sim}{\sigma} = \underset{\sim}{\sigma}^0 + \underset{\sim}{\sigma}', \qquad \underset{\sim}{\mu} = \underset{\sim}{\mu}^0 + \underset{\sim}{\mu}', \qquad \underset{\sim}{\gamma} = \underset{\sim}{\varepsilon}^0 + \underset{\sim}{\gamma}', \qquad \underset{\sim}{\kappa} = \underset{\sim}{\kappa}^0 + \underset{\sim}{\kappa}'.$$

We shall also give separated equations in $\Omega \times \mathbf{T}_+$:

— for φ^0_r, φ^0_z,

$$\overset{*}{\Box}{}^0_2\varphi^0_r - \tfrac{1}{2}\partial_z X_\theta = 0, \qquad \overset{*}{\Box}{}_2\varphi^0_z + \tfrac{1}{2}\left(r^{-1} + \partial_r\right) X_\theta = 0; \tag{2.304}$$

— for φ'_r, φ'_z,

$$\begin{aligned}
&\overset{*}{\Box}{}^0_2\Box^0_3(\Box^0_2\Box^0_4 + 4\alpha^2\nabla^2_0)\varphi'_r = [(\beta+\gamma-\varepsilon)\Box^0_2 - 4\alpha^2]\partial_r\hat{Y}\\
&\qquad - 2\alpha\overset{*}{\Box}{}^0_2\Box_3\partial_z\hat{X}_\theta - \Box^0_2\Box^0_3\hat{Y}_r,\\
&\overset{*}{\Box}{}_2\Box_3(\Box_2\Box_4 + 4\alpha^2\nabla^2)\varphi'_z = [(\beta+\gamma-\varepsilon)\Box_2 - 4\alpha^2]\partial_z\hat{Y}\\
&\qquad + 2\alpha\overset{*}{\Box}{}_2\Box_3\left(r^{-1} + \partial_r\right)\hat{X}_\theta - \Box_2\Box_3\hat{Y}_z,
\end{aligned} \tag{2.305}$$

where $\hat{Y} = \left(r^{-1} + \partial_r\right)\hat{Y}_r + \partial_z\hat{Y}_z$;

— for u'_θ,

$$\overset{*}{\Box}{}^0_2(\Box^0_2\Box^0_4 + 4\alpha^2\nabla^2_0)u'_\theta = 2\alpha(\partial_z\hat{Y}_r - \partial_r\hat{Y}_z) - \overset{*}{\Box}{}^0_2\Box^0_4\hat{X}_\theta. \tag{2.306}$$

The functions ζ_r, ζ_z satisfy in $\Omega \times \mathbf{T}_+$ the equations

$$\begin{aligned}
&2\overset{*}{\Box}{}^0_2\Box^0_3(\Box^0_2\Box^0_4 + 4\alpha^2\nabla^2_0)\zeta_r = [2\alpha\overset{*}{\Box}{}^0_4 - 2(\beta+\gamma-\varepsilon)\overset{*}{\Box}{}^0_2]\partial_r\hat{Y}\\
&\qquad + \overset{*}{\Box}{}^0_2\Box^0_3(2\hat{Y}_r + \overset{*}{\Box}{}^0_4\partial_z\hat{X}_\theta),\\
&2\overset{*}{\Box}{}_2\Box_3(\Box_2\Box_4 + 4\alpha^2\nabla^2)\zeta_z = [2\alpha\overset{*}{\Box}{}_4 - 2(\beta+\gamma-\varepsilon)\overset{*}{\Box}{}_2]\partial_z\hat{Y}\\
&\qquad + \overset{*}{\Box}{}_2\Box_3[2\hat{Y}_r - \overset{*}{\Box}{}_4\left(r^{-1} + \partial_r\right)\hat{X}_\theta].
\end{aligned} \tag{2.307}$$

In [68] we have solved the problem with the body couple (2.250) acting in the infinite space by using equations (2.292) and (2.304)–(2.306) and we have obtained the result (2.251). Let us note that if we assume the correspondence

$$(\varphi'_r, u'_\theta, \varphi'_z) \mapsto (\varphi_r, u_\theta, \varphi_z), \qquad (\hat{Y}_r, \hat{X}_\theta, \hat{Y}_z) \mapsto (\overset{*}{\Box}{}^0_2 Y_r, X_\theta, \overset{*}{\Box}{}_2 Y_z)$$

and remove $\overset{*}{\Box}{}^0_2$ or $\overset{*}{\Box}{}_2$, equations (2.298) and (2.305), (2.306) are transformed into equations (2.248) and (2.249), respectively.

2.3.6 Distortion field. Equations in displacements and rotations. Fundamental solutions and limiting results

In [49], we have derived the separated equations in displacements and rotations with a distortion field. In the method of superposition we use substitutions

$$X_\theta = -\overset{*}{\bar{R}}{}_3^0, \qquad \hat{X}_\theta = -\overset{**}{\bar{R}}{}_3^0, \tag{2.308}$$

$$\hat{Y}_r = -\hat{\bar{R}}{}_1^0, \qquad \hat{Y}_z = -\hat{\bar{R}}{}_2^0, \tag{2.309}$$

where we denote

$$\hat{\bar{R}}{}_1^0 = \overset{*}{\square}{}_2^0 \bar{R}_1^0 + \tfrac{1}{2}\overset{*}{\square}{}_4^0 \partial_z \overset{*}{\bar{R}}{}_3^0,$$

$$\hat{\bar{R}}{}_2^0 = \overset{*}{\square}{}_2 \bar{R}_2^0 - \tfrac{1}{2}\overset{*}{\square}{}_4 \left(r^{-1} + \partial_r\right) \overset{*}{\bar{R}}{}_3^0,$$

$$\bar{R}_1^0 = \sigma_{\theta z}^0 - \sigma_{z\theta}^0 + \partial_r \mu_{rr}^0 + \partial_z \mu_{zr}^0 + r^{-1}(\mu_{rr}^0 - \mu_{\theta\theta}^0),$$

$$\bar{R}_2^0 = \sigma_{r\theta}^0 - \sigma_{\theta r}^0 + \partial_r \mu_{rz}^0 + \partial_z \mu_{zz}^0 + r^{-1}\mu_{rz}^0.$$

The following relation holds:

$$\bar{R}_3^0 = \overset{*}{\bar{R}}{}_3^0 + \overset{**}{\bar{R}}{}_3^0, \tag{2.310}$$

where

$$\bar{R}_3^0 = \partial_r \sigma_{r\theta}^0 + \partial_z \sigma_{z\theta}^0 + r^{-1}(\sigma_{r\theta}^0 + \sigma_{\theta r}^0).$$

The sets of functions $\{\sigma_{r\theta}^0, \sigma_{\theta r}^0, \sigma_{\theta z}^0, \sigma_{z\theta}^0\}$ and $\{\mu_{rz}^0, \mu_{zr}^0, \mu_{rr}^0, \mu_{\theta\theta}^0, \mu_{zz}^0\}$ have the form of the right-hand sides of relations (2.269), written for the distortions $\gamma_{r\theta}^0, \gamma_{\theta r}^0, \gamma_{\theta z}^0, \gamma_{z\theta}^0$ and $\kappa_{rr}^0, \kappa_{\theta\theta}^0, \kappa_{zz}^0, \kappa_{rz}^0, \kappa_{zr}^0$. For example,

$$\sigma_{r\theta}^0 = (\mu + \alpha)\gamma_{r\theta}^0 + (\mu - \alpha)\gamma_{\theta r}^0 \qquad \text{etc.}$$

Hence, we obtain for expressions $\overset{*}{\bar{R}}{}_3^0$, $\overset{**}{\bar{R}}{}_3^0$ the formulae

$$\begin{aligned} \overset{*}{\bar{R}}{}_3^0 &= 2\mu[\partial_r \gamma_{(r\theta)}^0 + \partial_z \gamma_{(z\theta)}^0] + 4\mu r^{-1}\gamma_{(r\theta)}^0, \\ \overset{**}{\bar{R}}{}_3^0 &= 2\alpha[\partial_r \gamma_{[r\theta]}^0 + \partial_z \gamma_{[z\theta]}^0]. \end{aligned} \tag{2.311}$$

Now, from equation (2.292) and the analogy given by formula $(2.308)_1$, we get the classical equation in $\Omega \times \mathbf{T}_+$:

$$\overset{*}{\square}{}_2^0 u_\theta^0 = \overset{*}{\bar{R}}{}_3^0. \tag{2.312}$$

For the rotations φ_r^0, φ_z^0 we obtain equations from (2.304). By means of $(2.308)_2$ and (2.309), we obtain from equations (2.305) and (2.306) the complementary equations:

— for the rotations φ_r', φ_z' in $\Omega \times \mathbf{T}_+$,

$$\begin{aligned}
&\overset{*}{\square}{}_2^0\square_3^0(\square_2^0\square_4^0 + 4\alpha^2\nabla_0^2)\varphi_r' \\
&\quad = -[(\beta+\gamma-\varepsilon)\square_2^0 - 4\alpha^2]\partial_r[(r^{-1}+\partial_r)\,\hat{\bar{R}}_1^0 + \partial_z\hat{\bar{R}}_2^0] \\
&\qquad + 2\alpha\overset{*}{\square}{}_2^0\square_3^0\partial_z\,\overset{**}{\bar{R}}{}_3^0 + \square_2^0\square_3^0\hat{\bar{R}}_1^0, \\
&\overset{*}{\square}{}_2\square_3(\square_2\square_4 + 4\alpha^2\nabla^2)\varphi_z' \\
&\quad = -[(\beta+\gamma-\varepsilon)\square_2 - 4\alpha^2]\partial_z[(r^{-1}+\partial_r)\,\hat{\bar{R}}_1^0 + \partial_z\hat{\bar{R}}_2^0] \\
&\qquad - 2\alpha\overset{*}{\square}{}_2\square_3\,(r^{-1}+\partial_r)\,\overset{**}{\bar{R}}{}_3^0 + \square_2\square_3\hat{\bar{R}}_2^0;
\end{aligned} \tag{2.313}$$

— for the displacement u_θ' in $\Omega \times \mathbf{T}_+$,

$$\overset{*}{\square}{}_2^0(\square_2^0\square_4^0 + 4\alpha^2\nabla_0^2)u_\theta' = -2\alpha(\partial_z\hat{\bar{R}}_1^0 - \partial_r\hat{\bar{R}}_2^0) + \overset{*}{\square}{}_2^0\square_4^0\,\overset{**}{\bar{R}}{}_3^0. \tag{2.314}$$

Assuming the correspondence

$$(\hat{\bar{R}}_1^0, \overset{**}{\bar{R}}{}_3^0, \hat{\bar{R}}_2^0) \mapsto (\overset{*}{\square}{}_2^0\bar{R}_1^0, \bar{R}_3^0, \overset{*}{\square}{}_2\bar{R}_2^0) \tag{2.315}$$

and removing $\overset{*}{\square}{}_2^0$ or $\overset{*}{\square}{}_2$, we see that equations (2.313) and (2.314) are transformed into the respective equations below:

$$\begin{aligned}
\square_3^0(\square_2^0\square_4^0 + 4\alpha^2\nabla_0^2)\varphi_r &= 2\alpha\partial_z\square_3^0\bar{R}_3^0 + \square_2^0\square_3^0\bar{R}_1^0 \\
&\quad - [(\beta+\gamma-\varepsilon)\square_2^0 - 4\alpha^2]\partial_r\bar{R}^0, \\
\square_3(\square_2\square_4 + 4\alpha^2\nabla^2)\varphi_z &= -[(\beta+\gamma-\varepsilon)\square_2 - 4\alpha^2]\partial_z\bar{R}^0 \\
&\quad - 2\alpha\square_3[r^{-1}\partial_r(r\bar{R}_3^0)] + \square_2\square_3\bar{R}_2^0, \\
(\square_2^0\square_4^0 + 4\alpha^2\nabla_0^2)u_\theta &= 2\alpha(\partial_r\bar{R}_2^0 - \partial_z\bar{R}_1^0) + \square_4^0\bar{R}_3^0,
\end{aligned} \tag{2.316}$$

where $\bar{R}^0 = r^{-1}\partial_r(r\bar{R}_1^0) + \partial_z\bar{R}_2^0$.

These are equations (2.249) (obtained after separating the system of equations (2.248)), into which the distortion field should be introduced by making the substitution

$$Y_r = -\bar{R}_1^0, \qquad X_\theta = -\bar{R}_3^0, \qquad Y_z = -\bar{R}_2^0. \tag{2.317}$$

In [25] we have solved the problem with a distortion field in the infinite space, given by the formula

$$\kappa_{zz}^0 = \frac{\kappa_0^{zz}}{2\pi r}e^{-i\omega t}\delta(r)\delta(z) \tag{2.318}$$

using equations (2.312)–(2.314) and (2.316). The remaining distortions are equal to zero. By means of the Fourier-Hankel integral transformations we obtained closed formulae for rotations and displacements:

$$
\begin{aligned}
\varphi_r = & -\frac{\kappa_0 e^{-i\omega t}}{4\pi J\omega^2}\partial_r\left[2\gamma\partial_z^2\left(\frac{\sigma_4^2 A_2}{\lambda_1^2}\cdot\frac{e^{i\lambda_1 R}}{R}\right.\right.\\
& \left.\left.+\frac{\sigma_4^2 A_1}{\lambda_2^2}\cdot\frac{e^{i\lambda_2 R}}{R}-\sigma_3^2\cdot\frac{e^{i\lambda_3 R}}{R}\right)+\beta\sigma_3^2\cdot\frac{e^{i\lambda_3 R}}{R}\right],\\
\varphi_z = & \frac{\kappa_0 e^{-i\omega t}}{4\pi J\omega^2}\partial_z\left[2\gamma(\partial_r^2+r^{-1}\partial_r)\left(\frac{\sigma_4^2 A_2}{\lambda_1^2}\cdot\frac{e^{i\lambda_1 R}}{R}\right.\right.\\
& \left.\left.+\frac{\sigma_4^2 A_1}{\lambda_2^2}\cdot\frac{e^{i\lambda_2 R}}{R}-\sigma_3^2\cdot\frac{e^{i\lambda_3 R}}{R}\right)-(\beta+2\gamma)\sigma_3^2\cdot\frac{e^{i\lambda_3 R}}{R}\right],\\
u_\theta = & \frac{\kappa_0 e^{-i\omega t}}{4\pi J\omega^2}\cdot\frac{2\gamma s\sigma_4^2}{\lambda_1^2-\lambda_2^2}\partial_{zr}^2\left(\frac{e^{i\lambda_1 R}-e^{i\lambda_2 R}}{R}\right).
\end{aligned}
\tag{2.319}
$$

The notation in formulae (2.319) is the same as in (1.26)–(1.28), $R = \sqrt{r^2+z^2}$, $\kappa_0^{zz} \mapsto \kappa_0$. From formulae (2.319) one obtains many limiting results, e.g., taking $\alpha \to \infty$ and assuming (2.8), (2.10), we get from formula $(2.319)_3$ the following:

$$
u_\theta = \frac{\kappa_0(1+\eta)e^{-i\omega t}}{2\pi(\lambda_1^{\star 2}-\lambda_2^{\star 2})}\partial_{rz}^2\left(\frac{e^{i\lambda_1^\star R}-e^{i\lambda_2^\star R}}{R}\right). \tag{2.320}
$$

In turn, formulae $(2.319)_{1,2}$ go over into those which result from relations given by formulae $(2.253)_{2,3}$. It can be shown that the result obtained refers to couple-stress theory. In [25] all limiting cases for (2.319) referring to the theories (1)–(4), (7) and (8) (Fig. 1.1) were discussed.

2.3.7 Functions of displacements-rotations and the method of potentials

In [69] and [49] we have derived a representation for physical fields, which is expressed in terms of two solving functions $\psi^\star$, χ. We start our considerations from the following equations[19]:

— the system of equations of motion in $\Omega \times \mathbf{T}_+$,

$$
\begin{aligned}
&\sigma_{\theta z}-\sigma_{z\theta}+\partial_r\mu_{rr}+\partial_z\mu_{zr}+r^{-1}(\mu_{rr}-\mu_{\theta\theta})+Y_r = J\ddot{\varphi}_r,\\
&\sigma_{r\theta}-\sigma_{\theta r}+\partial_r\mu_{rz}+\partial_z\mu_{zz}+r^{-1}\mu_{rz}+Y_z = J\ddot{\varphi}_z,\\
&\partial_r\sigma_{r\theta}+\partial_z\sigma_{z\theta}+r^{-1}(\sigma_{r\theta}+\sigma_{\theta r})+X_\theta = \rho\ddot{u}_\theta;
\end{aligned}
\tag{2.321}
$$

— the compatibility equations in $\Omega \times \mathbf{T}$ [40],

$$
\begin{aligned}
&\partial_z(\gamma_{r\theta}+\gamma_{\theta r})+(r^{-1}-\partial_r)(\gamma_{z\theta}+\gamma_{\theta z}) = 0,\\
&\kappa_{rr}+\partial_r\gamma_{\theta z} = 0, \qquad \kappa_{\theta\theta}+r^{-1}\kappa_{\theta z} = 0,\\
&\partial_z\gamma_{\theta r}+r^{-1}\kappa_{z\theta}-\kappa_{zz}-\kappa_{\theta\theta} = 0,\\
&\kappa_{zr}+\partial_z\gamma_{\theta z} = 0, \qquad \partial_r(r\gamma_{\theta r})+\gamma_{r\theta}-r\kappa_{rz} = 0.
\end{aligned}
\tag{2.322}
$$

[19] Cf. the SEMP from Subsection 2.3.3. Equations (2.273), relations (2.267) and initial conditions (2.272) imply equations (2.322).

We express relations (2.322) in terms of stresses by using relations (2.274). Thus, $(2.322)_1$ takes the form

$$\partial_z(\sigma_{r\theta} + \sigma_{\theta r}) + (r^{-1} - \partial_r)(\sigma_{z\theta} + \sigma_{\theta z}) = 0. \tag{2.323}$$

Equation (2.323) can be satisfied identically by introducing the function $\Phi^\star$:

$$\sigma_{r\theta} + \sigma_{\theta r} = (\partial_r - r^{-1})\Phi^\star, \qquad \sigma_{\theta z} + \sigma_{z\theta} = \partial_z \Phi^\star. \tag{2.324}$$

Let us represent the vector $\mathbf{u} \overset{\text{def}}{=} (0, u_\theta, 0)$ in the form

$$\mathbf{u} = \operatorname{grad} \vartheta^\star + \operatorname{curl} \boldsymbol{\chi}, \qquad \operatorname{div} \boldsymbol{\chi} = 0, \tag{2.325}$$

where $\boldsymbol{\chi} \overset{\text{def}}{=} (\chi_r, 0, \chi_z)$.

The condition $(2.325)_2$ will be fulfilled for $\chi_r = \partial_z \chi$, $\chi_z = -r^{-1}\partial_r(r\chi)$. We obtain the following equation in $\Omega \times \mathbf{T}_+$:

$$\nabla_0^2 \chi = u_\theta. \tag{2.326}$$

The Stokes-Helmholtz decomposition for the vectors $\mathbf{X}$, $\mathbf{Y}$ leads to the following equations[20] in $\Omega \times \mathbf{T}$:

$$\begin{aligned} \nabla^2 \sigma &= r^{-1}\partial_r(rY_r) + \partial_z Y_z, \\ \nabla_0^2 \eta_\theta &= \partial_r Y_z - \partial_z Y_r, \\ \nabla_0^2 \chi^\star &= X_\theta. \end{aligned} \tag{2.327}$$

The equation of motion given by equation $(2.321)_3$ can be written in the form

$$\partial_r[-\sigma_{\theta r} + \partial_r \Phi^\star + r^{-1}\partial_r(r\bar{\chi})] + \partial_z[-\sigma_{\theta z} + \partial_z \Phi^\star + \partial_z \bar{\chi}] = 0, \tag{2.328}$$

where $\bar{\chi} = \chi^\star - \rho\ddot{\chi}$.

Equation (2.328) can be satisfied identically by introducing the function $\psi^\star$:

$$\begin{aligned} -\sigma_{\theta r} + r^{-1}\partial_r(r\bar{\chi}) + \partial_r \Phi^\star &= -\partial_z \psi^\star, \\ -\sigma_{\theta z} + \partial_z \bar{\chi} + \partial_z \Phi^\star &= \partial_r \psi^\star. \end{aligned} \tag{2.329}$$

---

[20] Assuming the decomposition

$$\mathbf{X} = \operatorname{grad} \vartheta + \operatorname{curl} \boldsymbol{\chi}^\star, \qquad \mathbf{Y} = \operatorname{grad} \sigma + \operatorname{curl} \boldsymbol{\eta},$$

where $\mathbf{X} \overset{\text{def}}{=} (0, X_\theta, 0)$, $\mathbf{Y} \overset{\text{def}}{=} (Y_r, 0, Y_z)$, $\boldsymbol{\chi}^\star \overset{\text{def}}{=} (\chi_r^\star, 0, \chi_z^\star)$, $\boldsymbol{\eta} \overset{\text{def}}{=} (0, \eta_\theta, 0)$, we obtain the relations:
(a) $X_\theta = \partial_z \chi_r^\star - \partial_r \chi_z^\star$,
(b) $Y_r = \partial_r \sigma - \partial_z \eta_\theta$, $Y_z = \partial_z \sigma + r^{-1}\partial_r(r\eta_\theta)$.
It holds $\operatorname{div} \boldsymbol{\eta} \equiv 0$. The condition $\operatorname{div} \boldsymbol{\chi}^\star = 0$ will be fulfilled if we assume that
(c) $\chi_r^\star = \partial_z \chi^\star$, $\chi_z^\star = -r^{-1}\partial_r(r\chi^\star)$.
Now, relations (a), (b) and (c) imply equation (2.327).

In this way, the stresses $\underset{\sim}{\sigma}$, and next, the strains $\underset{\sim}{\gamma}$ were determined by means of the functions $\Phi^\star$, $\psi^\star$, χ, the latter ones — by use of the constitutive relations (2.269). Note that from relations $(2.322)_{2-6}$ we determine the strains $\underset{\sim}{\kappa}$ by means of algebraic transformations, and then, using (2.269), we determine the stresses $\underset{\sim}{\mu}$, also in terms of the functions $\Phi^\star$, $\psi^\star$, χ. Making use of the geometric relations (2.267), we express the triple $(\varphi_r, u_\theta, \varphi_z)$ as follows:

$$
\begin{aligned}
2\mu\varphi_r &= \partial_r\psi^\star - \partial_z(\mu_1^2\Phi^\star + \bar{\chi}),\\
2\alpha\varphi_z &= \partial_z\psi^\star + r^{-1}\partial_r[r(\mu_1^2\Phi^\star + \bar{\chi})],\\
2\mu u_\theta &= \Phi^\star,
\end{aligned}
\tag{2.330}
$$

where $\mu_1^2 = \dfrac{\mu+\alpha}{2\mu}$.

The function $\Phi^\star$ can be eliminated from further considerations, since, from equations (2.326) and $(2.330)_3$, we get

$$
\Phi^\star = 2\mu\nabla_0^2\chi. \tag{2.331}
$$

It follows from our discussion that 7 out of 9 equations (2.321), (2.322) are identities. The remaining two equations of motion, namely equations $(2.321)_{1,2}$, are transformed into the following equations in $\Omega \times \mathbf{T}_+$:

$$
\begin{aligned}
&\partial_r(\square_3\psi^\star + 2\alpha\sigma) - \partial_z[(\square_2^0\square_4^0 + 4\alpha^2\nabla_0^2)\chi + \square_4^0\chi^\star + 2\alpha\eta_\theta] = 0,\\
&\partial_z(\square_3\psi^\star + 2\alpha\sigma)\\
&\qquad + (r^{-1} + \partial_r)\,[(\square_2^0\square_4^0 + 4\alpha^2\nabla_0^2)\chi + \square_4^0\chi^\star + 2\alpha\eta_\theta] = 0.
\end{aligned}
\tag{2.332}
$$

The representation for φ_r, u_θ, φ_z takes in $\Omega \times \mathbf{T}_+$ the form

$$
\begin{aligned}
2\alpha\varphi_r &= \partial_r\psi^\star - \partial_z(\square_2^0\chi + \chi^\star),\\
2\alpha\varphi_z &= \partial_z\psi^\star + (r^{-1} + \partial_r)\,(\square_2^0\chi + \chi^\star),\\
u_\theta &= \nabla_0^2\chi.
\end{aligned}
\tag{2.333}
$$

Satisfying equations (2.332) in a special way (the terms containing the functions $\psi^\star$ and χ, respectively, vanish independently of each other), we obtain the wave equations in $\Omega \times \mathbf{T}_+$:

$$
\begin{aligned}
&\square_3\psi^\star + 2\alpha\sigma = 0,\\
&(\square_2^0\square_4^0 + 4\alpha^2\nabla_0^2)\chi + \square_4^0\chi^\star + 2\alpha\eta_\theta = 0.
\end{aligned}
\tag{2.334}
$$

The functions $\psi^\star$, χ (introduced by means of stress-type equations) will be called functions of displacements-rotations, since, by substituting the representation (2.333) into the system of equations written in terms of displacements and rotations (2.248), we make the 3rd equation into an identity, whereas the 1st and 2nd equations go over into equations (2.332).

We have also considered action of the body loadings (2.250) in the infinite space, using (2.332) and (2.333). Integrating equations (2.332) in the form

$$\begin{aligned}&\nabla^2\square_3\psi^\star + 2\alpha[r^{-1}\partial_r(rY_r) + \partial_z Y_z] = 0,\\&\nabla_0^2(\square_2^0\square_4^0 + 4\alpha^2\nabla_0^2)\chi + \square_4^0 X_\theta + 2\alpha(\partial_r Y_z - \partial_z Y_r) = 0,\end{aligned}\tag{2.335}$$

and using Fourier (0.50) and Hankel (0.51) integral transformations, we obtain the following expressions for the functions $\psi^\star$, χ:

$$\begin{aligned}\psi^\star &= \frac{M_0\tau_0^2 e^{-i\omega t}}{8\pi\lambda_3^2}\partial_z\left(\frac{1-e^{i\lambda_3 R}}{R}\right),\\ \chi &= \frac{M_0 s e^{-i\omega t}}{4\pi J c_4^2}\partial_r\times\\ &\left[\frac{1}{\lambda_1^2(\lambda_1^2-\lambda_2^2)}\cdot\frac{e^{i\lambda_1 R}}{R} + \frac{1}{\lambda_2^2(\lambda_2^2-\lambda_1^2)}\cdot\frac{e^{i\lambda_2 R}}{R} + \frac{1}{\lambda_1^2\lambda_2^2}\left(\frac{1}{R}\right)\right],\end{aligned}\tag{2.336}$$

which, by means of (2.333) (for $\chi^\star \stackrel{\text{def}}{=} 0$), lead to (2.251). Notation is the same as in formulae (1.26)–(1.28), $R = \sqrt{r^2+z^2}$. It is interesting to connect the functions $\psi^\star$, χ with the functions $\varPsi$, $\varSigma$, used in the method of potentials. From the vector representation (1.147) and (1.151)–(1.153) we obtain for the triple $(\varphi_r, u_\theta, \varphi_z)$ the following formulae in $\varOmega \times \mathbf{T}_+$:

$$\begin{aligned}\varphi_r &= \partial_r\varSigma - \partial_z H_\theta,\\ \varphi_z &= \partial_z\varSigma + \left(r^{-1}+\partial_r\right)H_\theta,\\ u_\theta &= \partial_z\varPsi_r - \partial_r\varPsi_z.\end{aligned}\tag{2.337}$$

The condition $\left(r^{-1}+\partial_r\right)\varPsi_r + \partial_z\varPsi_z = 0$ is fulfilled by introducing the function $\varPsi$: $\varPsi_r = \partial_z\varPsi$, $\varPsi_z = -\left(r^{-1}+\partial_r\right)\varPsi$.

The functions $\varSigma$, $\varPsi_r$, $\varPsi_z$, H_θ are related through equations (1.151). Notice that our notation differs from the one used in the decomposition of the vector $\mathbf{X}$ in [49]: $X_\theta = \nabla_0^2\chi$.

From equation $(1.151)_1$ we obtain the following equation in $\varOmega \times \mathbf{T}_+$:

$$\square_2^0(\partial_z\varPsi_r - \partial_r\varPsi_z) - 2\alpha\nabla_0^2 H_\theta + \nabla_0^2\chi = 0,\tag{2.338}$$

from which

$$H_\theta = \frac{1}{2\alpha}(\square_2^0\varPsi + \chi).\tag{2.339}$$

In turn, from equation $(1.151)_2$ we have:

$$\begin{aligned}&\square_3^0\partial_r\varSigma - \square_4^0\partial_z H_\theta - 2\alpha\nabla_0^2\varPsi_r + \partial_r\sigma - \partial_z\eta_\theta = 0,\\&\square_3\partial_z\varSigma + \square_4\left(r^{-1}+\partial_r\right)H_\theta - 2\alpha\nabla^2\varPsi_z + \partial_z\sigma + (r^{-1}+\partial_r)\eta_\theta = 0.\end{aligned}\tag{2.340}$$

Expressing formulae (2.337) and (2.340) in terms of the function $\varPsi$, we finally get the representation

$$
\begin{aligned}
\varphi_r &= \partial_r \Sigma - \frac{1}{2\alpha}\partial_z(\Box_2^0 \Psi + \chi), \\
\varphi_z &= \partial_z \Sigma + \frac{1}{2\alpha}\left(r^{-1} + \partial_r\right)(\Box_2^0 \Psi + \chi), \\
u_\theta &= \nabla_0^2 \Psi
\end{aligned}
\tag{2.341}
$$

and the field equations

$$
\begin{aligned}
&2\alpha\partial_r(\Box_3 \Sigma + \sigma) - \partial_z[(\Box_2^0\Box_4^0 + 4\alpha^2\nabla_0^2)\Psi + \Box_4^0\chi + 2\alpha\eta_\theta] = 0, \\
&2\alpha\partial_z(\Box_3 \Sigma + \sigma) \\
&\quad + \left(r^{-1} + \partial_r\right)[(\Box_2^0\Box_4^0 + 4\alpha^2\nabla_0^2)\Psi + \Box_4^0\chi + 2\alpha\eta_\theta] = 0.
\end{aligned}
\tag{2.342}
$$

Note that after the following change in notation:

$$
\Psi \mapsto \chi, \qquad \Sigma \mapsto \frac{1}{2\alpha}\psi^\star, \qquad \chi \mapsto \chi^\star
$$

the representation (2.341) and equations (2.342) are transformed into the representation (2.333) and equations (2.332), respectively.

2.3.8 Potentials of Galerkin type

In [49], we introduced the generalized Galerkin vector in the form of two vectors[21]

$$
\mathbf{F} \stackrel{\text{def}}{=} (0, F_\theta, 0), \qquad \mathbf{M} \stackrel{\text{def}}{=} (M_r, 0, M_z),
\tag{2.343}
$$

with the following representation in $\Omega \times \mathbf{T}_+$:

$$
\begin{aligned}
u_\theta &= \Box_4^0 F_\theta - 2\alpha\Box_3^0(\partial_z M_r - \partial_r M_z), \\
\varphi_r &= 2\alpha\partial_z F_\theta + \Box_2^0\Box_3^0 M_r - [(\beta+\gamma-\varepsilon)\Box_2^0 - 4\alpha^2]\partial_r M, \\
\varphi_z &= -2\alpha\left(r^{-1} + \partial_r\right)F_\theta + \Box_2\Box_3 M_z - [(\beta+\gamma-\varepsilon)\Box_2 - 4\alpha^2]\partial_z M,
\end{aligned}
\tag{2.344}
$$

where $M = r^{-1}\partial_r(rM_r) + \partial_z M_z$.

Equations for the functions M_r, F_θ, M_z are obtained from equations (2.249) by means of (2.344):

$$
\begin{aligned}
&\Box_3^0(\Box_2^0\Box_4^0 + 4\alpha^2\nabla_0^2)M_r + Y_r = 0, \\
&\Box_3(\Box_2\Box_4 + 4\alpha^2\nabla^2)M_z + Y_z = 0, \\
&(\Box_2^0\Box_4^0 + 4\alpha^2\nabla_0^2)F_\theta + X_\theta = 0.
\end{aligned}
\tag{2.345}
$$

Let us consider the special case of body loadings given by (2.62). The loading X_z is connected with the triple $(u_r, \varphi_\theta, u_z)$; however, also the body moments Y_z appear in the problem. If we assume that only the component M_z does not

[21] Nowacki [192, p. 35] calls the vectors $\mathbf{F}$, $\mathbf{M}$ stress functions.

vanish identically in relation (2.343), we obtain the following representation for φ_r, u_θ, φ_z in $\Omega \times \mathbf{T}_+$:

$$\begin{aligned} u_\theta &= 2\alpha\Box_3^0\partial_r M_z, \\ \varphi_r &= -[(\beta+\gamma-\varepsilon)\Box_2^0 - 4\alpha^2]\partial_{rz}^2 M_z, \\ \varphi_z &= \left\{\Box_2\Box_3 - [(\beta+\gamma-\varepsilon)\Box_2 - 4\alpha^2]\partial_z^2\right\} M_z, \end{aligned} \tag{2.346}$$

where the stress function M_z satisfies

$$\Box_3(\Box_2\Box_4 + 4\alpha^2\nabla^2)M_z + Y_z = 0. \tag{2.347}$$

Another variant of introducing a stress function is connected with the body loading (2.65). Here, we obtain in $\Omega \times \mathbf{T}_+$ the representation

$$\begin{aligned} u_\theta &= -2\alpha\Box_3^0\partial_z M_r, \\ \varphi_r &= \left\{\Box_2^0\Box_3^0 - [(\beta+\gamma-\varepsilon)\Box_2^0 - 4\alpha^2]\partial_r\left(r^{-1}+\partial_r\right)\right\} M_r, \\ \varphi_z &= -[(\beta+\gamma-\varepsilon)\Box_2 - 4\alpha^2]\partial_z\left(r^{-1}+\partial_r\right) M_r, \end{aligned} \tag{2.348}$$

where M_r satisfies

$$\Box_3^0(\Box_2^0\Box_4^0 + 4\alpha^2\nabla_0^2)M_r + Y_r = 0. \tag{2.349}$$

The stress functions, M_r, M_z, are subject to the relation (we disregard body loadings):

$$\Box_3^0(\partial_r M_z + \partial_z M_r) = 0. \tag{2.350}$$

2.4 The second axially-symmetric problem. Elastostatics

2.4.1 Body loadings. Equations in displacements and rotations. Direct method and superposition method

The starting point for our considerations is given by the fundamental equations in displacements and rotations in Ω:

$$\begin{aligned} &[(\gamma+\varepsilon)\nabla_0^2 - 4\alpha]\varphi_r + (\beta+\gamma-\varepsilon)\partial_r\kappa - 2\alpha\partial_z u_\theta + Y_r = 0, \\ &[(\gamma+\varepsilon)\nabla^2 - 4\alpha]\varphi_z + (\beta+\gamma-\varepsilon)\partial_z\kappa + 2\alpha\left(r^{-1}+\partial_r\right)u_\theta + Y_z = 0, \\ &(\mu+\alpha)\nabla_0^2 u_\theta + 2\alpha(\partial_z\varphi_r - \partial_r\varphi_z) + X_\theta = 0, \end{aligned} \tag{2.351}$$

where $\kappa = r^{-1}\partial_r(r\varphi_r) + \partial_z\varphi_z$.

Let us formulate the first boundary value problem:

$$\varphi_r = \hat{\varphi}_r, \qquad \varphi_z = \hat{\varphi}_z, \qquad u_\theta = \hat{u}_\theta \qquad \text{on } \partial\Omega, \tag{2.352}$$

where $\hat{\varphi}_r$, $\hat{\varphi}_z$, $\hat{u}_\theta : \partial\Omega \to \mathbb{R}$ are given functions.

We introduce the substitution (2.290) to equations (2.351) and assume a solution in the form (2.291). Assuming $\zeta_r = \zeta_r^0 + \zeta_r'$, $\zeta_z = \zeta_z^0 + \zeta_z'$ for $\zeta_r^0 = \zeta_z^0 = 0$, we obtain from (2.351) the classical equations

$$\mu \nabla_0^2 u_\theta^0 + X_\theta = 0 \qquad \text{in } \Omega \tag{2.353}$$

with the boundary condition

$$u_\theta^0 = \hat{u}_\theta \qquad \text{on } \partial\Omega. \tag{2.354}$$

The rotations φ_r^0, φ_z^0 fulfill relations $(2.303)_{1,2}$ and the equations in Ω:

$$\begin{aligned} &\mu \nabla_0^2 \varphi_r^0 - \tfrac{1}{2}\partial_z X_\theta = 0, \\ &\mu \nabla^2 \varphi_z^0 + \tfrac{1}{2}\left(r^{-1} + \partial_r\right) X_\theta = 0. \end{aligned} \tag{2.355}$$

For the triple $(\varphi_r', u_\theta', \varphi_z')$ we give the final form of the system of equations in Ω:

$$\begin{aligned} &[(\gamma+\varepsilon)\nabla_0^2 - 4\alpha]\varphi_r' \\ &\qquad + (\beta+\gamma-\varepsilon)\partial_r \kappa' - 2\alpha\partial_z u_\theta' + \overset{*}{Y}_r = 0, \\ &[(\gamma+\varepsilon)\nabla^2 - 4\alpha]\varphi_z' \\ &\qquad + (\beta+\gamma-\varepsilon)\partial_z \kappa' + 2\alpha\left(r^{-1} + \partial_r\right) u_\theta' + \overset{*}{Y}_z = 0, \\ &(\mu+\alpha)\nabla_0^2 u_\theta' + 2\alpha(\partial_z \varphi_r' - \partial_r \varphi_z') + \hat{X}_\theta = 0, \end{aligned} \tag{2.356}$$

where it holds:

$$\begin{aligned} &\hat{X}_\theta = 0, \\ &\overset{*}{Y}_r = Y_r + \frac{\gamma+\varepsilon}{2\mu}\partial_z X_\theta, \\ &\overset{*}{Y}_z = Y_z - \frac{\gamma+\varepsilon}{2\mu}\left(r^{-1} + \partial_r\right) X_\theta. \end{aligned} \tag{2.357}$$

For the system of equations (2.356), the following boundary conditions are valid:

$$\varphi_r' = \hat{\varphi}_r - \varphi_r^0, \qquad \varphi_z' = \hat{\varphi}_z - \varphi_z^0, \qquad u_\theta' = 0 \qquad \text{on } \partial\Omega. \tag{2.358}$$

The quantities φ_r^0, φ_z^0 result from a solution of the classical problem. Now, from the system of equations (2.356), we derive the following separated equations:

— for the rotations φ_r', φ_z',

$$\begin{aligned}\nabla_0^2(l^2\nabla_0^2-1)(\nu_1^2\nabla_0^2-1)\varphi_r' &= -\frac{1}{2\mu}(\nu_1^2\nabla_0^2-1)\partial_z\hat{X}_\theta \\ &\quad -\frac{1}{4\alpha}(l^2\nabla_0^2-1)\partial_r\overset{*}{Y}-\frac{\mu+\alpha}{4\alpha\mu}(\nu_1^2\nabla_0^2-1)\partial_z(\partial_z\overset{*}{Y}_r-\partial_r\overset{*}{Y}_z), \\ \nabla^2(l^2\nabla^2-1)(\nu_1^2\nabla^2-1)\varphi_z' &= \frac{1}{2\mu}(\nu_1^2\nabla^2-1)\left(r^{-1}+\partial_r\right)\hat{X}_\theta \\ &\quad -\frac{1}{4\alpha}(l^2\nabla^2-1)\partial_z\overset{*}{Y} \\ &\quad +\frac{\mu+\alpha}{4\alpha\mu}(\nu_1^2\nabla^2-1)\left(r^{-1}+\partial_r\right)(\partial_z\overset{*}{Y}_r-\partial_r\overset{*}{Y}_z),\end{aligned} \tag{2.359}$$

where

$$\overset{*}{Y}=r^{-1}\partial_r(r\overset{*}{Y}_r)+\partial_z\overset{*}{Y}_z, \qquad \nu_1^2=\frac{2\gamma+\beta}{4\alpha}, \qquad l^2=\frac{(\gamma+\varepsilon)(\mu+\alpha)}{4\alpha\mu};$$

— for the displacement u_θ',

$$2\mu\nabla_0^2(l^2\nabla_0^2-1)u_\theta'=(\partial_z\overset{*}{Y}_r-\partial_r\overset{*}{Y}_z)-2(l^2\nabla_0^2-1)\hat{X}_\theta+\mu_0^{-1}\nabla_0^2\hat{X}_\theta, \tag{2.360}$$

where $\mu_0=\dfrac{2\mu}{\gamma+\varepsilon}$.

The functions ζ_r, ζ_z satisfy

$$\begin{aligned}4\alpha(\nu_1^2\nabla_0^2-1)(l^2\nabla_0^2-1)\zeta_r &= (\nu_1^2\nabla_0^2-1)\overset{*}{Y}_r+\mu_0^{-1}(\nu_1^2\nabla_0^2-1)\partial_z\hat{X}_\theta \\ &\quad +\left[\tfrac{1}{2}\mu_0^{-1}-(\nu_1^2-k^2)\right]\partial_r\overset{*}{Y}, \\ 4\alpha(\nu_1^2\nabla^2-1)(l^2\nabla^2-1)\zeta_z &= (\nu_1^2\nabla^2-1)\overset{*}{Y}_z \\ &\quad -\mu_0^{-1}(\nu_1^2\nabla^2-1)\left(r^{-1}+\partial_r\right)\hat{X}_\theta+\left[\tfrac{1}{2}\mu_0^{-1}-(\nu_1^2-k^2)\right]\partial_z\overset{*}{Y},\end{aligned} \tag{2.361}$$

where $k^2=\dfrac{\gamma+\varepsilon}{4\alpha}$.

Let us stress that by assuming the correspondence

$$(\varphi_r',u_\theta',\varphi_z')\mapsto(\varphi_r,u_\theta,\varphi_z), \qquad (\overset{*}{Y}_r,\hat{X}_\theta,\overset{*}{Y}_z)\mapsto(Y_r,X_\theta,Y_z), \tag{2.362}$$

we transform the system of equations (2.356) into (2.351), whereas equations (2.359), (2.360) go over into equations derived in [43]; the latter being equations for the triple $(\varphi_r,u_\theta,\varphi_z)$ obtained after separating the system (2.351). In [52] and [49] we gave formulae for the triple $(\varphi_r,u_\theta,\varphi_z)$, using equations (2.359) and (2.360), when in the infinite space a body moment of the following form acts:

$$Y_z=\frac{M_0}{2\pi r}\delta(r)\delta(z), \qquad Y_r=X_\theta=0. \tag{2.363}$$

Applying the Fourier transformations (0.50) with respect to the variable z and the Hankel transformations (0.51) with respect to the variable r, we arrive at formulae (2.252). We also solve the system of equations (2.351) of the form

$$\begin{aligned}
&\left[(\gamma+\varepsilon)\nabla_0^2-4\alpha\right]\varphi_r+(\beta+\gamma-\varepsilon)\partial_r\kappa-2\alpha\partial_z u_\theta=0,\\
&\left[(\gamma+\varepsilon)\nabla^2-4\alpha\right]\varphi_z+(\beta+\gamma-\varepsilon)\partial_z\kappa+2\alpha r^{-1}\partial_r(ru_\theta)=0,\\
&(\mu+\alpha)\nabla_0^2 u_\theta+2\alpha(\partial_z\varphi_r-\partial_r\varphi_z)=\tfrac{1}{2}\partial_r Y_z.
\end{aligned}\tag{2.364}$$

The following formulae are obtained:

$$\begin{aligned}
&16\pi\mu M_0^{-1}\varphi_r=\partial^2_{rz}\left(\frac{1-e^{-R/l}}{R}\right),\\
&16\pi\mu M_0^{-1}\varphi_z=-\left(r^{-1}+\partial_r\right)\partial_r\left(\frac{1-e^{-R/l}}{R}\right),\\
&8\pi\mu M_0^{-1}u_\theta=-\partial_r\left(\frac{1-e^{-R/l}}{R}\right)-\frac{\mu}{\mu+\alpha}\partial_r\left(\frac{e^{-R/l}}{R}\right),
\end{aligned}\tag{2.365}$$

which give the result of action of a unit classical moment in a micropolar medium;

$$R=\sqrt{r^2+z^2},\qquad l=\sqrt{\frac{(\gamma+\varepsilon)(\mu+\alpha)}{4\alpha\mu}}.$$

The need to find a solution of such a problem resulted from the discussion of limiting cases of formulae (2.252). When $\alpha\to 0^+$, these formulae do not lead directly to the classical result, but go over in the limit into the formulae obtained within the framework of a hypothetical medium, from which only by an analogy one can obtain the classical result

$$8\pi\mu M_0^{-1}u_\theta^0=-\partial_r\left(\frac{1}{R}\right).\tag{2.366}$$

One can also obtain the classical result (2.366) from formulae (2.252) in another manner. Namely, formulae (2.252) for $\alpha\to\infty$ $(l\to l^\star)$ are transformed into the formulae obtained in couple-stress theory

$$\begin{aligned}
&8\pi\mu M_0^{-1}u_\theta=-\partial_r\left(\frac{1-e^{-R/l^\star}}{R}\right),\\
&\varphi_r=-\tfrac{1}{2}\partial_z u_\theta,\\
&\varphi_z=\tfrac{1}{2}\left(r^{-1}+\partial_r\right)u_\theta.
\end{aligned}\tag{2.367}$$

Now, when $l^\star\to 0^+$, formulae (2.367) become the classical ones $[(2.367)_1\to(2.366)]$. The result (2.365) also has this property, i.e., for $\alpha\to\infty$, $l^\star\to 0^+$ it leads (by virtue of couple-stress theory) to the classical result. However, formulae (2.365) are also characterized by the fact that for $\alpha\to 0^+$ we obtain the classical result (2.366).

It should be noted that for $\alpha\to\infty$ the results (2.252) and (2.365) lead to the identical result (2.367), since in couple-stress theory the vector of reduced body loadings has the form: $\mathbf{F}=\mathbf{X}+\frac{1}{2}\operatorname{curl}\mathbf{Y}$.

2.4.2 Equations in displacements and rotations with a distortion field

In [49] we also gave uncoupled equations for the triple $(\varphi_r, u_\theta, \varphi_z)$ with a distortion field. Assuming the substitution (2.308) and

$$\overset{*}{Y}_r = -\overset{*}{\bar{R}}{}^0_1, \qquad \overset{*}{Y}_z = -\overset{*}{\bar{R}}{}^0_2, \tag{2.368}$$

where

$$\overset{*}{\bar{R}}{}^0_1 = \bar{R}^0_1 + \frac{\gamma+\varepsilon}{2\mu}\partial_z \overset{*}{\bar{R}}{}^0_3$$

and

$$\overset{*}{\bar{R}}{}^0_2 = \bar{R}^0_2 - \frac{\gamma+\varepsilon}{2\mu}\left(r^{-1}+\partial_r\right)\overset{*}{\bar{R}}{}^0_3,$$

we obtain equations (2.353), (2.355), (2.356), (2.359) and (2.360) with distortions (referring to the method of superposition). Further, assuming in equations (2.359) and (2.360) thus obtained (with a distortion field) the correspondence

$$(\overset{*}{\bar{R}}{}^0_1, \overset{**}{\bar{R}}{}^0_3, \overset{*}{\bar{R}}{}^0_2) \mapsto (\bar{R}^0_1, \bar{R}^0_3, \bar{R}^0_2), \tag{2.369}$$

we obtain in Ω uncoupled equations for $(\varphi_r, u_\theta, \varphi_z)$ with a distortion field, of the form:

$$\begin{aligned}
\nabla_0^2(l^2\nabla_0^2-1)(\nu_1^2\nabla_0^2-1)\varphi_r &= \frac{1}{2\mu}(\nu_1^2\nabla_0^2-1)\partial_z\bar{R}^0_3\\
&+\frac{1}{4\alpha}(l^2\nabla_0^2-1)\partial_r\bar{R}^0\\
&-\frac{\mu+\alpha}{4\alpha\mu}(\nu_1^2\nabla_0^2-1)\partial_z(\partial_r\bar{R}^0_2-\partial_z\bar{R}^0_1),\\
\nabla^2(l^2\nabla^2-1)(\nu_1^2\nabla^2-1)\varphi_z &= -\frac{1}{2\mu}(\nu_1^2\nabla^2-1)[r^{-1}\partial_r(r\bar{R}^0_3)]\\
&+\frac{1}{4\alpha}(l^2\nabla^2-1)\partial_z\bar{R}^0\\
&+\frac{\mu+\alpha}{4\alpha\mu}(\nu_1^2\nabla^2-1)\left\{r^{-1}\partial_r[r(\partial_r\bar{R}^0_2-\partial_z\bar{R}^0_1)]\right\},\\
2\mu\nabla_0^2(l^2\nabla_0^2-1)u_\theta &= (\partial_r\bar{R}^0_2-\partial_z\bar{R}^0_1)+2(k^2\nabla_0^2-1)\bar{R}^0_3,
\end{aligned} \tag{2.370}$$

where the symbols l^2, ν_1^2, k^2 have the same meaning as in Subsection 2.4.1, whereas the symbols $\bar{R}^0$, $\bar{R}^0_i$ $(i = 1, 2, 3)$ are the same as in Subsection 2.3.6. Equations (2.370) refer to the system of equations (2.351), in which the analogy (2.317) should be assumed. In [25] we used these equations to solve the problem of the infinite space with a distortion field of the form

$$\kappa_{zz}^{0} = \frac{\kappa_0^{zz}}{2\pi r}\delta(r)\delta(z). \tag{2.371}$$

The remaining distortions are equal to zero.

Applying the integral transformations of Fourier (0.50) and Hankel (0.51), we obtain the result in a closed form:

$$\begin{aligned}
\varphi_r = & -\frac{\kappa_0\gamma}{8\pi\alpha}\partial_r\left\{\partial_z^2\left[2\mu_1^2\left(\frac{1-e^{-R/l}}{R}\right)-\left(\frac{1-e^{-R/\nu_1}}{R}\right)\right]\right.\\
& \left.+\frac{2\alpha\beta}{\gamma(\beta+2\gamma)}\left(\frac{e^{-R/\nu_1}}{R}\right)\right\},\\
\varphi_z = & \frac{\kappa_0\gamma}{8\pi\alpha}\partial_z\left\{2\mu_1^2(r^{-1}\partial_r+\partial_r^2)\left(\frac{1-e^{-R/l}}{R}\right)\right.\\
& \left.+\partial_z^2\left(\frac{1-e^{-R/\nu_1}}{R}\right)-\frac{2\alpha\beta}{\gamma(\beta+2\gamma)}\left(\frac{e^{-R/\nu_1}}{R}\right)\right\},\\
u_\theta = & \frac{\kappa_0\gamma}{4\pi\mu}\partial_{rz}^2\left(\frac{1-e^{-R/l}}{R}\right),
\end{aligned} \tag{2.372}$$

where

$$\mu_1^2 = \frac{\mu+\alpha}{2\mu}, \qquad l = \sqrt{\frac{(\gamma+\varepsilon)(\mu+\alpha)}{4\alpha\mu}}, \qquad R = \sqrt{r^2+z^2}, \qquad \kappa_0^{zz} \mapsto \kappa_0.$$

Recall that ν_1 in the expression $\exp(-R/\nu_1)$ is determined by formula (1.189). Formulae (2.372) are a basis for deriving results for theories (4), (6) and (8) (Fig. 1.1). The very formulae (2.372) may be treated as a limiting case of formulae (2.319). To obtain them one must carry out a complex limiting procedure $\omega \to 0$.

2.4.3 Stress equations of Beltrami-Michell type and stress functions. Half-space problem

In [38], [40] and [49] we gave stress equations of the problem, which consist of:

— the equations of equilibrium in Ω,

$$\begin{aligned}
&\sigma_{\theta z} - \sigma_{z\theta} + \partial_r\mu_{rr} + \partial_z\mu_{zr} + r^{-1}(\mu_{rr} - \mu_{\theta\theta}) = 0,\\
&\sigma_{r\theta} - \sigma_{\theta r} + \partial_r\mu_{rz} + \partial_z\mu_{zz} + r^{-1}\mu_{rz} = 0,\\
&\partial_r\sigma_{r\theta} + \partial_z\sigma_{z\theta} + r^{-1}(\sigma_{r\theta} + \sigma_{\theta r}) = 0;
\end{aligned} \tag{2.373}$$

— the compatibility equations (2.322),
which, by using the constitutive relations (2.269), may be expressed in terms of stresses in Ω,

$$
\begin{aligned}
&\partial_z(\sigma_{r\theta}+\sigma_{\theta r})+(r^{-1}-\partial_r)(\sigma_{\theta z}+\sigma_{z\theta})=0,\\
&\frac{1}{2\gamma}(\mu_{rr}-\beta\kappa)+\partial_r\left[\frac{1}{4\mu}(\sigma_{z\theta}+\sigma_{\theta z})-\frac{1}{4\alpha}(\sigma_{z\theta}-\sigma_{\theta z})\right]=0,\\
&\frac{1}{2\gamma}(\mu_{\theta\theta}-\beta\kappa)+r^{-1}\left[\frac{1}{4\mu}(\sigma_{z\theta}+\sigma_{\theta z})-\frac{1}{4\alpha}(\sigma_{z\theta}-\sigma_{\theta z})\right]=0,\\
&\partial_z\left[\frac{1}{4\mu}(\sigma_{r\theta}+\sigma_{\theta r})-\frac{1}{4\alpha}(\sigma_{r\theta}-\sigma_{\theta r})\right]\\
&\qquad+r^{-1}\left[\frac{1}{4\mu}(\sigma_{z\theta}+\sigma_{\theta z})+\frac{1}{4\alpha}(\sigma_{z\theta}-\sigma_{\theta z})\right]\\
&\qquad-\frac{1}{2\gamma}(\mu_{\theta\theta}+\mu_{zz}-2\beta\kappa)=0,\\
&\frac{1}{4\gamma}(\mu_{zr}+\mu_{rz})+\frac{1}{4\varepsilon}(\mu_{zr}-\mu_{rz})\\
&\qquad+\partial_z\left[\frac{1}{4\mu}(\sigma_{z\theta}+\sigma_{\theta z})-\frac{1}{4\alpha}(\sigma_{z\theta}-\sigma_{\theta z})\right]=0,\\
&\frac{1}{2\mu}(\sigma_{r\theta}+\sigma_{\theta r})+r\partial_r\left[\frac{1}{4\mu}(\sigma_{r\theta}+\sigma_{\theta r})-\frac{1}{4\alpha}(\sigma_{r\theta}-\sigma_{\theta r})\right]\\
&\qquad-r\left[\frac{1}{4\gamma}(\mu_{zr}+\mu_{rz})-\frac{1}{4\varepsilon}(\mu_{zr}-\mu_{rz})\right]=0,
\end{aligned}
\tag{2.374}
$$

where

$$
\kappa=\frac{1}{2\gamma+3\beta}(\mu_{rr}+\mu_{\theta\theta}+\mu_{zz});
$$

— the boundary conditions on $\partial\Omega$,

$$
\begin{aligned}
&n_r\sigma_{r\theta}+n_z\sigma_{z\theta}=p_\theta, \qquad n_r\mu_{rr}+n_z\mu_{zr}=m_r,\\
&n_r\mu_{rz}+n_z\mu_{zz}=m_z,
\end{aligned}
\tag{2.375}
$$

where p_θ, m_r, $m_z:\partial\Omega\to\mathbb{R}$ are given functions, n_r, n_z are components of the unit outward vector $\mathbf{n}$ normal to the surface $\partial\Omega$, expressed in terms of a local basis associated with the assumed cylindrical coordinate system.

Equations (2.373)–(2.375) constitute an analogue of the classical Beltrami-Michell equations. From the stress equations of micropolar theory we can obtain the stress equations of the classical theory and of the theory of hypothetical medium [40]. From equations (2.373) and (2.374) one can derive equations for individual components of stress state, or for certain combinations of these components. Such equations, combined with equations (2.373)–(2.375), were also employed in [34] and [43] in solving the boundary value problem for the half-space. In [38], equations (2.373) and (2.374) are a basis for introducing the stress functions Φ, Ψ, $\chi^\star$, $\Omega^\star$, $\Delta^\star$, which are related to the stresses as follows:

$$\sigma_{\theta r} = \partial_r \Phi + \partial_z \Psi, \qquad \sigma_{r\theta} = -r^{-1}\Phi - \partial_z \Psi,$$

$$\sigma_{\theta z} = \partial_z \Phi - \partial_r \Psi, \qquad \sigma_{z\theta} = \partial_r \Psi,$$

$$\mu_{rr} = 2\Psi - \partial_z \Omega^\star, \qquad \mu_{zz} = 2\Psi - r^{-1}\partial_r(r\chi^\star),$$

$$\mu_{\theta\theta} = 2\Psi - \partial_z \Omega^\star - r\partial^2_{rz}\Delta^\star,$$

$$\mu_{zr} = -\Phi + \partial_r(\Omega^\star - \Delta^\star), \qquad \mu_{rz} = \Phi + \partial_z \chi^\star,$$

so that equations (2.373) and $(2.374)_1$ become identities. Five equations linking stress functions are obtained from equations $(2.374)_{2-6}$.

We shall determine the forms of those functions for the half-space problem. We start from the equations in Ω:

$$\begin{aligned} &\nabla_0^2(l^2\nabla_0^2 - 1)\Phi = 0, \qquad \nabla^2(\nu_1^2\nabla^2 - 1)\Psi = 0, \\ &\nabla_0^2(\nu_1^2\nabla_0^2 - 1)(l^2\nabla_0^2 - 1)\chi^\star = 0, \\ &\nabla^2(\nu_1^2\nabla^2 - 1)(l^2\nabla^2 - 1)(\Omega^\star - \Delta^\star) = 0, \end{aligned} \tag{2.376}$$

where

$$\nu_1^2 = \frac{2\gamma + \beta}{4\alpha}, \qquad l^2 = \frac{(\gamma + \varepsilon)(\mu + \alpha)}{4\alpha\mu}.$$

The separation of the expression $(\Omega^\star - \Delta^\star)$ is done by means of the equation

$$\partial_z \Delta^\star = -\frac{1}{r}\left(\frac{1}{a^0}\partial_z\Phi - \frac{\gamma}{\alpha}\partial_r\Psi\right), \tag{2.377}$$

where $a^0 = \dfrac{2\alpha\mu}{\gamma(\mu + \alpha)}$.

The stress functions are assumed in the following form:

$$\begin{aligned} &\Psi = \mathcal{H}_0\left[\!\left[(Ae^{-\xi z} + Be^{-\eta z}); \xi \to r\right]\!\right], \\ &\Phi = \mathcal{H}_1\left[\!\left[(2Ae^{-\xi z} + De^{-\rho z}); \xi \to r\right]\!\right], \\ &\chi^\star = \mathcal{H}_1\left[\!\left[(Ee^{-\xi z} + Fe^{-\eta z} + Ge^{-\rho z}); \xi \to r\right]\!\right], \\ &\Omega^\star - \Delta^\star = \mathcal{H}_0\left[\!\left[(He^{-\xi z} + Ke^{-\eta z} + Me^{-\rho z}); \xi \to r\right]\!\right], \end{aligned} \tag{2.378}$$

where

$$\eta = \sqrt{\xi^2 + \frac{1}{\nu_1^2}}, \qquad \rho = \sqrt{\xi^2 + \frac{1}{l^2}}.$$

The symbol $\mathcal{H}_\nu\left[\!\left[.\right]\!\right]$ denotes the Hankel transformation given by equation $(0.51)_2$ ($\nu = 0, 1$). We obtain the following formula for the function $\Delta^\star$:

$$\begin{aligned} \Delta^\star = -\frac{1}{r}\Bigg[&\frac{1}{a^0}\mathcal{H}_1\left[\!\left[(2Ae^{-\xi z} + De^{-\rho z}); \xi \to r\right]\!\right] \\ &- \frac{\gamma}{\alpha}\mathcal{H}_1\left[\!\left[\left(Ae^{-\xi z} + \frac{\xi}{\eta}Be^{-\eta z}\right); \xi \to r\right]\!\right]\Bigg]. \end{aligned} \tag{2.379}$$

In formulae (2.378) and (2.379), there appear nine constants A, B, D, E, F, G, H, K, M, which need to be determined. From the equations $(2.374)_{2-6}$, expressed in terms of the stress functions, we obtain:

$$E = 2\xi\left(\frac{\gamma}{2\mu}+\frac{1}{\xi^2}\right)A, \qquad H = 2\xi\left(\frac{\gamma}{2\mu}-\frac{1}{\xi^2}\right)A,$$

$$K = -\frac{\eta\gamma}{\alpha}B, \qquad F = -\frac{\xi\gamma}{\alpha}B,$$

$$G = \frac{\rho}{\eta}M = \frac{2\gamma\rho}{\gamma(\xi^2-\rho^2)+\varepsilon(\xi^2+\rho^2)}\left(\frac{\varepsilon}{\gamma a^0}\rho^2-1\right)D.$$

Stress functions are now expressed by means of three constants $A(\xi)$, $B(\xi)$ and $D(\xi)$, which may be determined from three boundary conditions (2.375). Using the stress functions (2.378) and (2.379), we have derived in [38] integral formulae for the stress fields concerning the half-space twisted on the boundary:

$$\sigma_{z\theta} = 0, \qquad \mu_{zr} = 0, \qquad \mu_{zz} = -m(r) \qquad \text{on } \partial\Omega, \tag{2.380}$$

where $m : \partial\Omega \to \mathbb{R}$ — a given function.

2.4.4 Functions of displacements-rotations

It was shown in [39], [40] and [49] that five stress functions Φ, Ψ, $\chi^\star$, $\Omega^\star$, $\Delta^\star$ may be reduced by means of eliminating $\chi^\star$, $\Omega^\star$ and $\Delta^\star$ to two functions of displacements-rotations $\Phi^\star$, $\Psi^\star$, which solve the problem[22]. In this way the following representation is obtained in Ω:

$$\begin{aligned} 2\alpha\varphi_r &= \partial_r\Psi^\star - \mu_1^2\partial_z\Phi^\star, \\ 2\alpha\varphi_z &= \partial_z\Psi^\star + \mu_1^2\left(r^{-1}+\partial_r\right)\Phi^\star, \\ 2\mu u_\theta &= \Phi^\star, \end{aligned} \tag{2.381}$$

where $\mu_1^2 = \dfrac{\mu+\alpha}{2\mu}$.

The functions of displacements-rotations $\Phi^\star$, $\Psi^\star$ must satisfy in Ω the differential relations

$$\begin{aligned} &2\partial_r(\nu_1^2\nabla^2-1)\Psi^\star - \partial_z(l^2\nabla_0^2-1)\Phi^\star = 0, \\ &2\partial_z(\nu_1^2\nabla^2-1)\Psi^\star + \left(r^{-1}+\partial_r\right)(l^2\nabla_0^2-1)\Phi^\star = 0, \end{aligned} \tag{2.382}$$

from which we obtain

[22] Love [145, pp. 274-276]. In this way, J. H. Mitchell introduces Love's function in the classical theory of elasticity, and Suchar [223] introduces a representation for displacements and rotations in the second plane problem of micropolar elastostatics.

$$\nabla^2(\nu_1^2\nabla^2 - 1)\Psi^\star = 0, \qquad \nabla_0^2(l^2\nabla_0^2 - 1)\Phi^\star = 0. \tag{2.383}$$

In [42], we dealt with stress equations (2.373)–(2.375) with body loading fields and a distortion field. Here, we give the final results. We obtain the following representation for body loadings in Ω:

$$\begin{aligned} 2\alpha\varphi_r &= \partial_r\Psi^\star - \partial_z(\mu_1^2\Phi^\star + \chi^\star), \\ 2\alpha\varphi_z &= \partial_z\Psi^\star + \left(r^{-1} + \partial_r\right)(\mu_1^2\Phi^\star + \chi^\star), \\ 2\mu u_\theta &= \Phi^\star. \end{aligned} \tag{2.384}$$

The functions $\Phi^\star$, $\Psi^\star$ satisfy the differential relations

$$\begin{aligned} &\partial_r[2(\nu_1^2\nabla^2 - 1)\Psi^\star + \sigma] \\ &\qquad - \partial_z[(l^2\nabla_0^2 - 1)\Phi^\star + 2(k^2\nabla_0^2 - 1)\chi^\star + \eta_\theta] = 0, \\ &\partial_z[2(\nu_1^2\nabla^2 - 1)\Psi^\star + \sigma] \\ &\qquad + \left(r^{-1} + \partial_r\right)[(l^2\nabla_0^2 - 1)\Phi^\star + 2(k^2\nabla_0^2 - 1)\chi^\star + \eta_\theta] = 0, \end{aligned} \tag{2.385}$$

where

$$\nu_1^2 = \frac{2\gamma + \beta}{4\alpha}, \qquad l^2 = \frac{(\gamma + \varepsilon)(\mu + \alpha)}{4\alpha\mu}, \qquad k^2 = \frac{\gamma + \varepsilon}{4\alpha}, \qquad \mu_1^2 = \frac{\mu + \alpha}{2\mu}.$$

The functions σ, η_θ, $\chi^\star$ are determined from equations (2.327). For the distortion field we obtain the representation

$$\begin{aligned} 2\alpha\varphi_r &= \partial_r\Psi^\star - \partial_z(\mu_1^2\Phi^\star + \chi^\star) - \sigma^0_{\theta z}, \\ 2\alpha\varphi_z &= \partial_z\Psi^\star + r^{-1}\partial_r[r(\mu_1^2\Phi^\star + \chi^\star)] + \sigma^0_{\theta r}, \\ 2\mu u_\theta &= \Phi^\star. \end{aligned} \tag{2.386}$$

Here, the functions $\Phi^\star$, $\Psi^\star$ satisfy the differential relations

$$\begin{aligned} &2\partial_r(\nu_1^2\nabla^2 - 1)\Psi^\star - \partial_z[(l^2\nabla_0^2 - 1)\Phi^\star - 2(k^2\nabla_0^2 - 1)\chi^\star] \\ &\quad = \bar{R}_1^0 + 2(\nu_1^2\nabla_0^2 - 1)\sigma^0_{\theta z} - 2(\nu_1^2 - k^2)\partial_z(\partial_z\sigma^0_{\theta z} + \partial_r\sigma^0_{\theta r}), \\ &2\partial_z(\nu_1^2\nabla^2 - 1)\Psi^\star + \left(r^{-1} + \partial_r\right)[(l^2\nabla_0^2 - 1)\Phi^\star - 2(k^2\nabla_0^2 - 1)\chi^\star] \\ &\quad = \bar{R}_2^0 - 2(k^2\nabla_0^2 - 1)\sigma^0_{\theta r} + 2(\nu_1^2 - k^2)\partial_z[r^{-1}\partial_r(r\sigma^0_{\theta z}) - \partial_z\sigma^0_{\theta r}], \end{aligned} \tag{2.387}$$

where the function $\chi^\star$ satisfies the equation

$$\nabla_0^2\chi^\star = r^{-1}\partial_r[r(\sigma^0_{r\theta} + \sigma^0_{\theta r})] + \partial_z(\sigma^0_{z\theta} + \sigma^0_{\theta z}).$$

The symbols $\bar{R}^0_\alpha$, $(\alpha = 1, 2)$, $\sigma^0_{\theta z}$, $\sigma^0_{z\theta}$, $\sigma^0_{\theta r}$, $\sigma^0_{r\theta}$ refer to the distortion field and have the same meaning as in Subsection 2.3.6. Let us consider, for example, the problem of the half-space Ω (2.119) with the loading

$$\sigma_{z\theta} = -p_\theta, \qquad \mu_{zr} = -m_r, \qquad \mu_{zz} = -m_z \qquad \text{on } \partial\Omega, \tag{2.388}$$

where functions p_θ, m_r, $m_z : \partial\Omega \to \mathbb{R}$ are given.

The form of the functions $\Phi^\star$, $\Psi^\star$ is chosen from equations (2.383) and (2.382) as follows:

$$\begin{aligned} \Phi^\star &= \mathcal{H}_1 \left[\!\left[(2Ce^{-\xi z} + Be^{-\rho z}); \xi \to r \right]\!\right], \\ \Psi^\star &= \mathcal{H}_0 \left[\!\left[(Ce^{-\xi z} + De^{-\eta z}); \xi \to r \right]\!\right], \end{aligned} \tag{2.389}$$

where

$$\rho = \sqrt{\xi^2 + l^{-2}}, \qquad \eta = \sqrt{\xi^2 + \nu_1^{-2}}.$$

The three constants $B(\xi)$, $C(\xi)$ and $D(\xi)$ are determined from three boundary conditions given by (2.388).

Consider now action of body loadings in the infinite space of the form (2.363). Here, we shall make use of equations (2.385), from which we obtain the equations

$$\begin{aligned} &2\nabla^2(\nu_1^2\nabla^2 - 1)\Psi^\star + r^{-1}\partial_r(rY_r) + \partial_z Y_z = 0, \\ &\nabla_0^2(l^2\nabla_0^2 - 1)\Phi^\star + 2(k^2\nabla_0^2 - 1)X_\theta + \partial_r Y_z - \partial_z Y_r = 0. \end{aligned} \tag{2.390}$$

By means of the Fourier (0.50) and Hankel (0.51) integral transformations we obtain the formulae

$$\begin{aligned} \Phi^\star &= -\frac{M_0}{4\pi}\partial_r\left(\frac{1 - e^{-R/l}}{R}\right), \\ \Psi^\star &= -\frac{M_0}{4\pi}\frac{z}{r}\partial_r\left(\frac{1 - e^{-R/\nu_1}}{R}\right), \end{aligned} \tag{2.391}$$

where the constant ν_1 is given by formula (1.189).

From formulae (2.384) (putting $\chi^\star \stackrel{\text{def}}{=} 0$) we obtain formulae (2.252) by use of (2.391). Let us make the following important remark concerning the representation (2.381) and equations (2.382)–(2.383). Namely, having performed the transformation

$$\Psi^\star \mapsto 2\alpha\Phi, \qquad \Phi^\star \mapsto -\frac{4\alpha\mu}{\mu + \alpha}\partial_r\Psi$$

we obtain from formula (2.381) in Ω the following:

$$\begin{aligned} \varphi_r &= \partial_r\Phi + \partial_{rz}^2\Psi, \\ \varphi_z &= \partial_z\Phi - (\nabla^2 - \partial_z^2)\Psi, \\ u_\theta &= -\frac{2\alpha}{\mu + \alpha}\partial_r\Psi. \end{aligned} \tag{2.392}$$

By integrating equation $(2.382)_1$ over r we obtain from (2.382) the equations

$$\begin{aligned}&(\nu_1^2\nabla_0^2-1)\Phi+\frac{\mu}{\mu+\alpha}(l^2\nabla_0^2-1)\partial_z\Psi=0,\\&(\nu_1^2\nabla^2-1)\partial_z\Phi-\frac{\mu}{\mu+\alpha}(l^2\nabla^2-1)\left(r^{-1}+\partial_r\right)\partial_r\Psi=0.\end{aligned}\tag{2.393}$$

This is the representation for potentials of Nowacki [188], derived directly from equations (2.351).

2.4.5 Method of potentials

In [49], we considered the vector equations (1.147) and (1.237)–(1.239). It follows from those equations that a scalar potential Σ and vectorial potentials $\boldsymbol{\Psi} \overset{\text{def}}{=} (\Psi_r, 0, \Psi_z)$, $\mathbf{H} \overset{\text{def}}{=} (0, H_\theta, 0)$ will appear in the problem. For the triple $(\varphi_r, u_\theta, \varphi_z)$ we obtain the following formulae in Ω:

$$\begin{aligned}&u_\theta=\partial_z\Psi_r-\partial_r\Psi_z, \qquad \varphi_r=\partial_r\Sigma-\partial_z H_\theta,\\&\varphi_z=\partial_z\Sigma+\left(r^{-1}+\partial_r\right)H_\theta.\end{aligned}\tag{2.394}$$

Equations for Σ, Ψ_r, Ψ_z, H_θ follow from equations (1.237)–(1.239). Taking into account the specific character of the second axial problem, one can show that its solution reduces to the determination of two functions Σ, H_θ. Let us introduce a function Ψ subject to the conditions

$$\left(r^{-1}+\partial_r\right)\Psi_r+\partial_z\Psi_z=0: \qquad \Psi_r=\partial_z\Psi, \qquad \Psi_z=-\left(r^{-1}+\partial_r\right)\Psi.$$

From equation $(1.237)_1$ we obtain the following equation in Ω:

$$(\mu+\alpha)\nabla_0^2(\partial_z\Psi_r-\partial_r\Psi_z)-2\alpha\nabla_0^2 H_\theta+\nabla_0^2\chi=0,\tag{2.395}$$

from which we find $\nabla_0^2\Psi$:

$$\nabla_0^2\Psi=\frac{2\alpha}{\mu+\alpha}H_\theta-\frac{1}{\mu+\alpha}\chi^\star.\tag{2.396}$$

From equation $(1.237)_2$ we obtain two equations linking Σ, H_θ, Ψ_r, Ψ_z, from which, eliminating $\nabla_0^2\Psi$ by means of formula (2.396), we get

$$\begin{aligned}&\partial_r[4\alpha(\nu_1^2\nabla^2-1)\Sigma+\sigma]\\&\quad-\partial_z\left[\frac{4\alpha\mu}{\mu+\alpha}(l^2\nabla_0^2-1)H_\theta-\frac{2\alpha}{\mu+\alpha}\chi^\star+\eta_\theta\right]=0,\\&\partial_z[4\alpha(\nu_1^2\nabla^2-1)\Sigma+\sigma]\\&\quad+\left(r^{-1}+\partial_r\right)\left[\frac{4\alpha\mu}{\mu+\alpha}(l^2\nabla_0^2-1)H_\theta-\frac{2\alpha}{\mu+\alpha}\chi^\star+\eta_\theta\right]=0.\end{aligned}\tag{2.397}$$

The functions σ, η_θ, $\chi^\star$ are determined from equations (2.327). The representation (2.394) now takes the form

$$\begin{aligned} \varphi_r &= \partial_r \Sigma - \partial_z H_\theta, \\ \varphi_z &= \partial_z \Sigma + \left(r^{-1} + \partial_r\right) H_\theta, \\ u_\theta &= \frac{2\alpha}{\mu+\alpha} H_\theta - \frac{1}{\mu+\alpha}\chi^\star. \end{aligned} \tag{2.398}$$

Separate equations for the functions Σ, H_θ (with the body loadings Y_r, X_θ, Y_z) have the form

$$\begin{aligned} &4\alpha\nabla^2(\nu_1^2\nabla^2 - 1)\Sigma + r^{-1}\partial_r(rY_r) + \partial_z Y_z = 0, \\ &2\mu\nabla_0^2(l^2\nabla_0^2 - 1)H_\theta + \frac{\mu+\alpha}{2\alpha}(\partial_r Y_z - \partial_z Y_r) - X_\theta = 0, \end{aligned} \tag{2.399}$$

where

$$\nu_1^2 = \frac{2\gamma+\beta}{4\alpha}, \qquad l^2 = \frac{(\gamma+\varepsilon)(\mu+\alpha)}{4\alpha\mu}.$$

Let us make an important remark. Recall that for the dynamical problem we have derived the same representation and equations: formulae (2.332) and (2.333), or formulae (2.341) and (2.342), starting from the equations of stress type (2.321) and (2.374), and further from the vector equations for potentials (1.147) and (1.151)–(1.153), respectively. We will show now that the dynamical representation (2.332) and (2.333) leads to a representation within the framework of elastostatics. Replacing the dynamical operators by the static ones in (2.332) and (2.333), we obtain:

$$\begin{aligned} 2\alpha\varphi_r &= \partial_r\psi^\star - \partial_z[(\mu+\alpha)\nabla_0^2\chi + \chi^\star], \\ 2\alpha\varphi_z &= \partial_z\psi^\star + \left(r^{-1} + \partial_r\right)[(\mu+\alpha)\nabla_0^2\chi + \chi^\star], \\ u_\theta &= \nabla_0^2\chi \end{aligned} \tag{2.400}$$

and the equations

$$\begin{aligned} &\partial_r[2(\nu_1^2\nabla^2 - 1)\psi^\star + \sigma] \\ &\qquad - \partial_z[2\mu\nabla_0^2(l^2\nabla_0^2 - 1)\chi + 2(k^2\nabla_0^2 - 1)\chi^\star + \eta_\theta] = 0, \\ &\partial_z[2(\nu_1^2\nabla^2 - 1)\psi^\star + \sigma] \\ &\qquad + \left(r^{-1} + \partial_r\right)[2\mu\nabla_0^2(l^2\nabla_0^2 - 1)\chi + 2(k^2\nabla_0^2 - 1)\chi^\star + \eta_\theta] = 0. \end{aligned} \tag{2.401}$$

Relation $(2.400)_3$ is a conclusion from the Stokes-Helmholtz decomposition of the vector $\mathbf{u}$. Moreover, in the method employing dynamical equations, we made use of the geometric relations (2.267), from which we obtained that:

$\gamma_{z\theta} + \gamma_{\theta z} = \partial_z u_\theta$. In turn, from the constitutive relations (2.269) we have: $\sigma_{z\theta} + \sigma_{\theta z} = 2\mu(\gamma_{\theta z} + \gamma_{z\theta})$. Since, from $(2.324)_2$, we have: $\sigma_{\theta z} + \sigma_{z\theta} = \partial_z \Phi^\star$ (where $\Phi^\star$ is introduced in such a way that the compatibility equation (2.323) is an identity), thus we get

$$\partial_z(2\mu u_\theta - \Phi^\star) = 0. \tag{2.402}$$

From equation (2.402) we obtain $2\mu u_\theta - \Phi^\star = C(r)$, where $C(r)$ is an arbitrary function of the real variable r. Assuming $\bar{\Phi}^\star = \Phi^\star + C(r)$, we are able to eliminate the function χ: $\bar{\Phi}^\star = 2\mu\nabla_0^2\chi$ from further considerations.

Now, from formulae (2.400) we obtain the following representation:

$$\begin{aligned} 2\alpha\varphi_r &= \partial_r\psi^\star - \partial_z(\mu_1^2\bar{\Phi}^\star + \chi^\star), \\ 2\alpha\varphi_z &= \partial_z\psi^\star + \left(r^{-1} + \partial_r\right)\left(\mu_1^2\bar{\Phi}^\star + \chi^\star\right), \\ 2\mu u_\theta &= \bar{\Phi}^\star \end{aligned} \tag{2.403}$$

and from (2.401) — the equations in Ω:

$$\begin{aligned} &\partial_r[2(\nu_1^2\nabla^2 - 1)\psi^\star + \sigma] \\ &\qquad - \partial_z[(l^2\nabla_0^2 - 1)\bar{\Phi}^\star + 2(k^2\nabla_0^2 - 1)\chi^\star + \eta_\theta] = 0, \\ &\partial_z[2(\nu_1^2\nabla^2 - 1)\psi^\star + \sigma] \\ &\qquad + \left(r^{-1} + \partial_r\right)[(l^2\nabla_0^2 - 1)\bar{\Phi}^\star + 2(k^2\nabla_0^2 - 1)\chi^\star + \eta_\theta] = 0, \end{aligned} \tag{2.404}$$

where

$$\mu_1^2 = \frac{\mu + \alpha}{2\mu}, \qquad k^2 = \frac{\gamma + \varepsilon}{4\alpha}.$$

Furthermore, note that changing the notation: $\bar{\Phi}^\star \mapsto \Phi^\star$ in formulae (2.403) and (2.404) (which means that equation (2.402) may be satisfied by assuming $2\mu u_\theta - \Phi^\star = 0$), we obtain the representation (2.384) and equations (2.385). Let us go further: if now, in formulae (2.403) and (2.404), we assume the correspondence

$$\psi^\star \mapsto 2\alpha\Sigma, \qquad \frac{\mu + \alpha}{2\mu}\bar{\Phi}^\star + \chi^\star \mapsto 2\alpha H_\theta,$$

then we obtain the representation (2.398) and equations (2.397). This procedure, together with our previous discussion, shows again that a basis for the derivation of representations and equations for potentials may be provided by vector equations (for potentials in the dynamical problem) (1.147) and (1.151). We shall return to this question when discussing the problem of plane state of strain.

2.4.6 Functions of Love type

In [49], we assumed the following relations in Ω for the stress vectors (2.343) and the triple $(\varphi_r, u_\theta, \varphi_z)$:

$$\begin{aligned}
2\mu u_\theta &= [-\mu_0^{-1}\nabla_0^2 + 2(l^2\nabla_0^2 - 1)]F_\theta - (\nu_1^2\nabla_0^2 - 1)(\partial_z M_r - \partial_r M_z),\\
2\mu\varphi_r &= (2\alpha)^{-1}\mu(l^2\nabla_0^2 - 1)\partial_r M\\
&\quad + \alpha_0^{-1}(\nu_1^2\nabla_0^2 - 1)\partial_z(\partial_z M_r - \partial_r M_z) + \partial_z F_\theta,\\
2\mu\varphi_z &= (2\alpha)^{-1}\mu(l^2\nabla^2 - 1)\partial_z M\\
&\quad - \alpha_0^{-1}(\nu_1^2\nabla^2 - 1)\left(r^{-1} + \partial_r\right)(\partial_z M_r - \partial_r M_z)\\
&\quad - \left(r^{-1} + \partial_r\right) F_\theta,
\end{aligned} \tag{2.405}$$

where

$$M = \left(r^{-1} + \partial_r\right) M_r + \partial_z M_z, \qquad \mu_0 = \frac{2\mu}{\gamma + \varepsilon}, \qquad \alpha_0 = \frac{2\alpha}{\mu + \alpha},$$
$$l^2 = \frac{(\gamma + \varepsilon)(\mu + \alpha)}{4\alpha\mu}, \qquad \nu_1^2 = \frac{2\gamma + \beta}{4\alpha}.$$

The functions M_r, F_θ, M_z satisfy the equations

$$\begin{aligned}
&\nabla^2(\nu_1^2\nabla^2 - 1)(l^2\nabla^2 - 1)M_z + Y_z = 0,\\
&\nabla_0^2(\nu_1^2\nabla_0^2 - 1)(l^2\nabla_0^2 - 1)M_r + Y_r = 0,\\
&\nabla_0^2(l^2\nabla_0^2 - 1)F_\theta + X_\theta = 0.
\end{aligned} \tag{2.406}$$

Let us consider body loadings of the form (2.62). In the problem considered, there appear body moments Y_z. Keeping only the function M_z which satisfies equation $(2.406)_1$, we obtain the representation

$$\begin{aligned}
2\mu u_\theta &= (\nu_1^2\nabla_0^2 - 1)\partial_r M_z,\\
2\mu\varphi_r &= [(2\alpha)^{-1}\mu(l^2\nabla_0^2 - 1) - \alpha_0^{-1}(\nu_1^2\nabla_0^2 - 1)]\partial_{rz}^2 M_z,\\
2\mu\varphi_z &= [(2\alpha)^{-1}\mu(l^2\nabla^2 - 1)\partial_z^2 + \alpha_0^{-1}(\nu_1^2\nabla^2 - 1)\left(r^{-1} + \partial_r\right)\partial_r]M_z.
\end{aligned} \tag{2.407}$$

The function M_z, selected in the second axially-symmetric problem in such a way, is an analogue of the generalized Love function F_z (formulae (2.166) and (2.167)). Assuming the relation

$$M_z = 16\alpha^2\mu\psi \tag{2.408}$$

we obtain a representation from (2.407), and from $(2.406)_1$ — an equation for the function ψ, which has been introduced to micropolar theory by Nowacki

[178]. Let us consider body loadings in the form (2.65). There are body couples Y_r associated with the investigated problem. Keeping only the function M_r, which satisfies $(2.406)_2$, we obtain the representation

$$\begin{aligned} 2\mu u_\theta &= -(\nu_1^2\nabla_0^2 - 1)\partial_z M_r, \\ 2\mu\varphi_r &= [(2\alpha)^{-1}\mu(l^2\nabla_0^2 - 1)\partial_r\left(r^{-1} + \partial_r\right) + \alpha_0^{-1}(\nu_1^2\nabla_0^2 - 1)\partial_z^2]M_r, \\ 2\mu\varphi_z &= [(2\alpha)^{-1}\mu(l^2\nabla^2 - 1)\partial_z\left(r^{-1} + \partial_r\right) \\ &\quad - \alpha_0^{-1}(\nu_1^2\nabla^2 - 1)\partial_z\left(r^{-1} + \partial_r\right)]M_r. \end{aligned} \tag{2.409}$$

The function M_r may be treated as a different variant of the generalized Love function in the second axial problem of elastostatics (as an analogue of the function F_r from the first axial problem, formulae (2.169) and (2.170)). For the problem without body loadings, the functions M_r, M_z satisfy

$$\begin{aligned} \nabla^2(\nu_1^2\nabla^2 - 1)(l^2\nabla^2 - 1)M_z = 0, \\ \nabla_0^2(\nu_1^2\nabla_0^2 - 1)(l^2\nabla_0^2 - 1)M_r = 0, \end{aligned} \tag{2.410}$$

respectively. The following relation between these functions holds:

$$(\nu_1^2\nabla_0^2 - 1)(\partial_r M_z + \partial_z M_r) = 0. \tag{2.411}$$

2.4.7 Problem of singularities of physical fields in the half-space twisted on the boundary

In [81] Dyszlewicz & Słotwińska studied the problem of singularities of displacements, rotations, force-stresses and couple-stresses in the half-space (2.119) twisted on the boundary by a concentrated moment. Two cases of twisting are considered:

(1°) classical twisting, which refers to a torque of value M, understood as the limiting case ($h \to 0$) of action of two pairs of tangent forces (Fig. 1.4);

(2°) non-classical (micropolar) twisting, referring to the problem in which there is a concentrated couple-stress $\mu_{11} = -M_\mu\delta(x_2)\delta(x_3)$ $(\sigma_{1i} = \mu_{1K} = 0)$ given on $\partial\Omega$.

We have the following conditions in the cylindrical system of coordinates:

$$\sigma_{z\theta} = \mu_{zr} = 0, \qquad \mu_{zz} = -\frac{M_\mu}{2\pi r}\delta(r)\delta(z),$$

where $M_\mu \in \mathbb{R}$.

Both problems are axially-symmetric, therefore the results are obtained for the rectangular system of coordinates $Ox_1x_2x_3$ as well as for the cylindrical coordinate system.

In Table 2.4 we present the results concerning the orders of singularities of physical fields ($R = \sqrt{x_ix_i} = \sqrt{r^2 + z^2}$).

Table 2.4.

Physical fields	(1)		(2)		(3)	
	(1°)	(2°)	(1°)	(2°)	(1°)	(2°)
u_θ	$O(R^{-2})$	$O(1)$	$O(1)$	$O(1)$	$O(R^{-2})$	–
φ_r, φ_z	$O(R^{-1})$	$O(R^{-1})$	$O(R^{-1})$	$O(R^{-1})$	$O(R^{-3})$	$O(R^{-1})$
$\sigma_{r\theta}$, $\sigma_{\theta r}$, $\sigma_{z\theta}$, $\sigma_{\theta z}$	$O(R^{-3})$	$O(R^{-1})$	$O(R^{-1})$	$O(R^{-1})$	$O(R^{-3})$	–
μ_{zz}, μ_{rr}, $\mu_{\theta\theta}$, μ_{zr}, μ_{rz}	$O(R^{-2})$	$O(R^{-2})$	$O(R^{-2})$	$O(R^{-2})$	–	$O(R^{-2})$

Key to the table: numbers (1) and (2) denote micropolar theory and couple-stress theory, respectively. Column (3) refers to the classical theory of elasticity for case (1°), and to hypothetical medium for case (2°)

Notation for stresses in the framework of couple-stress theory (2) should be understood as follows:

$$s_{r\theta} = s_{\theta r} = \tfrac{1}{2}(\sigma_{r\theta} + \sigma_{\theta r}), \qquad s_{z\theta} = s_{\theta z} = \tfrac{1}{2}(\sigma_{z\theta} + \sigma_{\theta z}); \tag{2.412}$$

$$\begin{aligned} &\mu_{rz} = m_{rz}, \qquad \mu_{zr} = m_{zr}, \\ &m_{rr} = \mu_{rr} - \hat{\mu}, \qquad m_{\theta\theta} = \mu_{\theta\theta} - \hat{\mu}, \qquad m_{zz} = \mu_{zz} - \hat{\mu}, \end{aligned} \tag{2.413}$$

where $\hat{\mu} = \frac{1}{3}(\mu_{rr} + \mu_{\theta\theta} + \mu_{zz})$.

In the problem of twisting the half-space discussed here, there is a visible influence of couple-stresses of couple-stress theory (2) on the phenomenon of singularities of physical fields. Moreover, let us note that in couple-stress theory (2), for both cases of twisting, (1°) and (2°), the respective physical quantities have identical orders of singularities (also, the respective coefficients of their intensities are the same). This observation is a part of more general conclusion, from which it follows that the couple-stress half-space does not distinguish the action of a classical moment M from that of a non-classical moment M_μ. Fields induced by the action of M and M_μ are identical (we assume $M = 1$, $M_\mu = 1$). The displacements u_i are expressed by the formula ($u_1 = 0$):

$$\begin{aligned} u_K = &- \epsilon_{1KN} \frac{M_\mu a^{0^*}}{2\pi^2 \mu} \int_0^\infty \int_0^\infty \frac{\xi_N}{\xi} \cdot \frac{1}{\Delta^{0^*}} \times \\ &[(\xi^2 + a^{0^*}) e^{-\xi x_1} - \xi^2 e^{-\rho^\star x_1}] \sin \xi_N x_N \cos \xi_K x_K \, d\xi_2 \, d\xi_3, \end{aligned} \tag{2.414}$$

where

$$\rho^\star = \sqrt{\xi^2 + \frac{1}{l^{\star 2}}}, \qquad a^{0^*} = \frac{1}{(1+\eta) l^{\star 2}}, \qquad \xi^2 = \xi_N \xi_N,$$

$\Delta^{0^*} = (\xi^2 + a^{0^*})^2 - \rho^\star \xi^3$, $K \neq N$ and we do not sum over K, N.

Formulae for the rotations φ_i are such that they satisfy the relation: $\varphi_i = \frac{1}{2}\epsilon_{ijk}u_{k,j}$. The stress fields s_{ij}, m_{ij} satisfy the constitutive relations ($s_{11} = 0$, $m_{23} = m_{32}$):

$$s_{ij} = \mu(u_{i,j} + u_{j,i}) + \lambda u_{k,k}\delta_{ij}, \qquad m_{ij} = 4\mu l^{\star 2}(\varphi_{j,i} + \eta\varphi_{i,j}), \tag{2.415}$$

where μ, λ, η, $l^\star$ are elasticity constants of couple-stress theory. Such result is obtained in couple-stress theory (2), both by using the equations of the theory, and as a limiting case of the results from micropolar theory:

$$\alpha \to \infty, \qquad \sqrt{\frac{\gamma+\varepsilon}{4\mu}} \to l^\star, \qquad \frac{\gamma-\varepsilon}{\gamma+\varepsilon} \to \eta.$$

This is illustrated by the diagram in Fig. 2.26.

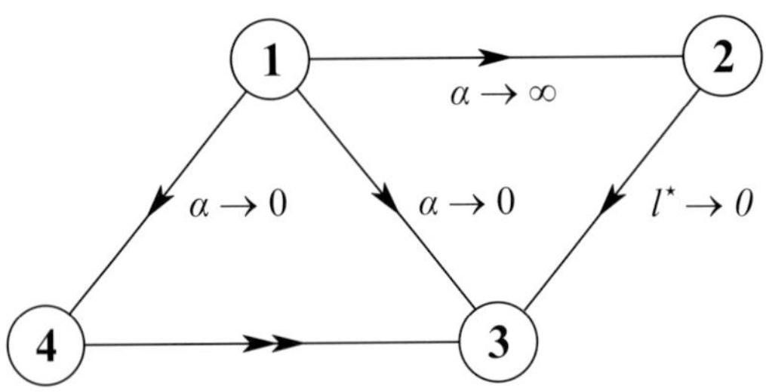

Fig. 2.26. Numbers (1), (2), (3) and (4) denote, respectively: micropolar theory, couple-stress theory, the classical theory of elasticity and the theory of a hypothetical medium

The case (1) $\xrightarrow{\alpha\to 0^+}$ (4) holds for the moment M_μ, while the case (1) $\xrightarrow{\alpha\to 0^+}$ (3) — for the moment M. By a certain analogy, we may construct the classical result (depicted by the double arrow on the diagram) from the results for (4). In turn, from micropolar theory (1) we obtain as $\alpha \to \infty$ (both for M and M_μ) the identical result as from couple-stress theory (2), which for $l^\star \to 0^+$ goes over into the classical result. Let us go back to Table 2.4. For force-stresses in the theory (1) (case (1°)) we obtain the formulae:

$$\begin{aligned}
\sigma_{r\theta} &= -\frac{M}{4\pi}\left(\frac{3r^2}{R^5} - \frac{2\alpha}{\mu+\alpha}\cdot\frac{1}{R^3}\right) + O(R^{-1}),\\
\sigma_{\theta r} &= -\frac{M}{4\pi}\left(3\,\frac{\mu-\alpha}{\mu+\alpha}\cdot\frac{r^2}{R^5} + \frac{2\alpha}{\mu+\alpha}\cdot\frac{1}{R^3}\right) + O(R^{-1}),\\
\sigma_{z\theta} &= -\frac{3M}{4\pi}\cdot\frac{zr}{R^5} + O(R^{-1}),\\
\sigma_{\theta z} &= -\frac{3M}{4\pi}\cdot\frac{\mu-\alpha}{\mu+\alpha}\cdot\frac{zr}{R^5} + O(R^{-1}),
\end{aligned} \tag{2.416}$$

from which it follows that the orders of singularities of those stresses are $O(R^{-3})$. Now observe that, constructing expressions (2.412), we find that the coefficient of singularity intensity (the coefficient standing by the singularity $O(R^{-3})$) depends on the elasticity constant α and for $\alpha \to \infty$ it tends to zero, thus reducing the orders of singularities of force-stresses in couple-stress theory (2).

For $\alpha \to 0$ we obtain the classical result:

$$\sigma_{r\theta} = \sigma_{\theta r} = -\frac{3M}{4\pi} \cdot \frac{r^2}{R^5}, \qquad \sigma_{z\theta} = \sigma_{\theta z} = -\frac{3M}{4\pi} \cdot \frac{zr}{R^5}. \tag{2.417}$$

Let us go back now to formula (2.414) from couple-stress theory. When $l^\star \to 0$, we obtain the classical result ($u_1 = 0$):

$$u_K = -\epsilon_{1KN} \frac{M}{2\pi^2\mu} \int_0^\infty \int_0^\infty \frac{\xi_N}{\xi} e^{-\xi x_1} \sin \xi_N x_N \cos \xi_K x_K \, d\xi_2 \, d\xi_3, \tag{2.418}$$

(where $K \neq N$ and we do not sum over K, N), which can be given in a closed form in the cylindrical system of coordinates:

$$u_\theta = \frac{M}{4\pi\mu} \cdot \frac{r}{R^3}. \tag{2.419}$$

Let us notice now that in micropolar theory (1) the solution for the displacements u_i (for case (1°)) with a singular part $u_i^s = O(R^{-2})$ has the following form in the cylindrical system of coordinates:

$$u_\theta = \frac{M}{4\pi} \cdot \frac{1}{\mu + \alpha} \cdot \frac{r}{R^3} + O(1). \tag{2.420}$$

The coefficient of singularity intensity depends on α and, when $\alpha \to 0$, $O(1) \to 0$ we obtain the classical result (2.419) from formula (2.420). However, when $\alpha \to \infty$, we obtain from (2.420) that in couple-stress theory u_θ is bounded at the point where the moment is applied, namely[23] $u_\theta = O(1)$. Nevertheless, in order to properly determine (or verify) singularities of rotations and stresses in couple-stress theory, one has to represent u_θ in the form ($u_\theta = u_\theta^s + O(1)$):

$$u_\theta = \frac{M}{4\pi\mu l^{\star 2}} \cdot \frac{1}{3 - \eta} \left[\frac{2}{1 + \eta} \cdot \frac{r}{R + z} + \frac{zr}{R(R + z)} \right] + O(1), \tag{2.421}$$

where the leading part of the solution u_θ^s is bounded at the point $r = z = 0$, but is not unique there. Now, let us note that formula (2.421) does not correspond with the classical formula (2.419), since, for $l^\star \to 0$, we obtain $|u_K^s| \to \infty$. Similar remarks refer to the rotations φ_i^s and the stresses s_{1K}^s,

[23] In [81] it was verified that for $\alpha \to \infty$, we obtain $u_\theta = O(1)$ from formula (2.420), since it holds $u_\theta^s \to 0$, $O(1) \to O(1)$.

s^s_{23}, s^s_{KK}. As far as the stresses m^s_{ij} are concerned, the constant $l^\star$ does not appear in the coefficient of singularity intensity. In general, let us note that the correspondence between the singular results of theories (1)–(4) (Fig. 2.26), by which we mean singular or leading parts of solutions, is connected with the coefficient of singularity intensity and does not follow the diagram in Fig. 2.26. Namely, the limit (2) $\xrightarrow{l^\star \to 0}$ (3) does not hold, and moreover, for the limit (1) $\xrightarrow{\alpha \to \infty}$ (2) the additional condition $\beta \to \infty$ is required (apart from: $\sqrt{\dfrac{\gamma + \varepsilon}{4\mu}} \to l^\star$, $\dfrac{\gamma - \varepsilon}{\gamma + \varepsilon} \to \eta$).

3. Two-dimensional problems

3.1 The first problem of plane strain state. Elastodynamics

3.1.1 Equations in displacements and rotations with a field of body loadings

From vector equations (1.1) for the plane strain state (causes and effects do not depend on the variable x_3) we obtain for the triple (u_α, φ_3) the following equations in $\Omega \times \mathbf{T}_+$:

$$\begin{aligned} &\Box_2 u_\alpha + (\lambda + \mu - \alpha) u_{\beta,\beta\alpha} + 2\alpha \epsilon_{\alpha\gamma} \varphi_{3,\gamma} + X_\alpha = 0, \\ &\Box_4 \varphi_3 + 2\alpha \epsilon_{\alpha\beta} u_{\beta,\alpha} + Y_3 = 0. \end{aligned} \tag{3.1}$$

The operators $\Box_i$ $(i = 1, \ldots, 4)$ contain the two-dimensional Laplace operator $\nabla_1^2(.) = (.)_{,\alpha\alpha}$ and $\epsilon_{\alpha\beta}$ is the permutation symbol. The vectors of body loadings have the form: $\mathbf{X} \overset{\text{def}}{=} (X_1, X_2, 0)$, $\mathbf{Y} \overset{\text{def}}{=} (0, 0, Y_3)$. In [49] a solution of the system of equations (3.1) is sought for by means of the superposition method. Let us formulate now the following initial-boundary conditions:

— boundary conditions on $\partial\Omega \times \mathbf{T}$,

$$p_\beta = \sigma_{\alpha\beta} n_\alpha, \qquad m_3 = \mu_{\alpha 3} n_\alpha \qquad \text{on } \partial\Omega_\sigma \times \mathbf{T}, \tag{3.2}$$

$$u_\alpha = f_\alpha, \qquad \varphi_3 = g_3 \qquad \text{on } \partial\Omega_u \times \mathbf{T}; \tag{3.3}$$

— initial conditions in $\Omega \times \{0\}$,

$$u_\alpha = h_\alpha, \qquad \varphi_3 = k_3, \qquad \dot{u}_\alpha = \psi_\alpha, \qquad \dot{\varphi}_3 = \chi_3, \tag{3.4}$$

where the functions p_β, $m_3 : \partial\Omega_\sigma \times \mathbf{T} \to \mathbb{R}$, f_α, $g_3 : \partial\Omega_u \times \mathbf{T} \to \mathbb{R}$ and h_α, ψ_α, k_3, $\chi_3 : \Omega \to \mathbb{R}$ are given.

To the system of equations (3.1) we introduce the substitution

$$\varphi_3 = \tfrac{1}{2} \epsilon_{\alpha\beta} u_{\beta,\alpha} - \zeta_3. \tag{3.5}$$

We obtain the following system of equations in $\Omega \times \mathbf{T}_+$:

$$\overset{*}{\Box}_2 u_\alpha + (\lambda+\mu) u_{\beta,\beta\alpha} + X_\alpha = 2\alpha \epsilon_{\alpha\gamma}\zeta_{3,\gamma},$$
$$\overset{*}{\Box}_4 \epsilon_{\alpha\beta} u_{\beta,\alpha} + 2Y_3 = 2\Box_4 \zeta_3. \tag{3.6}$$

The operators $\overset{*}{\Box}_2, \overset{*}{\Box}_4$ are obtained from operators $\Box_2, \Box_4$ by putting $\alpha = 0$. We assume the solution of the problem in the form: $u_\alpha = u_\alpha^0 + u'_\alpha$ and $\varphi_3 = \varphi_3^0 + \varphi'_3$, where the triple $(u_\alpha^0, \varphi_3^0)$ refers to the classical theory of elasticity. Such a result can be obtained by assuming $\zeta_3 = \zeta_3^0 + \zeta'_3$ and $Y_3 = Y_3^0 + Y'_3$. Now, for $\zeta_3^0 = 0$, we obtain from $(3.6)_1$ the classical equations

$$\overset{*}{\Box}_2 u_\alpha^0 + (\lambda+\mu) u_{\beta,\beta\alpha}^0 + X_\alpha = 0 \tag{3.7}$$

with the following initial-boundary conditions:

— boundary conditions on $\partial\Omega \times \mathbf{T}$,

$$\sigma_{\alpha\beta}^0 n_\alpha = p_\beta \qquad \text{on } \partial\Omega_\sigma \times \mathbf{T}, \tag{3.8}$$
$$u_\alpha^0 = f_\alpha \qquad \text{on } \partial\Omega_u \times \mathbf{T}; \tag{3.9}$$

— initial conditions in $\Omega \times \{0\}$,

$$u_\alpha^0 = h_\alpha, \qquad \dot{u}_\alpha^0 = \psi_\alpha. \tag{3.10}$$

The stresses $\sigma_{\alpha\beta}^0$ are determined from Hooke's law on $\partial\Omega \times \mathbf{T}$

$$\sigma_{\alpha\beta}^0 = 2\mu\varepsilon_{\alpha\beta}^0 + \lambda u_{\gamma,\gamma}^0 \delta_{\alpha\beta}, \qquad \sigma_{33}^0 = \lambda u_{\gamma,\gamma}^0, \tag{3.11}$$

where

$$\varepsilon_{\alpha\beta}^0 = \tfrac{1}{2}(u_{\alpha,\beta}^0 + u_{\beta,\alpha}^0) \tag{3.12}$$

holds in $\Omega \times \mathbf{T}_+$ and $\delta_{\alpha\beta}$ is the Kronecker delta.

From equation $(3.6)_2$ (for $Y_3^0 = 0$) we obtain the equation

$$2\overset{*}{\Box}_2 \varphi_3^0 + \epsilon_{\alpha\beta} X_{\beta,\alpha} = 0, \tag{3.13}$$

resulting also, which is important, from equation (3.7) since $2\varphi_3^0 = \epsilon_{\alpha\beta} u_{\beta,\alpha}^0$. From the system of equations given by (3.6), (3.7) and (3.13) we obtain for u'_α, ζ_3 the following:

$$\overset{*}{\Box}_2 u'_\alpha + (\lambda+\mu) u'_{\beta,\beta\alpha} + \hat{X}_\alpha = 2\alpha \epsilon_{\alpha\gamma}\zeta_{3,\gamma},$$
$$\overset{*}{\Box}_4 \epsilon_{\alpha\beta} u'_{\beta,\alpha} + 2\bar{Y}_3 - 2\Box_4\zeta_3 = 2(\gamma+\varepsilon)(c_4^{-2} - c_2^{\star -2})\ddot{\varphi}_3^0, \tag{3.14}$$

where

$$c_4^2 = \frac{\gamma+\varepsilon}{J}, \qquad c_2^{\star 2} = \frac{\mu}{\rho}, \qquad \hat{X}_\alpha = 0, \qquad \bar{Y}_3 = Y_3 - \frac{\gamma+\varepsilon}{2\mu}\epsilon_{\alpha\beta} X_{\beta,\alpha}.$$

With the system of equations (3.14) we associate the following initial-boundary conditions:

— boundary conditions on $\partial\Omega \times \mathbf{T}$,

$$\sigma'_{\alpha\beta} n_\alpha = 0, \qquad \mu'_{\alpha 3} n_\alpha = m_3 - m_3^0 \qquad \text{on } \partial\Omega_\sigma \times \mathbf{T}, \tag{3.15}$$

$$u'_\alpha = 0, \qquad \varphi'_3 = g_3 - \varphi_3^0 \qquad \text{on } \partial\Omega_u \times \mathbf{T}; \tag{3.16}$$

— initial conditions in $\Omega \times \{0\}$,

$$u'_\alpha = 0, \qquad \varphi'_3 = k_3 - \varphi_3^0, \qquad \dot{u}'_\alpha = 0, \qquad \dot{\varphi}'_3 = \chi_3 - \dot{\varphi}_3^0, \tag{3.17}$$

where

$$m_3^0 = \mu_{\alpha 3}^0 n_\alpha, \qquad \varphi_3^0 = \tfrac{1}{2}\epsilon_{\alpha\beta} u_{\beta,\alpha}^0, \qquad \kappa_{\alpha 3}^0 = \varphi_{3,\alpha}^0, \qquad \mu_{\alpha 3}^0 = (\gamma + \varepsilon)\kappa_{\alpha 3}^0,$$

$$\mu_{3\alpha}^0 = \frac{\gamma - \varepsilon}{\gamma + \varepsilon}\mu_{\alpha 3}^0, \qquad \dot{\varphi}_3^0 = \lim_{t \to 0} \frac{\partial \varphi_3^0(x_\alpha, t)}{\partial t}.$$

The stresses $\sigma'_{\alpha\beta}$, $\mu'_{\alpha 3}$ are determined from the constitutive relations in $\Omega \times \mathbf{T}$:

$$\sigma'_{\alpha\beta} = (\mu + \alpha)\gamma'_{\alpha\beta} + (\mu - \alpha)\gamma'_{\beta\alpha} + \lambda\gamma'_{\gamma\gamma}\delta_{\alpha\beta}, \qquad \mu'_{\alpha 3} = (\gamma + \varepsilon)\kappa'_{\alpha 3} \tag{3.18}$$

by means of the geometric relations in $\Omega \times \mathbf{T}_+$:

$$\gamma'_{\alpha\beta} = u'_{\beta,\alpha} - \epsilon_{\alpha\beta}\varphi'_3, \qquad \kappa'_{\alpha 3} = \varphi'_{3,\alpha}. \tag{3.19}$$

Equations (3.18) should be completed with the formulae: $\sigma'_{33} = \dfrac{\lambda}{2(\lambda + \mu)}\sigma'_{\alpha\alpha}$ and $\mu'_{3\alpha} = \dfrac{\gamma - \varepsilon}{\gamma + \varepsilon}\mu'_{\alpha 3}$.

The systems of equations (3.7) and (3.14) are coupled not only through the initial-boundary conditions (3.15)–(3.17). Let us note that in equation $(3.14)_2$ the classical rotations $\varphi_3^0 = \frac{1}{2}\epsilon_{\alpha\beta} u_{\beta,\alpha}^0$ appear in the operator $2(\gamma + \varepsilon)(c_4^{-2} - c_2^{\star -2})\partial_t^2$. For the stated initial-boundary value problem (3.1)–(3.4), the system of equations (3.1) can be solved by the superposition method in such a way that we first solve the classical equations (3.7) with conditions (3.8)–(3.10). Then, we substitute φ_3^0 thus obtained into equation $(3.14)_2$ and we solve the system of equations (3.14) with conditions (3.15)–(3.17). The solution of the problem is obtained in the form

$$\begin{aligned} u_\alpha &= u_\alpha^0 + u'_\alpha, & \varphi_3 &= \varphi_3^0 + \varphi'_3, & \sigma_{\alpha\beta} &= \sigma_{\alpha\beta}^0 + \sigma'_{\alpha\beta}, \\ \sigma_{33} &= \sigma_{33}^0 + \sigma'_{33}, & \mu_{\alpha 3} &= \mu_{\alpha 3}^0 + \mu'_{\alpha 3}, & \mu_{3\alpha} &= \mu_{3\alpha}^0 + \mu'_{3\alpha}, \\ \gamma_{\alpha\beta} &= \varepsilon_{\alpha\beta}^0 + \gamma'_{\alpha\beta}, & \kappa_{\alpha 3} &= \kappa_{\alpha 3}^0 + \kappa'_{\alpha 3}. \end{aligned}$$

In the sequel we shall give equations for the triple (u'_α, φ'_3) which do not contain the classical rotations φ_3^0. To this end we apply the operator $\overset{*}{\Box}_2$ to equation $(3.6)_2$ and we arrive at

$$\overset{*}{\Box}_2(\overset{*}{\Box}_4 \epsilon_{\alpha\beta} u_{\beta,\alpha} + 2Y_3 - 2\Box_4 \zeta_3) = 0. \tag{3.20}$$

Now, for $\zeta_3^0 = 0$ ($Y_3^0 = 0$), we obtain from (3.20) the equation

$$\overset{*}{\Box}_4(2\overset{*}{\Box}_2 \varphi_3^0 + \epsilon_{\alpha\beta} X_{\beta,\alpha}) = 0, \tag{3.21}$$

which will be satisfied identically by virtue of equation (3.13). Let us recall that equation (3.13) results from the classical equations (3.7).

Let us go over now to the complementary solution (u'_α, ζ_3). From equations (3.7) and $(3.6)_1$ we have obtained equation $(3.14)_1$, whereas from (3.20) and (3.21) we get

$$\overset{*}{\Box}_2(\overset{*}{\Box}_4 \epsilon_{\alpha\beta} u'_{\beta,\alpha} - 2\Box_4 \zeta_3) + 2\hat{Y}_3 = 0, \tag{3.22}$$

where $\hat{Y}_3 = \overset{*}{\Box}_2 Y_3 - \frac{1}{2}\overset{*}{\Box}_4 \epsilon_{\alpha\beta} X_{\beta,\alpha}$, in which φ_3^0 does not appear anymore.

Further, we eliminate the function ζ_3 from the system of equations $(3.14)_1$ and (3.22), obtaining a system of equations for u'_α, φ'_3 in $\Omega \times \mathbf{T}_+$ of the form

$$\begin{aligned} &\Box_2 u'_\alpha + (\lambda + \mu - \alpha) u'_{\beta,\beta\alpha} + 2\alpha \epsilon_{\alpha\gamma} \varphi'_{3,\gamma} + \hat{X}_\alpha = 0, \\ &\overset{*}{\Box}_2(\Box_4 \varphi'_3 + 2\alpha \epsilon_{\alpha\beta} u'_{\beta,\alpha}) + \hat{Y}_3 = 0. \end{aligned} \tag{3.23}$$

We associate the initial-boundary conditions (3.15)–(3.17) with the system of equations (3.23). Let us present the separated equations for u_α^0, φ_3^0, ζ_3, u'_α and φ'_3. The function φ_3^0 satisfies equation (3.13). From (3.7) we obtain

$$\Box_1 \overset{*}{\Box}_2 u_\alpha^0 = (\lambda + \mu) X_{\beta,\beta\alpha} - \Box_1 X_\alpha. \tag{3.24}$$

From the system of equations $(3.14)_1$ and (3.22) we obtain

$$2(\Box_2 \Box_4 + 4\alpha^2 \nabla_1^2)\zeta_3 = 2\hat{Y}_3 - \overset{*}{\Box}_4 \epsilon_{\alpha\beta} \hat{X}_{\beta,\alpha} \tag{3.25}$$

and

$$\begin{aligned} &\overset{*}{\Box}_2 \Box_1 (\Box_2 \Box_4 + 4\alpha^2 \nabla_1^2) u'_\alpha \\ &\quad = \overset{*}{\Box}_2 \left\{ [(\lambda + \mu - \alpha)\Box_4 - 4\alpha^2] \hat{X}_{\beta,\beta\alpha} - \Box_1 \Box_4 \hat{X}_\alpha \right\} \\ &\qquad + 2\alpha \Box_1 \epsilon_{\alpha\beta} \hat{Y}_{3,\beta}. \end{aligned} \tag{3.26}$$

In turn, from the system of equations (3.23) we obtain

$$\overset{*}{\Box}_2(\Box_2 \Box_4 + 4\alpha^2 \nabla_1^2)\varphi'_3 = -\Box_2 \hat{Y}_3 - 2\alpha \overset{*}{\Box}_2 \epsilon_{\delta\gamma} \hat{X}_{\delta,\gamma}, \tag{3.27}$$

where $\hat{X}_\alpha = 0$ and $\hat{Y}_3 = \overset{*}{\Box}_2 Y_3 - \frac{1}{2}\overset{*}{\Box}_4 \epsilon_{\alpha\beta} X_{\beta,\alpha}$. Let us make here another remark. Namely, assuming the correspondence

$$(u'_\alpha, \varphi'_3) \mapsto (u_\alpha, \varphi_3), \qquad (\hat{X}_\alpha, \hat{Y}_3) \mapsto (X_\alpha, \overset{*}{\Box}_2 Y_3),$$

we note that the system of equations (3.23) goes over into (3.1), and equations (3.26) and (3.27) (after removing the operator $\overset{*}{\square}_2$) are transformed into the equations

$$\square_1(\square_2\square_4 + 4\alpha^2\nabla_1^2)u_\alpha = [(\lambda+\mu-\alpha)\square_4 - 4\alpha^2]X_{\beta,\beta\alpha} + \square_1(2\alpha\epsilon_{\alpha\beta}Y_{3,\beta} - \square_4 X_\alpha), \tag{3.28}$$

$$(\square_2\square_4 + 4\alpha^2\nabla_1^2)\varphi_3 = -\square_2 Y_3 - 2\alpha\epsilon_{\delta\gamma}X_{\delta,\gamma},$$

which refer to the system of equations (3.1) in the sense that they result from uncoupling it.

3.1.2 Equations in displacements and rotations with a distortion field. Fundamental solutions and limiting cases

In [49] equations (3.1) with a distortion field are considered. Let us use here the method of superposition. Assuming the substitutions (an analogy of body forces):

$$X_\alpha = -\overset{*}{r}{}_\alpha^0, \qquad Y_3 = -r_3^0, \tag{3.29}$$

$$\hat{X}_\alpha = -\overset{**}{r}{}_\alpha^0, \qquad \hat{Y}_3 = -\hat{r}_3^0, \tag{3.30}$$

where $\hat{r}_3^0 = \overset{*}{\square}_2 r_3^0 - \frac{1}{2}\overset{*}{\square}_4\epsilon_{\alpha\beta}\overset{*}{r}{}_{\beta,\alpha}^0$, we obtain equations with a distortion field from equations (3.7), (3.13) and (3.23)–(3.27). We introduce the following notation here:

$$r_\alpha^0 = \overset{*}{r}{}_\alpha^0 + \overset{**}{r}{}_\alpha^0 = \sigma_{\delta\alpha,\delta}^0, \qquad r_3^0 = \epsilon_{\delta\gamma}\sigma_{\delta\gamma}^0 + u_{\varepsilon 3,\varepsilon}^0, \tag{3.31}$$

where

$$\begin{aligned}
\overset{*}{r}{}_1^0 &= (2\mu\gamma_{11}^0 + \lambda\gamma_{\beta\beta}^0)_{,1} + 2\mu\gamma_{(21),2}^0,\\
\overset{*}{r}{}_2^0 &= 2\mu\gamma_{(12),1}^0 + (2\mu\gamma_{22}^0 + \lambda\gamma_{\beta\beta}^0)_{,2},\\
\overset{**}{r}{}_1^0 &= 2\alpha\gamma_{[21],2}^0,\\
\overset{**}{r}{}_2^0 &= 2\alpha\gamma_{[12],1}^0.
\end{aligned}$$

The symbols $\sigma_{\alpha\beta}^0$ and $\mu_{\alpha 3}^0$ have the form of the constitutive relations (3.18), written for the distortions $\gamma_{\alpha\beta}^0$ and $\kappa_{\alpha 3}^0$.

If we assume now the correspondence

$$(u'_\alpha, \varphi'_3) \mapsto (u_\alpha, \varphi_3), \qquad (\overset{**}{r}{}_\alpha^0, \hat{r}_3^0) \mapsto (r_\alpha^0, \overset{*}{\square}_2 r_3^0)$$

in equations (3.26) and (3.27) with the substitution (3.30), then, removing the operator $\overset{*}{\square}_2$, we obtain the following equations in $\Omega \times \mathbf{T}_+$:

$$\begin{aligned}\Box_1(\Box_2\Box_4 + 4\alpha^2\nabla_1^2)u_\alpha &= -[(\lambda+\mu-\alpha)\Box_4 - 4\alpha^2]\sigma^0_{\varepsilon\beta,\varepsilon\beta\alpha}\\ &\quad + \Box_1[-2\alpha\epsilon_{\alpha\beta}(\epsilon_{\delta\gamma}\sigma^0_{\delta\gamma} + \mu^0_{\varepsilon3,\varepsilon})_{,\beta} + \Box_4\sigma^0_{\varepsilon\alpha,\varepsilon}],\\ (\Box_2\Box_4 + 4\alpha^2\nabla_1^2)\varphi_3 &= \Box_2(\epsilon_{\delta\gamma}\sigma^0_{\delta\gamma} + \mu^0_{\varepsilon3,\varepsilon}) + 2\alpha\epsilon_{\delta\gamma}\sigma^0_{\varepsilon\delta,\varepsilon\gamma}.\end{aligned} \tag{3.32}$$

Let us note that equations (3.32) are obtained from equations (3.28) by substituting

$$X_\alpha = -r^0_\alpha, \qquad Y_3 = -r^0_3. \tag{3.33}$$

In [83] we gave the fundamental solutions for displacements and rotations in the case of infinite space when harmonically-varying-in-time distortions of the form

$$(\gamma^0_{\alpha\beta}, \kappa^0_{\alpha3}) = (\gamma_0^{\alpha\beta}, \kappa_0^{\alpha3})e^{-i\omega t}\delta(x_1)\delta(x_2) \tag{3.34}$$

are given. As an example, consider the action of the distortions γ^0_{11} and κ^0_{13}. Applying the Fourier integral transformation (0.50) for $n=3$ (over the position variables x_α and time t) to equations (3.32), we obtain the following result:

— for the distortion γ^0_{11},

$$\begin{aligned}u_\alpha &= \frac{\gamma_0}{2\pi}e^{-i\omega t}\epsilon_{\alpha\beta}\Big\{\frac{2\mu}{\mu+\alpha}\cdot\frac{1}{\sigma_2^2}[A_1K_0(-ik_1r) + A_2K_0(-ik_2r)\\ &\quad - K_0(-i\sigma_1 r)]_{,\beta12} + a_{\alpha\beta}K_0(-i\sigma_1 r)_{,\alpha}\Big\},\\ \varphi_3 &= \frac{\gamma_0}{2\pi}e^{-i\omega t}\frac{1}{l^2(k_1^2-k_2^2)}[K_0(-ik_1r) - K_0(-ik_2r)]_{,12},\end{aligned} \tag{3.35}$$

where

$$\gamma_0^{11}\mapsto\gamma_0, \qquad a_{12}=-1, \qquad a_{21} = \frac{\lambda}{2\mu+\lambda}, \qquad l^2 = \frac{(\gamma+\varepsilon)(\mu+\alpha)}{4\alpha\mu},$$

$$A_1 = \frac{\sigma_2^2-k_2^2}{k_1^2-k_2^2}, \qquad A_2 = \frac{\sigma_2^2-k_1^2}{k_2^2-k_1^2},$$

$$k_1^2 = \tfrac{1}{2}\left(A - \sqrt{A^2 - 4\sigma_2^2k_4^2}\right), \qquad k_2^2 = \tfrac{1}{2}\left(A + \sqrt{A^2 - 4\sigma_2^2k_4^2}\right),$$

$$A = \sigma_2^2 + \sigma_4^2 + \eta_0^2 - \nu_0^2, \qquad \sigma_1 = \frac{\omega}{c_1}, \qquad \sigma_2 = \frac{\omega}{c_2}, \qquad \sigma_4 = \frac{\omega}{c_4},$$

$$c_1 = \sqrt{\frac{\lambda+2\mu}{\rho}}, \qquad c_2 = \sqrt{\frac{\mu+\alpha}{\rho}}, \qquad c_4 = \sqrt{\frac{\gamma+\varepsilon}{J}},$$

$$\nu_0^2 = \frac{4\alpha}{\gamma+\varepsilon}, \qquad \eta_0^2 = \frac{4\alpha^2}{(\mu+\alpha)(\gamma+\varepsilon)}, \qquad k_4^2 = \frac{J\omega^2 - 4\alpha}{\gamma+\varepsilon}.$$

The symbol $K_0(-i\hat{\beta}r)$ denotes the modified Bessel function of the second kind: $K_0(-i\hat{\beta}r) = \frac{1}{2}\pi i H_0^{(1)}(\hat{\beta}r)$, where $H_0^{(1)}(\hat{\beta}r)$ is the Hankel function of the first kind and of order zero, $r = \sqrt{x_\alpha x_\alpha}$ and $\hat{\beta} = (k_\alpha, \sigma_1)$ (we use the summation convention, but we do not sum over three identical indices);

— for the distortion κ_{13}^0,

$$
\begin{aligned}
u_\alpha &= \frac{\kappa_0}{2\pi} e^{-i\omega t} \frac{2\alpha}{\mu+\alpha} \cdot \frac{1}{k_1^2 - k_2^2} \epsilon_{\alpha\beta} [K_0(-ik_1 r) - K_0(-ik_2 r)]_{,\beta 1} \,, \\
\varphi_3 &= \frac{\kappa_0}{2\pi} e^{-i\omega t} \frac{1}{k_1^2 - k_2^2} (\nabla_1^2 + \sigma_2^2) [K_0(-ik_1 r) - K_0(-ik_2 r)]_{,1} \,,
\end{aligned}
\tag{3.36}
$$

where $\kappa_0^{13} \mapsto \kappa_0$.

Dyszlewicz & Wytrążek in [83] considered full matrices for the distortions $\gamma_{\alpha\beta}^0$ and $\kappa_{\alpha 3}^0$ and found fundamental solutions in the framework of micropolar theory for the displacements $u_{\alpha\gamma\varepsilon}^{\gamma^0}$, $u_{\alpha\gamma 3}^{\kappa^0}$ and the rotations $\varphi_{3\gamma\varepsilon}^{\gamma^0}$, $\varphi_{3\gamma 3}^{\kappa^0}$ (formulae (3.35) and (3.36) give $u_{\alpha 11}^{\gamma^0}$, $\varphi_{311}^{\gamma^0}$ and $u_{\alpha 13}^{\kappa^0}$, $\varphi_{313}^{\kappa^0}$, respectively).

On the basis of fundamental solutions we construct Green functions $U_{\alpha\delta\gamma}^{\gamma^0}$, $U_{\alpha\beta 3}^{\kappa^0}$, $\Phi_{3\delta\gamma}^{\gamma^0}$ and $\Phi_{3\beta 3}^{\kappa^0}$, which depend on $(x_1, x_2; x_1', x_2', \omega)$, where

$$
r = \sqrt{(x_1 - x_1')^2 + (x_2 - x_2')^2}.
$$

If now, in the region $\hat{\mathbb{R}}^2$ of the space $\mathbb{R}^2$, the given distortion field $\gamma_{\delta\gamma}^0(x_1', x_2')$, $\kappa_{\beta 3}^0(x_1', x_2')$ is a harmonic function of time, then the fields of displacements and rotations are determined by the method of Green functions:

$$
\begin{aligned}
u_\alpha(x_1, x_2, t) = e^{-i\omega t} \int_{\hat{\mathbb{R}}^2} & [\gamma_{\delta\gamma}^0(x_1', x_2') U_{\alpha\delta\gamma}^{\gamma^0}(x_1, x_2; x_1', x_2', \omega) \\
& + \kappa_{\beta 3}^0(x_1', x_2') U_{\alpha\beta 3}^{\kappa^0}(x_1, x_2; x_1', x_2', \omega)] \, dx_1' \, dx_2', \\
\varphi_3(x_1, x_2, t) = e^{-i\omega t} \int_{\hat{\mathbb{R}}^2} & [\gamma_{\delta\gamma}^0(x_1', x_2') \Phi_{3\delta\gamma}^{\gamma^0}(x_1, x_2; x_1', x_2', \omega) \\
& + \kappa_{\beta 3}^0(x_1', x_2') \Phi_{3\beta 3}^{\kappa^0}(x_1, x_2; x_1', x_2', \omega)] \, dx_1' \, dx_2'.
\end{aligned}
\tag{3.37}
$$

In [83] we put emphasis on obtaining fundamental solutions for u_α, φ_3 in the frameworks of the limiting theories (2)–(8) (Fig. 1.1) by use of their fundamental equations. Taking then the appropriate limits, we established correspondences between the results of different theories. In particular, in micropolar elastostatics (2) and for the limit (1) $\overset{\omega\to 0}{\longrightarrow}$ (2) (Fig. 1.1), the results of Suchar [221] have been obtained.

3.1.3 The method of potentials

In [49], we considered the vector equations (1.147) and (1.151)–(1.153) referring to the plane problem with the triple (u_α, φ_3). One can see that a scalar potential Φ and vector potentials $\boldsymbol{\Psi} \stackrel{\text{def}}{=} (0,0,\Psi_3)$ and $\mathbf{H} \stackrel{\text{def}}{=} (H_1,H_2,0)$ appear in the problem. We obtain the following relations in $\Omega \times \mathbf{T}_+$:

$$u_\alpha = \Phi_{,\alpha} + \epsilon_{\alpha\beta}\Psi_{3,\beta}, \qquad \varphi_3 = \epsilon_{\alpha\beta}H_{\beta,\alpha}. \tag{3.38}$$

The condition $\operatorname{div}\boldsymbol{\Psi} = 0$ is satisfied identically, while the condition $\operatorname{div}\mathbf{H} = 0$ implies that $H_{\alpha,\alpha} = 0$, which is fulfilled for the function H: $H_\alpha = \epsilon_{\beta\alpha}H_{,\beta}$. The functions Φ, Ψ_3, H_α are related by means of the following equations in $\Omega \times \mathbf{T}_+$:

$$\Box_4\epsilon_{\alpha\beta}H_{\beta,\alpha} - 2\alpha\nabla_1^2\Psi_3 + \nabla_1^2\eta = 0 \tag{3.39}$$

obtained from equation $(1.151)_2$, and

$$\Box_1\Phi_{,\alpha} + \Box_2\epsilon_{\alpha\beta}\Psi_{3,\beta} - 2\alpha\nabla_1^2H_\alpha + \vartheta_{,\alpha} + \epsilon_{\alpha\beta}\chi_{3,\beta} = 0, \tag{3.40}$$

which follows from equation $(1.151)_1$. The functions η, ϑ, χ_3 are connected with the decomposition (1.148) of body loading vectors of the form: $\mathbf{X} \stackrel{\text{def}}{=} (X_1,X_2,0)$ and $\mathbf{Y} \stackrel{\text{def}}{=} (0,0,Y_3)$. We obtain in $\Omega \times \mathbf{T}$:

$$X_\alpha = \vartheta_{,\alpha} + \epsilon_{\alpha\beta}\chi_{3,\beta}, \qquad Y_3 = \nabla_1^2\eta. \tag{3.41}$$

Going further, we eliminate the function Ψ_3 from further considerations by satisfying equation (3.39): $\Psi_3 = \frac{1}{2\alpha}(\Box_4H + \eta)$ for all arguments. From relations (3.38) we finally obtain the representation

$$u_\alpha = \Phi_{,\alpha} + \frac{1}{2\alpha}\epsilon_{\alpha\beta}(\Box_4H + \eta)_{,\beta}, \qquad \varphi_3 = \nabla_1^2H, \tag{3.42}$$

whereas by means of (3.40) we conclude that the expressions $2\alpha(\Box_1\Phi + \vartheta)$ and $(\Box_2\Box_4 + 4\alpha^2\nabla_1^2)H + \Box_2\eta + 2\alpha\chi_3$ are coupled harmonic functions, i.e. they satisfy the Laplace equation and the differential Cauchy-Riemann conditions.

We obtain the following equations in $\Omega \times \mathbf{T}_+$:

$$\begin{aligned} &\nabla_1^2\Box_1\Phi + X_{\alpha,\alpha} = 0, \\ &\nabla_1^2(\Box_2\Box_4 + 4\alpha^2\nabla_1^2)H + \Box_2Y_3 + 2\alpha\epsilon_{\beta\alpha}X_{\beta,\alpha} = 0 \end{aligned} \tag{3.43}$$

and the Cauchy-Riemann conditions:

$$2\alpha(\Box_1\Phi + \vartheta)_{,\alpha} + \epsilon_{\alpha\beta}[(\Box_2\Box_4 + 4\alpha^2\nabla_1^2)H + \Box_2\eta + 2\alpha\chi_3]_{,\beta} = 0. \tag{3.44}$$

Satisfying the Cauchy-Riemann conditions in a special way, we obtain the wave equations in $\Omega \times \mathbf{T}_+$:

$$\begin{aligned} &\Box_1\Phi + \vartheta = 0, \\ &(\Box_2\Box_4 + 4\alpha^2\nabla_1^2)H + \Box_2\eta + 2\alpha\chi_3 = 0. \end{aligned} \tag{3.45}$$

3.1.4 Wave equations in polar coordinates

In [49], we considered vector equations (1.1) in the system of polar coordinates. In order to obtain equations for the triples $(u_r, u_\theta, \varphi_z)$ and $(\varphi_r, \varphi_\theta, u_z)$, which represent the first and the second plane problems in polar coordinates (r, θ), respectively, one should assume that for the vector $\mathbf{a} \overset{\text{def}}{=} (a_r, a_\theta, a_z)$ the following relations hold:

$$
\begin{aligned}
\vec{\nabla}^2 \mathbf{a} &= \left(\nabla_0^2 a_r - \frac{2}{r^2} \partial_\theta a_\theta \,, \nabla_0^2 a_\theta + \frac{2}{r^2} \partial_\theta a_r \,, \nabla^2 a_z \right), \\
\operatorname{grad} \operatorname{div} \mathbf{a} &= \left(\partial_r a \,, \frac{1}{r} \partial_\theta a \,, 0 \right), \qquad (3.46) \\
\operatorname{curl} \mathbf{a} &= \left(\frac{1}{r} \partial_\theta a_z \,, -\partial_r a_z \,, \left(\frac{1}{r} + \partial_r \right) a_\theta - \frac{1}{r} \partial_\theta a_r \right),
\end{aligned}
$$

where the scalar Laplace operator

$$
\begin{aligned}
&\nabla^2 \overset{\text{def}}{=} \operatorname{div} \operatorname{grad} = \partial_r^2 + r^{-1} \partial_r + r^{-2} \partial_\theta^2, \\
&\nabla_0^2 = \nabla^2 - r^{-2}, \\
&a = r^{-1} \partial_r (r a_r) + r^{-1} \partial_\theta a_\theta .
\end{aligned}
$$

Disregarding body loadings, we obtain from (1.1) and (3.46) for the vectors

$$
\mathbf{u} \overset{\text{def}}{=} (u_r, u_\theta, 0), \qquad \boldsymbol{\varphi} \overset{\text{def}}{=} (0, 0, \varphi_z) \qquad (3.47)
$$

the following equations in $\Omega \times \mathbf{T}_+$:

$$
\begin{aligned}
&(\mu + \alpha) \left(\nabla_0^2 u_r - \frac{2}{r^2} \partial_\theta u_\theta \right) + (\lambda + \mu - \alpha) \partial_r e + 2\alpha \frac{1}{r} \partial_\theta \varphi_z = \rho \ddot{u}_r, \\
&(\mu + \alpha) \left(\nabla_0^2 u_\theta + \frac{2}{r^2} \partial_\theta u_r \right) + (\lambda + \mu - \alpha) \frac{1}{r} \partial_\theta e - 2\alpha \partial_r \varphi_z = \rho \ddot{u}_\theta, \qquad (3.48) \\
&[(\gamma + \varepsilon) \nabla^2 - 4\alpha] \varphi_z + 2\alpha \frac{1}{r} [\partial_r (r u_\theta) - \partial_\theta u_r] = J \ddot{\varphi}_z,
\end{aligned}
$$

where $e = \operatorname{div} \mathbf{u} = r^{-1} \partial_r (r u_r) + r^{-1} \partial_\theta u_\theta$.

We assume the representation

$$
u_r = \partial_r \Phi + \frac{1}{r} \partial_\theta \Psi, \qquad u_\theta = \frac{1}{r} \partial_\theta \Phi - \partial_r \Psi, \qquad (3.49)
$$

which we substitute into the system of equations (3.48). Thus we obtain a system of equations for the triple (Φ, Ψ, φ_z) in $\Omega \times \mathbf{T}_+$, namely

$$
\begin{aligned}
&\partial_r \Box_1 \Phi + \frac{1}{r} \partial_\theta (\Box_2 \Psi + 2\alpha \varphi_z) = 0, \\
&\frac{1}{r} \partial_\theta \Box_1 \Phi - \partial_r (\Box_2 \Psi + 2\alpha \varphi_z) = 0, \qquad (3.50) \\
&\Box_4 \varphi_z - 2\alpha \nabla^2 \Psi = 0.
\end{aligned}
$$

In the definitions of the operators $\Box_1, \Box_2, \Box_4$ we have now the Laplace operator ∇^2 for the polar coordinates (r, θ). Equations $(3.50)_{1,2}$ will be satisfied under the assumption that the expressions on which the operators ∂_r and $r^{-1}\partial_\theta$ act, should vanish independently. We arrive at

$$\Box_1 \Phi = 0, \qquad \Box_2 \Psi + 2\alpha\varphi_z = 0. \tag{3.51}$$

Now, equations $(3.50)_3$ and $(3.51)_2$ are coupled with each other. Uncoupling them, we obtain the equations

$$(\Box_2\Box_4 + 4\alpha^2\nabla^2)(\Psi, \varphi_z) = (0, 0). \tag{3.52}$$

The system of equations (3.48) has been transformed to wave equations $(3.51)_1$ and (3.52) with the representation (3.49).

3.1.5 SEMP

Let us return to the considerations in Subsection 2.1.6 and 2.3.3 and give here (for the plane problem) the stress equations of motion of Ignaczak type. The first plane problem is represented by the vectors:

$$\mathbf{u} \stackrel{\text{def}}{=} (u_1, u_2, 0), \qquad \boldsymbol{\varphi} \stackrel{\text{def}}{=} (0, 0, \varphi_3), \tag{3.53}$$

$$\mathbf{X} \stackrel{\text{def}}{=} (X_1, X_2, 0), \qquad \mathbf{Y} \stackrel{\text{def}}{=} (0, 0, Y_3). \tag{3.54}$$

We arrange the field equations in the following groups:

— the equations of motion in $\Omega \times \mathbf{T}_+$,

$$\begin{aligned} &\sigma_{\alpha\beta,\alpha} + X_\beta = \rho\ddot{u}_\beta, \\ &\epsilon_{\alpha\beta}\sigma_{\alpha\beta} + \mu_{\alpha 3,\alpha} + Y_3 = J\ddot{\varphi}_3, \end{aligned} \tag{3.55}$$

where matrices of stresses have the form

$$\sigma_{ji} \equiv \begin{bmatrix} \sigma_{11} & \sigma_{12} & 0 \\ \sigma_{21} & \sigma_{22} & 0 \\ 0 & 0 & \sigma_{33} \end{bmatrix}, \qquad \mu_{ji} \equiv \begin{bmatrix} 0 & 0 & \mu_{13} \\ 0 & 0 & \mu_{23} \\ \mu_{31} & \mu_{32} & 0 \end{bmatrix}.$$

There are 6 independent components of the stress state, since it holds

$$\sigma_{33} = \frac{\lambda}{2(\lambda + \mu)}\sigma_{\alpha\alpha}, \qquad \mu_{3\alpha} = \frac{\gamma - \varepsilon}{\gamma + \varepsilon}\mu_{\alpha 3};$$

— the geometric relations in $\Omega \times \mathbf{T}_+$,

$$\gamma_{\alpha\beta} = u_{\beta,\alpha} + \epsilon_{\beta\alpha}\varphi_3, \qquad \kappa_{\alpha 3} = \varphi_{3,\alpha}, \tag{3.56}$$

where the matrices

$$\gamma_{ji} \equiv \begin{bmatrix} \gamma_{11} & \gamma_{12} & 0 \\ \gamma_{21} & \gamma_{22} & 0 \\ 0 & 0 & 0 \end{bmatrix}, \qquad \kappa_{ji} \equiv \begin{bmatrix} 0 & 0 & \kappa_{13} \\ 0 & 0 & \kappa_{23} \\ 0 & 0 & 0 \end{bmatrix}$$

describe strain fields;

— the constitutive relations in $\Omega \times \mathbf{T}$,

$$\begin{aligned} \sigma_{\alpha\beta} &= (\mu+\alpha)\gamma_{\alpha\beta} + (\mu-\alpha)\gamma_{\beta\alpha} + \lambda\gamma_{\varepsilon\varepsilon}\delta_{\alpha\beta}, \\ \mu_{\alpha 3} &= (\gamma+\varepsilon)\kappa_{\alpha 3}, \end{aligned} \tag{3.57}$$

where μ, λ, α, γ, ε are elastic constants. Note that the constant β does not appear in the first plane problem. Substituting relations (3.57) to the equations of motion (3.55) and using relations (3.56), we see that equations (3.55) go over into the equations in displacements and rotations (3.1)[1].

Let us formulate the following initial-boundary conditions in $\overline{\Omega} \times \mathbf{T}$:

— the boundary conditions on $\partial\Omega \times \mathbf{T}$,

$$p_\beta = \sigma_{\alpha\beta} n_\alpha, \qquad m_3 = \mu_{\alpha 3} n_\alpha, \tag{3.58}$$

where p_β, $m_3 : \partial\Omega \times \mathbf{T} \to \mathbb{R}$ are given functions;

— the initial conditions in $\Omega \times \{0\}$,

$$u_\alpha = f_\alpha, \qquad \varphi_3 = f_3, \qquad \dot{u}_\alpha = g_\alpha, \qquad \dot{\varphi}_3 = g_3, \tag{3.59}$$

where the functions f_α, f_3, g_α, $g_3 : \Omega \to \mathbb{R}$ are given.

Using relations (3.56), we obtain from the equations of motion (3.55) the following equations in $\Omega \times \mathbf{T}_+$:

$$\begin{aligned} &\rho^{-1} R_{\alpha,\beta} + \epsilon_{\alpha\beta} J^{-1} R_3 = \ddot{\gamma}_{\beta\alpha}, \\ &J^{-1} R_{3,\alpha} = \ddot{\kappa}_{\alpha 3} \end{aligned} \tag{3.60}$$

where $R_\alpha = \sigma_{\beta\alpha,\beta} + X_\alpha$, $R_3 = \epsilon_{\alpha\beta}\sigma_{\alpha\beta} + \mu_{\alpha 3,\alpha} + Y_3$, $\epsilon_{11} = \epsilon_{22} = 0$ and $\epsilon_{12} = -\epsilon_{21} = 1$. We present the constitutive relations (3.57) in the equivalent form

$$\begin{aligned} \gamma_{\alpha\beta} &= \frac{1}{2\mu}\sigma_{(\alpha\beta)} + \frac{1}{2\alpha}\sigma_{[\alpha\beta]} - \frac{\nu}{2\mu}\sigma_{\varepsilon\varepsilon}\delta_{\alpha\beta}, \\ \kappa_{\alpha 3} &= \frac{1}{\gamma+\varepsilon}\mu_{\alpha 3}, \end{aligned} \tag{3.61}$$

[1] The compatibility equations in the form $\kappa_{23,1} - \kappa_{13,2} = 0$, $\kappa_{13} - \gamma_{21,1} + \gamma_{11,2} = 0$, $\kappa_{23} + \gamma_{12,2} - \gamma_{22,1} = 0$ will be then satisfied identically.

where $\sigma_{(\alpha\beta)} = \frac{1}{2}(\sigma_{\alpha\beta} + \sigma_{\beta\alpha})$, $\sigma_{[\alpha\beta]} = \frac{1}{2}(\sigma_{\alpha\beta} - \sigma_{\beta\alpha})$ and $\nu = \dfrac{\lambda}{2(\mu+\lambda)}$ is Poisson's ratio.

Introducing relations (3.61), we notice that equations (3.60) are transformed into

$$\begin{aligned}&\rho^{-1}R_{\alpha,\beta} + \epsilon_{\alpha\beta}J^{-1}R_3 - \frac{1}{2\mu}(\ddot{\sigma}_{(\alpha\beta)} - \nu\ddot{\sigma}_{\varepsilon\varepsilon}\delta_{\alpha\beta}) - \frac{1}{2\alpha}\ddot{\sigma}_{[\beta\alpha]} = 0,\\ &c_4^2 R_{3,\alpha} - \ddot{\mu}_{\alpha 3} = 0,\end{aligned} \tag{3.62}$$

where $c_4^2 = J^{-1}(\gamma + \varepsilon)$.

Within the framework of the first plane problem with the triple (u_1, u_2, φ_3), the stress equations of motion problem consists of the field equations (3.62), the boundary conditions (3.58) and the initial conditions of the following form in $\Omega \times \{0\}$:

$$\begin{aligned}\sigma_{\alpha\beta} &= \sigma^0_{\alpha\beta}, & \mu_{\alpha 3} &= \mu^0_{\alpha 3},\\ \dot{\sigma}_{\alpha\beta} &= \dot{\sigma}^0_{\alpha\beta}, & \dot{\mu}_{\alpha 3} &= \dot{\mu}^0_{\alpha 3},\end{aligned} \tag{3.63}$$

where

$$\sigma^0_{\alpha\beta} = (\mu+\alpha)\gamma^0_{\alpha\beta} + (\mu-\alpha)\gamma^0_{\beta\alpha} + \lambda\gamma^0_{\delta\delta}\delta_{\alpha\beta},$$

$$\mu^0_{\alpha 3} = (\gamma+\varepsilon)\kappa^0_{\alpha 3}, \qquad \gamma^0_{\alpha\beta} = f_{\beta,\alpha} - \epsilon_{\alpha\beta}f_3, \qquad \kappa^0_{\alpha 3} = f_{3,\alpha};$$

$$\dot{\sigma}^0_{\alpha\beta} = (\mu+\alpha)\dot{\gamma}^0_{\alpha\beta} + (\mu-\alpha)\dot{\gamma}^0_{\beta\alpha} + \lambda\dot{\gamma}^0_{\delta\delta}\delta_{\alpha\beta},$$

$$\dot{\mu}^0_{\alpha 3} = (\gamma+\varepsilon)\dot{\kappa}^0_{\alpha 3}, \qquad \dot{\gamma}^0_{\alpha\beta} = g_{\beta,\alpha} - \epsilon_{\alpha\beta}g_3, \qquad \dot{\kappa}^0_{\alpha 3} = g_{3,\alpha}.$$

The triple (u_1, u_2, φ_3) is obtained from the following formulae in $\Omega \times \mathbf{T}$:

$$\begin{aligned}u_\alpha &= f_\alpha + g_\alpha t + \rho^{-1}(t * R_\alpha),\\ \varphi_3 &= f_3 + g_3 t + J^{-1}(t * R_3),\end{aligned} \tag{3.64}$$

where

$$t * f(x_1, x_2, t) = \int_0^t (t-\tau) f(x_1, x_2, \tau)\, d\tau.$$

For sufficiently smooth functions describing physical fields and initial-boundary conditions, the initial-boundary value problems described by relations (3.55)–(3.59) and (3.58), (3.61)–(3.64) are equivalent (see Ignaczak [116]).

3.2 The first problem of plane strain state. Elastostatics

3.2.1 Equations in displacements-rotations with a field of body loadings

Let us consider the equations in displacements and rotations for the triple (u_α, φ_3) in $\Omega \times \mathbf{T}_+$:

$$\begin{aligned} &(\mu+\alpha)\nabla_1^2 u_\beta + (\lambda+\mu-\alpha)u_{\alpha,\alpha\beta} + 2\alpha\epsilon_{\beta\gamma}\varphi_{3,\gamma} + X_\beta = 0, \\ &[(\gamma+\varepsilon)\nabla_1^2 - 4\alpha]\varphi_3 + 2\alpha\epsilon_{\alpha\beta}u_{\beta,\alpha} + Y_3 = 0 \end{aligned} \tag{3.65}$$

with the following boundary conditions on $\partial\Omega$:

$$u_\alpha = f_\alpha, \qquad \varphi_3 = g_3, \tag{3.66}$$

where f_α, $g_3 : \partial\Omega \to \mathbb{R}$ are given functions.

We introduce the substitution (3.5) to the system of equations (3.65) and assume the solution of the form $u_\alpha = u_\alpha^0 + u_\alpha'$ and $\varphi_3 = \varphi_3^0 + \varphi_3'$. For $\zeta_3 = \zeta_3^0 + \zeta_3'$ and $Y_3 = Y_3^0 + Y_3'$ with $\zeta_3^0 = 0$ ($Y_3^0 = 0$), we obtain the classical equations in Ω:

$$\mu\nabla_1^2 u_\alpha^0 + (\lambda+\mu)u_{\beta,\beta\alpha}^0 + X_\alpha = 0 \tag{3.67}$$

with the classical boundary conditions

$$u_\alpha^0 = f_\alpha \qquad \text{on } \partial\Omega. \tag{3.68}$$

In turn, for the rotations φ_3^0 we obtain the equation

$$2\mu\nabla_1^2\varphi_3^0 + \epsilon_{\alpha\beta}X_{\beta,\alpha} = 0, \tag{3.69}$$

where $\varphi_3^0 = \frac{1}{2}\epsilon_{\alpha\beta}u_{\beta,\alpha}^0$.

Moreover, we obtain the complementary equations in Ω:

$$\begin{aligned} &(\mu+\alpha)\nabla_1^2 u_\alpha' + (\lambda+\mu-\alpha)u_{\beta,\beta\alpha}' + 2\alpha\epsilon_{\beta\gamma}\varphi_{3,\gamma}' + \hat{X}_\alpha = 0, \\ &[(\gamma+\varepsilon)\nabla_1^2 - 4\alpha]\varphi_3' + 2\alpha\epsilon_{\alpha\beta}u_{\beta,\alpha}' + \overset{*}{Y}_3 = 0 \end{aligned} \tag{3.70}$$

with the boundary conditions (3.66) and the boundary effect

$$u_\alpha' = 0, \qquad \varphi_3' = g_3 - \varphi_3^0 \qquad \text{on } \partial\Omega \tag{3.71}$$

being taken into account. The rotations φ_3^0 can be obtained from the solution of the classical problem, (3.67) and (3.68). Note that the reduced body loadings

$$\hat{X}_\alpha = 0, \qquad \overset{*}{Y}_3 = Y_3 - \frac{\gamma+\varepsilon}{2\mu}\epsilon_{\alpha\beta}X_{\beta,\alpha} \tag{3.72}$$

appear in equations (3.70). We shall still give separated equations for u_α^0, u_α' and φ_3'. From equation (3.67) we obtain

$$\mu\nabla_1^2\nabla_1^2 u_\alpha^0 = -\nabla_1^2 X_\alpha + \frac{\lambda+\mu}{2\mu+\lambda}X_{\beta,\beta\alpha}. \tag{3.73}$$

From the system of equations (3.70) we get the following equations[2] in Ω:

[2] In equations (3.74) we do not disregard differential operations on the zero vector field $\hat{X}_\alpha$. The explanation of this fact results from, among other reasons, the manner of introducing the separated equations (3.76). The problem is explained in detail in Section 4.4 and concerns this entire work.

$$
\begin{aligned}
2\mu\nabla_1^2\nabla_1^2(l^2\nabla_1^2-1)u'_\alpha &= \epsilon_{\alpha\beta}\left(\nabla_1^2\overset{*}{Y}_3+\frac{\gamma+\varepsilon}{2\mu}\nabla_1^2\epsilon_{\delta\gamma}\hat{X}_{\delta,\gamma}\right)_{,\beta} \\
&\quad -2\nabla_1^2(l^2\nabla_1^2-1)\hat{X}_\alpha+\frac{2(\lambda+\mu)}{\lambda+2\mu}(l^2\nabla_1^2-1)\hat{X}_{\beta,\beta\alpha}, \\
2\mu\nabla_1^2(l^2\nabla_1^2-1)\varphi'_3 &= -\frac{\mu+\alpha}{2\alpha}\nabla_1^2\overset{*}{Y}_3-\epsilon_{\delta\gamma}\hat{X}_{\delta,\gamma}.
\end{aligned}
\tag{3.74}
$$

In [43] and [49], the superposition method described by equations and relations (3.65)–(3.74) is extensively used for solving the first, the second and the thermoelastic boundary value problems.

Now, assuming the correspondence

$$
(\hat{X}_\alpha,\overset{*}{Y}_3)\mapsto(X_\alpha,Y_3) \tag{3.75}
$$

we note that equations (3.70) are transformed into equations (3.65) ($u'_\alpha\mapsto u_\alpha$, $\varphi'_3\mapsto\varphi_3$), while equations (3.74) go over into the respective equations given below in Ω:

$$
\begin{aligned}
2\mu\nabla_1^2\nabla_1^2(l^2\nabla_1^2-1)u_\alpha &= \epsilon_{\alpha\beta}\left(\nabla_1^2 Y_3+\frac{\gamma+\varepsilon}{2\mu}\nabla_1^2\epsilon_{\delta\gamma}X_{\delta,\gamma}\right)_{,\beta} \\
&\quad -2\nabla_1^2(l^2\nabla_1^2-1)X_\alpha+\frac{2(\lambda+\mu)}{\lambda+2\mu}(l^2\nabla_1^2-1)X_{\beta,\beta\alpha}, \\
2\mu\nabla_1^2(l^2\nabla_1^2-1)\varphi_3 &= -\frac{\mu+\alpha}{2\alpha}\nabla_1^2 Y_3-\epsilon_{\delta\gamma}X_{\delta,\gamma},
\end{aligned}
\tag{3.76}
$$

where $l^2=\dfrac{(\gamma+\varepsilon)(\mu+\alpha)}{4\alpha\mu}$.

Equations (3.76) are obtained after uncoupling the system of equations (3.65). In [49] we applied the Fourier integral transformations (0.50) for $n=2$ (over the position variables x_α) to equations (3.69), (3.73) and (3.74) and obtained the fundamental solutions for u_α, φ_3 when body loadings in the space $\mathbb{R}^2$ are given in the form

$$
(X_\alpha,Y_3)=(P_\alpha,M)\delta(x_1)\delta(x_2). \tag{3.77}
$$

In [43], we obtained the identical result by means of equations (3.76), e.g. for $X_\alpha=P_1\delta(x_1)\delta(x_2)\delta_{\alpha1}$, $Y_3=0$ we got the following formulae:

$$
\begin{aligned}
u_1 &= \frac{P_1}{4\pi\mu}\left\{2I_1+\mu_0^{-1}[I_1-I_2(l)]_{,22}+(1-\nu)I_{3,11}\right\}, \\
u_2 &= \frac{P_1}{4\pi\mu}\left\{-\mu_0^{-1}[I_1-I_2(l)]_{,12}+(1-\nu)I_{3,12}\right\}, \\
\varphi_3 &= -\frac{P_1}{4\pi\mu}[I_1-I_2(l)]_{,2},
\end{aligned}
\tag{3.78}
$$

where

$$\mu_0 = \frac{2\mu}{\gamma + \varepsilon}, \qquad l = \sqrt{\frac{(\gamma + \varepsilon)(\mu + \alpha)}{4\alpha\mu}}, \qquad \nu = \frac{\lambda}{2(\mu + \lambda)}.$$

Moreover, the symbols I_i ($i = 1, 2, 3$) denote:

$$\begin{aligned} I_1 &= \text{p.f. } (2\pi)^{-1} \int_{-\infty}^{\infty} \int_{-\infty}^{\infty} \underset{\sim}{\xi}^{-2} \exp(-i\xi_\alpha x_\alpha)\, d\xi_1\, d\xi_2 = -(\mathsf{C} + \ln r), \\ I_2(l) &= \text{p.f. } (2\pi)^{-1} \int_{-\infty}^{\infty} \int_{-\infty}^{\infty} (\underset{\sim}{\xi}^2 + l^{-2})^{-1} \exp(-i\xi_\alpha x_\alpha)\, d\xi_1\, d\xi_2 \\ &= K_0\left(\frac{r}{l}\right), \\ I_3 &= \text{p.f. } (2\pi)^{-1} \int_{-\infty}^{\infty} \int_{-\infty}^{\infty} \underset{\sim}{\xi}^{-4} \exp(-i\xi_\alpha x_\alpha)\, d\xi_1\, d\xi_2 = -\frac{r^2}{4} I_1, \end{aligned} \tag{3.79}$$

where $r = \sqrt{x_\alpha x_\alpha}$, C is the Euler's constant, $K_0\left(\frac{r}{l}\right)$ denotes the modified Bessel function of the second kind (Mac Donald's function), and $\underset{\sim}{\xi} = \sqrt{\xi_\alpha \xi_\alpha}$. The integrals (3.79) do not exist, if understood as improper integrals. They cannot be assigned the principal values in Cauchy's sense, either. It is possible, however, to extract their so-called finite parts (in formula (3.79) an abbreviation p.f. was used). This notion is known in the literature thanks to Hadamard [107]. The method of effective determination of formulae (3.79) was given by Ganowicz in [96], where he made use of the definition of the finite part in the sense of Nowacki [171, p. 145]. Fundamental solutions for displacements, rotations and stresses caused by the body loadings (3.77) were introduced by Nowacki[3], who employed the method of potentials and stress equations.

In [49], emphasis is put on the connection between the results of micropolar theory, couple-stress theory and the classical theory of elasticity. As an example, let us look at the formula for the stress σ_{22} caused by the body force $X_\alpha = P_1 \delta(x_1)\delta(x_2)\delta_{\alpha 1}$, $Y_3 = 0$. From the constitutive relations (3.57), the geometric relations (3.56) and formulae (3.78) we obtain the formula

$$\sigma_{22} = -\frac{P_1}{2\pi} \left\{ I_{1,1} + \frac{\gamma + \varepsilon}{2\mu} [I_1 - I_2(l)]_{,122} + (1 - \nu)^{-1} I_{3,111} \right\}. \tag{3.80}$$

When $\alpha \to \infty$, formula (3.80) is transformed into the formula of couple-stress theory:

$$s_{22} = -\frac{P_1}{2\pi} \left\{ I_{1,1} + 2l^{\star 2} [I_1 - I_2(l^\star)]_{,122} + (1 - \nu)^{-1} I_{3,111} \right\}. \tag{3.81}$$

[3] Nowacki [184].

However, when $\alpha \to 0^+$, formula (3.80) [and for $l^\star \to 0^+$, formula (3.81)] goes over into the classical formula:

$$\sigma_y = \frac{Fx}{4\pi(1-\nu)r^2}\left[(1+2\nu) - \frac{2x^2}{r^2}\right], \tag{3.82}$$

where $r^2 = x^2 + y^2$, $\nu = \dfrac{\lambda}{2(\mu+\lambda)}$, $F = P_1$; the notation in formula (3.82) is given after Sneddon & Berry [212, p. 130][4].

3.2.2 Distortion field. Fundamental solutions for displacements and rotations

In [49], we considered equations (3.65) with a distortion field. Assuming the substitution (3.29), $(3.30)_1$ (the so-called analogy of body loadings) and

$$\overset{*}{Y}_3 = -\overset{*0}{r}_3, \tag{3.83}$$

where

$$\overset{*0}{r}_3 = \overset{0}{r}_3 - \frac{\gamma+\varepsilon}{2\mu}\epsilon_{\alpha\beta}\overset{*0}{r}_{\beta,\alpha},$$

we obtain by the superposition method equations (3.67), (3.69), (3.70), (3.73) and (3.74), in which a distortion field is taken into account. The symbols used here and related to distortions have the same meaning as in Subsection 3.1.2. Let us consider equations (3.70), (3.74) with a distortion field (i.e., we substitute $\overset{*}{Y}_3 = -\overset{*0}{r}_3$ and $\hat{X}_\alpha = -\overset{**0}{r}_\alpha$) and assume the following correspondence:

$$(u'_\alpha, \varphi'_3) \mapsto (u_\alpha, \varphi_3), \qquad (\overset{**0}{r}_\alpha, \overset{*0}{r}_3) \mapsto (\overset{0}{r}_\alpha, \overset{0}{r}_3). \tag{3.84}$$

Equations (3.74) are transformed into

$$\begin{aligned} &2\mu\nabla_1^2\nabla_1^2(l^2\nabla_1^2 - 1)u_\alpha \\ &\quad = \epsilon_{\alpha\beta}\left[-\nabla_1^2(\epsilon_{\delta\gamma}\sigma^0_{\delta\gamma} + \mu^0_{\delta 3,\delta}) - \frac{\gamma+\varepsilon}{2\mu}\nabla_1^2\epsilon_{\delta\gamma}\sigma^0_{\varepsilon\delta,\varepsilon\gamma}\right]_{,\beta} \\ &\qquad + 2\nabla_1^2(l^2\nabla_1^2 - 1)\sigma^0_{\varepsilon\alpha,\varepsilon} - \frac{2(\lambda+\mu)}{\lambda+2\mu}(l^2\nabla_1^2 - 1)\sigma^0_{\varepsilon\beta,\varepsilon\beta\alpha}, \\ &2\mu\nabla_1^2(l^2\nabla_1^2 - 1)\varphi_3 = \frac{\mu+\alpha}{2\alpha}\nabla_1^2(\epsilon_{\alpha\beta}\sigma^0_{\alpha\beta} + \mu^0_{\alpha 3,\alpha}) + \epsilon_{\delta\gamma}\sigma^0_{\varepsilon\delta,\varepsilon\gamma}. \end{aligned} \tag{3.85}$$

In turn, the system of equations (3.70) goes over into

$$\begin{aligned} &(\mu+\alpha)\nabla_1^2 u_\alpha + (\lambda+\mu-\alpha)u_{\beta,\beta\alpha} + 2\alpha\epsilon_{\alpha\gamma}\varphi_{3,\gamma} = \overset{0}{r}_\alpha, \\ &[(\gamma+\varepsilon)\nabla_1^2 - 4\alpha]\varphi_3 + 2\alpha\epsilon_{\alpha\beta}u_{\beta,\alpha} = \overset{0}{r}_3. \end{aligned} \tag{3.86}$$

[4] See also Nowacki [175, p. 120].

Equations (3.85) are obtained after separating the system of equations (3.86). Let us note that also equations (3.85) may be obtained from equations (3.76) after assuming the substitution (3.33).
Using equations (3.85) and Fourier integral transformations (0.50) (for $n=2$), we gave in [83] fundamental solutions for displacements and rotations in the case when in the space $\mathbb{R}^2$ there acts a distortion field given by the formula

$$(\gamma^0_{\alpha\beta}, \kappa^0_{\alpha 3}) = (\gamma_0^{\alpha\beta}, \kappa_0^{\alpha 3})\delta(x_1)\delta(x_2). \tag{3.87}$$

It was shown that the result obtained in this manner for micropolar theories coincides with the result of Suchar [221] obtained by the method of potentials. This result constitutes a limiting case of the solution for the distortion (3.34). It was verified that such transition holds for $\omega \to 0$. In [83] we also discussed other limiting cases referring to the limiting theories (4), (6), (8) (Fig. 1.1). Consider, for example, action of the distortions γ^0_{11} and κ^0_{13}. We obtain the following formulae:

— for the distortion γ^0_{11},

$$\begin{aligned} u_1 &= -\frac{\gamma_0}{2\pi}\left[I_1 + \mu_0^{-1}(I_1 - I_2)_{,22} - \frac{1}{1-\nu} I_{3,22}\right]_{,1}, \\ u_2 &= \frac{\gamma_0}{2\pi}\left[I_1 + \mu_0^{-1}(I_1 - I_2)_{,11} + \frac{1}{1-\nu} I_{3,22}\right]_{,2}, \\ \varphi_3 &= \frac{\gamma_0}{2\pi}(I_1 - I_2)_{,12}, \end{aligned} \tag{3.88}$$

where

$$\mu_0 = \frac{2\mu}{\gamma+\varepsilon}, \qquad \nu = \frac{\lambda}{2(\mu+\lambda)}, \qquad I_2 = I_2(l) = K_0\left(\frac{r}{l}\right), \qquad \gamma_0^{11} \mapsto \gamma_0;$$

— for the distortion κ^0_{13},

$$\begin{aligned} u_\alpha &= \epsilon_{\beta\alpha}\frac{\kappa_0}{2\pi}\cdot\frac{\gamma+\varepsilon}{2\mu}[I_1 - I_2(l)]_{,\beta 1}, \\ \varphi_3 &= -\frac{\kappa_0}{2\pi} I_2(l)_{,1}, \\ \kappa_0^{13} &\mapsto \kappa_0. \end{aligned} \tag{3.89}$$

Formulae (3.88) and (3.89) are obtained from (3.35) and (3.36), respectively, by passing to the limit $\omega \to 0$. Let us consider the limiting case of formulae (3.88) for $\alpha \to \infty$. Assuming $\sqrt{\dfrac{\gamma+\varepsilon}{4\mu}} = l^\star$ and $\gamma_0 = \varepsilon_0$, we note that formulae (3.88) go over into the formulae:

$$
\begin{aligned}
u_1 &= -\frac{\varepsilon_0}{2\pi}[I_1 + 2l^{\star 2}(I_1 - I_2^\star)_{,22} - (1-\nu)^{-1}I_{3,22}]_{,1},\\
u_2 &= \frac{\varepsilon_0}{2\pi}[I_1 + 2l^{\star 2}(I_1 - I_2^\star)_{,11} + (1-\nu)^{-1}I_{3,22}]_{,2},\\
\varphi_3 &= \frac{\varepsilon_0}{2\pi}(I_1 - I_2^\star)_{,12},
\end{aligned}
\tag{3.90}
$$

where $I_2^\star = I_2(l^\star) = K_0\left(\frac{r}{l^\star}\right)$, $\nu = \dfrac{\lambda}{2(\mu+\lambda)}$.

Further, one considers equations of couple-stress theory of the form

$$
\mu\nabla_1^2 u_\alpha + (\lambda+\mu)u_{\beta,\beta\alpha} + \mu l^{\star 2}\nabla_1^2\epsilon_{\alpha\gamma}\varphi_{3,\gamma} + F_\alpha^0 = 0,
\tag{3.91}
$$

where

$$
\begin{aligned}
\varphi_3 &= \tfrac{1}{2}\epsilon_{\alpha\beta}u_{\beta,\alpha}, & F_\alpha^0 &= -(s_{\beta\alpha,\beta}^0 + \tfrac{1}{2}\epsilon_{\alpha\delta}m_{\gamma 3,\gamma\delta}^0),\\
m_{\alpha 3}^0 &= 4\mu l^{\star 2}\kappa_{\alpha 3}^0, & s_{\alpha\beta}^0 &= 2\mu\varepsilon_{\alpha\beta}^0 + \lambda\varepsilon_{\gamma\gamma}^0\delta_{\alpha\beta}.
\end{aligned}
$$

The symbols $\kappa_{\alpha 3}^0$, $\varepsilon_{\alpha\beta}^0$ determine the fields of the given distortions.

Separating equations (3.91), we obtain them in the form

$$
\begin{aligned}
2\mu\nabla_1^2\nabla_1^2(l^{\star 2}\nabla_1^2 - 1)u_\alpha &= -(1-\nu)^{-1}(l^{\star 2}\nabla_1^2 - 1)s_{\delta\gamma,\delta\gamma\alpha}^0\\
&\quad - 2l^{\star 2}\nabla_1^2\epsilon_{\alpha\beta}\epsilon_{\delta\gamma}F_{\gamma,\delta\beta}^0 - 2\nabla_1^2(l^{\star 2}\nabla_1^2 - 1)F_\alpha^0,\\
2\mu\nabla_1^2(l^{\star 2}\nabla_1^2 - 1)\varphi_3 &= \epsilon_{\alpha\beta}F_{\beta,\alpha}^0.
\end{aligned}
\tag{3.92}
$$

Now, for the distortion $\varepsilon_{11}^0 = \varepsilon_0^{11}\delta(x_1)\delta(x_2)$, one can show, by use of equations (3.92), that formulae (3.90) are solutions of equations (3.91) of couple-stress theory. In [83] we gave many remarks connected with correspondences of results obtained in different theories (1)–(8), (Fig. 1.1). The fundamental solutions for stresses may be determined by use of the constitutive relations for the problem with the triple (u_α, φ_3), or by means of stress equations[5]. Let us consider, for example, the formula for the stress σ_{11} caused by the distortion γ_{11}^0. Starting from formula

$$
\sigma_{11} = 2\mu u_{1,1} - 2\alpha\varphi_3 + \lambda u_{\alpha,\alpha}
\tag{3.93}
$$

and formulae (3.88) we obtain the formula

$$
\begin{aligned}
\sigma_{11} = &-\frac{\gamma_0}{2\pi}\{(2\mu+\lambda)\nabla_1^2 I_1 + (\gamma+\varepsilon)[I_1 - I_2(l)]_{,1122}\\
&+ 2\mu(1-\nu)^{-1}I_{3,2222}\}
\end{aligned}
\tag{3.94}
$$

[5] Nowacki [192, p. 316] applies the functions of Airy-Mindlin to determine the stress fields in the first plane problem with a distortion field.

where $\nabla_1^2(\cdot) = (\cdot)_{,\alpha\alpha}$. When $\alpha \to \infty$, formula (3.94) is transformed into the formula of couple-stress theory:

$$s_{11} = -\frac{\varepsilon_0^{11}}{2\pi}\Big\{(2\mu+\lambda)\nabla_1^2 I_1 + 4\mu l^{\star 2}[I_1 - I_2(l^\star)]_{,1122} + \frac{2\mu}{1-\nu} I_{3,2222}\Big\}. \tag{3.95}$$

On the other hand, when $\alpha \to 0^+$, formula (3.94) (and for $l^\star \to 0^+$, formula (3.95)) goes over into the classical formula:

$$\sigma_{11} = -\frac{\varepsilon_0^{11}}{2\pi}\left[(2\mu+\lambda)\nabla_1^2 I_1 + \frac{2\mu}{1-\nu} I_{3,2222}\right]. \tag{3.96}$$

Formulae defining stresses such as (3.94)–(3.96) may be the basis for qualitative analysis of the stress fields[6].

3.2.3 Stress equations of thermoelastostatics and displacement potentials in polar coordinates

In [54] and [49], we studied the stress equations in the polar coordinate system (r,θ) with temperature field taken into account. The stress state is described by the matrices

$$\underset{\sim}{\sigma} \overset{\text{def}}{=} \begin{bmatrix} \sigma_{rr} & \sigma_{r\theta} & 0 \\ \sigma_{\theta r} & \sigma_{\theta\theta} & 0 \\ 0 & 0 & \sigma_{zz} \end{bmatrix}, \qquad \underset{\sim}{\mu} \overset{\text{def}}{=} \begin{bmatrix} 0 & 0 & \mu_{rz} \\ 0 & 0 & \mu_{\theta z} \\ \mu_{zr} & \mu_{z\theta} & 0 \end{bmatrix}.$$

We impose the following field conditions on the components of stress state (we disregard here the fields of temperature and body loadings):

— the equations of equilibrium in Ω,

$$\begin{aligned} &\partial_r\sigma_{rr} + r^{-1}\partial_\theta\sigma_{\theta r} + r^{-1}(\sigma_{rr} - \sigma_{\theta\theta}) = 0, \\ &\partial_r\sigma_{r\theta} + r^{-1}\partial_\theta\sigma_{\theta\theta} + r^{-1}(\sigma_{r\theta} - \sigma_{\theta r}) = 0, \\ &\left(r^{-1} + \partial_r\right)\mu_{rz} + r^{-1}\partial_\theta\mu_{\theta z} + \sigma_{r\theta} - \sigma_{\theta r} = 0; \end{aligned} \tag{3.97}$$

— the compatibility equations in Ω,

$$\begin{aligned} &(1 + r\partial_r)\kappa_{\theta z} - \partial_\theta\kappa_{rz} = 0, \\ &\partial_\theta\gamma_{rr} - (1 + r\partial_r)\gamma_{\theta r} - \gamma_{r\theta} + r\kappa_{rz} = 0, \\ &\gamma_{rr} + \partial_\theta\gamma_{r\theta} - (1 + r\partial_r)\gamma_{\theta\theta} + r\kappa_{\theta z} = 0, \end{aligned} \tag{3.98}$$

where the matrices

[6] Stefaniak [219].

$$\underset{\sim}{\gamma} \stackrel{\text{def}}{=} \begin{bmatrix} \gamma_{rr} & \gamma_{r\theta} & 0 \\ \gamma_{\theta r} & \gamma_{\theta\theta} & 0 \\ 0 & 0 & 0 \end{bmatrix}, \qquad \underset{\sim}{\kappa} \stackrel{\text{def}}{=} \begin{bmatrix} 0 & 0 & \kappa_{rz} \\ 0 & 0 & \kappa_{\theta z} \\ 0 & 0 & 0 \end{bmatrix}$$

describe the strain state. Equations (3.98) should be expressed in terms of stresses by use of the constitutive relations in Ω:

$$\begin{aligned} \sigma_{rr} &= 2\mu\gamma_{rr} + \lambda(\gamma_{rr} + \gamma_{\theta\theta}), \\ \sigma_{\theta\theta} &= 2\mu\gamma_{\theta\theta} + \lambda(\gamma_{rr} + \gamma_{\theta\theta}), \\ \sigma_{r\theta} &= (\mu - \alpha)\gamma_{\theta r} + (\mu + \alpha)\gamma_{r\theta}, \\ \sigma_{\theta r} &= (\mu - \alpha)\gamma_{r\theta} + (\mu + \alpha)\gamma_{\theta r}, \\ \mu_{zr} &= (\gamma - \varepsilon)\kappa_{rz}, \qquad \mu_{z\theta} = (\gamma - \varepsilon)\kappa_{\theta z}. \end{aligned} \tag{3.99}$$

To the field equations (3.97) and (3.98) one should add the boundary conditions expressed in terms of stresses

$$\begin{aligned} &n_r\sigma_{rr} + n_\theta\sigma_{\theta r} = p_r, \qquad n_r\sigma_{r\theta} + n_\theta\sigma_{\theta\theta} = p_\theta, \\ &n_r\mu_{rz} + n_\theta\mu_{\theta z} = m_z, \end{aligned} \tag{3.100}$$

where p_r, p_θ, $m_z : \partial\Omega \to \mathbb{R}$ are the given loadings, n_r, n_θ are components of the outward unit vector $\mathbf{n}$ normal to $\partial\Omega$, which are expressed in a local basis connected with the assumed polar coordinate system.

According to the algorithm presented in Subsection 1.2.9, Table 1.3, the sum of the numbers of equations in (3.97) and (3.98) is the same as the number of independent components of stress state and equals 6. Independent components of the stress state are enumerated in relations (3.99), while the stress-state components σ_{zz}, μ_{rz} and $\mu_{\theta z}$ depend on the remaining components as follows:

$$\sigma_{zz} = \nu(\sigma_{rr} + \sigma_{\theta\theta}), \qquad (\mu_{rz}, \mu_{\theta z}) = \frac{\gamma + \varepsilon}{\gamma - \varepsilon}(\mu_{zr}, \mu_{z\theta}),$$

where $\nu = \dfrac{\lambda}{2(\mu + \lambda)}$.

Equations (3.97), together with the constitutive relations (3.99), (with the temperature field) and the geometric relations in Ω:

$$\begin{aligned} &\gamma_{rr} = \partial_r u_r, \qquad \gamma_{\theta\theta} = r^{-1}(\partial_\theta u_\theta + u_r), \\ &\gamma_{r\theta} = \partial_r u_\theta - \varphi_z, \qquad \gamma_{\theta r} = r^{-1}(\partial_\theta u_r - u_\theta) + \varphi_z, \\ &\kappa_{rz} = \partial_r \varphi_z, \qquad \kappa_{\theta z} = r^{-1}\partial_\theta \varphi_z \end{aligned} \tag{3.101}$$

lead to the equations of equilibrium in displacements and rotations for the triple $(u_r, u_\theta, \varphi_z)$ in Ω (the compatiblity equations (3.98) are satisfied identically):

$$
\begin{aligned}
&(\mu+\alpha)\left(\nabla_0^2 u_r - \frac{2}{r^2}\partial_\theta u_\theta\right)\\
&\qquad + (\lambda+\mu-\alpha)\partial_r e + 2\alpha\frac{1}{r}\partial_\theta\varphi_z = \nu_T\partial_r T,\\
&(\mu+\alpha)\left(\nabla_0^2 u_\theta + \frac{2}{r^2}\partial_\theta u_r\right)\\
&\qquad + (\lambda+\mu-\alpha)\frac{1}{r}\partial_\theta e - 2\alpha\partial_r\varphi_z = \nu_T\frac{1}{r}\partial_\theta T,\\
&[(\gamma+\varepsilon)\nabla^2 - 4\alpha]\varphi_z + 2\alpha\frac{1}{r}[\partial_r(r u_\theta) - \partial_\theta u_r] = 0,
\end{aligned}
\tag{3.102}
$$

where

$$
e = r^{-1}\partial_r(r u_r) + r^{-1}\partial_\theta u_\theta, \qquad \nabla^2 = \partial_r^2 + r^{-1}\partial_r + r^{-2}\partial_\theta^2,
$$

$$
\nu_T = (3\lambda + 2\mu)\alpha_t,
$$

α_t is the coefficient of linear thermal expansion of the medium, $\nabla_0^2 = \nabla^2 - r^{-2}$, μ, λ, α, ε, γ are the elasticity constants of a micropolar medium (the constant β does not appear in this plane problem).

In equations (3.102) we took into account the temperature field (denoted here by T). The function T satisfies the equation of heat conduction (1.174), i.e.:

$$
\left(\partial_r^2 + \frac{1}{r}\partial_r + \frac{1}{r^2}\partial_\theta^2\right) T(r,\theta) = -\frac{1}{\lambda_0} W(r,\theta), \qquad (r,\theta) \in \Omega \tag{3.103}
$$

and the boundary condition (e.g. $T = \vartheta$ on $\partial\Omega$), where $\vartheta : \partial\Omega \to \mathbb{R}$ is a given function. The function W determines the intensity of heat sources in the body, while λ_0 is a coefficient of heat conduction. In general, the following boundary conditions are important for the system of equations (3.102):

— the conditions (3.100) on $\partial\Omega_\sigma$,
— the conditions

$$
(u_r, u_\theta, \varphi_z) = (\hat{u}_r, \hat{u}_\theta, \hat{\varphi}_z) \qquad \text{on } \partial\Omega_u = \partial\Omega \setminus \partial\Omega_\sigma, \tag{3.104}
$$

where $\hat{u}_r$, $\hat{u}_\theta$, $\hat{\varphi}_z : \partial\Omega_u \to \mathbb{R}$ are given functions.

Let us introduce the following displacement representation in Ω:

$$
u_r = \partial_r\Phi + \frac{1}{r}\partial_\theta\Psi, \qquad u_\theta = \frac{1}{r}\partial_\theta\Phi - \partial_r\Psi. \tag{3.105}
$$

From the system of equations (3.102) we obtain now the following equations. Namely, the functions Φ, φ_z satisfy the equations

$$
\nabla^2(\nabla^2\Phi - mT) = 0, \qquad \nabla^2(l^2\nabla^2 - 1)\varphi_z = 0 \tag{3.106}
$$

and the following differential relations:

$$
\begin{aligned}
&\partial_r(\nabla^2\Phi - mT) + \frac{2\mu}{2\mu+\lambda}\cdot\frac{1}{r}\partial_\theta(l^2\nabla^2 - 1)\varphi_z = 0,\\
&\frac{1}{r}\partial_\theta(\nabla^2\Phi - mT) - \frac{2\mu}{2\mu+\lambda}\partial_r(l^2\nabla^2 - 1)\varphi_z = 0,
\end{aligned}
\tag{3.107}
$$

where $m = \dfrac{\nu_T}{2\mu+\lambda}$, $\quad l^2 = \dfrac{(\gamma+\varepsilon)(\mu+\alpha)}{4\alpha\mu}$.

The potential Ψ is determined from the equation

$$
\nabla^2\Psi = 2(k^2\nabla^2 - 1)\varphi_z, \tag{3.108}
$$

where $k^2 = \dfrac{\gamma+\varepsilon}{4\alpha}$. Recall that $\nabla^2 = \partial_r^2 + r^{-1}\partial_r + r^{-2}\partial_\theta^2$.

When solving the boundary value problem stated above one should express the components of stress state in terms of the functions Φ, Ψ, φ_z by means of relations (3.99) and (3.101).

3.2.4 Concentration of stresses. The problem of cylindrical inclusion. The case of a circular hole

In [54], [56] and [49] we solved and discussed the problem of a micropolar space in the plane strain state, uniformly stretched along one axis and containing a cylindrical hole. This problem is of great practical and technological importance due to the concentration of stresses on the boundary of the hole. The problem discussed is illustrated in Fig. 3.1 (p and q are different, constant body loadings).

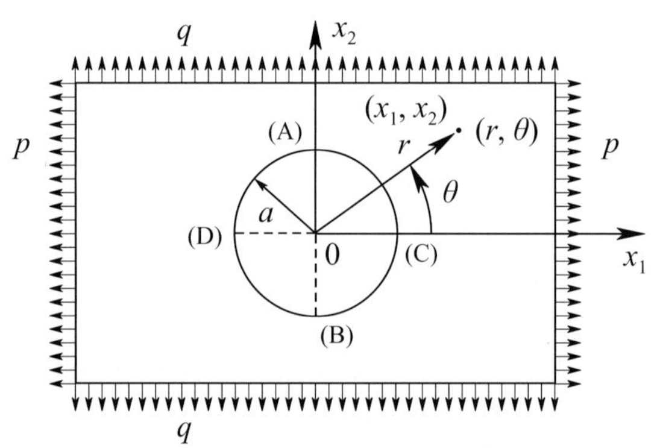

Fig. 3.1.

The problem is a generalization of the classical Kirsch problem to the micropolar theory of elasticity. We will present the solution of the problem in a closed form by using the representation (3.105) and equations (3.106)–(3.108) (for $T = 0$). In the polar coordinates (r,θ), we have at infinity $q \equiv 0$:

$$\begin{aligned} &\sigma_{rr} = p\cos^2\theta, \qquad \sigma_{\theta\theta} = p\sin^2\theta, \qquad \sigma_{r\theta} = \sigma_{\theta r} = -\tfrac{1}{2}p\sin 2\theta, \\ &\mu_{rz} = \mu_{z\theta} = 0. \end{aligned} \tag{3.109}$$

The form of the functions Φ, Ψ, φ_z satisfying equations (3.106) and (3.108), the differential equations (3.107) and the conditions (3.109) is the following:

$$\begin{aligned} \Phi &= E\ln r + \frac{p}{8(\lambda+\mu)}r^2 + \left[Ar^{-2} + \frac{\mu}{2(2\mu+\lambda)}C + \frac{p}{8\mu}r^2\right]\cos 2\theta, \\ \Psi &= \left[\tfrac{1}{2}C - D\frac{\gamma+\varepsilon}{2\mu}K_2\left(\frac{r}{l}\right)\right]\sin 2\theta, \\ \varphi_z &= \left[Cr^{-2} + DK_2\left(\frac{r}{l}\right)\right]\sin 2\theta, \end{aligned} \tag{3.110}$$

where $K_i(.)$ $(i = 0, 1, 2)$ — the modified Bessel functions of the second kind, and A, C, D, E denote arbitrary real constants which are determined from conditions on the boundary of hole. Namely, for $r = a$ (where a is the radius of the hole) and arbitrary θ, $\theta \in [0, 2\pi)$ (Fig. 3.1), one requires that the stresses σ_{rr}, $\sigma_{r\theta}$, μ_{rz} vanish:

$$\sigma_{rr} = \sigma_{r\theta} = \mu_{rz} = 0. \tag{3.111}$$

We limit our discussion of the solution to the vicinities of points where the concentration of circumferential stresses $\sigma_{\theta\theta}$ appears. Maximal values of those stresses for $r = a$ are obtained at points (A) and (B) in Fig. 3.1. Those stresses may be expressed by the concise formula

$$\sigma_{\theta\theta}^{\max} = 3p\frac{1+\frac{1}{3}G}{1+G}, \tag{3.112}$$

where

$$G = \frac{\dfrac{a_0}{l^2}}{1 + \dfrac{a^2}{4l^2} + \dfrac{a}{2l}\cdot\dfrac{K_0\left(\frac{a}{l}\right)}{K_1\left(\frac{a}{l}\right)}}.$$

The characteristic constants a_0, l are expressed as follows:

$$a_0 = \frac{(\gamma+\varepsilon)(2\mu+\lambda)}{4\mu(\mu+\lambda)}, \qquad l^2 = \frac{(\gamma+\varepsilon)(\mu+\alpha)}{4\alpha\mu}.$$

Limiting cases of formula (3.112) were also obtained. For $\alpha \to \infty$ we have

$$\sqrt{\frac{\gamma+\varepsilon}{4\mu}} = l^\star, \qquad l \to l^\star, \qquad a_0 = 2(1-\nu)l^{\star 2}, \qquad G \mapsto F,$$

where ν is the Poisson's ratio and $l^\star$ is the elasticity constant of couple-stress theory, and thus we obtain the result of Mindlin [157] from couple-stress theory:

$$\sigma_{\theta\theta}^{\max} = 3p\frac{1+\frac{1}{3}F}{1+F}, \tag{3.113}$$

where

$$F = \frac{8(1-\nu)}{4 + \dfrac{a^2}{l^{\star 2}} + \dfrac{2a}{l^\star} \cdot \dfrac{K_0\left(\frac{a}{l^\star}\right)}{K_1\left(\frac{a}{l^\star}\right)}}.$$

In turn, if $\alpha \to 0^+$ ($l \to \infty$, $G \to 0$), we obtain the classical result (Timoshenko & Goodier [228])[7]:

$$\sigma_{\theta\theta}^{\max} = 3p. \tag{3.114}$$

Assuming maximal circumferential stresses in the form of the formula $\sigma_{\theta\theta}^{\max} = \mathbf{C}p$, where $\mathbf{C}$ is the coefficient of stress concentration, we obtain $\mathbf{C} = 3$ in the classical theory. In couple-stress theory $\mathbf{C} = (3+F)/(1+F)$ and the concentration coefficient $\mathbf{C}$ depends on a geometric parameter a in addition to the elasticity constants μ, λ and $l^\star$ (for $l^\star \to 0$ we have $F \to 0$ and one obtains the classical result (3.114)). In micropolar theory we obtain:

$$\mathbf{C} = \frac{3+G}{1+G}. \tag{3.115}$$

The coefficient of stress concentration $\mathbf{C}$ depends on the material constants μ, λ and α of stress dimension, on constants γ, ε — of force dimension, and on a geometric parameter a of length dimension. Hence, one constructs for $\mathbf{C}$ a formula which depends on dimensionless parameters and which is richer in its structure than analogical formulae in the limiting theories. Namely, it is of the following form:

$$\mathbf{C} = \mathbf{C}\left(\frac{a}{l^\star}, \nu, r_1\right), \tag{3.116}$$

where $r_1 = \dfrac{\alpha}{\mu+\alpha} \in [0,1]$ is a new dimensionless parameter (as compared with couple-stress theory).

For the limiting values r_1: $r_1 = 0$ ($\alpha \to 0$) and $r_1 = 1$ ($\alpha \to \infty$) we obtain classical theory and couple-stress theory, repectively. Further explanations concerning formula (3.116) are the following: $\sqrt{\dfrac{\gamma+\varepsilon}{4\mu}} = l^\star$, $\dfrac{a}{l^\star} \in [0,\infty)$, ν — Poisson's ratio and $\nu \in \left[0, \frac{1}{2}\right]$. Careful study of the coefficient of stress concentration $\mathbf{C}$ carried out in [49] leads to the final conclusions:

[7] See also Sokołowski [215, p. 46].
Woźniak [244, p. 148] gives a solution for an infinite lattice-type homogenous disc with a circular hole (of radius many times larger than the length of disc connectors), expanded along one direction by forces of constant intensity. See also: Barański & Woźniak [12], Ariman & Kaloni [7], Niemish [164] and Eringen, Theory of Micropolar Elasticity, [in:] Fracture, Vol. II, edited by H. Liebowitz, Academic Press, New York 1968, p. 719.

1. In micropolar theory it is possible to modify the solution of the Kirsch-Mindlin problem by assuming suitable values of material coefficients, however this can be done in the region Δ only (Fig. 3.2).

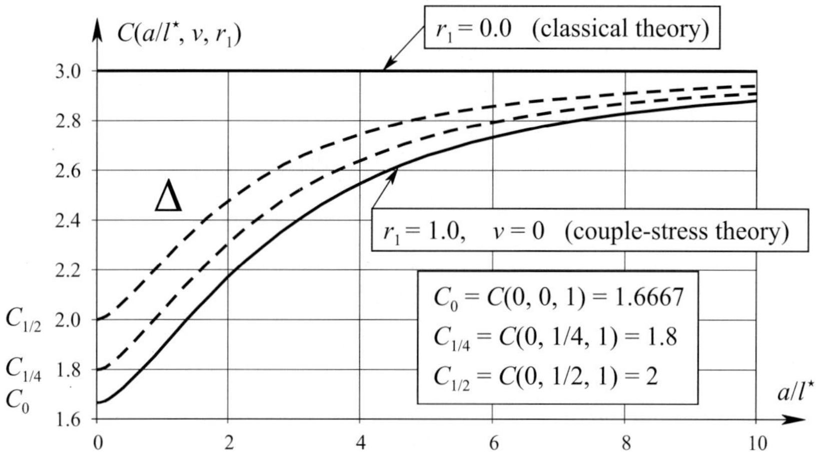

Fig. 3.2.

2. The following estimate of $\mathbf{C}$ holds:

$$(\forall \nu)(\forall a/l^{\star})(\forall r_1) \quad (1.6667 \leq \mathbf{C} \leq 3).$$

3.2.5 Stress concentration problem. Perfectly rigid cylindrical inclusion

In [55], [56] and [49] we solved and discussed the problem of stress concentration on the boundary of a perfectly rigid cylindrical inclusion in a micropolar space uniformly stretched along one axis. The problem concerns the plane strain state $\sigma_{11} = p$, $q = 0$ (Fig. 3.1).

We assume the solution for φ_z in the form of formula $(3.110)_3$. Using the representation (3.105) and formulae $(3.110)_{1,2}$, we obtain the following formulae for the displacements u_r, u_θ:

$$\begin{aligned} u_r &= \frac{p}{4(\lambda+\mu)} r + E r^{-1} \\ &\quad + \left[\frac{p}{4\mu} r + C r^{-1} - 2 A r^{-3} - 2 \frac{\gamma+\varepsilon}{2\mu} D r^{-1} K_2\left(\frac{r}{l}\right) \right] \cos 2\theta, \\ u_\theta &= -\left\{ \frac{p}{4\mu} r + \frac{\mu}{2\mu+\lambda} C r^{-1} + 2 A r^{-3} \right. \\ &\quad \left. + 2D \frac{\gamma+\varepsilon}{2\mu} \left[r^{-1} K_2\left(\frac{r}{l}\right) + \frac{1}{2l} K_1\left(\frac{r}{l}\right) \right] \right\} \sin 2\theta. \end{aligned} \tag{3.117}$$

The constants A, C, D, E are determined from the conditions for a prefectly rigid (in the micropolar sense) junction on the boundary of the inclusion; for $r = a$ and arbitrary θ it holds (Fig. 3.1):

$$u_r = u_\theta = \varphi_z = 0. \tag{3.118}$$

We are interested in normal stresses σ_{rr} which reach their maximal values on the boundary of inclusion at points (C) and (D), see Fig. 3.1. These stresses are expressed by the formula

$$\sigma_{rr}^{\max} = \frac{p}{2}\left[(3-2\nu) + \frac{G + 8(1-\nu)\dfrac{a_0}{l^2}}{(3-4\nu)G}\right], \tag{3.119}$$

where

$$G = (3-4\nu)\left[\frac{a}{l}\cdot\frac{K_0\left(\frac{a}{l}\right)}{K_1\left(\frac{a}{l}\right)} + 2\right] - \frac{2a_0}{l^2}.$$

Notation used here is identical with that in Subsection 3.2.4. Limiting cases of formula (3.119) were obtained. For $\alpha \to \infty$ ($l \to l^\star$, $a_0/l^2 \to 2(1-\nu)$, $G \to F$) we obtain the result of Banks & Sokołowski [10] from couple-stress theory:

$$\sigma_{rr}^{\max} = \frac{p}{2}\left[(3-2\nu) + \frac{F + 16(1-\nu)^2}{(3-4\nu)F}\right], \tag{3.120}$$

where

$$F = 2(1-2\nu) + (3-4\nu)\frac{a}{l^\star}\cdot\frac{K_0\left(\frac{a}{l^\star}\right)}{K_1\left(\frac{a}{l^\star}\right)}.$$

If $\alpha \to 0$ ($l \to \infty$, $G \to 2(3-4\nu)$), we obtain the classical result of Goodier [100]:

$$\sigma_{rr}^{\max} = \frac{p}{2}\cdot\frac{2(5-4\nu)(1-\nu)}{3-4\nu}. \tag{3.121}$$

Assuming maximal normal stresses in the form $\sigma_{rr}^{\max} = \mathbf{C}p$, we see that in classical theory the coefficient of stress concentration $\mathbf{C} = \dfrac{(5-4\nu)(1-\nu)}{3-4\nu}$ and it depends on the Poisson ratio ν, but it does not depend on the geometric parameter a. In turn, in couple-stress theory we have

$$\mathbf{C} = \tfrac{1}{2}\left[(3-2\nu) + \frac{F + 16(1-\nu)^2}{(3-4\nu)F}\right]$$

and the concentration coefficient depends on the elastic constants μ, λ, $l^\star$ and the radius a (when $l^\star \to 0$, we obtain the classical result). Finally, in micropolar theory we obtain

$$\mathbf{C} = \tfrac{1}{2}\left[(3-2\nu) + \frac{G + 8(1-\nu)\dfrac{a_0}{l^2}}{(3-4\nu)G}\right]. \tag{3.122}$$

The coefficient of stress concentration $\mathbf{C}$ depends on the parameters μ, λ, α, γ, ε, a. Hence, as in the case of a cylindrical hole, one builds up a formula giving the coefficient $\mathbf{C}$ in the form (3.116), whose structure is richer in dimensionless parameters than analogous formulae for classical and couple-stress theories[8]. Studying the coefficient of stress concentration $\mathbf{C} = \mathbf{C}\left(\frac{a}{l^\star}, \nu, r_1\right)$, we arrive at the following conclusions:

1. In micropolar theory it is possible to modify the solution of the problem of Goodier and Banks-Sokołowski by choosing the values of material coefficients, however this can be done in the region $\varDelta$ only (Fig. 3.3.).

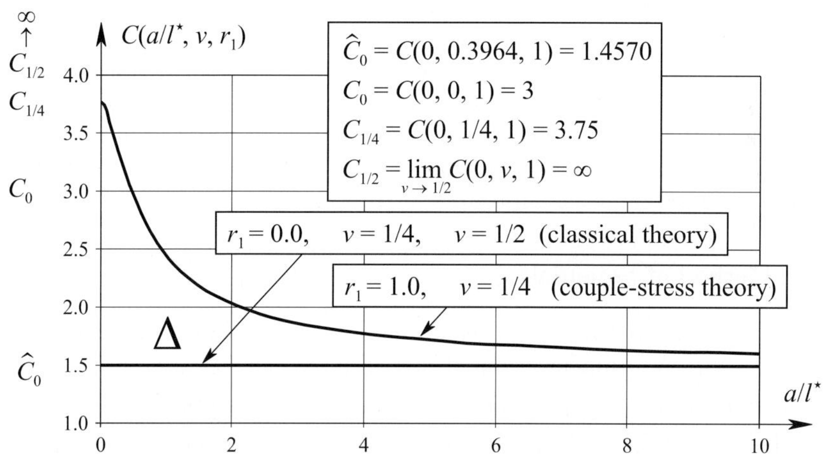

Fig. 3.3.

2. The coefficient of stress concentration may assume arbitrarily large values:

$$(\forall\nu)(\forall a/l^\star)(\forall r_1) \quad (1.4570 \le \mathbf{C}).$$

Let us make here a few remarks. It should be noted that the singularity $\mathbf{C} = \infty$ refers to the limiting case of micropolar theory $r_1 = 1$ and thus it concerns couple-stress theory. For $a/l^\star = 0$, $r_1 = 1$ and for $\nu \to 1/2$ we have $\mathbf{C} \to \infty$ ($\lim_{\nu\to 1/2} \mathbf{C}(0,\nu,1) = \infty$). In [49] we analysed the stress concentration coefficient $\mathbf{C}$ for all possible choices of dimensionless parameters which affect the stress concentration. This does not lead to any richer results than those given in the above conclusions. However, there is one interesting exception, concerning the coefficient of stress concentration in the form:

[8] See Sokołowski [215, p. 48].

$$\mathbf{C} = \mathbf{C}\left(\frac{a}{l}, \nu, \nu^{\star}\right),$$

where the dimensionless parameter $\nu^{\star} = \dfrac{a_0}{l^2} \in [0, 2]$.

In this case, micropolar theory does not exhibit the singularity $\mathbf{C} = \infty$ since the estimate $1.4570 \leq \mathbf{C} \leq 3$ holds. In both problems with stress concentration in micropolar theory discussed above one observes a reduced scale effect. Let us add that in both problems the solutions for the case of two-axial uniform stretching ($p = q$) and for the case of pure shear ($p = -q$) are obtained by the superposition method. Finally, let us make an important remark concerning the experimental verification of the theory. Kisiel [129] discusses a device for model investigations in geotechnology (it concerns here the Taylor-Schneebel's medium)[9] and points out that there is an analogy between the strain of that medium and the strain of the Cosserats medium. This analogy creates certain possibilities for designing an experiment and verifying micropolar theory experimentally by using the results presented here which concern cylindrical inclusion. According to [129], the essence of the results given here is that they are in a closed form and that the representation (3.105) used in solving the problems refers to displacements, which is of fundamental importance for an experiment based on mechanical measurements. One should use here research results in the area of model investigations[10].

3.2.6 Method of potentials

In [49], we analysed the vector equation (1.147) and (1.237)–(1.239) for the triple (u_α, φ_3). A formal procedure leads to the conclusion that in the problem we shall have the scalar potential Φ and the vector potentials $\boldsymbol{\Psi} \overset{\text{def}}{=} (0, 0, \Psi_3)$, and $\mathbf{H} \overset{\text{def}}{=} (H_1, H_2, 0)$, which are related to u_α and φ_3 through relations (3.38) with the condition $H_{\alpha,\alpha} = 0$. Equations for Φ, Ψ_3, H_α obtained from (1.148) and (1.239) are the following in Ω:

$$\begin{aligned}
&(2\mu + \lambda)\nabla_1^2\nabla_1^2\Phi = -X_{\alpha,\alpha}, \\
&4\alpha\mu\nabla_1^2\nabla_1^2(l^2\nabla_1^2 - 1)\Psi_3 = [(\gamma + \varepsilon)\nabla_1^2 - 4\alpha]\epsilon_{\alpha\beta}X_{\beta,\alpha} + 2\alpha\nabla_1^2 Y_3, \qquad (3.123) \\
&4\alpha\mu\nabla_1^2\nabla_1^2(l^2\nabla_1^2 - 1)H_\alpha = (\mu + \alpha)\nabla_1^2\epsilon_{\alpha\gamma}Y_{3,\gamma} - 2\alpha\epsilon_{\alpha\beta}\epsilon_{\delta\gamma}X_{\gamma,\delta\beta},
\end{aligned}$$

where $\epsilon_{\alpha\beta}$ is the permutation symbol, $l^2 = \dfrac{(\gamma + \varepsilon)(\mu + \alpha)}{4\alpha\mu}$, X_α, Y_3 are body forces and moments and $\nabla_1^2(.) = (.)_{,\alpha\alpha}$. The potentials Φ, Ψ_3 and H_α are coupled through equations which are obtained from (1.237), namely:

[9] See Kisiel, Dmitruk & Lysik [130, p. 291].

[10] See Dąbrowski [26] and a number of other works by him and his coworkers on model investigations.

$$
\begin{aligned}
&(2\mu+\lambda)\nabla_1^2\Phi_{,\alpha} \\
&\qquad + (\mu+\alpha)\nabla_1^2\epsilon_{\alpha\beta}\Psi_{3,\beta} - 2\alpha\nabla_1^2 H_\alpha + \vartheta_{,\alpha} + \epsilon_{\alpha\beta}\chi_{3,\beta} = 0,
\end{aligned}
\tag{3.124}
$$

$$
[(\gamma+\varepsilon)\nabla_1^2 - 4\alpha]\epsilon_{\alpha\beta}H_{\beta,\alpha} - 2\alpha\nabla_1^2\Psi_3 + \nabla_1^2\eta = 0.
$$

The functions η, ϑ, χ_3 are connected with the body loadings X_α, Y_3 by means of formula (3.41). We will show below that the representation (3.38) and equations (3.123) and (3.124) with the potentials Φ, Ψ_3, H_α can be simplified by eliminating the vector $\mathbf{H}$. First, we satisfy the condition $H_{\alpha,\alpha} = 0$ in Ω by introducing a new function H: $H_\alpha = \epsilon_{\beta\alpha}H_{,\beta}$. Now, from $(3.38)_2$ we have in Ω: $\varphi_3 = \nabla_1^2 H$. Eliminating H_α from equation $(3.124)_1$, we see that the latter leads to the differential relations for functions Φ, φ_3 in Ω:

$$
\begin{aligned}
&[(2\mu+\lambda)\nabla_1^2\Phi + \vartheta]_{,\alpha} \\
&\qquad + \epsilon_{\alpha\beta}\left[2\mu(l^2\nabla_1^2 - 1)\varphi_3 + \frac{\mu+\alpha}{2\alpha}\nabla_1^2\eta + \chi_3\right]_{,\beta} = 0.
\end{aligned}
\tag{3.125}
$$

The functions Φ, φ_3 satisfy the separate equations

$$
\begin{aligned}
\nabla_1^2\nabla_1^2\Phi &= -\frac{1}{2\mu+\lambda}X_{\alpha,\alpha}, \\
\nabla_1^2(l^2\nabla_1^2 - 1)\varphi_3 &= -\frac{\mu+\alpha}{4\alpha\mu}\nabla_1^2 Y_3 - \frac{1}{2\mu}\epsilon_{\beta\alpha}X_{\beta,\alpha}.
\end{aligned}
\tag{3.126}
$$

In turn, equation $(3.124)_2$ leads (after elimination of H_α) to the equation for the function Ψ_3:

$$
\nabla_1^2\Psi_3 = 2(k^2\nabla_1^2 - 1)\varphi_3 + \frac{1}{2\alpha}\nabla_1^2\eta,
\tag{3.127}
$$

where $k^2 = \dfrac{\gamma+\varepsilon}{4\alpha}$.

The representation for the displacements u_α has the form $(3.38)_1$, i.e.:

$$
u_\alpha = \Phi_{,\alpha} + \epsilon_{\alpha\beta}\Psi_{3,\beta} \qquad \text{in } \Omega.
\tag{3.128}
$$

Let us make here a few remarks. Disregarding body loadings and changing the notation ($\Psi_3 \mapsto \Psi$), we obtain from (3.125) the Cauchy-Riemann relations for the expressions: $\nabla_1^2\Phi$ and $\dfrac{2\mu}{2\mu+\lambda}(l^2\nabla_1^2 - 1)\varphi_3$ from the representation of Nowacki [188] derived directly from equations (3.65).

Another remark refers again to the vector equations (1.147) and (1.237)–(1.239) for the potentials Φ, Σ, $\boldsymbol{\Psi}$, $\mathbf{H}$. Constructing a representation for u_α, φ_3 by means of these equations enables us to choose a different approach. As in the first plane dynamical problem we can eliminate the vector $\boldsymbol{\Psi}$ from our considerations. Let us also disregard body loadings. Equation $(3.124)_2$ (with the function H) is satisfied in Ω if we assume Ψ_3 in the form

$$\Psi_3 = 2(k^2\nabla_1^2 - 1)H. \tag{3.129}$$

Eliminating H_α and Ψ_3 from equation $(3.124)_1$, we obtain the Cauchy-Riemann conditions for the expressions

$$(2\mu + \lambda)\nabla_1^2\Phi, \qquad 2\mu\nabla_1^2(l^2\nabla_1^2 - 1)H.$$

Thus, equations for Φ and H have the form

$$\nabla_1^2\nabla_1^2\Phi = 0, \qquad \nabla_1^2\nabla_1^2(l^2\nabla_1^2 - 1)H = 0. \tag{3.130}$$

In turn, the representation for u_α, φ_3 becomes

$$u_\alpha = \Phi_{,\alpha} + 2\epsilon_{\alpha\beta}(k^2\nabla_1^2 - 1)H_{,\beta}, \qquad \varphi_3 = \nabla_1^2 H. \tag{3.131}$$

Let us emphasize that the representation (3.131), the Cauchy-Riemann conditions and equations (3.130) result immediately from our considerations for the dynamical problem: see relations (3.42)–(3.44) (for $X_\alpha = Y_3 = 0$). We shall present in still another way the variant of the representation (3.131), the respective Cauchy-Riemann conditons, and equations (3.130) considered above. Assume the representation for the displacements u_α in the form $(3.131)_1$. The Cauchy-Riemann conditions take here the form

$$\nabla_1^2\Phi_{,\alpha} + \frac{2\mu}{2\mu + \lambda}\epsilon_{\alpha\beta}(l^2\nabla_1^2 - 1)\varphi_{3,\beta} = 0. \tag{3.132}$$

The functions Φ, φ_3 are determined from the following equations:

$$\nabla_1^2\nabla_1^2\Phi = 0, \qquad \nabla_1^2(l^2\nabla_1^2 - 1)\varphi_3 = 0. \tag{3.133}$$

On the other hand, the function H can be obtained from equation $(3.131)_2$, i.e. from the equation $\nabla_1^2 H = \varphi_3$.

Let us go back to the representation (3.131), equations (3.130) and the Cauchy-Riemann conditions for the expressions: $2\mu\nabla_1^2(l^2\nabla_1^2 - 1)H$ and $(2\mu + \lambda)\nabla_1^2\Phi$. The problem can be reduced to the determination of one unknown function H. Namely, assuming

$$\Phi_{,\alpha} + \frac{2\mu}{2\mu + \lambda}\epsilon_{\alpha\beta}(l^2\nabla_1^2 - 1)H_{,\beta} = 0, \tag{3.134}$$

we satisfy the Cauchy-Riemann conditions for all arguments and eliminate the potential Φ. The representation for u_α and φ_3 (3.131) takes the final form

$$\begin{aligned} u_\alpha &= (1-\nu)^{-1}\left[k^2\left(1 - \frac{\alpha}{\mu + \lambda}\right)\nabla_1^2 - 1\right]\epsilon_{\alpha\beta}H_{,\beta}, \\ \varphi_3 &= \nabla_1^2 H, \end{aligned} \tag{3.135}$$

where $\nu = \dfrac{\lambda}{2(\mu+\lambda)}$ and $k^2 = \dfrac{\gamma+\varepsilon}{4\alpha}$.

The function H is determined from the equation

$$\nabla_1^2\nabla_1^2(l^2\nabla_1^2 - 1)H = 0. \tag{3.136}$$

It should be noted that superimposing the operator $\nabla_1^2\nabla_1^2(l^2\nabla_1^2 - 1)$ on $(3.135)_1$ and the operator $\nabla_1^2(l^2\nabla_1^2 - 1)$ — on $(3.135)_2$, we obtain the equations

$$\nabla_1^2\nabla_1^2(l^2\nabla_1^2 - 1)u_\alpha = 0, \qquad \nabla_1^2(l^2\nabla_1^2 - 1)\varphi_3 = 0 \qquad \text{in } \Omega, \tag{3.137}$$

derived in [32] and [43] directly from equations (3.65).

3.2.7 Half-space problem. Application of the method of potentials

For the half-space Ω in the plane strain state:

$$\Omega = \{(x_1, x_2) : x_1 > 0 \wedge |x_2| < \infty\} \tag{3.138}$$

we shall formulate the following boundary value problem. One should solve the system of equations (3.65) with the boundary conditions ($x_1 = 0$):

$$u_\alpha = f_\alpha, \qquad \varphi_3 = g_3 \qquad \text{on } \partial\Omega, \tag{3.139}$$

where f_α, $g_3 : \partial\Omega \to \mathbb{R}$ are given functions. Moreover it is postulated that $u_\alpha \to 0$ and $\varphi_3 \to 0$ for $R = \sqrt{x_\alpha x_\alpha} \to \infty$. In [49], we discussed the solution of the half-space problem by using the equations and the representation obtained from the method of potentials[11]. We shall disregard the body loadings X_α and Y_3 in our considerations. We will briefly discuss here the procedure with the use of equations (3.123) and (3.124) as well as the representation (3.38) with the condition $H_{\alpha,\alpha} = 0$. Applying Fourier integral transformations (0.50) for $n = 1$ (over the variable x_2) to equations (3.123), we obtain ordinary differential equations for the transforms $\tilde{\Phi}$, $\tilde{\Psi}_3$, $\tilde{H}_\alpha$. Taking into account the regularity conditions at infinity (vanishing of physical fields), we find the general solutions of those equations in the form:

$$\begin{aligned} \tilde{\Phi} &= Ae^{-|\xi|x_1} + B|\xi|x_1e^{-|\xi|x_1}, \\ \tilde{\Psi}_3 &= (E_3 + F_3|\xi|x_1)e^{-|\xi|x_1} + G_3e^{-\rho x_1}, \\ \tilde{H}_\alpha &= (J_\alpha + K_\alpha|\xi|x_1)e^{-|\xi|x_1} + M_\alpha e^{-\rho x_1}, \end{aligned} \tag{3.140}$$

where $\rho = \sqrt{\xi^2 + \dfrac{1}{l^2}}$, $l^2 = \dfrac{(\gamma+\varepsilon)(\mu+\alpha)}{4\alpha\mu}$, ξ — the result of transformation of the variable x_2, and A, B, E_3, F_3, G_3, J_α, K_α, M_α $(\alpha = 1, 2)$ are

[11] Conditions (3.139) may be formulated for the first, second, mixed or thermoelastic boundary value problems.

constants which need to be determined. Satisfying the differential relations (3.124) and the relation $H_{\alpha,\alpha} = 0$ we eliminate 6 constants, since we obtain:

$$\begin{aligned} B &= \frac{2\mu}{2\mu+\lambda}\frac{1}{|\xi|}K_1, \qquad F_3 = \frac{2i}{\xi}K_1, \qquad G_3 = \frac{i\alpha_0}{\xi}M_1, \\ J_1 &= \frac{|\xi|K_1 - i\xi J_2}{|\xi|}, \qquad K_2 = i\frac{|\xi|}{\xi}K_1, \qquad M_2 = \frac{i\rho}{\xi}M_1, \end{aligned} \tag{3.141}$$

where $\alpha_0 = \dfrac{2\alpha}{\mu+\alpha}$ and i is the imaginary unit of the complex plane.

Using (3.140) and (3.141) and introducing three new constants

$$A_1 = -|\xi|A - i\xi E_3 + \frac{2\mu}{2\mu+\lambda}K_1, \qquad B_1 = \frac{2(\mu+\lambda)}{2\mu+\lambda}K_1, \qquad C_1 = \alpha_0 M_1$$

we uniquely determine diplacements and rotations from the representation (3.38), since we obtain in Ω:

$$\begin{aligned} u_1 &= \mathcal{F}^\star \left[\!\left[(A_1 + B_1|\xi|x_1)e^{-|\xi|x_1} + C_1 e^{-\rho x_1}; \xi \to x_2 \right]\!\right], \\ u_2 &= i\,\mathrm{sign}\,\xi\,\mathcal{F}^\star \left[\!\left[[A_1 - (3-4\nu)B_1 + B_1|\xi|x_1]e^{-|\xi|x_1} \right.\right. \\ &\quad \left.\left. + \frac{\rho}{|\xi|}C_1 e^{-\rho x_1}; \xi \to x_2 \right]\!\right], \\ \varphi_3 &= i\mu_0 \mathcal{F}^\star \left[\!\left[2a_0\xi B_1 e^{-|\xi|x_1} - \xi^{-1}C_1 e^{-\rho x_1}; \xi \to x_2 \right]\!\right], \end{aligned} \tag{3.142}$$

where $\mathrm{sign}\,\xi = +1$ for $\xi > 0$ (-1 for $\xi < 0$),

$$\mu_0 = \frac{2\mu}{\gamma+\varepsilon}, \qquad a_0 = \frac{(\gamma+\varepsilon)(2\mu+\lambda)}{4\mu(\mu+\lambda)}.$$

Three constants, namely A_1, B_1 and C_1, can be determined from the three boundary conditions (3.139). In the definitions of the constants A_1, B_1 and C_1, out of five previous constants A, E_3, J_2, K_1, M_1, only four have appeared. Namely, the constant J_2 was the one which did not appear, which means that we may assume in equation $(3.140)_3$ that $J_2 = 0$. It should be stressed that formulae (3.142) are a special case of formulae (1.220) and (1.221), in which we take $a_1 = b_1 = d_1 = \bar{a}_1 = \bar{b}_1 = \bar{d}_1 = 0$, $\mathcal{F}^\star_{(2)} \mapsto \mathcal{F}^\star$, $\xi_3 \mapsto 0$, $\xi_2 \mapsto \xi$ and $\xi \mapsto |\xi|$. Formulae (3.142) constitute also the result obtained in [43] by means of equations (3.137). Let us consider the same problem by use of the representation (3.131) and equations (3.132), (3.133). From equations (3.133) we obtain

$$\begin{aligned} \tilde{\Phi} &= (A + B|\xi|x_1)e^{-|\xi|x_1}, \\ \tilde{\varphi}_3 &= Fe^{-|\xi|x_1} + Ge^{-\rho x_1}. \end{aligned} \tag{3.143}$$

From equation $(3.131)_2$ we find

$$\tilde{H} = -\frac{1}{2|\xi|} F x_1 e^{-|\xi| x_1} + l^2 G e^{-\rho x_1}. \tag{3.144}$$

Four constants, namely A, B, F and G, remain to be determined. One of the constants F or B can be eliminated by use of relations (3.132), e.g. $B = -\frac{i\mu}{2\mu+\lambda} \cdot \frac{1}{\xi|\xi|} F$, and then the remaining three can be found from the three conditions (3.139) on $\partial\Omega$. Displacements are obtained from relations $(3.131)_1$. We may now use equations (3.130), from which we obtain $(3.143)_1$ and

$$\tilde{H} = (C + D|\xi| x_1) e^{-|\xi| x_1} + E e^{-\rho x_1}. \tag{3.145}$$

There are five constants A–E to be determined, from which we eliminate two by satysfying the Cauchy-Riemann conditions in the form of (3.134)[12]:

$$D = \frac{|\xi|}{\xi} \cdot \frac{2\mu + \lambda}{2\mu i} B,$$

$$A = \frac{2\mu i}{2\mu + \lambda} \cdot \frac{|\xi|}{\xi} C + (1 + 2\xi^2 l^2) B.$$

The remaining three constants can be found from three conditions on $\partial\Omega$. Displacements and rotations are determined from the representation (3.131). Finally, the stated problem may be solved by using one equation with function H, namely equation (3.136), from which we obtain (3.145). Displacements and rotations are determined by use of representation (3.135). The three constants are determined from three boundary conditions.

3.3 The second problem of plane strain state. Elastodynamics

3.3.1 Body loadings. Equations in displacements and rotations

Under the assumption of plane strain state (causes and effects do not depend on the position variable x_3) we obtain from the vector equations (1.1) the following system of equations in $\Omega \times \mathbf{T}_+$ for the triple (φ_α, u_3):

$$\begin{aligned} &\Box_4 \varphi_\alpha + (\beta + \gamma - \varepsilon)\varphi_{\beta,\beta\alpha} + 2\alpha \epsilon_{\alpha\gamma} u_{3,\gamma} + Y_\alpha = 0, \\ &\Box_2 u_3 + 2\alpha \epsilon_{\alpha\beta} \varphi_{\beta,\alpha} + X_3 = 0, \end{aligned} \tag{3.146}$$

[12] Satisfying the Cauchy-Riemann conditions for expressions $(2\mu + \lambda)\nabla_1^2 \Phi$ and $2\mu\nabla_1^2(l^2\nabla_1^2 - 1)H$ will not uniquely determine Φ and H. This does not change the fact that from the representation (3.131) we obtain unique (depending on three constants) fields u_α and φ_3.

where X_3, Y_α — body loadings, $\epsilon_{\alpha\beta}$ — the permutation symbol,

$$\Box_2 = (\mu+\alpha)\nabla_1^2 - \rho\partial_t^2,$$
$$\Box_4 = (\gamma+\varepsilon)\nabla_1^2 - 4\alpha - J\partial_t^2, \qquad \nabla_1^2(.) = (.)_{,\alpha\alpha}.$$

Let us formulate now the following initial-boundary conditions:

— the boundary conditions on $\partial\Omega \times \mathbf{T}$,

$$p_3 = \sigma_{\alpha 3} n_\alpha, \qquad m_\beta = \mu_{\alpha\beta} n_\alpha \qquad \text{on } \partial\Omega_\sigma \times \mathbf{T}, \tag{3.147}$$

$$u_3 = f_3, \qquad \varphi_\alpha = g_\alpha \qquad \text{on } \partial\Omega_u \times \mathbf{T}; \tag{3.148}$$

— initial conditions in $\Omega \times \{0\}$,

$$\begin{aligned} u_3 &= h_3, & \varphi_\alpha &= k_\alpha, \\ \dot{u}_3 &= \psi_3, & \dot{\varphi}_\alpha &= \chi_\alpha, \end{aligned} \tag{3.149}$$

where the functions p_3, $m_\beta : \partial\Omega_\sigma \times \mathbf{T} \to \mathbb{R}$, f_3, $g_\alpha : \partial\Omega_u \times \mathbf{T} \to \mathbb{R}$ and h_3, k_α, ψ_3, $\chi_\alpha : \Omega \to \mathbb{R}$ are given and n_α are the components of the outward unit vector $\mathbf{n}$ normal to the boundary $\partial\Omega$.

In [49], we studied the solution of the system of equations (3.146) with the initial-boundary conditions (3.147)–(3.149) by means of the superposition method. To this end we introduce the substitution

$$\varphi_\alpha = \tfrac{1}{2}\epsilon_{\alpha\beta} u_{3,\beta} - \zeta_\alpha \tag{3.150}$$

to the system of equations (3.146).

Then equations (3.146) are transformed into equations for u_3, ζ_α in $\Omega \times \mathbf{T}_+$, namely:

$$\begin{aligned} &\Box_4 \zeta_\alpha + (\beta+\gamma-\varepsilon)\zeta_{\beta,\beta\alpha} - \tfrac{1}{2}\overset{*}{\Box}_4 \epsilon_{\alpha\beta} u_{3,\beta} - Y_\alpha = 0, \\ &\overset{*}{\Box}_2 u_3 + 2\alpha\epsilon_{\alpha\beta}\zeta_{\alpha,\beta} + X_3 = 0, \end{aligned} \tag{3.151}$$

where $\overset{*}{\Box}_2 = \mu\nabla_1^2 - \rho\partial_t^2$ and $\overset{*}{\Box}_4 = (\gamma+\varepsilon)\nabla_1^2 - J\partial_t^2$.

The solution of the problem is assumed in the form

$$u_3 = u_3^0 + u_3', \qquad \varphi_\alpha = \varphi_\alpha^0 + \varphi_\alpha', \tag{3.152}$$

where the triple $(u_3^0, \varphi_\alpha^0)$ refers to the classical theory of elasticity, while the triple (u_3', φ_α') constitutes a complementary solution.

Such effect is obtained by assuming $\zeta_\alpha = \zeta_\alpha^0 + \zeta_\alpha'$ and $Y_\alpha = Y_\alpha^0 + Y_\alpha'$. Now, for $\zeta_\alpha^0 = 0$, we obtain from equation $(3.151)_2$ the classical equation

$$\overset{*}{\Box}_2 u_3^0 + X_3 = 0 \qquad \text{in } \Omega \times \mathbf{T}_+, \tag{3.153}$$

which should be connected with the following initial-boundary conditions:

— the boundary conditions on $\partial\Omega \times \mathbf{T}$,

$$\sigma^0_{\alpha 3} n_\alpha = p_3 \qquad \text{on } \partial\Omega_\sigma \times \mathbf{T}, \tag{3.154}$$

$$u^0_3 = f_3 \qquad \text{on } \partial\Omega_u \times \mathbf{T}; \tag{3.155}$$

— the initial conditions in $\Omega \times \{0\}$,

$$u^0_3 = h_3, \qquad \dot{u}^0_3 = \psi_3. \tag{3.156}$$

The stresses $\sigma^0_{\alpha 3}$ are determined from Hooke's law in $\Omega \times \mathbf{T}$:

$$\sigma^0_{\alpha 3} = 2\mu\varepsilon^0_{\alpha 3}, \tag{3.157}$$

where it holds in $\Omega \times \mathbf{T}_+$:

$$\varepsilon^0_{\alpha 3} = \tfrac{1}{2} u^0_{3,\alpha}, \qquad \sigma^0_{\alpha 3} = \sigma^0_{3\alpha}, \qquad \varepsilon^0_{\alpha 3} = \varepsilon^0_{3\alpha}. \tag{3.158}$$

We apply the operator $\overset{*}{\square}_2$ to the equation $(3.151)_1$, and now, for $\zeta^0_\alpha = 0$ $(Y^0_\alpha = 0)$, we obtain the equation in $\Omega \times \mathbf{T}_+$

$$\overset{*}{\square}_4(2\overset{*}{\square}_2\varphi^0_\alpha + \epsilon_{\alpha\beta}X_{3,\beta}) = 0. \tag{3.159}$$

Since in the classical theory of elasticity the following relationship holds in $\Omega \times \mathbf{T}_+$:

$$\varphi^0_\alpha = \tfrac{1}{2}\epsilon_{\alpha\beta}u^0_{3,\beta}, \tag{3.160}$$

thus equation (3.159) is satisfied identically, since from equation (3.153) we obtain the equation

$$\overset{*}{\square}_2\varphi^0_\alpha = \tfrac{1}{2}\epsilon_{\beta\alpha}X_{3,\beta} \qquad \text{in } \Omega \times \mathbf{T}_+. \tag{3.161}$$

Let us move on to the discussion of the solution of the complementary problem connected with the triple (u'_3, φ'_α). Applying the operator $\overset{*}{\square}_2$ to $(3.151)_1$, we obtain from the system of equations given by (3.151), (3.153) and (3.159) a system of equations for u'_3, ζ_α in $\Omega \times \mathbf{T}_+$, namely:

$$\begin{aligned} &\overset{*}{\square}_2 u'_3 + 2\alpha\epsilon_{\alpha\beta}\zeta_{\alpha,\beta} + \hat{X}_3 = 0,\\ &\overset{*}{\square}_2[\square_4\zeta_\alpha + (\beta+\gamma-\varepsilon)\zeta_{\beta,\beta\alpha} - \tfrac{1}{2}\overset{*}{\square}_4\epsilon_{\alpha\beta}u'_{3,\beta}] - \hat{Y}_\alpha = 0, \end{aligned} \tag{3.162}$$

where it holds

$$\hat{X}_3 = 0, \qquad \hat{Y}_\alpha = \overset{*}{\square}_2 Y_\alpha - \tfrac{1}{2}\overset{*}{\square}_4\epsilon_{\alpha\beta}X_{3,\beta}.$$

The system of equations (3.162) leads after uncoupling to the following equations for u'_3 and ζ_α:

$$\overset{*}{\square}_2(\square_2\square_4 + 4\alpha^2\nabla_1^2)u_3' = -2\alpha\epsilon_{\alpha\beta}\hat{Y}_{\alpha,\beta} - \square_4\overset{*}{\square}_2\hat{X}_3,$$

$$2\overset{*}{\square}_2\square_3(\square_2\square_4 + 4\alpha^2\nabla_1^2)\zeta_\alpha = [2\alpha\overset{*}{\square}_4 - 2(\beta+\gamma-\varepsilon)\overset{*}{\square}_2]\hat{Y}_{\beta,\beta\alpha} \qquad (3.163)$$

$$+\overset{*}{\square}_2\square_3(2\hat{Y}_\alpha - \overset{*}{\square}_4\epsilon_{\alpha\beta}\hat{X}_{3,\beta}).$$

Eliminating the functions ζ_α from (3.162), we obtain the following system of equations for u_3', φ_α':

$$\begin{aligned} &\square_2 u_3' + 2\alpha\epsilon_{\alpha\beta}\varphi_{\beta,\alpha}' + \hat{X}_3 = 0, \\ &\overset{*}{\square}_2[\square_4\varphi_\alpha' + (\beta+\gamma-\varepsilon)\varphi_{\beta,\beta\alpha}' + 2\alpha\epsilon_{\alpha\gamma}u_{3,\gamma}'] + \hat{Y}_\alpha = 0. \end{aligned} \qquad (3.164)$$

With the system of equations (3.164), as well as with (3.162), one should associate the following initial-boundary conditions:

— the boundary conditions on $\partial\Omega \times \mathbf{T}$,

$$\sigma_{\alpha 3}' n_\alpha = 0, \qquad \mu_{\alpha\beta}' n_\alpha = m_\beta - \overset{0}{m}_\beta \qquad \text{on } \partial\Omega_\sigma \times \mathbf{T}, \qquad (3.165)$$

$$u_3' = 0, \qquad \varphi_\alpha' = g_\alpha - \overset{0}{\varphi}_\alpha \qquad \text{on } \partial\Omega_u \times \mathbf{T}; \qquad (3.166)$$

— the initial conditions in $\Omega \times \{0\}$,

$$\begin{aligned} u_3' &= 0, & \varphi_\alpha' &= k_\alpha - \overset{0}{\varphi}_\alpha, \\ \dot{u}_3' &= 0, & \dot{\varphi}_\alpha' &= \chi_\alpha - \overset{0}{\varphi}_\alpha, \end{aligned} \qquad (3.167)$$

where

$$\overset{0}{m}_\beta = \overset{0}{\mu}_{\alpha\beta} n_\alpha, \qquad \overset{0}{\varphi}_\alpha = \tfrac{1}{2}\epsilon_{\alpha\beta}\overset{0}{u}_{3,\beta},$$

$$\overset{0}{\mu}_{\alpha\beta} = (\gamma+\varepsilon)\overset{0}{\kappa}_{\alpha\beta} + (\gamma-\varepsilon)\overset{0}{\kappa}_{\beta\alpha} + \beta\overset{0}{\kappa}_{\gamma\gamma}\delta_{\alpha\beta}, \qquad \overset{0}{\kappa}_{\alpha\beta} = \overset{0}{\varphi}_{\beta,\alpha},$$

and $\delta_{\alpha\beta}$ is the Kronecker delta.

The quantities $\overset{0}{m}_\beta$, $\overset{0}{\mu}_{\alpha\beta}$, $\overset{0}{\kappa}_{\alpha\beta}$ are expressed in terms of the rotations $\overset{0}{\varphi}_\alpha$, which can be obtained from the solution of the classical problem. In turn, the stresses $\sigma_{\alpha 3}'$, $\mu_{\alpha\beta}'$ are determined from the constitutive relations in $\Omega \times \mathbf{T}$:

$$\begin{aligned} \sigma_{\alpha 3}' &= (\mu+\alpha)\gamma_{\alpha 3}' + (\mu-\alpha)\gamma_{3\alpha}', \\ \mu_{\alpha\beta}' &= (\gamma+\varepsilon)\kappa_{\alpha\beta}' + (\gamma-\varepsilon)\kappa_{\beta\alpha}' + \beta\kappa_{\gamma\gamma}'\delta_{\alpha\beta} \end{aligned} \qquad (3.168)$$

by using the geometric relations in $\Omega \times \mathbf{T}_+$:

$$\gamma_{\alpha 3}' = u_{3,\alpha}' + \epsilon_{\alpha\beta}\varphi_\beta', \qquad \gamma_{3\alpha}' = \epsilon_{\beta\alpha}\varphi_\beta', \qquad \kappa_{\alpha\beta}' = \varphi_{\beta,\alpha}'. \qquad (3.169)$$

Relations (3.168) should be completed with the relations

$$\sigma'_{3\alpha} = (\mu+\alpha)\gamma'_{3\alpha} + (\mu-\alpha)\gamma'_{\alpha 3}, \qquad \mu'_{33} = \frac{\beta}{2(\gamma+\beta)}\mu'_{\alpha\alpha}.$$

A solution of the problem (3.146)–(3.149) is obtained in the form:

$$\begin{aligned} u_3 &= u^0_3 + u'_3, & \varphi_\alpha &= \varphi^0_\alpha + \varphi'_\alpha, & \sigma_{\alpha 3} &= \sigma^0_{\alpha 3} + \sigma'_{\alpha 3}, \\ \sigma_{3\alpha} &= \sigma^0_{3\alpha} + \sigma'_{3\alpha}, & \mu_{\alpha\beta} &= \mu^0_{\alpha\beta} + \mu'_{\alpha\beta}, & \mu_{33} &= \mu^0_{33} + \mu'_{33}, \\ \gamma_{\alpha 3} &= \varepsilon^0_{\alpha 3} + \gamma'_{\alpha 3}, & \gamma_{3\alpha} &= \varepsilon^0_{\alpha 3} + \gamma'_{3\alpha}, & \kappa_{\alpha\beta} &= \kappa^0_{\alpha\beta} + \kappa'_{\alpha\beta}. \end{aligned}$$

We shall still present an equation for the rotations φ'_α. From the system of equations (3.164) we obtain

$$\begin{aligned} \overset{*}{\square}_2\square_3(\square_2\square_4 + 4\alpha^2\nabla^2_1)\varphi'_\alpha &= [(\beta+\gamma-\varepsilon)\square_2 - 4\alpha^2]\hat{Y}_{\beta,\beta\alpha} \\ &\quad - \square_2\square_3\hat{Y}_\alpha + 2\alpha\overset{*}{\square}_2\square_3\epsilon_{\alpha\beta}\hat{X}_{3,\beta}, \end{aligned} \tag{3.170}$$

where $\square_3 = (\beta+2\gamma)\nabla^2_1 - 4\alpha - J\partial^2_t$.

Let us make here the following observations. Assuming the correspondence $(u'_3, \varphi'_\alpha) \mapsto (u_3, \varphi_\alpha)$ and $(\hat{X}_3, \hat{Y}_\alpha) \mapsto (X_3, \overset{*}{\square}_2 Y_\alpha)$, we transform equations $(3.163)_1$ and (3.170) (after removing the operator $\overset{*}{\square}_2$) into the following equations in $\Omega \times \mathbf{T}_+$, respectively:

$$\begin{aligned} (\square_2\square_4 + 4\alpha^2\nabla^2_1)u_3 &= -\square_4 X_3 - 2\alpha\epsilon_{\delta\gamma}Y_{\delta,\gamma}, \\ \square_3(\square_2\square_4 + 4\alpha^2\nabla^2_1)\varphi_\alpha &= [(\beta+\gamma-\varepsilon)\square_2 - 4\alpha^2]Y_{\beta,\beta\alpha} \\ &\quad + \square_3(2\alpha\epsilon_{\alpha\beta}X_{3,\beta} - \square_2 Y_\alpha). \end{aligned} \tag{3.171}$$

Equations (3.171) are obtained after uncoupling the system of equations (3.146). The system of equations (3.146) shows a structural similarity to the system of equations (3.1). Hence equations (3.171) can be obtained from equations (3.28):

$$\begin{aligned} &[u_\alpha, \varphi_3, X_\alpha, Y_3, (\lambda+\mu-\alpha), \alpha, \nabla^2_1, \square_1, \square_2, \square_4] \\ &\quad \mapsto [\varphi_\alpha, u_3, Y_\alpha, X_3, (\beta+\gamma-\varepsilon), \alpha, \nabla^2_1, \square_3, \square_4, \square_2]. \end{aligned}$$

3.3.2 Equations in displacements and rotations with a distortion field. Fundamental solutions and limiting cases

In [49], we analysed equations (3.146) with a distortion field. We shall give here the respective equations using the superposition method. Assuming the following substitutions:

$$Y_\alpha = -\bar{r}^{\,0}_{\alpha}, \qquad X_3 = -\overset{*}{r}{}^{0}_{3}, \tag{3.172}$$

$$\hat{X}_3 = -\overset{**}{r}{}^{0}_{3}, \qquad \hat{Y}_\alpha = -\hat{r}^{0}_{\alpha}, \tag{3.173}$$

where $\hat{r}^0_\alpha = \overset{*}{\square}_2 \bar{r}^{\,0}_\alpha - \frac{1}{2}\overset{*}{\square}_4 \epsilon_{\alpha\beta} \overset{*}{r}{}^{0}_{3,\beta}$, we obtain equations with a distortion field from equations (3.153), (3.161)–(3.164) and (3.170).

We introduce the following notation:

$$\bar{r}^{\,0}_{\alpha} = \epsilon_{\gamma\alpha}(\sigma^0_{3\gamma} - \sigma^0_{\gamma 3}) + \mu^0_{\beta\alpha,\beta}, \qquad \bar{r}^{\,0}_{3} = \overset{*}{r}{}^{0}_{3} + \overset{**}{r}{}^{0}_{3} = \sigma^0_{\varepsilon 3,\varepsilon}, \tag{3.174}$$

where $\overset{*}{r}{}^{0}_{3} = 2\mu\gamma^0_{(\alpha 3),\alpha}$ and $\overset{**}{r}{}^{0}_{3} = 2\alpha\gamma^0_{[\alpha 3],\alpha}$. The symbols $\sigma^0_{\alpha 3}$, $\sigma^0_{3\alpha}$, $\mu^0_{\alpha\beta}$ have the form of the constitutive relations (3.168), written for the distortions $\gamma^0_{\alpha 3}$, $\gamma^0_{3\alpha}$ and $\kappa^0_{\alpha\beta}$, e.g. for $\sigma^0_{3\alpha}$ we obtain

$$\sigma^0_{3\alpha} = (\mu + \alpha)\gamma^0_{3\alpha} + (\mu - \alpha)\gamma^0_{\alpha 3}.$$

Parentheses (.) and brackets [.] denote symmetric and antisymmetric parts of the tensor, respectively. If now in equations $(3.163)_1$ and (3.170) (for which we assume the substitution (3.173)) we assume the correspondence $(u_3', \varphi_\alpha') \mapsto (u_3, \varphi_\alpha)$ and $(\hat{r}^0_\alpha, \overset{**}{r}{}^{0}_{3}) \mapsto (\overset{*}{\square}_2 \bar{r}^{\,0}_{\alpha}, \bar{r}^{\,0}_{3})$, then, removing the operator $\overset{*}{\square}_2$, we obtain the following uncoupled equations with a distortion field in $\Omega \times \mathbf{T}_+$:

$$\begin{aligned}
&(\square_2\square_4 + 4\alpha^2\nabla_1^2)u_3 \\
&\quad = \square_4 \sigma^0_{\delta 3,\delta} + 2\alpha\epsilon_{\delta\alpha}[\epsilon_{\gamma\delta}(\sigma^0_{3\gamma} - \sigma^0_{\gamma 3}) + \mu^0_{\beta\delta,\beta}]_{,\alpha}, \\
&\square_3(\square_2\square_4 + 4\alpha^2\nabla_1^2)\varphi_\alpha \\
&\quad = -[(\beta + \gamma - \varepsilon)\square_2 - 4\alpha^2][\epsilon_{\gamma\beta}(\sigma^0_{3\gamma} - \sigma^0_{\gamma 3}) + \mu^0_{\delta\beta,\delta}]_{,\beta\alpha} \\
&\qquad + \square_3\left\{-2\alpha\epsilon_{\alpha\beta}\sigma^0_{\delta 3,\delta\beta} + \square_2[\epsilon_{\gamma\alpha}(\sigma^0_{3\gamma} - \sigma^0_{\gamma 3}) + \mu^0_{\beta\alpha,\beta}]\right\}.
\end{aligned} \tag{3.175}$$

Let us note that equations (3.175) are obtained from equations (3.171) by means of the substitution

$$Y_\alpha = -\bar{r}^{\,0}_{\alpha}, \qquad X_3 = -\bar{r}^{\,0}_{3}. \tag{3.176}$$

Dyszlewicz & Wytrążek considered in [84] the action of harmonically-varying-in-time distortion field in the infinite space of the form:

$$[(\gamma^0_{\alpha 3}, \gamma^0_{3\alpha}), \kappa^0_{\alpha\beta}] = [(\gamma_0^{\alpha 3}, \gamma_0^{3\alpha}), \kappa_0^{\alpha\beta}]e^{-i\omega t}\delta(x_1)\delta(x_2). \tag{3.177}$$

Using equations (3.175) and applying Fourier integral transformations (0.50) for $n = 3$, fundamental solutions in the closed form were given for displacements and rotations. Full matrices for the distortions $\gamma^0_{\alpha 3}$, $\gamma^0_{3\alpha}$ and $\kappa^0_{\alpha\beta}$ were considered. On the basis of micropolar theory (1) we found and discussed fundamental solutions for the limiting theories (2)–(8) (Fig. 1.1). In particular, for the limit (1)$\overset{\omega\to 0}{\longrightarrow}$(2) (Fig. 1.1) we obtained results in the framework of micropolar elastostatics which were given by Gajl & Suchar [95]. Let us consider, for example, the action of the distortions κ_{11}^0 and γ_{13}^0. We obtain the following formulae:

— for κ_{11}^0 (the remaining distortions are equal to zero),

$$\begin{aligned}
\varphi_1 &= \frac{\kappa_0}{2\pi}e^{-i\omega t}\left\{\frac{2\gamma}{\gamma+\varepsilon}\cdot\frac{1}{k_4^2}[B_1K_0(-ik_1r)+B_2K_0(-ik_2r)\right.\\
&\quad\left.-K_0(-ik_3r)]_{,22}-K_0(-ik_3r)\right\}_{,1},\\
\varphi_2 &= -\frac{\kappa_0}{2\pi}e^{-i\omega t}\left\{\frac{2\gamma}{\gamma+\varepsilon}\cdot\frac{1}{k_4^2}[B_1K_0(-ik_1r)+B_2K_0(-ik_2r)\right.\\
&\quad\left.-K_0(-ik_3r)]_{,11}+\frac{\beta}{2\gamma+\beta}K_0(-ik_3r)\right\}_{,2},\\
u_3 &= \frac{\kappa_0}{2\pi}e^{-i\omega t}\frac{\gamma}{\mu l^2(k_1^2-k_2^2)}[K_0(-ik_1r)-K_0(-ik_2r)]_{,12},
\end{aligned} \tag{3.178}$$

where

$$\kappa_0^{11}\mapsto\kappa_0,\qquad k_3=\lambda_3=\sqrt{\frac{J\omega^2-4\alpha}{2\gamma+\beta}},$$

$$B_1=\frac{k_4^2-k_2^2}{k_1^2-k_2^2},\qquad B_2=\frac{k_4^2-k_1^2}{k_2^2-k_1^2};$$

— for γ_{13}^0 (the remaining distortions are equal to zero),

$$\begin{aligned}
\varphi_\alpha &= \frac{\gamma_0}{2\pi}e^{-i\omega t}\left\{\frac{2\alpha}{\gamma+\varepsilon}\cdot\frac{1}{k_4^2}\epsilon_{\alpha\beta}[C_1K_0(-ik_1r)+C_2K_0(-ik_2r)\right.\\
&\quad\left.-K_0(-ik_3r)]_{,1\beta}+a_{\alpha1}K_0(-ik_3r)\right\},\\
u_3 &= \frac{\gamma_0}{2\pi}e^{-i\omega t}\frac{1}{k_1^2-k_2^2}(\nabla_1^2+k_4^2+\eta_0^2)[K_0(-ik_1r)-K_0(-ik_2r)]_{,1},
\end{aligned} \tag{3.179}$$

where

$$\gamma_0^{13}\mapsto\gamma_0,\qquad a_{11}=0,\qquad a_{21}=\frac{2\alpha}{2\gamma+\beta},$$

$$C_1=\frac{k_2^2}{k_2^2-k_1^2},\qquad C_2=\frac{k_1^2}{k_1^2-k_2^2}.$$

The remaining notations used in formulae (3.178) and (3.179) are the same as those in Subsection 3.1.2. Let us yet consider the limiting case of formulae

(3.178) for $\alpha \to \infty$. Then, one should assume the limits[13]

$$\sqrt{\frac{\gamma+\varepsilon}{4\mu}} \to l^\star, \qquad \frac{\gamma-\varepsilon}{\gamma+\varepsilon} \to \eta, \qquad J \to 0,$$

where $l^\star$ and η are the elasticity constants of couple-stress theory.

Formulae (3.178) go over into the formulae:

$$\begin{aligned} \varphi_\alpha &= \frac{\kappa_0}{2\pi} e^{-i\omega t}(1+\eta)\frac{1}{k_1^{\star 2}-k_2^{\star 2}}\epsilon_{\alpha\beta}[K_0(-ik_1^\star r) - K_0(-ik_2^\star r)]_{,12\beta}, \\ u_3 &= \frac{\kappa_0}{\pi} e^{-i\omega t}(1+\eta)\frac{1}{k_1^{\star 2}-k_2^{\star 2}}[K_0(-ik_1^\star r) - K_0(-ik_2^\star r)]_{,12}, \end{aligned} \tag{3.180}$$

where

$$k_1^{\star 2} = \frac{1}{2l^{\star 2}}\left(-1-\sqrt{1+\frac{4\rho\omega^2}{\mu}l^{\star 2}}\right),$$

$$k_2^{\star 2} = \frac{1}{2l^{\star 2}}\left(-1+\sqrt{1+\frac{4\rho\omega^2}{\mu}l^{\star 2}}\right).$$

This is a result concerning couple-stress theory. The explanation of this fact is the following. From the equations of couple-stress theory in the form

$$\overset{*}{\Box}_2 u_3 + 2\mu l^{\star 2}\nabla_1^2 \epsilon_{\delta\gamma}\varphi_{\gamma,\delta} + F_3^0 = 0, \qquad \varphi_\alpha = \tfrac{1}{2}\epsilon_{\alpha\beta}u_{3,\beta}, \tag{3.181}$$

where

$$\overset{*}{\Box}_2 = \mu\nabla_1^2 - \rho\partial_t^2, \qquad F_3^0 = -(s_{\delta 3,\delta}^0 + \tfrac{1}{2}\epsilon_{\gamma\beta}m_{\varepsilon\beta,\varepsilon\gamma}^0),$$

$$m_{\varepsilon\beta}^0 = 4\mu l^{\star 2}(\kappa_{\beta\varepsilon}^0 + \eta\kappa_{\varepsilon\beta}^0), \qquad s_{\delta 3}^0 = 2\mu\varepsilon_{\delta 3}^0,$$

we obtain the equations

$$\begin{aligned} [\nabla_1^2(l^{\star 2}\nabla_1^2 - 1) + \hat{c}_2^{-2}\partial_t^2]u_3 &= \frac{1}{\mu}F_3^0, \\ [\nabla_1^2(l^{\star 2}\nabla_1^2 - 1) + \hat{c}_2^{-2}\partial_t^2]\varphi_\alpha &= \frac{1}{2\mu}\epsilon_{\alpha\beta}F_{3,\beta}^0, \end{aligned} \tag{3.182}$$

[13] The condition $J \to 0$ has a physical motivation. Assuming it, we are motivated by the fact that the inerti term (the product of the rotational inertia and the second time derivative of a certain rotation angle) plays an essential role in the equations of motion in structural theories (see Woźniak [244]), and in the theory of continuos micropolar medium of the second order (see Nowacki [192]). This factor does not appear, however, in the equations of motion of couple-stress theory (see Sokołowski [215, p. 81]).

where $\hat{c}_2 = \sqrt{\mu/\rho}$ and $\kappa^0_{\alpha\beta}$, $\varepsilon^0_{\alpha 3}$ determine the fields of given distortions. Integrating equations (3.182) for $\kappa^0_{11} = \kappa^{11}_0 \exp(-i\omega t)\delta(x_1)\delta(x_2)$, we arrive at formulae (3.180) for which formula $(3.181)_2$ holds. Assuming that $\nabla^2_1(\ln r) = 2\pi\delta(x_1)\delta(x_2)$, we see that equation $(3.181)_1$ is satisfied identically and, moreover, for $r = \sqrt{x_\alpha x_\alpha} \to \infty$ we have $\varphi_\alpha \to 0$ and $u_3 \to 0$.

3.3.3 Functions of displacements and rotations

In [49], we derived a representation for the fields of displacements, rotations, force-stresses and couple-stressess. This representation is expressed in terms of two potentials $\psi^\star$ and χ. We start our considerations with the following equations:

— the equations of motion in $\Omega \times \mathbf{T}_+$,

$$\begin{aligned} &\epsilon_{\alpha\beta}(\sigma_{3\alpha} - \sigma_{\alpha 3}) + \mu_{\alpha\beta,\alpha} + Y_\beta = J\ddot{\varphi}_\beta, \\ &\sigma_{\alpha 3,\alpha} + X_3 = \rho\ddot{u}_3; \end{aligned} \tag{3.183}$$

— the compatibility equations in $\Omega \times \mathbf{T}$,

$$\kappa_{\alpha\beta} + \epsilon_{\delta\beta}\gamma_{3\delta,\alpha} = 0, \qquad \kappa_{\alpha\alpha} + \epsilon_{\delta\beta}\gamma_{\beta 3,\delta} = 0. \tag{3.184}$$

We proceed as follows. Using the Stokes-Helmholtz representation we express the displacement u_3 in terms of the function χ:

$$u_3 = \nabla^2_1\chi \qquad \text{in } \Omega \times \mathbf{T}_+. \tag{3.185}$$

We assume a similar decomposition for body loadings in $\Omega \times \mathbf{T}$:

$$Y_\alpha = \sigma_{,\alpha} + \epsilon_{\alpha\beta}\eta_{3,\beta}, \qquad X_3 = \nabla^2_1\chi^\star. \tag{3.186}$$

Further, we satisfy for all arguments the equation of motion $(3.183)_2$ the compatibility equations (3.184), expressed previously in terms of stresses by means of the constitutive relations (3.168). We analyse further the equations of motion $(3.183)_1$, from which we eliminate the rotations φ_α by using the geometric relations in $\Omega \times \mathbf{T}_+$:

$$\gamma_{\alpha 3} = u_{3,\alpha} + \epsilon_{\alpha\beta}\varphi_\beta, \qquad \gamma_{3\alpha} = \epsilon_{\beta\alpha}\varphi_\beta, \qquad \kappa_{\alpha\beta} = \varphi_{\beta,\alpha}. \tag{3.187}$$

Finally, we obtain the following representation for displacements and rotations, namely formula (3.185) and the formula

$$2\alpha\varphi_\alpha = \psi^\star_{,\alpha} + \epsilon_{\alpha\beta}(\Box_2\chi + \chi^\star)_{,\beta} \qquad \text{in } \Omega \times \mathbf{T}_+. \tag{3.188}$$

Moreover, we obtain the Cauchy-Riemann conditions for the expressions: $\Box_3\psi^\star + 2\alpha\sigma$ and $(\Box_2\Box_4 + 4\alpha^2\nabla^2_1)\chi + \Box_4\chi^\star + 2\alpha\eta_3$ in $\Omega \times \mathbf{T}_+$:

$$(\Box_3\psi^\star + 2\alpha\sigma)_{,\alpha} + \epsilon_{\alpha\beta}\left[(\Box_2\Box_4 + 4\alpha^2\nabla_1^2)\chi + \Box_4\chi^\star + 2\alpha\eta_3\right]_{,\beta} = 0. \quad (3.189)$$

It means that the following equations are valid in $\Omega \times \mathbf{T}_+$:

$$\begin{aligned} &\nabla_1^2\Box_3\psi^\star + 2\alpha Y_{\alpha,\alpha} = 0, \\ &\nabla_1^2(\Box_2\Box_4 + 4\alpha^2\nabla_1^2)\chi + \Box_4 X_3 + 2\alpha\epsilon_{\alpha\beta}Y_{\alpha,\beta} = 0. \end{aligned} \quad (3.190)$$

Satisfying the Cauchy-Riemann conditions (3.189) in a special way, we obtain the wave equations in $\Omega \times \mathbf{T}_+$:

$$\begin{aligned} &\Box_3\psi^\star + 2\alpha\sigma = 0, \\ &(\Box_2\Box_4 + 4\alpha^2\nabla_1^2)\chi + \Box_4\chi^\star + 2\alpha\eta_3 = 0. \end{aligned} \quad (3.191)$$

We call the functions $\psi^\star$, χ used in the representations (3.185) and (3.188) functions of displacements-rotations, since by substituting (3.185) and (3.188) to the system of equations (3.146) we obtain the Cauchy-Riemann conditions (3.189).

3.3.4 Method of potentials

In [49], we analyse the vector equations (1.147) and (1.151)–(1.153) with respect to the second plane problem with the triple (u_3, φ_α). A scalar potential Σ and vector potentials of the form $\boldsymbol{\Psi} \stackrel{\text{def}}{=} (\Psi_1, \Psi_2, 0)$ and $\mathbf{H} \stackrel{\text{def}}{=} (0, 0, H_3)$ will appear in the problem. From (1.147) we obtain the following relations in $\Omega \times \mathbf{T}_+$:

$$u_3 = \epsilon_{\alpha\beta}\Psi_{\beta,\alpha}, \qquad \varphi_\alpha = \Sigma_{,\alpha} + \epsilon_{\alpha\beta}H_{3,\beta}. \quad (3.192)$$

The condition $\operatorname{div}\mathbf{H} = 0$ is satisfied identically, while the condition $\operatorname{div}\boldsymbol{\Psi} = 0$ implies that $\Psi_{\alpha,\alpha} = 0$, which can be fulfilled by introducing a new scalar function Ψ through the substitution $\Psi_\alpha = \epsilon_{\beta\alpha}\Psi_{,\beta}$. The functions Σ, H_3, Ψ_α are connected with each other by means of the following equations in $\Omega\times\mathbf{T}_+$. From equation $(1.151)_1$ we obtain

$$\Box_2\epsilon_{\alpha\beta}\Psi_{\beta,\alpha} - 2\alpha\nabla_1^2 H_3 + \nabla_1^2\chi^\star = 0. \quad (3.193)$$

In turn, from equation $(1.151)_2$ we get

$$\Box_3\Sigma_{,\alpha} + \Box_4\epsilon_{\alpha\beta}H_{3,\beta} - 2\alpha\nabla_1^2\Psi_\alpha + \sigma_{,\alpha} + \epsilon_{\alpha\beta}\eta_{3,\beta} = 0. \quad (3.194)$$

The functions σ, $\chi^\star$, η_3 are connected with the decomposition (1.148) of body loading vectors $\mathbf{X} \stackrel{\text{def}}{=} (0, 0, X_3)$ and $\mathbf{Y} \stackrel{\text{def}}{=} (Y_1, Y_2, 0)$, and they are given by equations (3.186). The representation (3.192) can be simplified by eliminating the vector $\mathbf{H}$, since from equation (3.193) we obtain $2\alpha H_3 = \Box_2\Psi + \chi^\star$ by substituting $\Psi_\alpha = \epsilon_{\beta\alpha}\Psi_{,\beta}$. Finally, the representation (3.192) will be expressed in terms of the functions Σ, Ψ as follows:

$$\begin{aligned} \varphi_\alpha &= \Sigma_{,\alpha} + \frac{1}{2\alpha}\epsilon_{\alpha\beta}(\Box_2\Psi + \chi^\star)_{,\beta}, \\ u_3 &= \nabla_1^2\Psi. \end{aligned} \tag{3.195}$$

From equation (3.194) we obtain the Cauchy-Riemann conditions for the expressions:

$$2\alpha(\Box_3\Sigma + \sigma), \qquad (\Box_2\Box_4 + 4\alpha^2\nabla_1^2)\Psi + \Box_4\chi^\star + 2\alpha\eta_3.$$

It holds in $\Omega \times \mathbf{T}_+$:

$$2\alpha(\Box_3\Sigma + \sigma)_{,\alpha} + \epsilon_{\alpha\beta}\left[(\Box_2\Box_4 + 4\alpha^2\nabla_1^2)\Psi + \Box_4\chi^\star + 2\alpha\eta_3\right]_{,\beta} = 0. \tag{3.196}$$

It should be noted that formulae (3.185), (3.188) and (3.195) as well as (3.189) and (3.196) refer to the same representation and to the same equations. It suffices to assume in formulae (3.185), (3.188) and (3.189) the correspondence $\chi \mapsto \Psi$ and $\psi^\star \mapsto 2\alpha\Sigma$.

3.3.5 Rotation potentials in polar coordinates

From vector equations (1.1) we obtain (in the polar coordinate system (r,θ)) for the triple $(\varphi_r, \varphi_\theta, u_z)$ the following equations in $\Omega \times \mathbf{T}_+$[14]:

$$\begin{aligned} &(\gamma+\varepsilon)\left(\nabla_0^2\varphi_r - \frac{2}{r^2}\partial_\theta\varphi_\theta\right) \\ &\qquad - 4\alpha\varphi_r + (\beta+\gamma-\varepsilon)\partial_r\kappa + 2\alpha\frac{1}{r}\partial_\theta u_z + Y_r = J\ddot{\varphi}_r, \\ &(\gamma+\varepsilon)\left(\nabla_0^2\varphi_\theta + \frac{2}{r^2}\partial_\theta\varphi_r\right) \\ &\qquad - 4\alpha\varphi_\theta + (\beta+\gamma-\varepsilon)\frac{1}{r}\partial_\theta\kappa - 2\alpha\partial_r u_z + Y_\theta = J\ddot{\varphi}_\theta, \\ &(\mu+\alpha)\nabla^2 u_z + 2\alpha\frac{1}{r}[\partial_r(r\varphi_\theta) - \partial_\theta\varphi_r] + X_z = \rho\ddot{u}_z, \end{aligned} \tag{3.197}$$

where

$$\kappa = \frac{1}{r}\partial_r(r\varphi_r) + \frac{1}{r}\partial_\theta\varphi_\theta, \qquad \nabla^2 = \partial_r^2 + r^{-1}\partial_r + r^{-2}\partial_\theta^2, \qquad \nabla_0^2 = \nabla^2 - \frac{1}{r^2}.$$

[14] Taking into account the fact that for the vector $\mathbf{a} \stackrel{\text{def}}{=} (a_r, a_\theta, a_z)$ we have

$$\begin{aligned} &\vec{\nabla}^2\mathbf{a} = (\nabla_0^2 a_r - 2r^{-2}\partial_\theta a_\theta, \nabla_0^2 a_\theta + 2r^{-2}\partial_\theta a_r, \nabla^2 a_z), \\ &\operatorname{curl}\mathbf{a} = (r^{-1}\partial_\theta a_z, -\partial_r a_z, (r^{-1} + \partial_r)\, a_\theta - r^{-1}\partial_\theta a_r), \\ &\operatorname{grad}\operatorname{div}\mathbf{a} = (\partial_r a, r^{-1}\partial_\theta a, 0), \end{aligned}$$

where $a = r^{-1}\partial_r(ra_r) + r^{-1}\partial_\theta a_\theta$, we obtain equations (3.197) and, independently, equations (3.102) for the triple $(u_r, u_\theta, \varphi_z)$ from the vector equations (1.1).

In [49], we introduce rotation potentials in $\Omega \times \mathbf{T}_+$ to the system of equations (3.197):

$$\varphi_r = \partial_r \Sigma + \frac{1}{r}\partial_\theta H, \qquad \varphi_\theta = \frac{1}{r}\partial_\theta \Sigma - \partial_r H. \tag{3.198}$$

If we disregard the body loadings Y_r, Y_θ, X_z, the system of equations (3.197) is transformed into a system of equations for the functions Σ, H, u_z in $\Omega \times \mathbf{T}_+$, namely:

$$\begin{aligned} &\partial_r \Box_3 \Sigma + \frac{1}{r}\partial_\theta(\Box_4 H + 2\alpha u_z) = 0, \\ &\frac{1}{r}\partial_\theta \Box_3 \Sigma - \partial_r(\Box_4 H + 2\alpha u_z) = 0, \\ &\Box_2 u_z - 2\alpha \nabla^2 H = 0. \end{aligned} \tag{3.199}$$

The operators $\Box_2$, $\Box_3$, $\Box_4$ are determined here by the Laplacian ∇^2 in coordinates (r, θ). If we assume that in equations $(3.199)_{1,2}$ the expressions acted upon by the operators ∂_r, $\frac{1}{r}\partial_\theta$ vanish independently of each other, we obtain the equations

$$\Box_3 \Sigma = 0, \qquad \Box_4 H + 2\alpha u_z = 0 \qquad \text{in } \Omega \times \mathbf{T}_+. \tag{3.200}$$

Equations $(3.199)_3$ and $(3.200)_2$ are coupled with each other. Separating them, we obtain the equations

$$(\Box_2 \Box_4 + 4\alpha^2 \nabla^2)(u_z, H) = (0, 0) \qquad \text{in } \Omega \times \mathbf{T}_+. \tag{3.201}$$

Thus, the system of equations (3.197) has been replaced by the system of wave equations $(3.200)_1$ and (3.201).

3.3.6 SEMP

Let us consider the following mixed initial-boundary value problem for the second plane problem represented by the displacement vector $\mathbf{u}$ and the rotation vector $\boldsymbol{\varphi}$ of the forms

$$\mathbf{u} \stackrel{\text{def}}{=} (0, 0, u_3), \qquad \boldsymbol{\varphi} \stackrel{\text{def}}{=} (\varphi_1, \varphi_2, 0) \tag{3.202}$$

and by the vectors of body loadings

$$\mathbf{X} \stackrel{\text{def}}{=} (0, 0, X_3), \qquad \mathbf{Y} \stackrel{\text{def}}{=} (Y_1, Y_2, 0). \tag{3.203}$$

We arrange the field equations in the following groups:

— the equations of motion in $\Omega \times \mathbf{T}_+$,

$$\begin{aligned} &\epsilon_{\alpha\beta}(\sigma_{3\alpha} - \sigma_{\alpha 3}) + \mu_{\alpha\beta,\alpha} + Y_\beta = J\ddot{\varphi}_\beta, \\ &\sigma_{\alpha 3,\alpha} + X_3 = \rho \ddot{u}_3, \end{aligned} \tag{3.204}$$

where $\epsilon_{\alpha\beta}$ — the permutation symbol, ρ — density, J — rotational inertia of the medium, $\sigma_{\alpha 3}$, $\sigma_{3\alpha}$, $\mu_{\alpha\beta}$ and μ_{33} — stress fields. There are eight independent components of the stress state, since the following formula holds

$$\mu_{33} = \frac{\beta}{2(\gamma+\beta)}\mu_{\alpha\alpha};$$

— the geometric relations in $\Omega \times \mathbf{T}_+$,

$$\gamma_{\alpha 3} = u_{3,\alpha} + \epsilon_{\alpha\beta}\varphi_\beta, \qquad \gamma_{3\alpha} = \epsilon_{\beta\alpha}\varphi_\beta, \qquad \kappa_{\alpha\beta} = \varphi_{\beta,\alpha}, \tag{3.205}$$

where $\gamma_{\alpha 3}$, $\gamma_{3\alpha}$ and $\kappa_{\alpha\beta}$ are strain fields;
— the constitutive relations in $\Omega \times \mathbf{T}$,

$$\begin{aligned} (\sigma_{\alpha 3}, \sigma_{3\alpha}) &= (\mu+\alpha)(\gamma_{\alpha 3}, \gamma_{3\alpha}) + (\mu-\alpha)(\gamma_{3\alpha}, \gamma_{\alpha 3}), \\ \mu_{\alpha\beta} &= (\gamma+\varepsilon)\kappa_{\alpha\beta} + (\gamma-\varepsilon)\kappa_{\beta\alpha} + \beta\kappa_{\delta\delta}\delta_{\alpha\beta}, \end{aligned} \tag{3.206}$$

where $\delta_{\alpha\beta}$ is the Kronecker delta and μ, α, β, γ, ε are material constants (the constant λ does not appear in the second plane problem).

Substituting relations (3.206) to the equations of motion (3.204) and using relations (3.205), we transform (3.204) into equations (3.146) for the triple $(\varphi_1, \varphi_2, u_3)$[15].

Now, to equations (3.204)–(3.206) we add the initial-boundary conditions in $\overline{\Omega} \times \mathbf{T}$:

— the boundary conditions on $\partial\Omega \times \mathbf{T}$,

$$p_3 = \sigma_{\alpha 3} n_\alpha, \qquad m_\beta = \mu_{\alpha\beta} n_\alpha \qquad \text{on } \partial\Omega_\sigma \times \mathbf{T}, \tag{3.207}$$

$$u_3 = \hat{u}_3, \qquad \varphi_\alpha = \hat{\varphi}_\alpha \qquad \text{on } \partial\Omega_u \times \mathbf{T}, \tag{3.208}$$

where p_3, $m_\beta : \partial\Omega_\sigma \times \mathbf{T} \to \mathbb{R}$ and $\hat{u}_3$, $\hat{\varphi}_\alpha : \partial\Omega_u \times \mathbf{T} \to \mathbb{R}$ are given functions, $\partial\Omega = \partial\Omega_\sigma \cup \partial\Omega_u$, $\partial\Omega_\sigma \cap \partial\Omega_u = \emptyset$, and n_α are components of an outward unit vector $\mathbf{n}$ normal to $\partial\Omega_\sigma$;
— the initial conditions in $\Omega \times \{0\}$,

$$\begin{aligned} \varphi_\alpha &= h_\alpha, & u_3 &= h_3, \\ \dot{\varphi}_\alpha &= l_\alpha, & \dot{u}_3 &= l_3, \end{aligned} \tag{3.209}$$

where h_α, h_3, l_α, $l_3 : \Omega \to \mathbb{R}$ are given functions.

Relations (3.204)–(3.209) [or, (3.146) and (3.205)–(3.209)] describe the mixed initial-boundary value problem for the second problem of the plane strain state. Now, we shall formulate the above problem in an equivalent, stress-type form[16]. Using relations (3.205), we obtain from the equations of motion (3.204) the following equations in $\Omega \times \mathbf{T}_+$:

---

[15] The compatibility equations (3.184) will be satisfied for all arguments.

[16] See Ignaczak [116] and Gurtin [105, p. 222].

$$
\begin{aligned}
&J^{-1}\epsilon_{\beta\delta}R_{\beta}=\ddot{\gamma}_{3\delta},\\
&J^{-1}R_{\beta,\delta}=\ddot{\kappa}_{\delta\beta},\\
&J^{-1}\epsilon_{\delta\beta}R_{\beta}+\rho^{-1}R_{3,\delta}=\ddot{\gamma}_{\delta 3},
\end{aligned}
\tag{3.210}
$$

where $R_{\beta}=\epsilon_{\alpha\beta}(\sigma_{3\alpha}-\sigma_{\alpha 3})+\mu_{\alpha\beta,\alpha}+Y_{\beta}$ and $R_3=\sigma_{\alpha 3,\alpha}+X_3$.

Solving relations (3.206) with respect to strains, we obtain

$$
\begin{aligned}
&\gamma_{\alpha 3}=\frac{1}{2\mu}\sigma_{(\alpha 3)}+\frac{1}{2\alpha}\sigma_{[\alpha 3]},\qquad \gamma_{3\alpha}=\frac{1}{2\mu}\sigma_{(\alpha 3)}-\frac{1}{2\alpha}\sigma_{[\alpha 3]},\\
&\kappa_{\alpha\beta}=\frac{1}{2\gamma}\mu_{(\alpha\beta)}+\frac{1}{2\varepsilon}\mu_{[\alpha\beta]}-\frac{\hat{\nu}}{2\gamma}\mu_{\varepsilon\varepsilon}\delta_{\alpha\beta},
\end{aligned}
\tag{3.211}
$$

where

$$
\hat{\nu}=\frac{\beta}{2(\gamma+\beta)},\qquad \sigma_{(\alpha 3)}=\tfrac{1}{2}(\sigma_{\alpha 3}+\sigma_{3\alpha}),\qquad \sigma_{[\alpha 3]}=\tfrac{1}{2}(\sigma_{\alpha 3}-\sigma_{3\alpha}),
$$

$$
\mu_{(\alpha\beta)}=\tfrac{1}{2}(\mu_{\alpha\beta}+\mu_{\beta\alpha}),\qquad \mu_{[\alpha\beta]}=\tfrac{1}{2}(\mu_{\alpha\beta}-\mu_{\beta\alpha}).
$$

Using relations (3.211), we transform equations (3.210) into the equations

$$
\begin{aligned}
&J^{-1}\epsilon_{\beta\delta}R_{\beta}-\frac{1}{2\mu}\ddot{\sigma}_{(\delta 3)}-\frac{1}{2\alpha}\ddot{\sigma}_{[3\delta]}=0,\\
&J^{-1}R_{\beta,\delta}-\frac{1}{2\gamma}(\ddot{\mu}_{(\delta\beta)}-\hat{\nu}\ddot{\mu}_{\varepsilon\varepsilon}\delta_{\alpha\beta})-\frac{1}{2\varepsilon}\ddot{\mu}_{[\delta\beta]}=0,\\
&J^{-1}\epsilon_{\delta\beta}R_{\beta}+\rho^{-1}R_{3,\delta}-\frac{1}{2\mu}\ddot{\sigma}_{(\delta 3)}-\frac{1}{2\alpha}\ddot{\sigma}_{[\delta 3]}=0.
\end{aligned}
\tag{3.212}
$$

Having integrated the equations of motion (3.204) with respect to time t and taken into account the initial conditions (3.209), we obtain formulae for the triple $(\varphi_1,\varphi_2,u_3)$ in $\Omega\times\mathbf{T}$:

$$
\begin{aligned}
&\varphi_{\alpha}=h_{\alpha}+l_{\alpha}t+J^{-1}(t*R_{\alpha}),\\
&u_3=h_3+l_3t+\rho^{-1}(t*R_3),
\end{aligned}
\tag{3.213}
$$

where

$$
t*f(x_1,x_2,t)=\int_0^t(t-\tau)f(x_1,x_2,\tau)\,d\tau.
$$

The SEMP[17] for the second mixed plane problem consists of the following relations: the field equations (3.212); the initial conditions in $\Omega\times\{0\}$:

$$
\begin{aligned}
&\sigma_{\alpha 3}=\sigma^0_{\alpha 3},\qquad \sigma_{3\alpha}=\sigma^0_{3\alpha},\qquad \mu_{\alpha\beta}=\mu^0_{\alpha\beta},\\
&\dot{\sigma}_{\alpha 3}=\dot{\sigma}^0_{\alpha 3},\qquad \dot{\sigma}_{3\alpha}=\dot{\sigma}^0_{3\alpha},\qquad \dot{\mu}_{\alpha\beta}=\dot{\mu}^0_{\alpha\beta},
\end{aligned}
\tag{3.214}
$$

[17] See footnote 8 in Subsection 2.1.6 (SEMP — Stress Equations of Motion Problem).

where initial values for fields of stresses and velocities of those fields are obtained by using (3.205), (3.206) and (3.209), namely:

$$(\sigma^0_{\alpha 3}, \sigma^0_{3\alpha}) = (\mu + \alpha)(\gamma^0_{\alpha 3}, \gamma^0_{3\alpha}) + (\mu - \alpha)(\gamma^0_{3\alpha}, \gamma^0_{\alpha 3}),$$

$$\mu^0_{\alpha\beta} = (\gamma + \varepsilon)\kappa^0_{\alpha\beta} + (\gamma - \varepsilon)\kappa^0_{\beta\alpha} + \beta\kappa^0_{\delta\delta}\delta_{\alpha\beta},$$

$$\gamma^0_{\alpha 3} = h_{3,\alpha} + \epsilon_{\alpha\beta}h_\beta, \qquad \gamma^0_{3\alpha} = \epsilon_{\beta\alpha}h_\beta, \qquad \kappa^0_{\alpha\beta} = h_{\beta,\alpha},$$

$$(\dot\sigma^0_{\alpha 3}, \dot\sigma^0_{3\alpha}) = (\mu + \alpha)(\dot\gamma^0_{\alpha 3}, \dot\gamma^0_{3\alpha}) + (\mu - \alpha)(\dot\gamma^0_{3\alpha}, \dot\gamma^0_{\alpha 3}),$$

$$\dot\mu^0_{\alpha\beta} = (\gamma + \varepsilon)\dot\kappa^0_{\alpha\beta} + (\gamma - \varepsilon)\dot\kappa^0_{\beta\alpha} + \beta\dot\kappa^0_{\delta\delta}\delta_{\alpha\beta},$$

$$\dot\gamma^0_{\alpha 3} = l_{3,\alpha} + \epsilon_{\alpha\beta}l_\beta, \qquad \dot\gamma^0_{3\alpha} = \epsilon_{\beta\alpha}l_\beta, \qquad \dot\kappa^0_{\alpha\beta} = l_{\beta,\alpha};$$

as well as the following boundary conditions which we obtain from conditions (3.207), (3.208) and (3.213), namely:

$$\sigma_{\alpha 3}n_\alpha = p_3, \qquad \mu_{\alpha\beta}n_\alpha = m_\beta \qquad \text{on } \partial\Omega_\sigma \times \mathbf{T}, \tag{3.215}$$

and

$$\begin{aligned} &h_3 + l_3 t + \rho^{-1}(t * R_3) = \hat u_3, \\ &h_\alpha + l_\alpha t + J^{-1}(t * R_\alpha) = \hat\varphi_\alpha \qquad \text{on } \partial\Omega_u \times \mathbf{T}. \end{aligned} \tag{3.216}$$

Now, let the system of functions $\sigma_{\alpha 3}$, $\sigma_{3\alpha}$, $\mu_{\alpha\beta}$ be a solution of the initial-boundary value problem described by relations (3.212), (3.214)–(3.216). Then, evaluating the systems of functions $\gamma_{\alpha 3}$, $\gamma_{3\alpha}$, $\kappa_{\alpha\beta}$ and φ_α, u_3 by using relations (3.211) and (3.213), respectively[18], we conclude that for sufficiently smooth physical fields the initial-boundary value problems described by relations (3.204)–(3.209) and (3.211)–(3.216) are equivalent.

3.4 The second problem of plane strain state. Elastostatics

3.4.1 Equations in displacements and rotations with body loadings. Fundamental solutions

In [49] ,we studied equations in displacements and rotations for the triple (u_3, φ_α) in Ω:

[18] The compatibility equations (3.184) are implied by relations (3.205), (3.209) and (3.210) and will be satisfied identically.

$$\begin{aligned}&[(\gamma+\varepsilon)\nabla_1^2-4\alpha]\varphi_\beta+(\beta+\gamma-\varepsilon)\varphi_{\alpha,\alpha\beta}+2\alpha\epsilon_{\beta\gamma}u_{3,\gamma}+Y_\beta=0,\\&(\mu+\alpha)\nabla_1^2u_3+2\alpha\epsilon_{\delta\beta}\varphi_{\beta,\delta}+X_3=0\end{aligned}\tag{3.217}$$

with the following boundary conditions on $\partial\Omega$:

$$u_3=f_3,\qquad \varphi_\alpha=g_\alpha,\tag{3.218}$$

where f_3, $g_\alpha:\partial\Omega\to\mathbb{R}$ are given functions and $\nabla_1^2(.)=(.)_{,\alpha\alpha}$.

To the system of equations (3.217) we introduce the substitution (3.150) with the function ζ_α. The system of equations (3.217) is transformed into the following:

$$\begin{aligned}&[(\gamma+\varepsilon)\nabla_1^2-4\alpha]\zeta_\alpha+(\beta+\gamma-\varepsilon)\zeta_{\beta,\beta\alpha}\\&\qquad-\tfrac12(\gamma+\varepsilon)\nabla_1^2\epsilon_{\alpha\beta}u_{3,\beta}-Y_\alpha=0,\\&\mu\nabla_1^2u_3+2\alpha\epsilon_{\alpha\beta}\zeta_{\alpha,\beta}+X_3=0.\end{aligned}\tag{3.219}$$

We assume the solution of the problem (3.217)–(3.219) in the form of the superposition (3.152). Such superposition of solution can be obtained for $\zeta_\alpha=\zeta_\alpha^0+\zeta_\alpha'$ by putting $\zeta_\alpha^0=0$. From equation $(3.219)_2$ we obtain the classical equation

$$\mu\nabla_1^2u_3^0+X_3=0\tag{3.220}$$

with the boundary condition

$$u_3^0=f_3\qquad\text{on }\partial\Omega.\tag{3.221}$$

From equation $(3.219)_1$ ($Y_\alpha=Y_\alpha^0+Y_\alpha'$) for $\zeta_\alpha^0=0$ and $Y_\alpha^0=0$ we obtain an equation for the rotations φ_α^0 in the form

$$2\mu\nabla_1^2\varphi_\alpha^0+\epsilon_{\alpha\beta}X_{3,\beta}=0\qquad\text{in }\Omega.\tag{3.222}$$

Let us note that equation (3.222) will be satisfied identically since it results also from the classical equation (3.220) ($\varphi_\alpha^0=\frac12\epsilon_{\alpha\beta}u_{3,\beta}^0$). Moreover, from (3.219), (3.220) and (3.222) we obtain the equations

$$\begin{aligned}&[(\gamma+\varepsilon)\nabla_1^2-4\alpha]\zeta_\alpha+(\beta+\gamma-\varepsilon)\zeta_{\beta,\beta\alpha}\\&\qquad-\tfrac12(\gamma+\varepsilon)\nabla_1^2\epsilon_{\alpha\beta}u_{3,\beta}'-\overset{*}{Y}_\alpha=0,\\&\mu\nabla_1^2u_3'+2\alpha\epsilon_{\alpha\beta}\zeta_{\alpha,\beta}+\hat{X}_3=0,\end{aligned}\tag{3.223}$$

where $\hat{X}_3=0$ and $\overset{*}{Y}_\alpha=Y_\alpha-\dfrac{\gamma+\varepsilon}{2\mu}\epsilon_{\alpha\beta}X_{3,\beta}$.

With the system of equations (3.223) we associate boundary conditions in the form

$$u_3' = 0, \qquad \varphi_\alpha' = g_\alpha - \varphi_\alpha^0 \qquad \text{on } \partial\Omega. \tag{3.224}$$

The rotations φ_α^0 are obtained from the solution of the classical boundary value problem given by (3.220) and (3.221). Separating the system of equations (3.223), we arrive at

$$\begin{aligned}
&2\mu\nabla_1^2(l^2\nabla_1^2 - 1)u_3' = \epsilon_{\alpha\beta}\overset{*}{Y}_{\beta,\alpha} - 2(l^2\nabla_1^2 - 1)\hat{X}_3 + \mu_0^{-1}\nabla_1^2\hat{X}_3,\\
&4\alpha(\nu_1^2\nabla_1^2 - 1)(l^2\nabla_1^2 - 1)\zeta_\alpha = \left[\tfrac{1}{2}\mu_0^{-1} - (\nu_1^2 - k^2)\right]\overset{*}{Y}_{\beta,\beta\alpha}\\
&\qquad + (\nu_1^2\nabla_1^2 - 1)\overset{*}{Y}_\alpha - \mu_0^{-1}(\nu_1^2\nabla_1^2 - 1)\epsilon_{\alpha\beta}\hat{X}_{3,\beta},
\end{aligned} \tag{3.225}$$

where

$$l^2 = \frac{(\gamma+\varepsilon)(\mu+\alpha)}{4\alpha\mu}, \qquad \nu_1^2 = \frac{2\gamma+\beta}{4\alpha}, \qquad k^2 = \frac{\gamma+\varepsilon}{4\alpha}, \qquad \mu_0 = \frac{2\mu}{\gamma+\varepsilon}.$$

Eliminating the function ζ_α from the system of equations (3.223), we obtain the following system of equations for the complementary solution (u_3', φ_α'):

$$\begin{aligned}
&(\mu+\alpha)\nabla_1^2 u_3' + 2\alpha\epsilon_{\alpha\beta}\varphi'_{\beta,\alpha} + \hat{X}_3 = 0,\\
&[(\gamma+\varepsilon)\nabla_1^2 - 4\alpha]\varphi_\alpha' + (\beta+\gamma-\varepsilon)\varphi'_{\beta,\beta\alpha} + 2\alpha\epsilon_{\alpha\gamma}u'_{3,\gamma} + \overset{*}{Y}_\alpha = 0
\end{aligned} \tag{3.226}$$

with the boundary conditions (3.224). The function φ_α' satisfies the equation

$$\begin{aligned}
&\nabla_1^2(l^2\nabla_1^2 - 1)(\nu_1^2\nabla_1^2 - 1)\varphi_\alpha' = -\frac{1}{4\alpha}(l^2\nabla_1^2 - 1)\overset{*}{Y}_{\beta,\beta\alpha}\\
&\qquad + \epsilon_{\alpha\varepsilon}\left[-\frac{\mu+\alpha}{4\alpha\mu}(\nu_1^2\nabla_1^2 - 1)\epsilon_{\delta\gamma}\overset{*}{Y}_{\delta,\gamma} + \frac{1}{2\mu}(\nu_1^2\nabla_1^2 - 1)\hat{X}_3\right]_{,\varepsilon}.
\end{aligned} \tag{3.227}$$

It should be stressed that by assuming the following correspondence

$$(u_3', \varphi_\alpha') \mapsto (u_3, \varphi_\alpha), \qquad (\hat{X}_3, \overset{*}{Y}_\alpha) \mapsto (X_3, Y_\alpha) \tag{3.228}$$

in equations $(3.225)_1$ and (3.227), we obtain separated equations for u_3, φ_α, which were derived in [43] from the system of equations (3.217). Thus, using equations (3.220), (3.222), (3.225) and (3.227), we have determined the fundamental solutions for displacements and rotations in the case of action of body loadings in the space $\mathbb{R}^2$, which are of the form

$$(X_3, Y_\alpha) = (P, M_\alpha)\delta(x_1)\delta(x_2). \tag{3.229}$$

Applying the Fourier integral transformations (0.50) for $n = 2$, we obtain the following results:

— for $X_3 = P\delta(x_1)\delta(x_2)$, $Y_\alpha = 0$:

$$\begin{aligned} u_3 &= \frac{P}{4\pi\mu}[2I_1 - \alpha_0 I_2(l)], \\ \varphi_\alpha &= \frac{P}{4\pi\mu}\epsilon_{\alpha\beta}[I_1 - I_2(l)]_{,\beta}; \end{aligned} \tag{3.230}$$

— for $Y_1 = M_1\delta(x_1)\delta(x_2)$, $X_3 = Y_2 = 0$:

$$\begin{aligned} u_3 &= -\frac{M_1}{4\pi\mu}[I_1 - I_2(l)]_{,2}, \\ \varphi_1 &= -\frac{M_1}{2\pi}\left\{\frac{1}{4\alpha}[I_1 - I_2(\nu_1)]_{,11} + \frac{1}{2\mu\alpha_0}[I_1 - I_2(l)]_{,22}\right\}, \\ \varphi_2 &= \frac{M_1}{8\pi\mu}\left[I_1 + \frac{\mu}{\alpha}I_2(\nu_1) - \frac{2}{\alpha_0}I_2(l)\right]_{,12}, \end{aligned} \tag{3.231}$$

where $\alpha_0 = \dfrac{2\alpha}{\mu + \alpha}$ and the symbols I_1, I_2 are determined by formula (3.79).

In the expression $I_2(\nu_1) = K_0(r/\nu_1)$, the constant ν_1 is determined by formula (1.189). It should be emphasized that these results were derived by Nowacki [182] by means of the method of elastic potentials.

3.4.2 Equations in displacements and rotations with a distortion field. Fundamental solutions

In [49], we analysed equations (3.217) with a distortion field taken into account. Assuming the substitution (3.172) and

$$\hat{X}_3 = -\overset{**}{\bar{r}}{}^0_3, \qquad \overset{*}{Y}_\alpha = -\overset{*}{\bar{r}}{}^0_\alpha \tag{3.232}$$

where

$$\overset{*}{\bar{r}}{}^0_\alpha = \bar{r}^0_\alpha - \frac{\gamma + \varepsilon}{2\mu}\epsilon_{\alpha\beta}\overset{*}{\bar{r}}{}^0_{3,\beta},$$

we obtain equations (3.220), (3.222), $(3.225)_1$, (3.226) and (3.227) by means of the superposition method. In these equations, a distortion field has already been taken into account. The symbols $\bar{r}^0_\alpha$, $\overset{*}{\bar{r}}{}^0_3$, $\overset{**}{\bar{r}}{}^0_3$, $\bar{r}^0_3$ have the same meaning as in formula (3.174). Let us consider equations $(3.225)_1$ and (3.227) with a distortion field (i.e., we substitute (3.232) to $(3.225)_1$ and (3.227)). Assuming now the correspondence $(u_3', \varphi_\alpha') \mapsto (u_3, \varphi_\alpha)$, $(\overset{*}{\bar{r}}{}^0_\alpha, \overset{**}{\bar{r}}{}^0_3) \mapsto (\bar{r}^0_\alpha, \bar{r}^0_3)$, we obtain the following equations in Ω:

$$
\begin{aligned}
&2\mu\nabla_1^2(l^2\nabla_1^2-1)u_3\\
&\quad=2(k^2\nabla_1^2-1)\sigma^0_{\beta3,\beta}+\epsilon_{\delta\alpha}\left[\epsilon_{\gamma\delta}(\sigma^0_{3\gamma}-\sigma^0_{\gamma3})+\mu^0_{\beta\delta,\beta}\right]_{,\alpha},\\
&\nabla_1^2(l^2\nabla_1^2-1)(\nu_1^2\nabla_1^2-1)\varphi_\alpha\\
&\quad=\frac{1}{4\alpha}(l^2\nabla_1^2-1)\left[\epsilon_{\gamma\beta}(\sigma^0_{3\gamma}-\sigma^0_{\gamma3})+\mu^0_{\delta\beta,\delta}\right]_{,\beta\alpha}\\
&\qquad+\epsilon_{\alpha\varepsilon}\left\{\frac{\mu+\alpha}{4\alpha\mu}(\nu_1^2\nabla_1^2-1)\epsilon_{\delta\tau}\left[\epsilon_{\gamma\delta}(\sigma^0_{3\gamma}-\sigma^0_{\gamma3})+\mu^0_{\beta\delta,\beta}\right]_{,\tau}\right.\\
&\qquad\left.-\frac{1}{2\mu}(\nu_1^2\nabla_1^2-1)\sigma^0_{\beta3,\beta}\right\}_{,\varepsilon},
\end{aligned}
\tag{3.233}
$$

where

$$
l^2=\frac{(\gamma+\varepsilon)(\mu+\alpha)}{4\alpha\mu},\qquad \nu_1^2=\frac{2\gamma+\beta}{4\alpha},\qquad k^2=\frac{\gamma+\varepsilon}{4\alpha}.
$$

The symbols $\sigma^0_{\alpha3}$, $\sigma^0_{3\alpha}$, $\mu^0_{\alpha\beta}$ have the same meaning as in Subsection 3.3.2 of this chapter. Equations (3.233) are obtained after separating the system of equations (3.217) and performing the substitution (3.176). In [84], we found the fundamental solution for φ_α and u_3 in the case of action, in the space $\mathbb{R}^2$, of a distortion field given by the formula

$$
[(\gamma^0_{\alpha3},\gamma^0_{3\alpha}),\kappa^0_{\alpha\beta}]=[(\gamma_0^{\alpha3},\gamma_0^{3\alpha}),\kappa_0^{\alpha\beta}]\delta(x_1)\delta(x_2). \tag{3.234}
$$

For example, let us consider the action of the distortions κ^0_{11} and γ^0_{13}. Applying the direct and the inverse Fourier integral transformations (0.50) (for $n=2$) to equations (3.233), we arrive at the following results:

— for $\kappa^0_{\alpha\beta}=\kappa_0^{\alpha\beta}\delta_{\alpha1}\delta_{\beta1}\delta(x_1)\delta(x_2)$, $\gamma^0_{\alpha3}=\gamma^0_{3\alpha}=0$:

$$
\begin{aligned}
\varphi_\alpha=&-\frac{\kappa_0}{2\pi}\left\{[I_2(\nu_1)]_{,\alpha}+\frac{\gamma}{2\alpha}[I_1-I_2(\nu_1)]_{,22}\right.\\
&\left.-\epsilon_{\alpha\beta}(a^0)^{-1}[I_1-I_2(l)]_{,\beta12}\right\},\\
u_3=&\ \frac{\kappa_0\gamma}{2\pi\mu}[I_1-I_2(l)]_{,12};
\end{aligned}
\tag{3.235}
$$

— for $\gamma^0_{\alpha3}=\gamma_0^{\alpha3}\delta_{\alpha1}\delta(x_1)\delta(x_2)$, $\gamma^0_{3\alpha}=\kappa^0_{\alpha\beta}=0$:

$$
\varphi_\alpha=-\frac{\gamma_0}{4\pi}[I_1-I_2(\nu_1)]_{,\alpha2},\qquad u_3=-\frac{\gamma_0}{2\pi}I_{1,1}, \tag{3.236}
$$

where $\kappa_0^{\alpha\beta}\mapsto\kappa_0$, $\gamma_0^{\alpha3}\mapsto\gamma_0$, $\delta_{\alpha\beta}$ — Kronecker's delta, $\delta(.)$ — Dirac's delta distribution, $a^0=\dfrac{2\alpha\mu}{\gamma(\mu+\alpha)}$, $l=\sqrt{\dfrac{(\gamma+\varepsilon)(\mu+\alpha)}{4\alpha\mu}}$, the constant ν_1 is given by formula (1.189) and the symbols I_1, I_2 are determined by formula (3.79).

In [84], full distortion matrices: $\gamma^0_{\alpha 3}$, $\gamma^0_{3\alpha}$, $\kappa^0_{\alpha\beta}$ were considered. The formulae of micropolar theory (2) were obtained and on that basis formulae for the limiting theories (4), (6) and (8) (Fig. 1.1) were derived. The very results for micropolar theory (2) constitute a limiting case of fundamental solutions for u_3, φ_α when the distortion field is given by formula (3.177). Passing to the limit $\omega \to 0$, we obtain the results of micropolar elastostatics ((1) $\to$ (2), Fig. 1.1), which are compatible with the results of Gajl & Suchar [95] obtained by the method of potentials. In that sense, formulae (3.235) and (3.236) result from formulae (3.178) and (3.179), respectively. Stress fields may be found by using the constitutive relations and the geometric relations for the second plane problem. It should be noted that Nowacki [192, p. 319] applies the stress equations to determine stress fields with a distortion field.

3.4.3 Method of potentials

In [49], we consider the vector equations (1.147) and (1.237)–(1.239) for the case of plane problem with the triple (φ_α, u_3). A formal procedure leads us to the conclusion that a scalar potential Σ and vector potentials $\boldsymbol{\Psi} \stackrel{\text{def}}{=} (\Psi_1, \Psi_2, 0)$, $\mathbf{H} \stackrel{\text{def}}{=} (0, 0, H_3)$ will appear in the problem, which are related to the triple (φ_α, u_3) in the following manner:

$$u_3 = \epsilon_{\alpha\beta}\Psi_{\beta,\alpha}, \qquad \varphi_\alpha = \Sigma_{,\alpha} + \epsilon_{\alpha\beta}H_{3,\beta} \qquad \text{in } \Omega. \tag{3.237}$$

Moreover, the following condition must be fulfilled:

$$\Psi_{\alpha,\alpha} = 0 \qquad \text{in } \Omega. \tag{3.238}$$

Equations for the functions Σ, Ψ_α, H_3 are obtained from equations (1.238) and (1.239):

$$\begin{aligned} &4\alpha\nabla_1^2(\nu_1^2\nabla_1^2 - 1)\Sigma = -Y_{\alpha,\alpha}, \\ &4\alpha\mu\nabla_1^2\nabla_1^2(l^2\nabla_1^2 - 1)\Psi_\alpha = \left[(\gamma+\varepsilon)\nabla_1^2 - 4\alpha\right]\epsilon_{\alpha\gamma}X_{3,\gamma} \\ &\qquad - 2\alpha\epsilon_{\alpha\beta}\epsilon_{\delta\gamma}Y_{\gamma,\delta\beta}, \\ &4\alpha\mu\nabla_1^2\nabla_1^2(l^2\nabla_1^2 - 1)H_3 = \nabla_1^2[(\mu+\alpha)\epsilon_{\alpha\beta}Y_{\beta,\alpha} + 2\alpha X_3], \end{aligned} \tag{3.239}$$

where $\epsilon_{\alpha\beta}$ is the permutation symbol, $l^2 = \dfrac{(\gamma+\varepsilon)(\mu+\alpha)}{4\alpha\mu}$, $\nu_1^2 = \dfrac{2\gamma+\beta}{4\alpha}$, $\nabla_1^2(.) = (.)_{,\alpha\alpha}$ and X_3, Y_α are body loadings.

The potentials Σ, Ψ_α, H_3 are coupled by means of equations (1.237). We obtain here

$$
\begin{aligned}
&(\mu+\alpha)\nabla_1^2\epsilon_{\alpha\beta}\Psi_{\beta,\alpha} - 2\alpha\nabla_1^2 H_3 + \nabla_1^2\chi^\star = 0,\\
&4\alpha(\nu_1^2\nabla_1^2 - 1)\Sigma_{,\alpha} + 4\alpha(k^2\nabla_1^2 - 1)\epsilon_{\alpha\beta}H_{3,\beta} - 2\alpha\nabla_1^2\Psi_\alpha \\
&\qquad + \sigma_{,\alpha} + \epsilon_{\alpha\beta}\eta_{3,\beta} = 0,
\end{aligned}
\tag{3.240}
$$

where $k^2 = \dfrac{\gamma+\varepsilon}{4\alpha}$.

The functions σ, $\chi^\star$, η_3 are given by formula (3.186). It will be shown that the representation (3.237) and equations (3.238)–(3.240) can be simplified by eliminating the vector $\boldsymbol{\Psi}$ from our considerations. First, we fulfil the field condition (3.238) by introducing a new function Ψ: $\Psi_\alpha = -\epsilon_{\alpha\beta}\Psi_{,\beta}$. From $(3.237)_1$ we obtain $u_3 = \nabla_1^2\Psi$. Now, equation $(3.240)_1$ takes the form

$$
\nabla_1^2[(\mu+\alpha)u_3 - 2\alpha H_3 + \chi^\star] = 0,
\tag{3.241}
$$

from which we obtain

$$
u_3 = \frac{2\alpha}{\mu+\alpha}H_3 - \frac{1}{\mu+\alpha}\chi^\star.
\tag{3.242}
$$

It remains to consider equation $(3.240)_2$. Eliminating the expression $\nabla_1^2\Psi_\alpha$ from $(3.240)_2$, we arrive at the conclusion that the expressions

$$
\begin{aligned}
&4\alpha(\nu_1^2\nabla_1^2 - 1)\Sigma + \sigma,\\
&\frac{4\alpha\mu}{\mu+\alpha}(l^2\nabla_1^2 - 1)H_3 - \frac{2\alpha}{\mu+\alpha}\chi^\star + \eta_3
\end{aligned}
$$

are coupled harmonic functions. Thus we obtain the Cauchy-Riemann conditions

$$
\begin{aligned}
&[4\alpha(\nu_1^2\nabla_1^2 - 1)\Sigma + \sigma]_{,\alpha}\\
&\qquad + \epsilon_{\alpha\beta}\left[\frac{4\alpha\mu}{\mu+\alpha}(l^2\nabla_1^2 - 1)H_3 - \frac{2\alpha}{\mu+\alpha}\chi^\star + \eta_3\right]_{,\beta} = 0
\end{aligned}
\tag{3.243}
$$

and equations for the functions Σ, H_3 in Ω, namely:

$$
\begin{aligned}
&4\alpha\nabla_1^2(\nu_1^2\nabla_1^2 - 1)\Sigma + Y_{\alpha,\alpha} = 0,\\
&2\mu\nabla_1^2(l^2\nabla_1^2 - 1)H_3 + \frac{\mu+\alpha}{2\alpha}\epsilon_{\alpha\beta}Y_{\alpha,\beta} - X_3 = 0.
\end{aligned}
\tag{3.244}
$$

The representation for φ_α, u_3 is expressed by formulae $(3.237)_2$ and (3.242). We shall make here the following observations. Thus, let us emphasize that in the dynamical problem, starting from equations (3.183), (3.184) and the vector equations for potentials, (1.147) and (1.151)–(1.153), we derived (in two different ways) the same representation (formulae (3.185) and (3.188) or (3.195)) and equations (formulae (3.189) or (3.196)). We will show now that the representation $(3.237)_2$ and (3.242), as well as equations (3.243) and (3.244), result from the dynamical problem. Replacing in formulae (3.195), (3.196) the dynamical operators by the static ones, we obtain in Ω the representation

$$\begin{aligned} 2\alpha\varphi_\alpha &= \psi^\star_{,\alpha} + \epsilon_{\alpha\beta}[(\mu+\alpha)\nabla_1^2\chi + \chi^\star]_{,\beta}, \\ u_3 &= \nabla_1^2\chi \end{aligned} \tag{3.245}$$

and the Cauchy-Riemann conditions

$$\begin{aligned} &[2(\nu_1^2\nabla_1^2 - 1)\psi^\star + \sigma]_{,\alpha} \\ &\quad + \epsilon_{\alpha\beta}\left[2\mu(l^2\nabla_1^2 - 1)\nabla_1^2\chi + 2(k^2\nabla_1^2 - 1)\chi^\star + \eta_3\right]_{,\beta} = 0. \end{aligned} \tag{3.246}$$

Recall that $\psi^\star$, χ are the functions sought for, while σ, $\chi^\star$, η_3 are connected with the body loadings Y_α, X_3 by means of equations (3.186). In the dynamical problem, in the procedure based on equations (3.183) and (3.184), relation $(3.245)_2$ follows from the Stokes-Helmholtz decomposition of the vector $\mathbf{u}$. Further, one eliminates the function χ since we arrive at the relation

$$u_3 = \frac{1}{2\mu}\Phi^\star, \tag{3.247}$$

and now, we obtain from $(3.245)_2$ and (3.247):

$$\Phi^\star = 2\mu\nabla_1^2\chi. \tag{3.248}$$

The function $\Phi^\star$ has been introduced in such a way that one of the compatibility equations (3.184), expressed in terms of stresses as follows:

$$(\sigma_{31} + \sigma_{13})_{,2} - (\sigma_{23} + \sigma_{32})_{,1} = 0,$$

is satisfied identically: $\sigma_{3\alpha} + \sigma_{\alpha 3} = \Phi^\star_{,\alpha}$. Using the geometric equations (3.187), we obtain $(2\mu u_3 - \Phi^\star)_{,\alpha} = 0$, which leads to relation (3.247). If we now assume that equations (3.245) and (3.246) are a result of dynamical considerations based on the equations for potentials (1.147) and (1.151), then elimination of the function χ by assuming relation (3.248) is quite obvious, since in formulae (3.245) and (3.246) χ appears only under the symbol of the Laplacian ∇_1^2. Thus we obtain from relations (3.245)–(3.248) the representation

$$\begin{aligned} 2\alpha\varphi_\alpha &= \psi^\star_{,\alpha} + \epsilon_{\alpha\beta}(\mu_1^2\Phi^\star + \chi^\star)_{,\beta}, \\ 2\mu u_3 &= \Phi^\star \end{aligned} \tag{3.249}$$

and the Cauchy-Riemann conditions

$$
\begin{aligned}
&[2(\nu_1^2\nabla_1^2 - 1)\psi^\star + \sigma]_{,\alpha} \\
&\qquad + \epsilon_{\alpha\beta}\left[(l^2\nabla_1^2 - 1)\Phi^\star + 2(k^2\nabla_1^2 - 1)\chi^\star + \eta_3\right]_{,\beta} = 0,
\end{aligned}
\tag{3.250}
$$

where

$$
\mu_1^2 = \frac{\mu+\alpha}{2\mu}, \qquad k^2 = \frac{\gamma+\varepsilon}{4\alpha}, \qquad l^2 = \frac{(\gamma+\varepsilon)(\mu+\alpha)}{4\alpha\mu}, \qquad \nu_1^2 = \frac{2\gamma+\beta}{4\alpha}.
$$

Now, disregarding body loadings and adopting the following change of notation:

$$
\psi^\star \mapsto 2\alpha\Phi, \qquad \Phi^\star \mapsto \frac{4\alpha\mu}{\mu+\alpha}\Psi
$$

we obtain from (3.249) and (3.250) the representation and the equations of Nowacki [188], derived directly from equations (3.217):

$$
\begin{aligned}
&\varphi_\alpha = \Phi_{,\alpha} + \epsilon_{\alpha\beta}\Psi_{,\beta}, \qquad u_3 = \frac{2\alpha}{\mu+\alpha}\Psi, \\
&(\nu_1^2\nabla_1^2 - 1)\Phi_{,\alpha} + \frac{\mu}{\mu+\alpha}\epsilon_{\alpha\beta}(l^2\nabla_1^2 - 1)\Psi_{,\beta} = 0, \\
&\nabla_1^2(\nu_1^2\nabla_1^2 - 1)\Phi = 0, \qquad \nabla_1^2(l^2\nabla_1^2 - 1)\Psi = 0.
\end{aligned}
\tag{3.251}
$$

If one adopts a change of notation in formulae (3.249) and (3.250), namely $(\psi^\star, \Phi^\star) \mapsto (\Phi, \Psi)$ and $(\sigma, \chi^\star, \eta_3) \mapsto (p, \varphi, q)$, then we obtain the result of Suchar [223], derived by means of stress equations. If we assume the correspondence

$$
\psi^\star \mapsto 2\alpha\Sigma, \qquad (\mu_1^2\Phi^\star + \chi^\star) \mapsto 2\alpha H_3
$$

then the representation (3.249) and equations (3.250) go over into the representation $(3.237)_2$, (3.242) and equations (3.243), respectively.

3.4.4 Potentials in polar coordinates

Disregarding body loadings, we obtain from the vector equations (1.155) the following equations for the triple $(\varphi_r, \varphi_\theta, u_z)$ in the polar coordinate system in Ω:

$$
\begin{aligned}
&(\gamma+\varepsilon)(\nabla_0^2\varphi_r - \frac{2}{r^2}\partial_\theta\varphi_\theta) \\
&\qquad - 4\alpha\varphi_r + (\beta+\gamma-\varepsilon)\partial_r\kappa + 2\alpha r^{-1}\partial_\theta u_z = 0, \\
&(\gamma+\varepsilon)(\nabla_0^2\varphi_\theta + \frac{2}{r^2}\partial_\theta\varphi_r) \\
&\qquad - 4\alpha\varphi_\theta + (\beta+\gamma-\varepsilon)r^{-1}\partial_\theta\kappa - 2\alpha\partial_r u_z = 0, \\
&(\mu+\alpha)\nabla^2 u_z + 2\alpha r^{-1}[\partial_r(r\varphi_\theta) - \partial_\theta\varphi_r] = 0,
\end{aligned}
\tag{3.252}
$$

where

$$\kappa = r^{-1}\partial_r(r\varphi_r) + r^{-1}\partial_\theta\varphi_\theta, \quad \nabla^2 = \partial_r^2 + r^{-1}\partial_r + r^{-2}\partial_\theta^2, \quad \nabla_0^2 = \nabla^2 - r^{-2}.$$

In [49], we assume the following representation for rotations:

$$\varphi_r = \partial_r\Sigma + \frac{1}{r}\partial_\theta H, \qquad \varphi_\theta = \frac{1}{r}\partial_\theta\Sigma - \partial_r H. \tag{3.253}$$

Equations $(3.252)_3$ and (3.253) yield

$$u_z = \frac{2\alpha}{\mu+\alpha}H \qquad \text{in } \Omega. \tag{3.254}$$

From the remaining two equations $(3.252)_{1,2}$ we obtain equations coupling the potentials Σ, H in Ω:

$$\begin{aligned} -\partial_r(\nu_1^2\nabla^2 - 1)\Sigma &= \frac{\mu}{\mu+\alpha}\cdot\frac{1}{r}\partial_\theta(l^2\nabla^2 - 1)H, \\ \frac{1}{r}\partial_\theta(\nu_1^2\nabla^2 - 1)\Sigma &= \frac{\mu}{\mu+\alpha}\partial_r(l^2\nabla^2 - 1)H. \end{aligned} \tag{3.255}$$

The functions Σ, H satisfy the equations

$$\nabla^2(\nu_1^2\nabla^2 - 1)\Sigma = 0, \qquad \nabla^2(l^2\nabla^2 - 1)H = 0, \tag{3.256}$$

where

$$l^2 = \frac{(\gamma+\varepsilon)(\mu+\alpha)}{4\alpha\mu}, \qquad \nu_1^2 = \frac{2\gamma+\beta}{4\alpha}.$$

In the mixed boundary value problem we add to equations (3.252)–(3.256) the following boundary conditions:

$$\begin{aligned} n_r\sigma_{rz} + n_\theta\sigma_{\theta z} &= p_z, \qquad & n_r\mu_{rr} + n_\theta\mu_{\theta r} &= m_r, \\ n_r\mu_{r\theta} + n_\theta\mu_{\theta\theta} &= m_\theta \qquad & \text{on } \partial\Omega_\sigma; & \end{aligned} \tag{3.257}$$

$$\varphi_r = \hat{\varphi}_r, \qquad \varphi_\theta = \hat{\varphi}_\theta, \qquad u_z = \hat{u}_z \qquad \text{on } \partial\Omega_u, \tag{3.258}$$

where p_z, m_r, $m_\theta : \partial\Omega_\sigma \to \mathbb{R}$ and $\hat{\varphi}_r$, $\hat{\varphi}_\theta$, $\hat{u}_z : \partial\Omega_u \to \mathbb{R}$ are given functions expressed by means of the constitutive relations and the triple $(\varphi_r, \varphi_\theta, u_z)$ in terms of the potentials Σ, H.

3.4.5 Half-space problem

For the half-space Ω in the plane strain state:

$$\Omega = \{(x_1, x_2) : x_1 > 0 \wedge |x_2| < \infty\} \tag{3.259}$$

we formulate the following boundary value problem. The system of equations (3.217) should be solved (for $Y_\alpha = X_3 = 0$) with the boundary conditions (for $x_1 = 0$)

$$u_3 = f_3, \qquad \varphi_\alpha = g_\alpha \qquad \text{on } \partial\Omega, \tag{3.260}$$

where $f_3,\ g : \partial\Omega \to \mathbb{R}$ are given functions[19]. In addition, we assume that for $R = \sqrt{x_\alpha x_\alpha} \to \infty$, $u_3 \to 0$ and $\varphi_\alpha \to 0$. We shall discuss the solution of boundary value problem for the half-space using equations (3.239) and (3.240) as well as the representation (3.237), known from the method of potentials.

Applying the Fourier transformation (0.50), (for $Y_\alpha = X_3 = 0$) with respect to the variable x_2, to equations (3.239), we obtain ordinary differential equations for the transforms $\tilde{\Sigma}$, $\tilde{\Psi}_\alpha$, $\tilde{H}_3$, namely:

$$\begin{aligned} &(\mathbb{D}_1^2 - \eta^2)(\mathbb{D}_1^2 - \xi^2)\tilde{\Sigma} = 0, \\ &(\mathbb{D}_1^2 - \xi^2)^2(\mathbb{D}_1^2 - \rho^2)(\tilde{\Psi}_\alpha, \tilde{H}_3) = (0, 0), \end{aligned} \tag{3.261}$$

where

$$\eta = \sqrt{\xi^2 + \frac{1}{\nu_1^2}}, \qquad \rho = \sqrt{\xi^2 + \frac{1}{l^2}}, \qquad l^2 = \frac{(\gamma+\varepsilon)(\mu+\alpha)}{4\alpha\mu},$$
$$\nu_1^2 = \frac{2\gamma+\beta}{4\alpha}, \qquad \mathbb{D}_1 \overset{\text{def}}{=} \frac{d}{dx_1}$$

and ξ is the transformation parameter with respect to the variable x_2. General solutions of equations (3.261), adapted to the half-space Ω, have the form:

$$\begin{aligned} &\tilde{\Sigma} = Ce^{-|\xi|x_1} + De^{-\eta x_1}, \\ &\tilde{\Psi}_\alpha = (E_\alpha + F_\alpha|\xi|x_1)e^{-|\xi|x_1} + G_\alpha e^{-\rho x_1}, \\ &\tilde{H}_3 = (J_3 + K_3|\xi|x_1)e^{-|\xi|x_1} + M_3 e^{-\rho x_1}, \end{aligned} \tag{3.262}$$

where C, D, E_α, F_α, G_α, J_3, K_3 and M_3 are integration constants to be defined.

Satisfying the transformed field conditions (3.238) and (3.240) (for $Y_\alpha = X_3 = 0$) we eliminate six constants:

$$\begin{aligned} &C = -\frac{i\xi J_3 + \xi^2 F_1}{|\xi|}, \qquad G_2 = i\frac{\rho}{\xi}G_1, \qquad K_3 = 0, \\ &E_1 = F_1 - \frac{i\xi}{|\xi|}E_2, \qquad F_2 = \frac{i\xi}{|\xi|}F_1, \qquad M_3 = -i\mu_0\xi^{-1}G_1, \end{aligned} \tag{3.263}$$

where $\mu_0 = \dfrac{2\mu}{\gamma+\varepsilon}$ and $i = \sqrt{-1}$.

There remain five constants to be determined, namely D, E_2, F_1, G_1, J_3.

[19] In [49], we also consider boundary conditions formulated in terms of stresses.

Using relations (3.262), (3.263) and the Fourier transformations (0.50), with changed notation: $\bar{A}_1 = \xi^2 F_1$, $\bar{B}_1 = -\eta D$, $\bar{C}_1 = -\mu_0 G_1$, we can uniquely determine displacements and rotations from the representation (3.237), and thus we obtain in Ω:

$$
\begin{aligned}
u_3 &= \mathcal{F}^\star \left[\!\left[i\frac{1}{\xi}\left(2\bar{A}_1 e^{-|\xi|x_1} + \alpha_0 \bar{C}_1 e^{-\rho x_1}\right); \xi \to x_2 \right]\!\right], \\
\varphi_1 &= \mathcal{F}^\star \left[\!\left[\left(\bar{A}_1 e^{-|\xi|x_1} + \bar{B}_1 e^{-\eta x_1} + \bar{C}_1 e^{-\rho x_1}\right); \xi \to x_2 \right]\!\right], \\
\varphi_2 &= \mathcal{F}^\star \left[\!\left[i\left(\frac{\xi}{|\xi|}\bar{A}_1 e^{-|\xi|x_1} + \frac{\xi}{\eta}\bar{B}_1 e^{-\eta x_1} + \frac{\rho}{\xi}\bar{C}_1 e^{-\rho x_1}\right); \xi \to x_2 \right]\!\right],
\end{aligned}
\tag{3.264}
$$

where $\alpha_0 = \dfrac{2\alpha}{\mu + \alpha}$.

Three constants, $\bar{A}_1$, $\bar{B}_1$ and $\bar{C}_1$, are determined from the three boundary conditions (3.260). Let us note that in the definition of the constants $\bar{A}_1$, $\bar{B}_1$ and $\bar{C}_1$, the constants E_2, J_3 did not appear. This means that in formulae (3.262) we can assume[20] $E_2 = J_3 = 0$.

One may also arrive at the result (3.264) by using equations (3.243), (3.244) and the representation given by $(3.237)_2$ and (3.242) (with no body loadings). It should be emphasized that formulae (3.264) are a special case of formulae (1.220) and (1.221), in which one should assume $a_1 = b_1 = d_1 = \bar{a}_1 = \bar{b}_1 = \bar{d}_1 = 0$, $\mathcal{F}^\star_{(2)} \mapsto \mathcal{F}^\star$, $\xi_3 = 0$, $\xi_2 \mapsto \xi$, $\xi \mapsto |\xi|$. The result (3.264) is also obtained in [43] by means of the equations

$$
\begin{aligned}
&\nabla_1^2 (l^2 \nabla_1^2 - 1) u_3 = 0, \\
&\nabla_1^2 (l^2 \nabla_1^2 - 1)(\nu_1^2 \nabla_1^2 - 1)\varphi_\alpha = 0.
\end{aligned}
\tag{3.265}
$$

3.4.6 The problem of singularities of physical fields in the half-space loaded on the boundary

The problem of singularities of stress fields in the plane strain state (3.259) with discontinous loadings on the boundary was investigated by Bogy & Sternberg [15], Muki & Sternberg [161], in the framework of couple-stress theory and by Dyszlewicz & Matysiak [72]–[75], Kokot & Matysiak [132] in the framework of micropolar theory. Dyszlewicz & Słotwińska [79] extended

[20] Applying the vector equations with potentials Φ, Σ, $\boldsymbol{\Psi}$, $\mathbf{H}$, (1.238) and (1.239), adapted to the solution of three-dimensional and two-dimensional boundary value problems (for plane and axially-symetric strain states), we determine uniquely only the physical fields (of displacements, rotations, stresses and strains). In the method of potentials of Nowacki (see, e.g., [188]), equations with potentials for two-dimensional problems are derived from displacement-rotation equations (for the first and second plane and axially-symmetric problems, respectively) in such a way that we obtain also unique potentials in the solutions of the boundary value problems.

and completed the investigations and conclusions concerning the singularities of physical fields in the half-space for the second problem of plane strain state. First of all, we compared the solutions obtained in the framework of micropolar theory with the analogous solutions of the limiting theories, which were lacking in the literature. The following boundary conditions are considered for the half-space Ω (3.259):

$$\sigma_{13} = -S\delta(x_2), \qquad \mu_{11} = \mu_{12} = 0 \qquad \text{on } \partial\Omega. \tag{3.266}$$

Making use of formulae (3.264), one determines a solution for the displacement u_3 and the rotations φ_α implied by the boundary conditions (3.266). The force-stresses $\sigma_{\alpha 3}$, $\sigma_{3\alpha}$ and the couple-stresses $\mu_{\alpha 3}$, μ_{33} were determined by means of the constitutive relations in Ω:

$$\begin{aligned}
\sigma_{\alpha 3} &= (\mu + \alpha)\gamma_{\alpha 3} + (\mu - \alpha)\gamma_{3\alpha}, \\
\sigma_{3\alpha} &= (\mu + \alpha)\gamma_{3\alpha} + (\mu - \alpha)\gamma_{\alpha 3}, \\
\mu_{\alpha\beta} &= (\gamma + \varepsilon)\kappa_{\alpha\beta} + (\gamma - \varepsilon)\kappa_{\beta\alpha} + \beta\kappa_{\gamma\gamma}\delta_{\alpha\beta}, \\
\mu_{33} &= \frac{\beta}{2(\gamma + \beta)}\mu_{\alpha\alpha}
\end{aligned} \tag{3.267}$$

and the geometric relations in Ω:

$$\gamma_{\alpha 3} = u_{3,\alpha} + \epsilon_{\alpha\beta}\varphi_\beta, \qquad \gamma_{3\alpha} = \epsilon_{\beta\alpha}\varphi_\beta, \qquad \kappa_{\alpha\beta} = \varphi_{\beta,\alpha}.$$

The respective formulae are obtained in the form of single Fourier integrals, which are complex functions. Extracting the real parts of solutions, one performs for the integrands certain asymptotic expansions. In this manner, from improper integrals, determining u_3, φ_α, $\sigma_{\alpha 3}$, $\sigma_{3\alpha}$, $\mu_{\alpha 3}$, μ_{33} and converging in the entire half-plane $\overline{\Omega} = \{(x_1, x_2) : x_1 \geq 0, |x_2| < \infty\}$ except maybe the point $(0,0)$, one retrieves explicitly the singular parts, which depend on the behaviour of integrands at the point $(0,0;\xi)$ for $\xi \to \infty$. We mean here those expressions in the integrands which for $x_1 = x_2 = 0$ are of order $O(\xi^{-1})$ or higher when $\xi \to \infty$. The symbol $O(\xi^{-n})$, where $n \in \mathbb{N}$, denotes an expression, which at $\xi \to \infty$ behaves like the function $g(\xi) = 1/\xi^n$. The well-known method of analysis of singularities of physical fields is due to Sternberg and was discussed in [15] and [161] in detail. The suggestion to carry it over to the area of micropolar theory was made by Olesiak[21] and it was applied and discussed in detail in [72]–[75] and [132]. Let us write formulae for displacements and rotations in the form: $u_3 = u_3^s + O(1)$, $\varphi_\alpha = \varphi_\alpha^s + O(1)$, where u_3^s, φ_α^s are singular parts of a solution for u_3, φ_α, the symbol $O(1)$ denotes the part of a solution which is bounded at $(0,0)$:

[21] In 1970–1973, at the Institute of Mechanics at the University of Warsaw, Z. Olesiak conducted a seminar on the mechanics of continuous media with microstructure.

$$u_3 = -\frac{S}{\pi\mu}\log r + \frac{\alpha}{\mu+\alpha}\cdot\frac{S}{\pi A}\left[\frac{1}{\mu}\left(a^0 - \frac{1}{2l^2}\right) + \frac{1}{\mu+\alpha}\left(\frac{2\alpha}{\gamma} - \frac{1}{2\nu_1^2}\right)\right]\log r + O(1), \tag{3.268}$$

$$\varphi_\alpha = \epsilon_{\beta\alpha}\frac{S}{2\pi\mu}\cdot\frac{x_\beta}{r^2} + \epsilon_{\alpha\beta}\frac{S\hat{B}}{2(\mu+\alpha)\pi A}\cdot\frac{x_\beta}{r^2} + O(1),$$

where

$$r = \sqrt{x_\alpha x_\alpha}, \qquad \epsilon_{12} = 1, \qquad \epsilon_{21} = -1, \qquad \epsilon_{11} = \epsilon_{22} = 0,$$

$$\nu_1^2 = \frac{2\gamma+\beta}{4\alpha}, \qquad S \in \mathbb{R}, \qquad a^0 = \frac{2\alpha\mu}{\gamma(\mu+\alpha)}, \qquad l^2 = \frac{(\gamma+\varepsilon)(\mu+\alpha)}{4\alpha\mu},$$

$$\hat{B} = \left(a^0 - \frac{1}{2l^2}\right)\frac{\mu+\alpha}{\mu} + \left(\frac{2\alpha}{\gamma} - \frac{1}{2\nu_1^2}\right),$$

$$A = \frac{2\alpha\mu}{\mu+\alpha}\cdot\frac{\gamma^2+\gamma\beta+3\varepsilon\gamma+2\varepsilon\beta}{\gamma(\gamma+\varepsilon)(\beta+2\gamma)}.$$

Now, let us make a few remarks concerning the limiting theories. In accordance with the discussion in the Introduction (formula (0.29)), the boundary conditions (3.266) take in couple-stress theory the following form:

$$s_{13} = -\overset{a}{\sigma}_{13} - S\delta(x_2), \qquad m_{11} = -\hat{\mu}, \qquad m_{12} = 0 \qquad \text{on } \partial\Omega \tag{3.269}$$

and by virtue of equations (0.28) (adapted to the second plane problem in elastostatics) imply the field of displacements

$$u_3 = \frac{S}{\pi\mu}\int_0^\infty \left\{\frac{1}{\xi}e^{-\xi x_1} - \frac{\xi}{\Delta^{0\star}}\left[(\xi^2 + a^{0\star} - \xi\rho^\star)e^{-\xi x_1} + a^{0\star}e^{-\rho^\star x_1}\right]\right\}\times \cos\xi x_2\, d\xi, \tag{3.270}$$

where

$$a^{0\star} = \frac{1}{(1+\eta)l^{\star 2}}, \qquad \Delta^{0\star} = (\xi^2 + a^{0\star})^2 - \rho^\star\xi^3, \qquad \rho^\star = \sqrt{\xi^2 + \frac{1}{l^{\star 2}}}.$$

For the field of rotations φ_α we obtain formulae satisfying the relation

$$\varphi_\alpha = \frac{S}{2\pi\mu}\int_0^\infty \epsilon_{\alpha\beta}\tilde{u}_{3,\beta}\, d\xi, \tag{3.271}$$

where $\tilde{u}_3$ is the integrand from formula (3.270). Moreover, we obtain the following results for stresses:

— force-stresses:

$$s_{\alpha 3} = -\frac{S}{\pi}\int_0^\infty e^{-\xi x_1} a_{\alpha 3}\, d\xi + \frac{S}{\pi}\int_0^\infty \frac{\xi}{\Delta^{0\star}}(\xi^2 + a^{0\star} - \xi\rho^\star)\times \left[\xi e^{-\xi x_1} + \frac{b_{\alpha 3} a^{0\star} e^{-\rho^\star x_1}}{\xi^2 + a^{0\star} - \xi\rho^\star}\right] a_{\alpha 3}\, d\xi, \tag{3.272}$$

where $a_{13} = \cos\xi x_2$, $a_{23} = \sin\xi x_2$, $b_{13} = \rho^\star$, $b_{23} = \xi$, and we do not sum over α:

— couple-stresses:

$$m_{11} = \frac{2S}{\pi a^{0\star}}\int_0^\infty \xi e^{-\xi x_1}\sin\xi x_2\, d\xi - \frac{2S}{\pi a^{0\star}}\int_0^\infty \frac{\xi^2}{\Delta^{0\star}}\left[(\xi^2 + a^{0\star} - \xi\rho^\star)\xi e^{-\xi x_1} + \rho^\star a^{0\star} e^{-\rho^\star x_1}\right]\times \sin\xi x_2\, d\xi,$$

$$m_{\alpha\beta} = -\frac{2S}{\pi a^{0\star}}\int_0^\infty \xi e^{-\xi x_1}\cos\xi x_2\, d\xi + \frac{2S}{\pi a^{0\star}}\int_0^\infty \frac{\xi}{\Delta^{0\star}}\left[\xi^2(\xi^2 + a^{0\star} - \xi\rho^\star)e^{-\xi x_1} + (\xi^2 + c_{\alpha\beta} a^{0\star})a^{0\star} e^{-\rho^\star x_1}\right]\cos\xi x_2\, d\xi, \tag{3.273}$$

where $\alpha \neq \beta$, $c_{12} = 1$, $c_{21} = \eta$.

Moreover, it holds

$$s_{\alpha 3} = s_{3\alpha}, \qquad m_{11} = -m_{22}, \qquad m_{33} = 0. \tag{3.274}$$

Using the asymptotic expansions

$$\begin{aligned}
\rho^\star &= \xi + \frac{1}{2l^{\star 2}}\cdot\frac{1}{\xi} - \frac{1}{8l^{\star 4}}\cdot\frac{1}{\xi^3} + O(\xi^{-5}), \qquad \xi\to\infty,\\
e^{-\rho^\star x_1} &= e^{-\xi x_1}\left[1 - \frac{x_1}{2l^{\star 2}}\cdot\frac{1}{\xi} + \frac{x_1^2}{8l^{\star 4}}\cdot\frac{1}{\xi^2} + O(\xi^{-3})\right], \qquad \xi\to\infty,\\
\Delta^{0\star} &= A^\star\xi^2 + B^\star + O(\xi^{-2}), \qquad \xi\to\infty,\\
\frac{1}{\Delta^{0\star}} &= \frac{1}{A^\star}\cdot\frac{1}{\xi^2} - \frac{B^\star}{A^{\star 2}}\cdot\frac{1}{\xi^4} + O(\xi^{-6}), \qquad \xi\to\infty,
\end{aligned} \tag{3.275}$$

where

$$A^\star = \frac{3-\eta}{1+\eta}\cdot\frac{1}{2l^{\star 2}}, \qquad B^\star = \frac{8 + (1+\eta)^2}{8l^{\star 4}(1+\eta)^2},$$

we arrive, by virtue of formulae (3.270)–(3.274), at the conclusion that the quantities u_3, φ_α, $s_{\alpha 3}$, m_{11}, m_{22}, m_{12}, $s_{3\alpha}$, m_{33} are bounded at $(0,0)$ (the point of application of a concentrated load S) and only the couple-stress m_{21} exhibits the logarithmic singularity:

$$m_{21} = \frac{2S}{\pi} \cdot \frac{1-\eta}{3-\eta} \log r + O(1), \tag{3.276}$$

(η is the constant of couple-stress theory and $|\eta| < 1$).

In the classical theory of elasticity, the boundary conditions (3.266) take the form

$$\sigma_{13} = -S\delta(x_2) \qquad \text{on } \partial\Omega \tag{3.277}$$

and imply the displacement field[22]:

$$u_3 = \frac{S}{\pi\mu} \int_0^\infty \frac{1}{\xi} e^{-\xi x_1} \cos \xi x_2 \, d\xi. \tag{3.278}$$

For the rotation fields φ_α formula (3.271) remains valid, where

$$\tilde{u}_3 = \xi^{-1} \exp(-\xi x_1) \cos \xi x_2.$$

The stress tensor is symmetric, i.e. $\sigma_{\alpha 3} = \sigma_{3\alpha}$. We obtain the formula

$$\sigma_{\alpha 3} = -\frac{S}{\pi} \int_0^\infty e^{-\xi x_1} d_{\alpha 3} \, d\xi, \tag{3.279}$$

where $d_{13} = \cos \xi x_2$ and $d_{23} = \sin \xi x_2$. Since it holds:

$$\int_0^\infty \frac{1}{\xi} e^{-\xi x_1} \cos \xi x_2 \, d\xi = -\log r + O(1), \text{ for } r \to 0, \tag{3.280}$$

(see Sternberg [15]), hence the displacement u_3 can be expressed by the formula

$$u_3 = -\frac{S}{\pi\mu} \log r + O(1). \tag{3.281}$$

In turn, the rotations φ_α and the stresses $\sigma_{\alpha 3}$ can be given in the form of closed formulae:

$$\varphi_\alpha = -\epsilon_{\alpha\beta} \frac{S}{2\pi\mu} \cdot \frac{x_\beta}{r^2}, \qquad \sigma_{\alpha 3} = -\frac{S}{\pi} \cdot \frac{x_\alpha}{r^2}. \tag{3.282}$$

[22] With this divergent integral are connected singularities of displacement field in the classical theory of elasticity, in the framework of the half-space problem in plane strain state subjected to the action of forces concentrated on the boundary. Compare the problem of Boussinesq-Flamant: Nowacki [175, p. 127], Fung [94, p. 268]. See also the works of Sternberg (e.g., [15]). In general, one proves that in the static plane problem for a semi-infinite elastic body with a non-zero resultant load, one obtains displacements which diverge logarithmically at infinity (see Muschelishvili [162, §36]).

Within the framework of couple-stress theory, integral formulae (3.270)–(3.274) can be also obtained from the respective integral formulae of micropolar theory for $\alpha \to \infty$. Namely, one should take the following limits: $\sqrt{\dfrac{\gamma+\varepsilon}{4\mu}} \to l^\star$, $\dfrac{\gamma-\varepsilon}{\gamma+\varepsilon} \to \eta$ and, in the case of force-stresses, separate the symmetric part (there is no need however to separate the deviatoric part from the couple-stresses $\mu_{\alpha\beta}$, since $\hat{\mu} \xrightarrow{\alpha\to\infty} 0$). If $\alpha \to 0$, then the integral formulae of micropolar theory for the displacement u_3 and force stresses are transformed into the classical integral formulae (3.278) and (3.279), respectively. Couple-stresses converge to zero in the limit. Also, the rotations φ_α converge to zero and do not lead to the classical result (3.271). However, integral formulae of couple-stress theory (3.270)–(3.272) go over for $l^\star \to 0$ as a whole into the respective formulae of the classical theory, (3.271), (3.278) and (3.279) (couple-stresses from formula (3.273) converge to zero). Let us go over to a short discussion of singular parts of solutions[23]. The results determining the orders of singularities of the physical fields are given in Table 3.1.

Table 3.1.

Physical quantities	(1)	(2)	(3)
u_3	$O(\log r)$	$O(1)$	$O(\log r)$
φ_α	$O(R^{-1})$	$O(1)$	$O(R^{-1})$
$\sigma_{\alpha 3},\ \sigma_{3\alpha}$	$O(R^{-1})$	$O(1)$	$O(R^{-1})$
μ_{21}	$O(R^{-2})$	$O(\log r)$	–
μ_{33}	$O(1)$	0	–
$\mu_{11},\ \mu_{22},\ \mu_{12}$	$O(R^{-2})$	$O(1)$	–

Numbers (1), (2) and (3) denote: micropolar theory (E-N), couple-stress theory (K-M) and the classical theory of elasticity (H), respectively

From Table 3.1 one can see compatibility of results (concerning singular parts of the solution) of the theories (1) and (3), and a considerable influence of couple-stresses in the theory (2) on the phenomenon of singularities of physical fields.

[23] See [79], [81]. For the models (E-N), (K-M), (H) and (HM), an exhaustive discussion of the results (for singular and regular parts) concerning the problem of a half-space subjected to a concentrated loading on the boundary has been presented in those works. Therefore, we limit our discussion here to certain conclusions connected with the singular parts of solutions.

A discussion of the limits for singular parts in the theories (1), (2) and (3) is quite interesting. We give here the following relations determined for singular parts:

1°. (1) $\xrightarrow{\alpha\to 0}$ (3) for $\overset{s}{u}_3$, $\overset{s}{\sigma}_{\alpha 3}$, $\overset{s}{\sigma}_{3\alpha}$. The rotations φ_α and the couple-stresses $\mu_{\alpha\beta}$, μ_{33} converge to zero.

2°. (1) $\xrightarrow{\alpha\to\infty,\beta\to\infty}$ (2), however let us recall that in the theory (2) one ses the symmetric part of the force-stress tensor ($s_{\alpha 3} = s_{3\alpha}$) and the deviatoric part of the couple-stresses $m_{\alpha\beta}$, m_{33}. Only the stress m_{21} shows the logarithmic singularity. The couple-stresses $\mu_{\alpha\beta}$ ($\alpha \neq \beta$) in the theory (1) are expressed by the formula

$$\mu_{\alpha\beta} = -\frac{S\gamma}{\pi\mu} \cdot \frac{x_1^2 - x_2^2}{r^4} + \frac{\gamma}{\mu+\alpha} \cdot \frac{S\hat{B}}{\pi A} \cdot \frac{x_1^2 - x_2^2}{r^4} + O(\log r)$$

and for the iterated limit ($\alpha \to \infty$, $\beta \to \infty$) we obtain for m_{21} formula (3.276), while for m_{12} we obtain:

$$O(\log r) \to 0, \qquad \overset{s}{\mu}_{12} \to 0, \qquad (m_{12} = O(1));$$

3°. As to the singular parts of results in the framework of the theory (2) for $l^\star \to 0$, the correspondence with the theory (3) is not preserved at all. However, recall that the full correspondence (2) $\xrightarrow{l^\star\to 0}$ (3) is preserved for the results in the integral form.

4. Hemitropic medium

4.1 Vector equations. Elastodynamics

4.1.1 Equations in displacements and rotations with body vectors. The method of direct integration

Dyszlewicz and Kołodziej [66], [65], [49] considered vector equations written in terms of displacements and rotations for an isotropic hemitropic medium in $\Omega \times \mathbf{T}_+$, namely:

$$\begin{aligned} &\vec{\Box}_2 \mathbf{u} + (\lambda + \mu - \alpha)\,\mathrm{grad}\,\mathrm{div}\,\mathbf{u} + 2\alpha\,\mathrm{curl}\,\boldsymbol{\varphi} \\ &\qquad + (\chi + \nu_0)\vec{\nabla}^2 \boldsymbol{\varphi} + (\chi + \kappa - \nu_0)\,\mathrm{grad}\,\mathrm{div}\,\boldsymbol{\varphi} + \mathbf{X} = \mathbf{0}, \\ &\vec{\Box}_4 \boldsymbol{\varphi} + (\beta + \gamma - \varepsilon)\,\mathrm{grad}\,\mathrm{div}\,\boldsymbol{\varphi} + 2\alpha\,\mathrm{curl}\,\mathbf{u} \\ &\qquad + 4\nu_0\,\mathrm{curl}\,\boldsymbol{\varphi} + (\chi + \nu_0)\vec{\nabla}^2 \mathbf{u} + (\chi + \kappa - \nu_0)\,\mathrm{grad}\,\mathrm{div}\,\mathbf{u} + \mathbf{Y} = \mathbf{0}, \end{aligned} \tag{4.1}$$

where $\mathbf{u}$, $\boldsymbol{\varphi}$ are the vectors of displacement and rotation, respectively, $\mathbf{X}$, $\mathbf{Y}$ are the body loadings vectors, $\mathbf{0}$ is the zero vector,

$$\vec{\Box}_2 = (\mu + \alpha)\vec{\nabla}^2 - \rho\partial_t^2, \qquad \vec{\Box}_4 = (\gamma + \varepsilon)\vec{\nabla}^2 - 4\alpha - J\partial_t^2,$$

$$\vec{\nabla}^2 \overset{\mathrm{def}}{=} \mathrm{grad}\,\mathrm{div} - \mathrm{curl}\,\mathrm{curl},$$

∂_t denotes the time derivative, ρ is the density, J — rotational inertia of the medium, and μ, λ, α, β, γ, ε, χ, ν_0, κ are material constants.

From the system of equations (4.1) we derived the following separated equations. Namely, for $e = \mathrm{div}\,\mathbf{u}$ and $z = \mathrm{div}\,\boldsymbol{\varphi}$ we obtain the equations in $\Omega \times \mathbf{T}_+$

$$\begin{aligned} \mathcal{L}e &= (\kappa + 2\chi)\nabla^2\,\mathrm{div}\,\mathbf{Y} - \Box_3\,\mathrm{div}\,\mathbf{X}, \\ \mathcal{L}z &= (\kappa + 2\chi)\nabla^2\,\mathrm{div}\,\mathbf{X} - \Box_1\,\mathrm{div}\,\mathbf{Y}, \end{aligned} \tag{4.2}$$

where $\mathcal{L}$ is a scalar operator of the form

$$\mathcal{L} = \Box_1\Box_3 - (\kappa + 2\chi)^2\nabla^4, \tag{4.3}$$

and $\square_1 = (\lambda + 2\mu)\nabla^2 - \rho\partial_t^2$, $\square_3 = (\beta + 2\gamma)\nabla^2 - 4\alpha - J\partial_t^2$, $\nabla^2 \stackrel{\text{def}}{=} \operatorname{div}\operatorname{grad}$, the bi-Laplacian operator $\nabla^4 = \nabla^2\nabla^2$.

Furthermore, we obtain the following equations in $\Omega \times \mathbf{T}_+$:

— for the displacement vector $\mathbf{u}$,

$$\begin{aligned}
&\vec{\mathcal{L}}\Big\{\vec{\mathcal{D}}^2 + 16[\nu_0\vec{\square}_2 - \alpha(\chi+\nu_0)\vec{\nabla}^2]^2\vec{\nabla}^2\Big\}\mathbf{u} \\
&\quad = \vec{\mathcal{L}}\Big\{\vec{\mathcal{D}} - 4[\nu_0\vec{\square}_2 - \alpha(\chi+\nu_0)\vec{\nabla}^2]\operatorname{curl}\Big\}\times \\
&\qquad [(\chi+\nu_0)\vec{\nabla}^2\mathbf{Y} - \vec{\square}_4\mathbf{X} + 2\alpha\operatorname{curl}\mathbf{Y} - 4\nu_0\operatorname{curl}\mathbf{X}] \\
&\qquad + \vec{\mathcal{D}}[(\lambda+\mu-\alpha)\vec{\square}_4 - (\chi+\nu_0)(\kappa+\chi-\nu_0)\vec{\nabla}^2 - 4\alpha^2]\times \\
&\qquad \operatorname{grad}\operatorname{div}[\vec{\square}_3\mathbf{X} - (\kappa+2\chi)\vec{\nabla}^2\mathbf{Y}] \\
&\qquad - 16[\nu_0\vec{\square}_2 - \alpha(\chi+\nu_0)\vec{\nabla}^2]^2\operatorname{grad}\operatorname{div}[\vec{\square}_3\mathbf{X} - (\kappa+2\chi)\vec{\nabla}^2\mathbf{Y}] \\
&\qquad + \vec{\mathcal{D}}[(\kappa+\chi-\nu_0)\vec{\square}_4 - (\beta+\gamma-\varepsilon)(\chi+\nu_0)\vec{\nabla}^2]\times \\
&\qquad \operatorname{grad}\operatorname{div}[\vec{\square}_1\mathbf{Y} - (\kappa+2\chi)\vec{\nabla}^2\mathbf{X}],
\end{aligned} \tag{4.4}$$

where $\vec{\mathcal{L}}$, $\vec{\mathcal{D}}$ are vector operators of the form:

$$\begin{aligned}
\vec{\mathcal{L}} &= \vec{\square}_1\vec{\square}_3 - (\kappa+2\chi)^2\vec{\nabla}^4, \\
\vec{\mathcal{D}} &= \vec{\square}_2\vec{\square}_4 - (\chi+\nu_0)^2\vec{\nabla}^4 + 4\alpha^2\vec{\nabla}^2,
\end{aligned} \tag{4.5}$$

where, in turn,

$$\vec{\square}_1 = (\lambda+2\mu)\vec{\nabla}^2 - \rho\partial_t^2, \qquad \vec{\square}_3 = (\beta+2\gamma)\vec{\nabla}^2 - 4\alpha - J\partial_t^2,$$

$$\vec{\nabla}^4 = \vec{\nabla}^2\vec{\nabla}^2;$$

— for the rotation vector $\boldsymbol{\varphi}$,

$$\begin{aligned}
&\vec{\mathcal{L}}\Big\{\vec{\mathcal{D}}^2 + 16[\nu_0\vec{\square}_2 - \alpha(\chi+\nu_0)\vec{\nabla}^2]^2\vec{\nabla}^2\Big\}\boldsymbol{\varphi} \\
&\quad = \vec{\mathcal{L}}\Big\{\vec{\mathcal{D}} - 4[\nu_0\vec{\square}_2 - \alpha(\chi+\nu_0)\vec{\nabla}^2]\operatorname{curl}\Big\}\times \\
&\qquad [(\chi+\nu_0)\vec{\nabla}^2\mathbf{X} - \vec{\square}_2\mathbf{Y} + 2\alpha\operatorname{curl}\mathbf{X}] \\
&\qquad + \vec{\mathcal{D}}[(\beta+\gamma-\varepsilon)\vec{\square}_2 - (\chi+\nu_0)(\kappa+\chi-\nu_0)\vec{\nabla}^2 - 4\alpha^2]\times \\
&\qquad \operatorname{grad}\operatorname{div}[\vec{\square}_1\mathbf{Y} - (\kappa+2\chi)\vec{\nabla}^2\mathbf{X}] \\
&\qquad - 16[\nu_0\vec{\square}_2 - \alpha(\chi+\nu_0)\vec{\nabla}^2]^2\operatorname{grad}\operatorname{div}[\vec{\square}_1\mathbf{Y} - (\kappa+2\chi)\vec{\nabla}^2\mathbf{X}] \\
&\qquad + \vec{\mathcal{D}}[(\kappa+\chi-\nu_0)\vec{\square}_2 - (\lambda+\mu-\alpha)(\chi+\nu_0)\vec{\nabla}^2]\times \\
&\qquad \operatorname{grad}\operatorname{div}[\vec{\square}_3\mathbf{X} - (\kappa+2\chi)\vec{\nabla}^2\mathbf{Y}].
\end{aligned} \tag{4.6}$$

It should be stressed that the derived vector equations (4.4) and (4.6) have been transformed (it concerns their right-hand sides) in such a way that connecting them (for $\kappa = \chi = \nu_0 = 0$) with analogous equations of micropolar theory, (1.19) and (1.20), is straightforward. Now, the vector equations (1.19), (1.20) and (1.17), (1.18) in micropolar theory correspond in a simple way to the equations of a hemitropic medium. Moreover, the correspondence of these equations with the equations concerning three- and two-dimensional problems (in the coordinate systems $Ox_1x_2x_3$ and Ox_1x_2, (r, θ, z), (r, θ), respectively) presented here, is preserved. This problem has an interesting instructive aspect and that is why it will be discussed in a separate subsection.

4.1.2 The vector of Galerkin-Cauchy type

Let us write the vector equations (4.4) and (4.6) in the operator form:

$$\begin{aligned} &\vec{\mathcal{L}}\Big\{\vec{\mathcal{D}}^2 + 16[\nu_0 \vec{\Box}_2 - \alpha(\chi+\nu_0)\vec{\nabla}^2]^2 \vec{\nabla}^2\Big\}\mathbf{u} = \vec{\mathcal{L}}_{Xu}\mathbf{X} + \vec{\mathcal{L}}_{Yu}\mathbf{Y}, \\ &\vec{\mathcal{L}}\Big\{\vec{\mathcal{D}}^2 + 16[\nu_0 \vec{\Box}_2 - \alpha(\chi+\nu_0)\vec{\nabla}^2]^2 \vec{\nabla}^2\Big\}\boldsymbol{\varphi} = \vec{\mathcal{L}}_{X\varphi}\mathbf{X} + \vec{\mathcal{L}}_{Y\varphi}\mathbf{Y}, \end{aligned} \tag{4.7}$$

where the operators $\vec{\mathcal{L}}_{Xu}$, $\vec{\mathcal{L}}_{X\varphi}$, $\vec{\mathcal{L}}_{Yu}$, $\vec{\mathcal{L}}_{Y\varphi}$ are derived from equations (4.4) and (4.6), taking into account that the indices u, φ determine the equations for $\mathbf{u}$, $\boldsymbol{\varphi}$, respectively, whereas the indices X, Y — the vectors $\mathbf{X}$, $\mathbf{Y}$, to which the operator is referred. In [49], we introduced the following representation in $\Omega \times \mathbf{T}_+$:

$$\begin{aligned} &\mathbf{u} = -\vec{\mathcal{L}}_{Xu}\mathbf{F} - \vec{\mathcal{L}}_{X\varphi}\mathbf{M}, \qquad \boldsymbol{\varphi} = -\vec{\mathcal{L}}_{X\varphi}\mathbf{F} - \vec{\mathcal{L}}_{Y\varphi}\mathbf{M}, \\ &\vec{\mathcal{L}}_{X\varphi} = \vec{\mathcal{L}}_{Yu}, \end{aligned} \tag{4.8}$$

where the vectors $\mathbf{F}$, $\mathbf{M}$ satisfy the equations

$$\begin{aligned} &\vec{\mathcal{L}}\Big\{\vec{\mathcal{D}}^2 + 16[\nu_0 \vec{\Box}_2 - \alpha(\chi+\nu_0)\vec{\nabla}^2]^2 \vec{\nabla}^2\Big\}\mathbf{F} + \mathbf{X} = \mathbf{0}, \\ &\vec{\mathcal{L}}\Big\{\vec{\mathcal{D}}^2 + 16[\nu_0 \vec{\Box}_2 - \alpha(\chi+\nu_0)\vec{\nabla}^2]^2 \vec{\nabla}^2\Big\}\mathbf{M} + \mathbf{Y} = \mathbf{0}. \end{aligned} \tag{4.9}$$

The integral representation (4.8) and equations (4.9) with vectors $\mathbf{F}$, $\mathbf{M}$ constitute a generalization of the vectors of Galerkin-Iacovache type of micropolar elastostodynamics[1]; for $\kappa = \chi = \nu_0 = 0$, we obtain from (4.8) and (4.9) the representation (1.137) and equations (1.138), respectively. The problem of completness of the representation (4.8) has not been studied yet and still remains open.

[1] Gurtin [105, p. 235] calls the analog of $(4.8)_1$ in the classical theory of elasticity — the representation of Cauchy-Kovalevski-Somigliana. Nowacki [180, p. 505] calls the classical representation

$$\mathbf{u} = \frac{\lambda + 2\mu}{\mu}\left(\vec{\Box}_1^2\boldsymbol{\varphi} - \frac{\lambda+\mu}{\lambda+2\mu}\,\mathrm{grad\,div}\,\boldsymbol{\varphi}\right)$$

the solution of Iacovache (in elastostatics $\boldsymbol{\varphi}$ becomes the Galerkin vector).

4.2 Three-dimensional problems. Elastostatics

4.2.1 Vector equations in displacements and rotations. Separated equations

In [66] and [49], we considered a system of equations in displacements and rotations for an isotropic hemitropic medium within the framework of thermoelastostatics. In elastostatics it has the following form in Ω:

$$\begin{aligned}
&(\mu+\alpha)\vec{\nabla}^2\mathbf{u} + (\lambda+\mu-\alpha)\,\mathrm{grad\,div}\,\mathbf{u} + 2\alpha\,\mathrm{curl}\,\boldsymbol{\varphi}\\
&\qquad + (\chi+\nu_0)\vec{\nabla}^2\boldsymbol{\varphi} + (\kappa+\chi-\nu_0)\,\mathrm{grad\,div}\,\boldsymbol{\varphi} + \mathbf{X} = \mathbf{0},\\
&[(\gamma+\varepsilon)\vec{\nabla}^2 - 4\alpha]\boldsymbol{\varphi} + (\beta+\gamma-\varepsilon)\,\mathrm{grad\,div}\,\boldsymbol{\varphi} + 2\alpha\,\mathrm{curl}\,\mathbf{u}\\
&\qquad + 4\nu_0\,\mathrm{curl}\,\boldsymbol{\varphi} + (\chi+\nu_0)\vec{\nabla}^2\mathbf{u} + (\kappa+\chi-\nu_0)\,\mathrm{grad\,div}\,\mathbf{u} + \mathbf{Y} = \mathbf{0}.
\end{aligned} \tag{4.10}$$

We shall separate the system of equations (4.10). For $e = \mathrm{div}\,\mathbf{u}$ and $z = \mathrm{div}\,\boldsymbol{\varphi}$, we obtain the equations

$$\begin{aligned}
&\nabla^2 L^2 e = (\kappa+2\chi)\nabla^2\,\mathrm{div}\,\mathbf{Y} - [(\beta+2\gamma)\nabla^2 - 4\alpha]\,\mathrm{div}\,\mathbf{X},\\
&L^2 z = (\kappa+2\chi)\,\mathrm{div}\,\mathbf{X} - (\lambda+2\mu)\,\mathrm{div}\,\mathbf{Y},
\end{aligned} \tag{4.11}$$

where L^2 is a scalar operator of the form

$$L^2 = [(\lambda+2\mu)(\beta+2\gamma) - (\kappa+2\chi)^2]\nabla^2 - 4\alpha(\lambda+2\mu). \tag{4.12}$$

The equation for the displacement vector $\mathbf{u}$ has the form

$$\begin{aligned}
&\vec{\nabla}^4[\vec{\mathrm{D}}^4 + 16(\nu_0\mu-\alpha\chi)^2\vec{\nabla}^2]\vec{\mathrm{L}}^2\mathbf{u}\\
&\quad = -\,8(\nu_0\mu-\alpha\chi)\vec{\mathrm{L}}^2\vec{\nabla}^2\,\mathrm{grad}(\alpha\,\mathrm{div}\,\mathbf{Y} - 2\nu_0\,\mathrm{div}\,\mathbf{X})\\
&\qquad + [\vec{\mathrm{D}}^4 + 16(\nu_0\mu-\alpha\chi)^2\vec{\nabla}^2]\times\\
&\qquad \mathrm{grad}\Big\{(\kappa+2\chi)\nabla^2\,\mathrm{div}\,\mathbf{Y} - [(\beta+2\gamma)\nabla^2 - 4\alpha]\,\mathrm{div}\,\mathbf{X}\Big\}\\
&\qquad + \vec{\mathrm{L}}^2[\vec{\mathrm{D}}^2\,\mathrm{curl} + 4(\nu_0\mu-\alpha\chi)\vec{\nabla}^2]\times\\
&\qquad \Big[\!\Big[\mathrm{curl}\Big\{[(\gamma+\varepsilon)\vec{\nabla}^2 - 4\alpha]\mathbf{X} - (\chi+\nu_0)\vec{\nabla}^2\mathbf{Y}\Big\}\\
&\qquad + 2\alpha\vec{\nabla}^2\mathbf{Y} - 4\nu_0\vec{\nabla}^2\mathbf{X}\Big]\!\Big],
\end{aligned} \tag{4.13}$$

where $\vec{\nabla}^4 = \vec{\nabla}^2\vec{\nabla}^2$, $\vec{\mathrm{D}}^4 = \vec{\mathrm{D}}^2\vec{\mathrm{D}}^2$, and $\vec{\mathrm{L}}^2$, $\vec{\mathrm{D}}^2$ are vector operators of the form

$$\begin{aligned}
&\vec{\mathrm{L}}^2 = [(\lambda+2\mu)(\beta+2\gamma) - (\kappa+2\chi)^2]\vec{\nabla}^2 - 4\alpha(\lambda+2\mu),\\
&\vec{\mathrm{D}}^2 = [(\mu+\alpha)(\gamma+\varepsilon) - (\chi+\nu_0)^2]\vec{\nabla}^2 - 4\alpha\mu,
\end{aligned} \tag{4.14}$$

where $\vec{\nabla}^2$ — the vector Laplace operator.

For the rotation vector $\boldsymbol{\varphi}$ we obtain

$$\begin{aligned}
&\overrightarrow{\nabla}^2[\overrightarrow{\mathrm{D}}^4 + 16(\nu_0\mu - \alpha\chi)^2\overrightarrow{\nabla}^2]\overrightarrow{\mathrm{L}}^2\boldsymbol{\varphi} \\
&\quad = -8\alpha(\nu_0\mu - \alpha\chi)\overrightarrow{\mathrm{L}}^2 \operatorname{grad}\operatorname{div}\mathbf{X} \\
&\qquad + [\overrightarrow{\mathrm{D}}^4 + 16(\nu_0\mu - \alpha\chi)^2\overrightarrow{\nabla}^2]\times \\
&\qquad \operatorname{grad}[(\kappa + 2\chi)\operatorname{div}\mathbf{X} - (\lambda + 2\mu)\operatorname{div}\mathbf{Y}] \\
&\qquad + 4(\nu_0\mu - \alpha\chi)\overrightarrow{\mathrm{L}}^2\Big\{\overrightarrow{\nabla}^2\operatorname{curl}[(\mu+\alpha)\mathbf{Y} - (\chi + \nu_0)\mathbf{X}] + 2\alpha\overrightarrow{\nabla}^2\mathbf{X}\Big\} \\
&\qquad + \overrightarrow{\mathrm{L}}^2\overrightarrow{\mathrm{D}}^2\operatorname{curl}\Big\{\operatorname{curl}[(\mu+\alpha)\mathbf{Y} - (\chi+\nu_0)\mathbf{X}] + 2\alpha\mathbf{X}\Big\}.
\end{aligned} \tag{4.15}$$

For $\kappa = \chi = \nu_0 = 0$, equations (4.13) and (4.15) are transformed into equations (1.171) and (1.172).

4.2.2 Generalized vectors of Galerkin type

In [49], we write the vector equations (4.13) and (4.15) in the operator form:

$$\begin{aligned}
\overrightarrow{\nabla}^4[\overrightarrow{\mathrm{D}}^4 + 16(\nu_0\mu - \alpha\chi)^2\overrightarrow{\nabla}^2]\overrightarrow{\mathrm{L}}^2\mathbf{u} &= \overrightarrow{\mathrm{L}}_{Xu}\mathbf{X} + \overrightarrow{\mathrm{L}}_{Yu}\mathbf{Y}, \\
\overrightarrow{\nabla}^2[\overrightarrow{\mathrm{D}}^4 + 16(\nu_0\mu - \alpha\chi)^2\overrightarrow{\nabla}^2]\overrightarrow{\mathrm{L}}^2\boldsymbol{\varphi} &= \overrightarrow{\mathrm{L}}_{X\varphi}\mathbf{X} + \overrightarrow{\mathrm{L}}_{Y\varphi}\mathbf{Y}
\end{aligned} \tag{4.16}$$

and assume the following representation[2] :

$$\begin{aligned}
\mathbf{u} &= -\overrightarrow{\mathrm{L}}_{Xu}\mathbf{F} - \overrightarrow{\mathrm{L}}_{X\varphi}\mathbf{M}, \\
\boldsymbol{\varphi} &= -\overrightarrow{\nabla}^2\overrightarrow{\mathrm{L}}_{X\varphi}\mathbf{F} - \overrightarrow{\mathrm{L}}_{Y\varphi}\mathbf{M},
\end{aligned} \tag{4.17}$$

where the operators $\overrightarrow{\mathrm{L}}_{Xu}$, $\overrightarrow{\mathrm{L}}_{X\varphi}$, $\overrightarrow{\mathrm{L}}_{Yu}$, $\overrightarrow{\mathrm{L}}_{Y\varphi}$ are determined from equations (4.13) and (4.15) according to the rule given in Subsection 4.1.2. Making use of (4.16) and (4.17), we obtain equations satisfied by the vectors $\mathbf{F}$, $\mathbf{M}$ in Ω:

$$\begin{aligned}
\overrightarrow{\nabla}^4[\overrightarrow{\mathrm{D}}^4 + 16(\nu_0\mu - \alpha\chi)^2\overrightarrow{\nabla}^2]\overrightarrow{\mathrm{L}}^2\mathbf{F} + \mathbf{X} &= \mathbf{0}, \\
\overrightarrow{\nabla}^2[\overrightarrow{\mathrm{D}}^4 + 16(\nu_0\mu - \alpha\chi)^2\overrightarrow{\nabla}^2]\overrightarrow{\mathrm{L}}^2\mathbf{M} + \mathbf{Y} &= \mathbf{0}.
\end{aligned} \tag{4.18}$$

The representation (4.17) and equations (4.18) constitute analogues of the representation (1.227) and equations (1.228), respectively, of micropolar elastostatics, which refer to the generalized vector of Galerkin type.

---

[2] In the representation (4.17) we took into account that the operators $\overrightarrow{\mathrm{L}}_{X\varphi}$ and $\overrightarrow{\mathrm{L}}_{Yu}$ are related by means of the following formula:

$$\overrightarrow{\mathrm{L}}_{Yu} = \overrightarrow{\nabla}^2\overrightarrow{\mathrm{L}}_{X\varphi}.$$

For $\kappa = \chi = \nu_0 = 0$, one can obtain (1.227) and (1.228) from (4.17) and (4.18), respectively. Let us add that completness of both the representation (4.17) and the representation in micropolar theory (1.227) has not been studied.

4.2.3 The problem of isotropic hemitropic half-space

In [66] and [49], we considered (in the Cartesian coordinate system $Ox_1x_2x_3$) the problem of the half-space (1.209) with the boundary conditions (1.210) and (1.213). The starting point for our considerations is given by equations (4.13) and (4.15) (without body vectors), which are transformed into the following form in Ω:

$$\begin{aligned}&\vec{\nabla}^2[(l_h^2\vec{\nabla}^2 - 1)^2 + \hat{k}^2\vec{\nabla}^2](\nu_h^2\vec{\nabla}^2 - 1)\boldsymbol{\varphi} = \mathbf{0},\\&\vec{\nabla}^4[(l_h^2\vec{\nabla}^2 - 1)^2 + \hat{k}^2\vec{\nabla}^2](\nu_h^2\vec{\nabla}^2 - 1)\mathbf{u} = \mathbf{0},\end{aligned} \tag{4.19}$$

where

$$\vec{\nabla}^2 \equiv (\nabla^2, \nabla^2, \nabla^2), \qquad \nabla^2(.) = (.)_{,ii}, \qquad \hat{k} = \frac{\nu_0\mu - \alpha\chi}{\alpha\mu}.$$

Moreover, we introduced the notation:

$$l_h^2 = l^2 - \frac{(\chi + \nu_0)^2}{4\alpha\mu}, \qquad \nu_h^2 = \nu_1^2 - \frac{(\kappa + 2\chi)^2}{4\alpha(\lambda + 2\mu)}, \tag{4.20}$$

where $l^2 = \dfrac{(\gamma + \varepsilon)(\mu + \alpha)}{4\alpha\mu}$, $\nu_1^2 = \dfrac{2\gamma + \beta}{4\alpha}$.

Let us justify this notation by invoking the energy inequalities (0.17) and the discussion carried out in the Introduction, concerning the expression for the free energy (0.16) of an isotropic hemitropic medium. Taking into account the matrices of material constants [I]–[IV], we obtain, by virtue of Sylvester's theorem[3],

$$\begin{aligned}&\Delta_1 = \lambda + 2\mu > 0,\\&\Delta_2 = (\lambda + 2\mu)(\beta + 2\gamma) - (\kappa + 2\chi)^2 > 0.\end{aligned} \tag{4.21}$$

Consider the determinants Δ_7, Δ_8:

$$\Delta_7 = (\mu + \alpha)\Delta_6 > 0 \Longrightarrow (\mu + \alpha) > 0,$$

$$\Delta_8 = 4\alpha\mu\Delta_6 > 0 \Longrightarrow \alpha\mu > 0.$$

Thus it holds

[3] See Mostowski & Stark [160, p. 349]; Δ_i ($i = 1, 2, \ldots, 18$) denotes the determinant of a square $i \times i$ matrix. For each i, the inequality must hold: $\Delta_i > 0$.

$$\mu > 0, \qquad \alpha > 0. \tag{4.22}$$

Consider also the determinant Δ_{12}:

$$\begin{aligned} \Delta_{12} &= [(\mu+\alpha)(\gamma+\varepsilon) - (\chi+\nu_0)^2]\Delta_{10} > 0 \\ &\Longrightarrow (\mu+\alpha)(\gamma+\varepsilon) - (\chi+\nu_0)^2 > 0. \end{aligned} \tag{4.23}$$

Inequalities (4.21)–(4.23) justify the notation for l_h^2, ν_h^2 in formula (4.20).

In view of Boggio's theorem, we assume the solution of equations (4.19) in the form

$$\mathbf{u} = \mathbf{u}_1 + \mathbf{u}_2 + \mathbf{u}_3, \qquad \boldsymbol{\varphi} = \boldsymbol{\varphi}_1 + \boldsymbol{\varphi}_2 + \boldsymbol{\varphi}_3, \tag{4.24}$$

where the vectors $\mathbf{u}_i$, $\boldsymbol{\varphi}_i$ satisfy in Ω the equations

$$\begin{aligned} &(\nu_h^2 \overrightarrow{\nabla}^2 - 1)(\mathbf{u}_1, \boldsymbol{\varphi}_1) = (\mathbf{0}, \mathbf{0}), \\ &\overrightarrow{\nabla}^2(\overrightarrow{\nabla}^2 \mathbf{u}_3, \boldsymbol{\varphi}_3) = (\mathbf{0}, \mathbf{0}) \end{aligned} \tag{4.25}$$

and

$$[(l_h^2 \overrightarrow{\nabla}^2 - 1)^2 + \hat{k}^2 \overrightarrow{\nabla}^2](\mathbf{u}_2, \boldsymbol{\varphi}_2) = (\mathbf{0}, \mathbf{0}). \tag{4.26}$$

Having applied the Fourier integral transformations (0.50) for $n = 2$ (over the variables x_2, x_3) to equations (4.25), we obtain for the half-space Ω (1.209) the following formulae (in the form of transforms)[4]:

$$\begin{aligned} \tilde{\boldsymbol{\varphi}}_1 &= \mathbf{A}_1 e^{-\eta_h x_1}, \qquad & \tilde{\mathbf{u}}_1 &= \mathbf{B}_1 e^{-\eta_h x_1}, \\ \tilde{\boldsymbol{\varphi}}_3 &= \mathbf{A}_3 e^{-\xi x_1}, \qquad & \tilde{\mathbf{u}}_3 &= (\mathbf{B}_3 + \mathbf{B}_4 x_1) e^{-\xi x_1}, \end{aligned} \tag{4.27}$$

where $\eta_h = \sqrt{\xi^2 + \nu_h^{-2}}$, $\xi = \sqrt{\xi_N \xi_N}$, and $\mathbf{A}_1$, $\mathbf{A}_3$, $\mathbf{B}_1$, $\mathbf{B}_3$, $\mathbf{B}_4$ — constant vectors which need to be determined.

Recall that in the case of a micropolar centrosymmetric medium, the equations analogous to equations (4.19) for the vectors $\mathbf{u}$, $\boldsymbol{\varphi}$ have the form

$$\overrightarrow{\nabla}^2(l^2 \overrightarrow{\nabla}^2 - 1)[\overrightarrow{\nabla}^2 \mathbf{u}, (\nu_1^2 \overrightarrow{\nabla}^2 - 1)\boldsymbol{\varphi}] = (\mathbf{0}, \mathbf{0}), \tag{4.28}$$

where $l^2 = \dfrac{(\gamma+\varepsilon)(\mu+\alpha)}{4\alpha\mu}$, $\nu_1^2 = \dfrac{2\gamma+\beta}{4\alpha}$.

For the considered boundary value problem for the half-space Ω, equations (4.28) lead to the formulae

$$\begin{aligned} \tilde{\mathbf{u}} &= (\mathbf{A} + \mathbf{B} x_1) e^{-\xi x_1} + \mathbf{C} e^{-\rho x_1}, \\ \tilde{\boldsymbol{\varphi}} &= \mathbf{D} e^{-\xi x_1} + \mathbf{E} e^{-\eta x_1} + \mathbf{F} e^{-\rho x_1}, \end{aligned} \tag{4.29}$$

where $\rho = \sqrt{\xi^2 + l^{-2}}$, $\eta = \sqrt{\xi^2 + \nu_1^{-2}}$.

---

[4] In the sequel, partial solutions for the vectors $\mathbf{u}$, $\boldsymbol{\varphi}$ (e.g., the solution of the equation $(\nu_h^2 \overrightarrow{\nabla}^2 - 1)\mathbf{u}_1 = \mathbf{0}$ of the form $\tilde{\mathbf{u}}_1 = \mathbf{B}_1 \exp(-\eta_h x_1)$) will be simply called solutions.

The constant vectors **A**–**F** need to be determined. Integrals of the same type as in formulae (4.29) appear in $(4.27)_{3,4}$. The operator $(\nu_h^2 \overrightarrow{\nabla}^2 - 1)$ constitutes a generalization of the operator $(\nu_1^2 \overrightarrow{\nabla}^2 - 1)$, since the following implication holds: $(\kappa \to 0 \wedge \chi \to 0) \Longrightarrow [(\nu_h^2 \overrightarrow{\nabla}^2 - 1) \to (\nu_1^2 \overrightarrow{\nabla}^2 - 1)]$, thus a generalization of the solution $\mathbf{E}e^{-\eta x_1}$ for the vector φ in a centrosymmetric medium is given by formula $(4.27)_{1,2}$ for **u** and φ in a hemitropic medium. The solutions $(\mathbf{C}, \mathbf{F})e^{-\rho x_1}$ in formulae (4.29) are implemented by the operator $(l^2 \overrightarrow{\nabla}^2 - 1)$. A generalization of this operator to a hemitropic medium is the operator

$$(l_h^2 \overrightarrow{\nabla}^2 - 1) : (\chi \to 0 \wedge \nu_0 \to 0) \Longrightarrow [(l_h^2 \overrightarrow{\nabla}^2 - 1) \to (l^2 \overrightarrow{\nabla}^2 - 1)],$$

which is a part of the more complex one:

$$\overrightarrow{\mathrm{L}}^h = [(l_h^2 \overrightarrow{\nabla}^2 - 1)^2 + \hat{k}^2 \overrightarrow{\nabla}^2].$$

When searching for a counterpart of the solution $(\mathbf{C}, \mathbf{F})e^{-\rho x_1}$ for a hemitropic medium, one needs to verify if the equation

$$l_h^4 (\overrightarrow{\nabla}^2)^2 + (\hat{k}^2 - 2 l_h^2) \overrightarrow{\nabla}^2 + 1 = 0, \tag{4.30}$$

quadratic with respect to the Laplacian $\overrightarrow{\nabla}^2$ and with real-valued coefficients, has real-valued roots. The answer is negative since the discriminant Δ of equation (4.30) is negative:

$$\Delta = (\mu + \alpha) \hat{k}^2 [\mu \nu_0^2 + \alpha \chi^2 - \alpha \mu (\gamma + \varepsilon)] < 0. \tag{4.31}$$

To justify inequality (4.31), we consider the determinant $\Delta_9 = \Delta_6 W_3$, where

$$W_3 = \begin{vmatrix} \mu + \alpha & \mu - \alpha & \chi + \nu_0 \\ \mu - \alpha & \mu + \alpha & \chi - \nu_0 \\ \chi + \nu_0 & \chi - \nu_0 & \gamma + \varepsilon \end{vmatrix} = 4[\alpha \mu (\gamma + \varepsilon) - \mu \nu_0^2 - \alpha \chi^2].$$

From the condition $\Delta_9 > 0$ we obtain the inequality

$$\alpha \mu (\gamma + \varepsilon) - \mu \nu_0^2 - \alpha \chi^2 > 0. \tag{4.32}$$

Since $\hat{k}^2 > 0$, hence, taking into account inequalities (4.22) and (4.32), we obtain inequality (4.31). Thus, the operator $\overrightarrow{\mathrm{L}}^h = [(l_h^2 \overrightarrow{\nabla}^2 - 1)^2 + \hat{k}^2 \overrightarrow{\nabla}^2]$ can be only represented in the form: $(\overrightarrow{\nabla}^2 - \lambda)(\overrightarrow{\nabla}^2 - \bar{\lambda})$, where λ, $\bar{\lambda}$ are conjugate complex roots of equation (4.30). Therefore, in the solution for the hemitropic half-space, we will not obtain a solution of the type $(\mathbf{C}_1, \mathbf{F}_1)e^{-\rho_h x_1}$, i.e. an analogue of the solution of the type $(\mathbf{C}, \mathbf{F})e^{-\rho x_1}$, where $\rho_h = \sqrt{\xi^2 + l_h^{-2}}$ and $\mathbf{C}_1$, $\mathbf{F}_1$ are constant vectors. Having applied the Fourier transformation

$(0.50)_1$ to equations (4.26), we obtain linear homogeneous ordinary differential equations of order four with constant coefficients (for the transforms $\tilde{\mathbf{u}}_2$, $\tilde{\boldsymbol{\varphi}}_2$) in the form

$$\begin{aligned}&\Big\{l_h^4\mathbb{D}_1^4+[\hat{k}^2-2l_h^2(1+l_h^2\xi^2)]\mathbb{D}_1^2\\&\qquad+(l_h^4\xi^4+2l_h^2\xi^2-\hat{k}^2\xi^2+1)\Big\}\,\tilde{\mathbf{u}}_2=0,\\&\Big\{l_h^4\mathbb{D}_1^4+[\hat{k}^2-2l_h^2(1+l_h^2\xi^2)]\mathbb{D}_1^2\\&\qquad+(l_h^4\xi^4+2l_h^2\xi^2-\hat{k}^2\xi^2+1)\Big\}\,\tilde{\boldsymbol{\varphi}}_2=0,\end{aligned}\tag{4.33}$$

where $\mathbb{D}_1 \overset{\text{def}}{=} \dfrac{d}{dx_1}$.

The characteristic polynomial of this equations has four complex roots: $p+iq$, $p-iq$, $-p+iq$, $-p-iq$, where $p,\ q\in\mathbb{R}$ and are equal to[5]

$$\begin{aligned}p&=2^{-1/2}\Big\{[(\xi^2+l_h^{-2})^2-\hat{k}^2\xi^2l_h^{-4}]^{1/2}+(\xi^2+l_h^{-2}-\tfrac{1}{2}\hat{k}^2l_h^{-4})\Big\}^{1/2},\\q&=2^{-1/2}\Big\{[(\xi^2+l_h^{-2})^2-\hat{k}^2\xi^2l_h^{-4}]^{1/2}-(\xi^2+l_h^{-2}-\tfrac{1}{2}\hat{k}^2l_h^{-4})\Big\}^{1/2}.\end{aligned}\tag{4.34}$$

It means that the solution for the vectors $\mathbf{u}_2$, $\boldsymbol{\varphi}_2$ (for the half-space Ω) is obtained in the form

$$(\tilde{\boldsymbol{\varphi}}_2,\tilde{\mathbf{u}}_2)=[(\mathbf{A}_5,\mathbf{B}_7)\cos qx_1+(\mathbf{A}_6,\mathbf{B}_8)\sin qx_1]e^{-px_1},\tag{4.35}$$

where the constant vectors $\mathbf{A}_5$, $\mathbf{A}_6$, $\mathbf{B}_7$, $\mathbf{B}_8$ need to be determined.
It should be stressed that the solutions of the type

$$(\sin qx_1,\cos qx_1)\exp(-px_1),$$

appearing in formula (4.35), do not appear in a micropolar centrosymmetric medium. Moreover, note that the following implication holds:

$$(\chi\to 0\wedge \nu_0\to 0)\Longrightarrow(q\to 0\wedge p\to\rho),$$

and the solutions (4.35) are transformed into the solutions for a centrosymmetric medium. Further implications resulting from the presence of the operator $\overrightarrow{\mathrm{L}}^h=[(l_h^2\overrightarrow{\nabla}^2-1)^2+\hat{k}\overrightarrow{\nabla}^2]$ in the equations for a hemitropic medium, as a generalization of the operator $(l^2\overrightarrow{\nabla}^2-1)$ (which appears in the equations of a micropolar centrosymmetric medium), will be discussed on the basis of [70].

[5] It can be verified that the characteristic equation of (4.33) is a bi-quadratic equation and that, after reduction to a quadratic one (with real coefficients), we obtain:

$$\Delta=\hat{k}^2(\hat{k}^2-4l_h^2)<0,$$

due to inequality (4.32).

4.2.4 Potentials of Nowacki in elastostatics and thermoelastostatics

In [49], we introduced the potentials (1.147):

$$\begin{aligned} \mathbf{u} &= \operatorname{grad} \Phi + \operatorname{curl} \boldsymbol{\Psi}, \qquad \operatorname{div} \boldsymbol{\Psi} = 0, \\ \varphi &= \operatorname{grad} \Sigma + \operatorname{curl} \mathbf{H}, \qquad \operatorname{div} \mathbf{H} = 0, \end{aligned} \tag{4.36}$$

into the system of vector equations (4.10), where we assumed $\mathbf{X} \mapsto \mathbf{X}^\star$ and $\mathbf{Y} \mapsto \mathbf{Y}^\star$. As a result, we obtain the following vector equations for Φ, Σ; $\boldsymbol{\Psi}$, $\mathbf{H}$:

$$\begin{aligned} &\vec{\nabla}^2 \Big\{ \operatorname{grad}[(2\mu + \lambda)\Phi + (2\chi + \kappa)\Sigma] \\ &\quad + \operatorname{curl}[(\mu + \alpha)\boldsymbol{\Psi} + (\chi + \nu_0)\mathbf{H}] - 2\alpha \mathbf{H} \Big\} + \mathbf{X}^\star = \mathbf{0}, \\ &\operatorname{grad}[4\alpha(\nu_1^2 \nabla^2 - 1)\Sigma + (2\chi + \kappa)\nabla^2 \Phi] \\ &\quad + \operatorname{curl}[4\alpha(k^2 \vec{\nabla}^2 - 1)\mathbf{H} + (\chi + \nu_0)\vec{\nabla}^2 \boldsymbol{\Psi}] \\ &\quad - \vec{\nabla}^2 (2\alpha \boldsymbol{\Psi} + 4\nu_0 \mathbf{H}) + \mathbf{Y}^\star = \mathbf{0} \qquad \text{in } \Omega, \end{aligned} \tag{4.37}$$

where $\nu_1^2 = \dfrac{2\gamma + \beta}{4\alpha}$, $k^2 = \dfrac{\gamma + \varepsilon}{4\alpha}$.

It holds

$$\mathbf{X}^\star = \mathbf{X} - \eta \operatorname{grad} \theta, \qquad \mathbf{Y}^\star = \mathbf{Y} - \zeta \operatorname{grad} \theta, \tag{4.38}$$

where η, ζ are thermoelastic material constants, $\mathbf{X}$, $\mathbf{Y}$ are the vectors of body loadings, and $\theta = T - T_0$ defines the temperature field subject to the equation of heat conduction in Ω:

$$\operatorname{div} \operatorname{grad} \theta = -\frac{W}{\lambda_0}, \tag{4.39}$$

where, in turn, W describes the action of heat sources, and λ_0 denotes the coefficient of heat conduction. Furthermore, we separate the system of equations (4.37) and obtain a scalar system of equations coupling only the scalar potentials Φ, Σ, of the following form in Ω:

$$\begin{aligned} &\nabla^2 \nabla^2 [(2\mu + \lambda)\Phi + (2\chi + \kappa)\Sigma] + \operatorname{div} \mathbf{X}^\star = 0, \\ &\nabla^2 [4\alpha(\nu_1^2 \nabla^2 - 1)\Sigma + (2\chi + \kappa)\nabla^2 \Phi] + \operatorname{div} \mathbf{Y}^\star = 0, \end{aligned} \tag{4.40}$$

in which the temperature field θ appears. Separated equations for Φ, Σ have the form

$$\begin{aligned} &\nabla^2 L^2 \Sigma = (2\chi + \kappa) \operatorname{div} \mathbf{X}^\star - (2\mu + \lambda) \operatorname{div} \mathbf{Y}^\star, \\ &\nabla^2 \nabla^2 L^2 \Phi = (2\chi + \kappa) \operatorname{div} \mathbf{Y}^\star - 4\alpha(\nu_1^2 \nabla^2 - 1) \operatorname{div} \mathbf{X}^\star. \end{aligned} \tag{4.41}$$

The operators $\vec{\nabla}^2$, ∇^2 are the vector- and the scalar Laplace operator, respectively, the operator L^2 is given by formula (4.12). Furthermore, from the system of equations (4.37), we obtain a system of vector equations (in which there is no temperature field) coupling the vector potentials $\boldsymbol{\Psi}$, $\mathbf{H}$ in Ω:

$$\begin{aligned}
&\vec{\nabla}^2\Big\{\vec{\nabla}^2[(\mu+\alpha)\boldsymbol{\Psi}+(\chi+\nu_0)\mathbf{H}]+2\alpha\,\mathrm{curl}\,\mathbf{H}\Big\}=\mathrm{curl}\,\mathbf{X},\\
&\vec{\nabla}^2[4\alpha(k^2\vec{\nabla}^2-1)\mathbf{H}\\
&\qquad+(\chi+\nu_0)\vec{\nabla}^2\boldsymbol{\Psi}+\mathrm{curl}(2\alpha\boldsymbol{\Psi}+4\nu_0\mathbf{H})]=\mathrm{curl}\,\mathbf{Y}.
\end{aligned}\tag{4.42}$$

We shall also give the separated equations for the vectors $\boldsymbol{\Psi}$, $\mathbf{H}$ (here, under the assumption that there exists a potential of the body vector fields). Namely, the potentials $\boldsymbol{\Psi}$, $\mathbf{H}$ satisfy

$$\begin{aligned}
&\vec{\nabla}^2\vec{\nabla}^2[\vec{\mathrm{D}}^4+16(\nu_0\mu-\alpha\chi)^2\vec{\nabla}^2]\boldsymbol{\Psi}=\mathbf{0},\\
&\vec{\nabla}^2\vec{\nabla}^2[\vec{\mathrm{D}}^4+16(\nu_0\mu-\alpha\chi)^2\vec{\nabla}^2]\mathbf{H}=\mathbf{0}.
\end{aligned}\tag{4.43}$$

The operator $\vec{\mathrm{D}}^4=\vec{\mathrm{D}}^2\vec{\mathrm{D}}^2$, where $\vec{\mathrm{D}}^2$ is given by equation $(4.14)_2$. Applications of relations and equations (4.36)–(4.43), known from the method of potentials, shall be discussed in the following subsections.

4.2.5 Method of potentials. Certain solutions in $\mathbb{R}^3$

By means of equations (4.41) and (4.43) (in the rectangular system of coordinates $Ox_1x_2x_3$), we consider in [65] and [49] a particular problem referring to a given temperature field or a body loadings field in the infinite space.

1. The temperature field: from equations (4.43) we obtain $\mathbf{H}\equiv\mathbf{0}$ and $\boldsymbol{\Psi}\equiv\mathbf{0}$. Equations (4.41) take the form

$$\begin{aligned}
&\nabla^2(\nu_h^2\nabla^2-1)\Sigma=\tau_1\nabla^2\theta,\\
&\nabla^2\nabla^2(\nu_h^2\nabla^2-1)\Phi=(\tau_2\nabla^2-\tau_3)\nabla^2\theta,
\end{aligned}\tag{4.44}$$

where $\nabla^2(.)=(.)_{,ii}$,

$$\tau_1=\frac{\zeta}{4\alpha}-\eta\tau_4,\qquad \tau_2=\frac{\eta(2\gamma+\beta)}{4\alpha(\lambda+2\mu)}-\zeta\tau_4,\qquad \tau_3=\frac{\eta}{\lambda+2\mu}.$$

The constant ν_h^2 is expressed by formula $(4.20)_2$, whereas τ_4 is given by

$$\tau_4=\frac{2\chi+\kappa}{4\alpha(\lambda+2\mu)}.\tag{4.45}$$

The temperature θ satisfies the Poisson equation:

$$\nabla^2\theta = -\frac{W}{\lambda_0}. \tag{4.46}$$

Displacements and rotations are determined from the formulae $u_i = \Phi_{,i}$, $\varphi_i = \Sigma_{,i}$.

Consider two cases of temperature fields:

A. Thermal inclusion

$$\theta = \hat{T}_0\delta(x_1)\delta(x_2)\delta(x_3). \tag{4.47}$$

In this case equation $(4.44)_1$ for Σ is an equation of Helmholtz type:

$$(\nu_h^2\nabla^2 - 1)\Sigma = \tau_1\hat{T}_0\delta(\mathbf{x}) \tag{4.48}$$

with the solution

$$\Sigma = -\frac{\tau_1\hat{T}_0}{4\pi\nu_h^2}\cdot\frac{e^{-R/\nu_h}}{R}, \tag{4.49}$$

where $R = \sqrt{x_i x_i}$ and $\delta(\mathbf{x})$ denotes the three-dimensional Dirac delta distribution: $\delta(x_1)\delta(x_2)\delta(x_3)$.

The equation for Φ given by equation $(4.44)_2$ takes the form

$$\nabla^2(\nu_h^2\nabla^2 - 1)\Phi = (\tau_2\nabla^2 - \tau_3)\hat{T}_0\delta(\mathbf{x}) \tag{4.50}$$

with the solution

$$\Phi = -\frac{\hat{T}_0}{4\pi}\left[\frac{\tau_2}{\nu_h^2}\cdot\frac{e^{-R/\nu_h}}{R} + \tau_3\left(\frac{1 - e^{-R/\nu_h}}{R}\right)\right]. \tag{4.51}$$

For $(\nu_0 \to 0 \wedge \chi \to 0 \wedge \kappa \to 0 \wedge \zeta \to 0)$ it holds:

$$\tau_1 \to 0 \wedge \eta \to \nu_T \wedge \tau_2\nu_h^{-2} \to \tau_3 \to m$$

and we obtain $\Sigma \equiv 0$, $\varphi_i \equiv 0$ and then

$$\Phi = -\frac{m\hat{T}_0}{4\pi}\cdot\frac{1}{R}, \qquad u_i = \Phi_{,i}, \tag{4.52}$$

where $m = \dfrac{\nu_T}{2\mu + \lambda}$, $\nu_T = (3\lambda + 2\mu)\alpha_t$, and α_t is the coefficient of linear thermal expansion of the medium.

Formula (4.52) gives the result for a micropolar centrosymmetric medium and it is at the same time the result for the Hookean solid;

B. Point heat source

$$W = W_0\delta(x_1)\delta(x_2)\delta(x_3). \tag{4.53}$$

Equations (4.44) take the form

$$\begin{aligned}\nabla^2(\nu_h^2\nabla^2 - 1)\Sigma &= -\frac{\tau_1 W_0}{\lambda_0}\delta(\mathbf{x}),\\ \nabla^2\nabla^2(\nu_h^2\nabla^2 - 1)\Phi &= -\frac{W_0}{\lambda_0}(\tau_2\nabla^2 - \tau_3)\delta(\mathbf{x})\end{aligned} \tag{4.54}$$

and lead to the following result:

$$\begin{aligned}\Sigma &= -\frac{\tau_1 W_0}{4\pi\lambda_0}\left(\frac{1 - e^{-R/\nu_h}}{R}\right),\\ \Phi &= -\frac{W_0}{4\pi\lambda_0}\left[-\tfrac{1}{2}\tau_3 R + (\tau_2 - \tau_3\nu_h^2)\left(\frac{1 - e^{-R/\nu_h}}{R}\right)\right].\end{aligned} \tag{4.55}$$

For $\nu_0 = \chi = \kappa = \zeta = 0$ we have $\tau_1 = 0$, $\tau_2 = \tau_3\nu_h^2$. Thus we obtain the result for a micropolar centrosymmetric medium and also for the Hookean solid ($\Sigma \equiv 0$, $\varphi_i \equiv 0$):

$$\Phi = \frac{mW_0}{8\pi\lambda_0}R, \qquad u_i = \Phi_{,i}. \tag{4.56}$$

Note that in the case of a hemitropic medium, a rotation field φ_i appears in both cases **A** and **B**.

2. Potential field of body loadings: from equations (4.43) we obtain $\mathbf{H} \equiv \mathbf{0}$ and $\boldsymbol{\Psi} \equiv \mathbf{0}$. For the case $\mathbf{Y} \equiv \mathbf{0}$, we get from (4.41) the equations

$$\begin{aligned}&(\nu_h^2\nabla^2 - 1)\Sigma = \tau_4\varphi,\\ &(\lambda + 2\mu)\nabla^2(\nu_h^2\nabla^2 - 1)\Phi = -(\nu_1^2\nabla^2 - 1)\varphi,\end{aligned} \tag{4.57}$$

where φ is the potential of the body forces field $\mathbf{X}$ and $\nu_1^2 = \dfrac{2\gamma + \beta}{4\alpha}$. When $\mathbf{X} \equiv \mathbf{0}$, we obtain from (4.41) equations of Helmholtz type:

$$(\nu_h^2\nabla^2 - 1)(\Sigma, \Phi) = \left(-\frac{1}{4\alpha}, \tau_4\right)\psi, \tag{4.58}$$

where ψ is the potential of the body moments field $\mathbf{Y}$, with a solution in the form[6]

$$(\Sigma, \Phi) = \left(\frac{1}{16\pi\alpha\nu_h^2}, -\frac{\tau_4}{4\pi\nu_h^2}\right)\int_\Omega \frac{\psi(\mathbf{x}')e^{-|\mathbf{x}-\mathbf{x}'|/\nu_h}}{|\mathbf{x}-\mathbf{x}'|}\,d\Omega(\mathbf{x}'), \tag{4.59}$$

where $\mathbf{x} \overset{\text{def}}{=} (x_1, x_2, x_3)$, $\mathbf{x}' \overset{\text{def}}{=} (x_1', x_2', x_3')$.

[6] See Nowacki [192, p. 292].

3. The potentials φ, ψ satisfy the Poisson equation $\nabla^2(\varphi,\psi)=(\overset{\circ}{X},\overset{\circ}{Y})\delta(\mathbf{x})$:

(a) $\operatorname{div}\mathbf{X} = \overset{\circ}{X}\,\delta(x_1)\delta(x_2)\delta(x_3), \qquad \psi\equiv 0.$
We obtain

$$\begin{aligned}\Sigma &= \frac{\tau_4\,\overset{\circ}{X}}{4\pi}\left(\frac{1-e^{-R/\nu_h}}{R}\right),\\ \Phi &= \frac{\overset{\circ}{X}}{4\pi}\left[\frac{1}{2(\lambda+2\mu)}R - 4\alpha\tau_4^2\left(\frac{1-e^{-R/\nu_h}}{R}\right)\right].\end{aligned} \tag{4.60}$$

The following implication holds: $[(\chi\to 0\wedge\nu_0\to 0\wedge\kappa\to 0)\Longrightarrow(\tau_4\to 0)]$ and we arrive at the result for both the micropolar centrosymmetric medium and Hooke's medium ($\Sigma\equiv 0$, $\varphi_i\equiv 0$):

$$\Phi = \frac{\overset{\circ}{X}}{8\pi(\lambda+2\mu)}R, \qquad u_i = \Phi_{,i}; \tag{4.61}$$

(b) $\operatorname{div}\mathbf{Y} = \overset{\circ}{Y}\,\delta(x_1)\delta(x_2)\delta(x_3), \ \ \varphi\equiv 0.$
We obtain

$$\Sigma = -\frac{\overset{\circ}{Y}}{16\pi\alpha}\left(\frac{1-e^{-R/\nu_h}}{R}\right), \qquad \Phi = \frac{\tau_4\,\overset{\circ}{Y}}{4\pi}\left(\frac{1-e^{-R/\nu_h}}{R}\right), \tag{4.62}$$

where $\overset{\circ}{X}$, $\overset{\circ}{Y}$ are constants.

When $\chi=\nu_0=\kappa=0$, it holds $\nu_h\to\nu_1$ (and it is the constant given by formula (1.189)), and the micropolar centrosymmetric medium behaves as the hypothetical medium ($\Phi\equiv 0$, $u_i\equiv 0$):

$$\Sigma = -\frac{\overset{\circ}{Y}}{16\pi\alpha}\left(\frac{1-e^{-R/\nu_1}}{R}\right), \qquad \varphi_i = \Sigma_{,i}. \tag{4.63}$$

4.2.6 Hypothetical hemitropic medium

In [65] and [49], we consider certain limiting cases of an isotropic hemitropic medium, understood in the mathematical sense, which have a physical motivation. Such media will be called hypothetical media. Here, we shall consider the cases of a hypothetical hemitropic medium.

1. $\nu_0\mu=\alpha\chi$:
equations for displacement and rotation vectors with body loadings are obtained from equations (4.13) and (4.15). Consider here the homogeneous equations

$$\begin{aligned}\overrightarrow{\nabla}^2(l_h^2\overrightarrow{\nabla}^2-1)(\nu_h^2\overrightarrow{\nabla}^2-1)\overrightarrow{\nabla}^2\mathbf{u}&=\mathbf{0},\\ \overrightarrow{\nabla}^2(l_h^2\overrightarrow{\nabla}^2-1)(\nu_h^2\overrightarrow{\nabla}^2-1)\boldsymbol{\varphi}&=\mathbf{0}.\end{aligned} \tag{4.64}$$

We can see that in this case the operator $(l_h^2\overrightarrow{\nabla}^2 - 1)$, which is a generalization of $(l^2\overrightarrow{\nabla}^2 - 1)$, appears in the explicit form. However, a micropolar (non-centrosymmetric) medium has 8, and not 9, constants: α, β, γ, ε, μ, λ as well as κ and ν_0 or χ. For the problem of the half-space Ω (1.209) the solution in the transformed region will have the form

$$\begin{aligned}\tilde{\varphi} &= \mathbf{A}_1 e^{-\xi x_1} + \mathbf{A}_2 e^{-\eta_h x_1} + \mathbf{A}_3 e^{-\rho_h x_1},\\ \tilde{\mathbf{u}} &= (\mathbf{B}_1 + \mathbf{B}_2 x_1)e^{-\xi x_1} + \mathbf{B}_3 e^{-\eta_h x_1} + \mathbf{B}_4 e^{-\rho_h x_1},\end{aligned} \tag{4.65}$$

where the vectors $\mathbf{A}_1$–$\mathbf{A}_3$, $\mathbf{B}_1$–$\mathbf{B}_4$ should be determined.

In the solution for $\tilde{\mathbf{u}}$, a qualitative difference between formulae (4.65) and (4.29) constitutes the solution $\mathbf{B}_3\exp(-\eta_h x_1)$. For a centrosymmetric medium it holds:

$$\begin{array}{llll}\eta_h \to \eta, & \rho_h \to \rho, & & \\ \mathbf{A}_1 \mapsto \mathbf{D}, & \mathbf{A}_2 \mapsto \mathbf{E}, & \mathbf{A}_3 \mapsto \mathbf{F}, & \\ \mathbf{B}_1 \mapsto \mathbf{A}, & \mathbf{B}_2 \mapsto \mathbf{B}, & \mathbf{B}_3 \mapsto \mathbf{0}, & \mathbf{B}_4 \mapsto \mathbf{C}\end{array}$$

and formulae (4.65) are transformed into (4.29).

2. $\nu_0 \to 0 \wedge \chi \to 0$:

within the framework of elastostatics a medium is described by 7 elasticity constants: α, β, γ, ε, μ, λ and κ. In the problem of thermoelastostatics we have, in addition, the constants ζ, η.

Equations separated from the temperature field have the following form:

— for the vector $\mathbf{u}$ in Ω,

$$\begin{aligned}&4\alpha\mu(\lambda+2\mu)\overrightarrow{\nabla}^4(l^2\overrightarrow{\nabla}^2 - 1)(\hat{\nu}_h^2\overrightarrow{\nabla}^2 - 1)\mathbf{u} = \mu(l^2\overrightarrow{\nabla}^2 - 1)\times\\ &\quad \operatorname{grad}\left[\kappa\nabla^2\left(\operatorname{div}\mathbf{Y} + \frac{\zeta W}{\lambda_0}\right) - 4\alpha(\nu_1^2\nabla^2 - 1)\left(\operatorname{div}\mathbf{X} + \frac{\eta W}{\lambda_0}\right)\right]\\ &\quad + (\lambda+2\mu)(\hat{\nu}_h^2\overrightarrow{\nabla}^2 - 1)\operatorname{curl}\left[4\alpha(k^2\overrightarrow{\nabla}^2 - 1)\operatorname{curl}\mathbf{X} + 2\alpha\overrightarrow{\nabla}^2\mathbf{Y}\right],\end{aligned} \tag{4.66}$$

where

$$\begin{aligned}&\hat{\nu}_h^2 = \nu_1^2 - \kappa\hat{\tau}_4, \ \nu_1^2 = \frac{2\gamma+\beta}{4\alpha},\\ &\hat{\tau}_4 = \frac{\kappa}{4\alpha(\lambda+2\mu)}, \ l^2 = \frac{(\gamma+\varepsilon)(\mu+\alpha)}{4\alpha\mu},\end{aligned}$$

$\overrightarrow{\nabla}^4 = \overrightarrow{\nabla}^2\overrightarrow{\nabla}^2$, and $\overrightarrow{\nabla}^2$, ∇^2 are the vector- and the scalar Laplace operators, respectively, $k^2 = \dfrac{\gamma+\varepsilon}{4\alpha}$;

— for the vector φ in Ω,

$$
\begin{aligned}
&4\alpha\mu(\lambda+2\mu)\vec{\nabla}^2(l^2\vec{\nabla}^2-1)(\hat{\nu}_h^2\vec{\nabla}^2-1)\boldsymbol{\varphi} \\
&\quad= \mu(l^2\vec{\nabla}^2-1)\,\mathrm{grad}\Bigg[\kappa\left(\mathrm{div}\,\mathbf{X}+\frac{\eta W}{\lambda_0}\right) \\
&\qquad-(\lambda+2\mu)\left(\mathrm{div}\,\mathbf{Y}+\frac{\zeta W}{\lambda_0}\right)\Bigg] \\
&\qquad+(\lambda+2\mu)(\hat{\nu}_h^2\vec{\nabla}^2-1)\,\mathrm{curl}\,[(\mu+\alpha)\,\mathrm{curl}\,\mathbf{Y}+2\alpha\mathbf{X}]\,.
\end{aligned}
\tag{4.67}
$$

Consider the following fundamental solutions for $\mathbf{u}$, $\boldsymbol{\varphi}$ in the case of action of a temperature field and a field of body loadings in $\mathbb{R}^3$ (in the rectangular coordinate system $Ox_1x_2x_3$):

(a) a point heat source (4.53); using Fourier transformations (0.50), we obtain from equations (4.66) and (4.67) the following formulae:

$$
\begin{aligned}
\varphi_i &= -\frac{\hat{\tau}_1 W_0}{4\pi\lambda_0}\left(\frac{1-e^{-R/\nu_h}}{R}\right)_{,i}, \\
u_i &= -\frac{W_0}{4\pi\lambda_0}\left[-\tfrac{1}{2}\tau_3 R+(\hat{\tau}_2-\tau_3\hat{\nu}_h^2)\left(\frac{1-e^{-R/\hat{\nu}_h}}{R}\right)\right]_{,i},
\end{aligned}
\tag{4.68}
$$

where

$$
\hat{\tau}_1=\frac{\zeta}{4\alpha}-\eta\hat{\tau}_4,\quad \hat{\tau}_2=\frac{\eta(2\gamma+\beta)}{4\alpha(\lambda+2\mu)}-\zeta\hat{\tau}_4,\quad \tau_3=\frac{\eta}{\lambda+2\mu},\quad R=\sqrt{x_ix_i}.
$$

Note that formulae (4.68) are a special case of (4.55).

(b) body loadings of the form (1.182). Let us write equations (4.66) and (4.67) in the operator form ($\eta=\zeta=0$):

$$
\begin{aligned}
\vec{\nabla}^4(l^2\vec{\nabla}^2-1)(\hat{\nu}_h^2\vec{\nabla}^2-1)\mathbf{u} &= \vec{\mathrm{N}}_{Xu}\mathbf{X}+\vec{\mathrm{N}}_{Yu}\mathbf{Y}, \\
\vec{\nabla}^2(l^2\vec{\nabla}^2-1)(\hat{\nu}_h^2\vec{\nabla}^2-1)\boldsymbol{\varphi} &= \vec{\mathrm{N}}_{X\varphi}\mathbf{X}+\vec{\mathrm{N}}_{Y\varphi}\mathbf{Y}.
\end{aligned}
\tag{4.69}
$$

We introduce vectors of Galerkin type $\hat{\mathbf{F}}$, $\hat{\mathbf{M}}$, assuming the representation

$$
\begin{aligned}
\mathbf{u} &= -\vec{\mathrm{N}}_{Xu}\hat{\mathbf{F}}-\vec{\mathrm{N}}_{X\varphi}\hat{\mathbf{M}}, \\
\boldsymbol{\varphi} &= -\vec{\nabla}^2\vec{\mathrm{N}}_{X\varphi}\hat{\mathbf{F}}-\vec{\mathrm{N}}_{Y\varphi}\hat{\mathbf{M}},
\end{aligned}
\tag{4.70}
$$

where the operators $\vec{\mathrm{N}}_{Xu}$, $\vec{\mathrm{N}}_{X\varphi}$, $\vec{\mathrm{N}}_{Y\varphi}$ are determined from equations (4.66) and (4.67) by putting $\eta=\zeta=0$ and using formulae (4.69). The vectors $\hat{\mathbf{F}}$, $\hat{\mathbf{M}}$ satisfy the equations

$$
\begin{aligned}
\vec{\nabla}^4(l^2\vec{\nabla}^2-1)(\hat{\nu}_h^2\vec{\nabla}^2-1)\hat{\mathbf{F}}+\mathbf{X} &= \mathbf{0}, \\
\vec{\nabla}^2(l^2\vec{\nabla}^2-1)(\hat{\nu}_h^2\vec{\nabla}^2-1)\hat{\mathbf{M}}+\mathbf{Y} &= \mathbf{0}.
\end{aligned}
\tag{4.71}
$$

Now, we obtain for the body force ($X_i = P_0\delta(\mathbf{x})\delta_{is}$, $Y_i = 0$) the following expression ($\hat{M}_i \equiv 0$):

$$4\pi\hat{F}_i^{(s)} = P_0\left[\frac{1}{2}R + (l^2 + \hat{\nu}_h^2)\frac{1}{R} - \frac{l^4}{l^2 - \hat{\nu}_h^2}\cdot\frac{e^{-R/l}}{R} + \frac{\hat{\nu}_h^4}{l^2 - \hat{\nu}_h^2}\cdot\frac{e^{-R/\hat{\nu}_h}}{R}\right]\delta_{is}. \tag{4.72}$$

In turn, for the body moment ($Y_i = M_0\delta(\mathbf{x})\delta_{is}$, $X_i = 0$) we obtain ($\hat{F}_i \equiv 0$):

$$4\pi\hat{M}_i^{(s)} = M_0\left(\frac{1}{R} - \frac{l^2}{l^2 - \hat{\nu}_h^2}\cdot\frac{e^{-R/l}}{R} + \frac{\hat{\nu}_h^2}{l^2 - \hat{\nu}_h^2}\cdot\frac{e^{-R/\hat{\nu}_h}}{R}\right)\delta_{is}. \tag{4.73}$$

Displacements and rotations are obtained in the form of the respective formulae:

— for the body force ($P_0 = 1$),

$$\begin{aligned} 4\pi\varphi_i^{(s)} &= \hat{\tau}_4\left(\frac{1 - e^{-R/\hat{\nu}_h}}{R}\right)_{,is} + \frac{1}{2\mu}\epsilon_{ijs}\left(\frac{1 - e^{-R/l}}{R}\right)_{,j}, \\ 8\pi\mu u_i^{(s)} &= -\frac{1}{2(1-\nu)}R_{,is} + 2\left(\frac{1}{R}\right)\delta_{is} \\ &\quad - \mu_0^{-1}\epsilon_{ijk}\epsilon_{krs}\left(\frac{1 - e^{-R/l}}{R}\right)_{,rj} \\ &\quad - 8\alpha\mu\hat{\tau}_4^2\left(\frac{1 - e^{-R/\hat{\nu}_h}}{R}\right)_{,is}; \end{aligned} \tag{4.74}$$

— for the body moment ($M_0 = 1$),

$$\begin{aligned} 8\pi\mu\varphi_i^{(s)} &= -(2\alpha)^{-1}\mu\left(\frac{1 - e^{-R/\hat{\nu}_h}}{R}\right)_{,is} \\ &\quad + \alpha_0^{-1}\epsilon_{ijk}\epsilon_{krs}\left(\frac{1 - e^{-R/l}}{R}\right)_{,rj}, \\ 4\pi u_i^{(s)} &= \hat{\tau}_4\left(\frac{1 - e^{-R/\hat{\nu}_h}}{R}\right)_{,is} + \frac{1}{2\mu}\epsilon_{ijs}\left(\frac{1 - e^{-R/l}}{R}\right)_{,j}, \end{aligned} \tag{4.75}$$

where

$$R = \sqrt{x_i x_i}, \qquad \mu_0 = \frac{2\mu}{\gamma + \varepsilon}, \qquad \alpha_0 = \frac{2\alpha}{\mu + \alpha}, \qquad l = \frac{1}{\sqrt{\mu_0\alpha_0}},$$

ν is Poisson's ratio ($\nu = \frac{\lambda}{2(\mu+\lambda)}$). Identical forms of formulae $(4.74)_1$ and $(4.75)_2$ follows from the Betti-Maxwell theorem on reciprocity of works. It is important to stress that for $\kappa \to 0$ we have $\hat{\nu}_h \to \nu_1$ (the constant ν_1 given by formula (1.189), i.e. $\nu_1 = \sqrt{\frac{2\gamma+\beta}{4\alpha}}$), $\hat{\tau}_4 \to 0$ and formulae (4.74) and (4.75) are transformed into formulae (1.187) and (1.188), respectively.

4.2.7 Boussinesq's problem. Method of potentials. Singularities of physical fields in the half-space

In [80] and [65], we considered the hypothetical isotropic hemitropic half-space Ω (1.209) in the rectangular coordinate system $Ox_1x_2x_3$. In the Boussinesq problem, the following boundary conditions are given:

$$\sigma_{11} = -P\delta(x_2)\delta(x_3), \qquad \sigma_{1N} = \mu_{1j} = 0 \qquad \text{on } \partial\Omega. \tag{4.76}$$

Moreover, it is postulated that the physical fields vanish for $R \to \infty$. In order to solve the problem, we applied the method of potentials: formulae (4.36)–(4.43) for $\mathbf{X}^\star \stackrel{\text{def}}{=} \mathbf{0}$ and $\mathbf{Y}^\star \stackrel{\text{def}}{=} \mathbf{0}$; moreover, we put $\nu_0 = \chi = 0$. Then, the scalar potentials Σ, Φ satisfy the equations

$$\nabla^2(\hat{\nu}_h^2\nabla^2 - 1)(\nabla^2\Phi, \Sigma) = (0,0) \qquad \text{in } \Omega. \tag{4.77}$$

For the vector potentials $\mathbf{H}$, $\boldsymbol{\Psi}$ we arrive at one equation, namely

$$\nabla^2\nabla^2(l^2\nabla^2 - 1)(H_i, \Psi_i) = (0,0) \qquad \text{in } \Omega, \tag{4.78}$$

where $\nabla^2(.) = (.)_{,ii}$.

By means of the Fourier transformations (0.50) we choose from equations (4.77) and (4.78) the form of potentials in the transformed region as follows:

$$\begin{aligned}
\tilde{\Phi} &= (A + B\xi x_1)e^{-\xi x_1} + Ne^{-\hat{\eta}_h x_1}, \\
\tilde{\Sigma} &= Ce^{-\xi x_1} + De^{-\hat{\eta}_h x_1}, \\
\tilde{\Psi}_i &= (E_i + F_i\xi x_1)e^{-\xi x_1} + G_i e^{-\rho x_1}, \\
\tilde{H}_i &= (J_i + K_i\xi x_1)e^{-\xi x_1} + M_i e^{-\rho x_1},
\end{aligned} \tag{4.79}$$

where

$$\hat{\eta}_h = \sqrt{\xi^2 + \frac{1}{\hat{\nu}_h^2}}, \qquad \xi^2 = \xi_N\xi_N,$$

the notation for $\hat{\nu}_h^2$ is the same as in Subsection 4.2.6,

$$\rho = \sqrt{\xi^2 + \frac{1}{l^2}} \qquad \text{and} \qquad l^2 = (\alpha_0\mu_0)^{-1}.$$

The solutions $\tilde{\Phi}$, $\tilde{\Sigma}$; $\tilde{\Psi}_i$, $\tilde{H}_i$ depend on 23 constants A–D, N, E_i–G_i, J_i, K_i, M_i. We shall reduce the number of these constants to 10 by satisfying equations (4.37) (for $X_i^\star = Y_i^\star = 0$, $\chi = \nu_0 = 0$) and the conditions from (4.36): $\Psi_{i,i} = 0$ and $H_{i,i} = 0$. Such potentials (depending on 10 constants) lead to unique displacements u_i and rotations φ_i by means of the representation (4.36) for u_i, φ_i. The procedure is similar to that described for the micropolar centrosymmetric half-space (see Subsection 1.2.7).

We obtain the following formulae:

$$
\begin{aligned}
\tilde{u}_1 &= (A_1 + B_1\xi x_1)e^{-\xi x_1} + C_1 e^{-\rho x_1} - \frac{\kappa}{2\mu+\lambda}\bar{B}_1 e^{-\hat{\eta}_h x_1},\\
\tilde{u}_K &= i\left\{\frac{\xi_K}{\xi}[A_1 - (3-4\nu)B_1] - 2\epsilon_{1KL}\xi_L\xi^{-2}\bar{A}_1 + \xi_K B_1 x_1\right\}e^{-\xi x_1}\\
&\quad - i\frac{\kappa}{2\mu+\lambda}\xi_K\hat{\eta}_h^{-1}\bar{B}_1 e^{-\hat{\eta}_h x_1} + i(\rho\xi_K C_1 - \alpha_0\epsilon_{1KL}\xi_L\bar{C}_1)\xi^{-2}e^{-\rho x_1},\\
\tilde{\varphi}_1 &= \left[\bar{A}_1 + \frac{\kappa}{2\alpha}(1-2\nu)B_1\xi^2\right]e^{-\xi x_1} + \bar{B}_1 e^{-\hat{\eta}_h x_1} + \bar{C}_1 e^{-\rho x_1},\\
\tilde{\varphi}_K &= i\left[\xi_K\xi^{-1}\bar{A}_1 - \epsilon_{1KL}\xi_L\, 2(1-\nu)B_1 + \frac{\kappa}{2\alpha}(1-2\nu)\xi_K\xi B_1\right]e^{-\xi x_1}\\
&\quad + i\xi_K\hat{\eta}_h^{-1}\bar{B}_1 e^{-\hat{\eta}_h x_1} + i(\rho\xi_K\bar{C}_1 + \mu_0\epsilon_{1KL}\xi_L C_1)\xi^{-2}e^{-\xi x_1},
\end{aligned}
\tag{4.80}
$$

where, in the expressions: $(3-4\nu)$, $(1-\nu)$, $(1-2\nu)$ we have Poisson's ratio ν, $\alpha_0 = 2\alpha/(\mu+\alpha)$, $\mu_0 = 2\mu/(\gamma+\varepsilon)$, the indices K, L assume values 2 and 3. In the solution (4.80), 6 new integration constants appear, A_1–C_1, $\bar{A}_1$–$\bar{C}_1$, which can be uniquely determined from 6 boundary conditions (4.76). It should be noted that in [80] it is shown that from (4.80) we can arrive at the results of the limiting theories: micropolar theory (E-N), couple-stress theory (K-M), the classical theory of elasticity (H) and the theory of a hypothetical medium (HM). For example, for $\kappa \to 0$, formulae (4.80) are transformed into formulae (1.220) and (1.221) for the centrosymmetric micropolar half-space:

$$
a_1 = b_1 = d_1 = \bar{a}_1 = \bar{b}_1 = \bar{d}_1 = 0,\ \hat{\eta}_h \to \eta = \sqrt{\xi^2 + \nu_1^{-2}},\ \hat{\nu}_h^2 \to \nu_1^2 = \frac{2\gamma+\beta}{4\alpha}.
$$

For $\alpha \to \infty$, we have: $(\nu_h^2 \to 0) \Longrightarrow (\eta_h^2 \to \infty)$.

Introducing a new notation:

$$
A_2 = \frac{i\xi_2}{\xi}[A_1 - (3-4\nu)B_1] - \frac{i\xi_3}{\xi^2}2\bar{A}_1, \qquad C_2 = i\left(\rho^\star\frac{\xi_2}{\xi^2}C_1 - \frac{2\xi_3}{\xi^2}\bar{C}_1\right),
$$

where $\rho^\star = \sqrt{\xi^2 + \dfrac{1}{l^{\star 2}}}$, we obtain from formulae $(4.80)_{1,2}$ for $\tilde{u}_i$ the following formulae:

$$
\begin{aligned}
\tilde{u}_1 &= (A_1 + B_1\xi x_1)e^{-\xi x_1} + C_1 e^{-\rho^\star x_1},\\
\tilde{u}_2 &= (A_2 + i\xi_2 B_1 x_1)e^{-\xi x_1} + C_2 e^{-\rho^\star x_1},\\
\tilde{u}_3 &= \frac{1}{\xi_3}[iA_1\xi - \xi_2 A_2 - i(3-4\nu)B_1\xi]e^{-\xi x_1} + i\xi_3 B_1 x_1 e^{-\xi x_1}\\
&\quad + \frac{1}{\xi_3}(i\rho^\star C_1 - \xi_2 C_2)e^{-\rho^\star x_1},
\end{aligned}
\tag{4.81}
$$

which refer to the couple-stress half-space. Five constants A_1, B_1, C_1, A_2, C_2 are determined from five boundary conditions of the form

$$
\begin{aligned}
s_{11} &= -P\delta(x_2)\delta(x_3), & s_{12} &= \tfrac{1}{2}(m_{j3,j} - m_{11,3}), \\
s_{13} &= \tfrac{1}{2}(m_{j2,j} - m_{11,2}), & m_{1K} &= 0 \quad \text{on } \partial\Omega,
\end{aligned}
\tag{4.82}
$$

where s_{ij}, m_{ij} are the symmetric and the deviatoric parts of the stress tensors σ_{ij}, μ_{ij}, respectively. In [80], we derived formulae for displacements, rotations, force-stresses and couple-stresses in the form of double Fourier integrals in the Bousinessq problem (on the boundary $\partial\Omega$ of the hypothetical hemitropic half-space Ω, the conditions (4.76) are given)[7]. We also discussed the result obtained. Thus, for $\kappa \to 0$, we get the formulae for a micropolar centrosymmetric medium. In turn, for $\alpha \to 0$, we obtain the classical result for the Boussinesq problem for u_i, but the rotations $\varphi_i \to 0$. We arrive at the complete classical result when we evaluate the iterated limit: $\alpha \to \infty$, $l^\star \to 0$. It is interesting that for $\alpha \to \infty$ one obtains the formulae of couple-stress theory from the result for a hemitropic (hypothetical) medium. It confirms, in particular, a more general observation, namely that couple-stress theory cannot be generalized by discarding the postulate of centrosymmetry.

The integral formulae obtained for the hypothetical hemitropic half-space were a basis for the investigation of singularities of physical fields at the point of application of a vertical concentrated force. In Table 4.1 we present the final results concerning the types and the orders of singularities of physical fields and analogous results for the limiting theories.

Table 4.1.

Physical fields	(1)	(2)	(3)	(4)
u_i	$O(R^{-1})$	$O(R^{-1})$	$O(R^{-1})$	$O(R^{-1})$
φ_1	$O(R^{-1})$	0	0	0
φ_N	$O(R^{-1})$	$O(1)$	$O(1)$	$0(R^{-2})$
σ_{ij}	$O(R^{-2})$	$O(R^{-2})$	$O(R^{-2})$	$O(R^{-2})$
μ_{11}	$O(R^{-2})$	0	0	–
remaining μ_{ij}	$O(R^{-2})$	$O(R^{-1})$	$O(R^{-1})$	–

Numbers (1)–(4) denote: hypothetical hemitropic medium, micropolar centrosymmetric medium, couple-stress medium, classical medium, respectively. For couple-stress theory (3) the symbols σ_{ji}, μ_{ji} are understood as s_{ji}, m_{ji}, respectively

[7] Dyszlewicz & Kołodziej [65]. In this paper we give more general formulae for u_i and φ_i, for $n = 9$ elasticity constants. However, it should be stressed that the expressions are quite complex and difficult for further discussion.

In [80], we analysed in detail the problem of correspondence between singular parts of a solution within the frameworks of the theories (1)–(4). Such correspondence depends on the fact whether the coefficients of singularities intensities k^0 depend on the material constants, which influence the limiting procedures of passage from one theory to another (for the theory (1) it is κ and α, for (2) — α, and for (3) — $l^\star$). We will present here a few selected final conclusions[8]. For the purpose of better illustration we will make use of the diagram in Fig. 4.1.

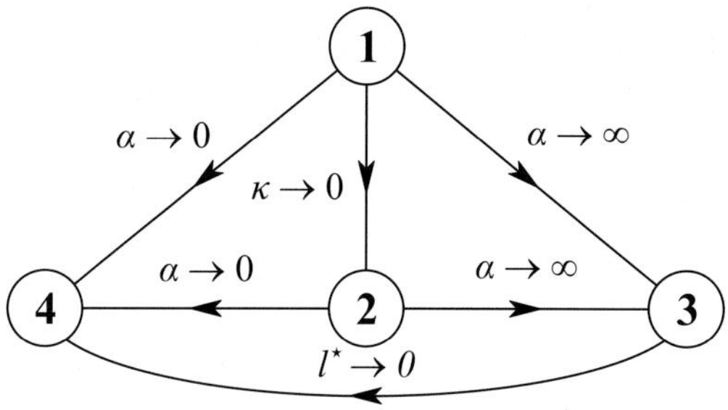

Fig. 4.1.

1. From the general solution for the half-space in the theory (1), given by formulae (4.80) one can derive analogous solutions for the theories (2), (3) and (4) (as in Fig. 4.1, keeping the rules of changing the notation for constants, introduced in [80]).

2. From the results (written in the integral form) for physical fields in the theory (1) one can obtain integral representations for the results obtained in the theories (2), (3) and (4): here, one should take into account that for $((1),(2)) \xrightarrow{\alpha\to\infty} (3)$ we assume $\sqrt{\dfrac{\gamma+\varepsilon}{4\mu}} \to l^\star$, $\dfrac{\gamma-\varepsilon}{\gamma+\varepsilon} \to \eta$ and from the stresses σ_{ji}, μ_{ji} we extract s_{ji}, m_{ji}, respectively.

3. From the singular parts of solutions in the theory (1) one obtains the singular parts of solutions in the theories (2), (3) and (4), since in the theory (1) the coefficient of singularity intensity $k^0 = k^0(\kappa, \alpha)$, whereas in the theory (2) $k^0 = k^0(\alpha)$. Recall that in the case of singular parts of solutions, the correspondence $(3) \xrightarrow{l^\star\to 0} (4)$ is not preserved since $k^0 = k^0(\nu)$ (ν — Poisson's ratio) and does not depend on $l^\star$. An exception is furnished by the expressions φ^s_K, for which $k^0 = k^0(l^\star)$, but for $l^\star \to 0$ we obtain the peculiar result[9]: $\varphi^s_2 \to -\infty$, $\varphi^s_3 \to +\infty$.

[8] See Dyszlewicz & Słotwińska [80]. This paper contains a detailed discussion of the results and their singular parts for the Boussinesq problem, within the framework of the theories (1)–(4).

[9] See [80], [72], [74] and [50]. In the cylindrical system of coordinates, we obtain in couple-stress theory (3) the following results

4. All (non-zero) components of physical fields appear in the framework of the theory (1), and they are singular at the point of application of a concentrated force (Table 4.1).

4.3 One-dimensional problems of elastostatics and thermoelastostatics

4.3.1 The half-space problem

Dyszlewicz & Kołodziej [70] studied the one-dimensional problem of the isotropic hemitropic half-space $\Omega = \{x_1 : x_1 > 0\}$. The existence of a solution of the boundary value problem[10], implies displacements and rotations on the boundary $\partial\Omega$ (for $x_1 = 0$) of the half-space. Moreover, one requires that displacements and rotations be finite for $x_1 \to \infty$. The following boundary conditions on $\partial\Omega$ are considered:

$$u_1 = U_1, \qquad \varphi_1 = \Phi_1, \tag{4.83}$$

where U_1, $\Phi_1 : \partial\Omega \to \mathbb{R}$ are given functions. With the boundary conditions (4.83) we associate a system of ordinary differential equations for the pair (u_1, φ_1) of the following form in Ω:

$$\begin{aligned} &(\lambda + 2\mu)\mathbb{D}_1^2 u_1 + (\kappa + 2\chi)\mathbb{D}_1^2 \varphi_1 = 0, \\ &[(\beta + 2\gamma)\mathbb{D}_1^2 - 4\alpha]\varphi_1 + (\kappa + 2\chi)\mathbb{D}_1^2 u_1 = 0, \end{aligned} \tag{4.84}$$

in which we disregarded both the temperature- and the body loading fields. The system of equations (4.84) is one of the two systems which are obtained from (4.10) under the assumption that physical fields depend only on the variable x_1. Also the deformations γ_{11}, γ_{KL}, κ_{11} $(K \neq L)$ and stresses σ_{11}, σ_{MN}, μ_{11}, μ_{MN} are associated with the system of equations (4.84).

After uncoupling, the system of equations (4.84) leads to the equations

$$(\nu_h^2 \mathbb{D}_1^2 - 1)(\mathbb{D}_1^2 u_1, \varphi_1) = (0, 0), \tag{4.85}$$

where

$$\mathbb{D}_1 \stackrel{\text{def}}{=} \frac{d}{dx_1}, \quad \nu_h^2 = \nu_1^2 - (\kappa + 2\chi)\tau_4, \quad \nu_1^2 = \frac{2\gamma + \beta}{4\alpha}, \quad \tau_4 = \frac{2\chi + \kappa}{4\alpha(\lambda + 2\mu)}.$$

$$\varphi_\theta = \overset{s}{\varphi_\theta} + O(1),$$

where $\overset{s}{\varphi_\theta} = k^0(l^\star)\frac{r}{R}$, $k^0(l^\star) = \frac{P}{8\pi\mu}\frac{1}{l^{\star 2}}\frac{\beta^2}{2\beta^2 - 1}$, $\beta^2 = \frac{2(1-\nu)}{1-2\nu}$ and $\nu = \frac{\lambda}{2(\mu + \lambda)}$. For $l^\star \to 0$, $\overset{s}{\varphi_\theta} \to \infty$.

[10] See Gawinecki [99].

The following expressions are general solutions of equations (4.85):

$$u_1 = A_1 e^{x_1/\nu_h} + B_1 e^{-x_1/\nu_h} + E_1 x_1 + F_1, \; \varphi_1 = \bar{A}_1 e^{x_1/\nu_h} + \bar{B}_1 e^{-x_1/\nu_h}, \quad (4.86)$$

where A_1, B_1, E_1, F_1, $\bar{A}_1$, $\bar{B}_1 \in \mathbb{R}$.

For the boundary value problem stated at the beginning of this section, the integration constants A_1, B_1, E_1, F_1, $\bar{A}_1$, $\bar{B}_1$ (and hence also all physical fields) will be uniquely determined if the system of equations (4.84) is satisfied and the boundary conditions (4.83) hold, under the assumption of finitness of the displacement u_1 and the rotation φ_1 at infinity[11]. We arrive at the following result:

$$u_1 = U_1 + 4\alpha\tau_4 \Phi_1 (1 - e^{-x_1/\nu_h}), \qquad \varphi_1 = \Phi_1 e^{-x_1/\nu_h}. \qquad (4.87)$$

The fields of displacements and rotations in a hemitropic medium are illustrated by the diagrams in Fig. 4.2 and Fig. 4.3 (the curves denoted by (1)).

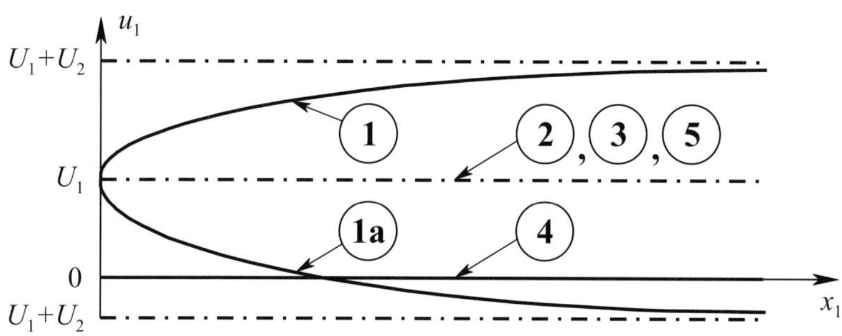

Fig. 4.2.

We also present the diagrams of analogous solutions for limiting theories, denoted, respectively: (2) — centrosymmetric micropolar medium (E-N), (3)

[11] In [70], we assumed the condition of finitness of the displacement u_1 and rotation φ_1 at infinity due to the fact that if we assume the condition
(a) $(u_1, \varphi_1) \to (0,0)$ for $x_1 \to \infty$, the boundary value problem will be ill-posed. In fact, notice that by satisfying the system of equations (4.84), we obtain
(b) $(A_1, B_1) = -\dfrac{\kappa + 2\gamma}{\lambda + 2\mu}(\bar{A}_1, \bar{B}_1)$, $4\alpha\tau_4 = \dfrac{\kappa + 2\chi}{\lambda + 2\mu}$. From (4.86), (a) and (b) we obtain
(c) $A_1 = \bar{A}_1 = E_1 = F_1 = 0$. Now, using (4.86), (b) and (c), and assuming
(d) $\Phi_1 + \dfrac{\lambda + 2\mu}{\kappa + 2\chi} U_1 \neq 0$, we obtain from the boundary conditions (4.83) a contradictory system of equations for the constant $\bar{B}_1$, namely
(e) $\bar{B}_1 = \Phi_1$, $\quad -\dfrac{\kappa + 2\chi}{\lambda + 2\mu}\bar{B}_1 = U_1$. The condition (a) makes sense if
(f) $\Phi_1 + \dfrac{\lambda + 2\mu}{\kappa + 2\chi} U_1 = 0$ and is included in the solution given by formula (4.87).
Then, it holds:
(g) $u_1 = U_1 \exp(-x_1/\nu_h)$, $\qquad \varphi_1 = \Phi_1 \exp(-x_1/\nu_h)$.

— classical theory (H), (4) — hypothetical medium (HM), (5) — couple-stress theory (K-M).

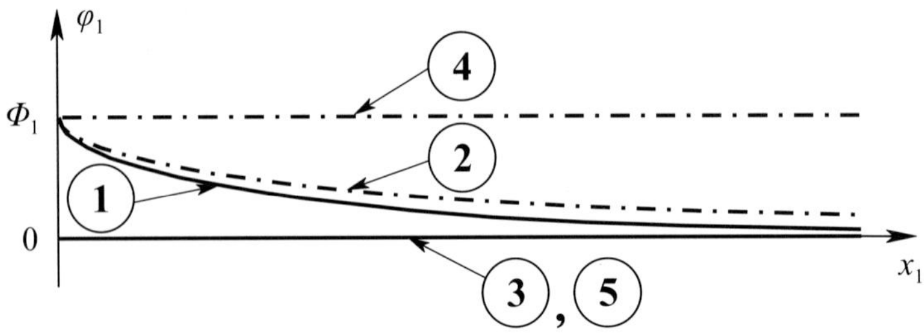

Fig. 4.3.

The quantities U_1, Φ_1 are given on $\partial\Omega$, $U_0 = 4\alpha\tau_4\Phi_1$, where τ_4 is the same as in formula (4.85). Let us make a few remarks. In a hemitropic medium (1) the rotation Φ_1 influences the value of the displacement u_1, which increases to the limiting value $U_1 + U_0$ as depth increases. Figure 4.2 was made under the assumption that the quantities Φ_1 and $(\kappa + 2\chi)$ have equal signs[12]. For $x_1 > 0$, the rotation φ_1 in a hemitropic medium (1) is smaller than the rotation in a micropolar centrosymmetric medium (2), since the following inequality holds: $\nu_h^2 - \nu_1^2 < 0$ $(\Phi_1 > 0)$. Consider the next case of boundary conditions in displacements and rotations:

$$u_N = U_N, \qquad \varphi_N = \Phi_N \qquad \text{on } \partial\Omega, \tag{4.88}$$

where $U_N, \Phi_N : \partial\Omega \to \mathbb{R}$ are given functions $(N = 2, 3)$.

Conditions (4.88) are connected with the system of ordinary differential equations for u_N, φ_N, obtained (in addition to (4.84)) from the system of equations (4.10) for $X_i = Y_i = 0$ in Ω, namely:

$$\begin{aligned} &(\mu + \alpha)\mathbb{D}_1^2 u_K + 2\alpha\epsilon_{1LK}\mathbb{D}_1\varphi_L + (\chi + \nu_0)\mathbb{D}_1^2\varphi_K = 0, \\ &[(\gamma + \varepsilon)\mathbb{D}_1^2 - 4\alpha]\varphi_K + 2\alpha\epsilon_{1LK}\mathbb{D}_1 u_L \\ &\qquad + (\chi + \nu_0)\mathbb{D}_1^2 u_K + 4\nu_0\epsilon_{1LK}\varphi_L = 0. \end{aligned} \tag{4.89}$$

Uncoupling the system of equations (4.89), we obtain the equations

$$[(l_h^2\mathbb{D}_1^2 - 1)^2 + \hat{k}^2\mathbb{D}_1^2](\mathbb{D}_1^2 u_N, \mathbb{D}_1\varphi_N) = (0, 0), \tag{4.90}$$

where

[12] For $U_1 > 0$ (as in Fig. 4.2), if we assume the sign of Φ_1 opposite to that of the constant $\kappa + 2\chi$, the displacement u_1 in the half-space is not of constant sign (curve (1a)).

$$l_h^2 = l^2 - \frac{(\chi+\nu_0)^2}{4\alpha\mu}, \qquad \hat{k}^2 = \left(\frac{\nu_0\mu - \alpha\chi}{\alpha\mu}\right)^2, \qquad l^2 = \frac{(\gamma+\varepsilon)(\mu+\alpha)}{4\alpha\mu}.$$

General solutions of equations (4.90) have the form:

$$\begin{aligned} u_N &= F_N + E_N x_1 + C_N e^{-\hat{p}x_1}\cos\hat{q}x_1 + D_N e^{-\hat{p}x_1}\sin\hat{q}x_1 \\ &\quad + G_N e^{\hat{p}x_1}\cos\hat{q}x_1 + H_N e^{\hat{p}x_1}\sin\hat{q}x_1, \\ \varphi_N &= \bar{F}_N + \bar{C}_N e^{-\hat{p}x_1}\cos\hat{q}x_1 + \bar{D}_N e^{-\hat{p}x_1}\sin\hat{q}x_1 + \bar{G}_N e^{\hat{p}x_1}\cos\hat{q}x_1 \\ &\quad + \bar{H}_N e^{\hat{p}x_1}\sin\hat{q}x_1, \end{aligned} \tag{4.91}$$

where $\hat{p} = \frac{1}{2}l_h^{-2}(4l_h^2 - \hat{k}^2)^{1/2}$ and $\hat{q} = \frac{1}{2}\hat{k}l_h^{-2}$.

The expressions $\hat{p},\hat{q} \in \mathbb{R}$, since we can verify that the following inequality holds:

$$(4l_h^2 - \hat{k}^2) = (\mu+\alpha)[\alpha\mu(\gamma+\varepsilon) - (\alpha\chi^2 + \mu\nu_0^2)] > 0$$

in view of inequalities (4.22) and (4.32).

To justify formulae (4.91), let us note that in equations (4.90) we deal with the operator $\overrightarrow{\mathrm{L}}^h = [(l_h^2\overrightarrow{\nabla}^2 - 1)^2 + \hat{k}^2\overrightarrow{\nabla}^2]$ discussed in Subsection 4.2.3. In the rectangular system of coordinates $Ox_1x_2x_3$, for the one-dimensional problem with the variable x_1 it holds: $\overrightarrow{\nabla}^2 \equiv (\nabla^2, \nabla^2, \nabla^2)$, where

$$\nabla^2(.) = (.)_{,ii} = (.)_{,11} = \mathbb{D}_1^2.$$

The discussion of the quadratic equation (4.30) (where now $\overrightarrow{\nabla}^2 \to \mathbb{D}_1^2$) remains in force. This implies that solutions of the type

$$(\sin\hat{q}x_1, \cos\hat{q}x_1)\exp(-\hat{p}x_1)$$

appear in (4.91). We can uniquely determine the integration constants C_N, $\bar{C}_N$, $\bar{D}_N$, D_N, E_N, F_N, $\bar{F}_N$, G_N, $\bar{G}_N$, H_N, $\bar{H}_N$ by using formulae (4.91), satisfying the system of equations (4.89) and the boundary conditions (4.88), and assuming finitness of the displacements u_N and rotations φ_N for $x_1 \to +\infty$. We obtain the following result:

$$\begin{aligned} u_N &= U_N + (a\Phi_N + b\epsilon_{1NK}\Phi_K)(1 - e^{-\hat{p}x_1}\cos\hat{q}x_1) \\ &\quad + (b\Phi_N - a\epsilon_{1NK}\Phi_K)e^{-\hat{p}x_1}\sin\hat{q}x_1, \\ \varphi_N &= \Phi_N e^{-\hat{p}x_1}\cos\hat{q}x_1 + \epsilon_{1NK}\Phi_K e^{-\hat{p}x_1}\sin\hat{q}x_1, \end{aligned} \tag{4.92}$$

where

$$a = \chi\mu^{-1}, \qquad b = s\hat{p}l_h^2 \qquad \text{and} \qquad s = \alpha_0 = \frac{2\alpha}{\mu+\alpha}.$$

A complete analysis of formulae (4.92) is difficult. In [70], we made the following important remark. If, on $\partial\Omega$, only the rotation Φ_2 ($\Phi_3 = U_N = 0$) is given, then (4.92) leads to the formulae:

$$
\begin{aligned}
(u_2, u_3) &= (a, -b)\Phi_2 + (a^2+b^2)^{1/2}\Phi_2 e^{-\hat{p}x_1}(\sin, \cos)(\hat{q}x_1 - \psi),\\
(\varphi_2, \varphi_3) &= (1, -1)\Phi_2 e^{-\hat{p}x_1}(\cos, \sin)\hat{q}x_1,
\end{aligned}
\tag{4.93}
$$

where $\cos\psi = b(a^2+b^2)^{-1/2}$ and $\sin\psi = a(a^2+b^2)^{-1/2}$.

In contrast to the limiting media (2)–(5), in a hemitropic medium (1) the diagrams for the displacements u_N and rotations φ_N are of oscillatory character. In [70], we presented such curves. As an example, we will give here the plot of $\varphi_3(x_1)$ (see Fig. 4.4).

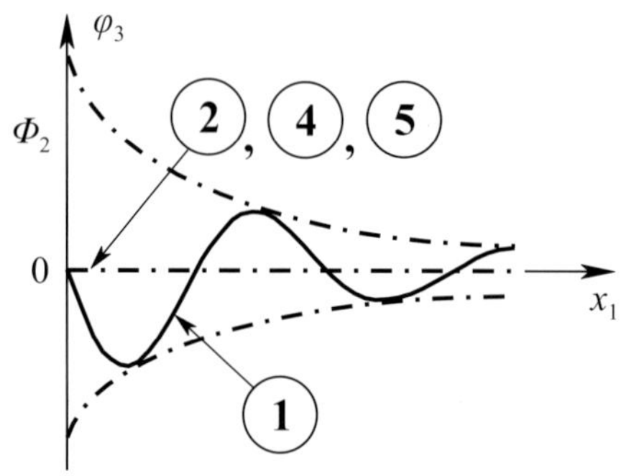

Fig. 4.4.

The rotations Φ_2, given on $\partial\Omega$, produce the rotation φ_3 only in the case of a hemitropic medium (1).

4.3.2 The problem of a layer with temperature field

In [70], we considered the isotropic elastic hemitropic layer

$$\Omega = \{(x_1, x_2, x_3) : 0 < x_1 < a_1 \wedge |x_2| < \infty \wedge |x_3| < \infty\} \tag{4.94}$$

with one rigidly fixed (in the micropolar sense) boundary:

$$u_1 = 0, \qquad \varphi_1 = 0 \qquad \text{for } x_1 = 0 \tag{4.95}$$

the second boundary being free:

$$\sigma_{11} = 0, \qquad \mu_{11} = 0 \qquad \text{for } x_1 = a_1. \tag{4.96}$$

The layer is heated in the following way:

$$\theta = \theta_1 \;\; \text{for } x_1 = 0 \qquad \text{and} \qquad \theta = \theta_2 \;\; \text{for } x_1 = a_1, \tag{4.97}$$

where the functions θ_1, θ_2 are given. The function θ, describing the temperature field, satisfies in Ω the equation

$$\mathbb{D}_1^2\theta = 0. \tag{4.98}$$

The stated thermoelastic boundary value problem is connected with the system of equations (4.84) (for u_1, φ_1), in which we take into account the temperature field:

$$\begin{aligned}(\lambda + 2\mu)\mathbb{D}_1^2 u_1 + (\kappa + 2\chi)\mathbb{D}_1^2\varphi_1 &= \eta\mathbb{D}_1\theta,\\ [(\beta + 2\gamma)\mathbb{D}_1^2 - 4\alpha]\varphi_1 + (\kappa + 2\chi)\mathbb{D}_1^2 u_1 &= \eta\mathbb{D}_1\theta.\end{aligned} \tag{4.99}$$

We will give here a solution of the boundary value problem (4.94)–(4.99) when $\theta_1 = 0$ and $\theta_2 \neq 0$:

$$\begin{aligned} u_1 &= -4\alpha\tau_4\varphi_1 + \frac{\eta\theta_2}{2(\lambda+2\mu)a_1}x_1^2,\\ \varphi_1 &= \frac{\tau\theta_2}{4\alpha a_1}\left\{\frac{1}{\cosh\dfrac{a_1}{\nu_h}}\left[\frac{a_1}{\nu_h}\sinh\frac{x_1}{\nu_h} + \cosh\frac{x_1 - a_1}{\nu_h}\right] - 1\right\},\end{aligned} \tag{4.100}$$

where $\tau = \zeta - 4\alpha\eta\tau_4$, ζ, η are thermoelastic material constants, τ_4 is given by formula (4.45), a_1 is layer's thickness, and $\nu_h = [\nu_1^2 - (\kappa + 2\chi)\tau_4]^{1/2}$, $\nu_1^2 = \dfrac{2\gamma + \beta}{4\alpha}$. In formula (4.100) and in the sequel, by sinh, cosh and atanh we denote hyperbolic functions. In Fig. 4.5 we present the plot of $\varphi_1(x_1)$ under the assumption that $\tau > 0$.

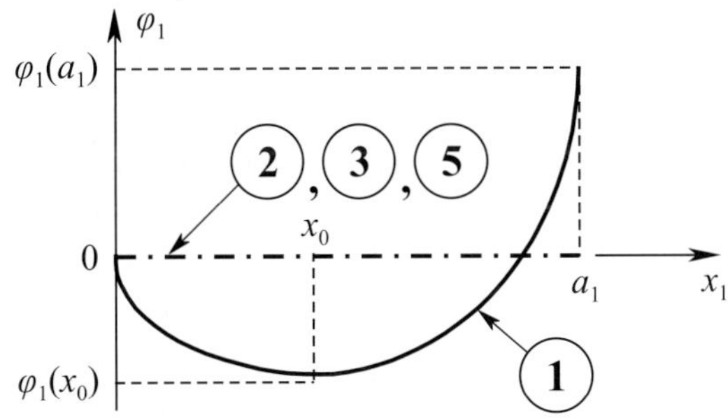

Fig. 4.5.

Notation in Fig. 4.5: (1), (2), (3) and (5) have the same meaning as in Subsection 4.3.1. The rotation φ_1 reaches the extremal values for two values of the variable x_1: namely for $x_1 = x_0$ and $x_1 = a_1$. The value x_0 is given by

$$x_0 = \nu_h \operatorname{atanh}\left[\frac{\sinh\dfrac{a_1}{\nu_h} - \dfrac{a_1}{\nu_h}}{\cosh\dfrac{a_1}{\nu_h}}\right]$$

and does not depend on thermoelastic properties of the material ζ, η, but only on the thickness a_1 of the layer and on the characteristic constant ν_h. We shall give now a solution of the boundary value problem when $\theta_1 \neq 0$ and $\theta_2 = 0$:

$$u_1 = -4\alpha\tau_4\varphi_1 - \frac{\eta\theta_1}{2a_1(\lambda+2\mu)}(x_1^2 - 2a_1x_1),$$
$$\varphi_1 = \frac{\tau\theta_1}{4\alpha a_1}\left[1 - \frac{\cosh\dfrac{x_1-a_1}{\nu_h}}{\cosh\dfrac{a_1}{\nu_h}}\right]. \tag{4.101}$$

The plot of $\varphi_1(x_1)$ for $\tau > 0$ is given in Fig. 4.6.

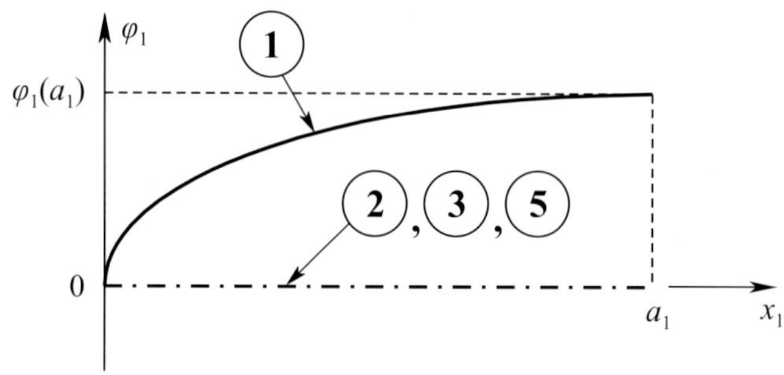

Fig. 4.6.

The rotation φ_1 increases with the distance from the heated boundary, the latter being fixed. In [70], we also considered the problem of the layer (4.94) with surface and body loadings. The conclusions presented there indicate that the deformation of a hemitropic medium under the action of surface loadings, temperature field and body loadings exhibits singularities which do not appear in the limiting theories (2)–(5).

4.4 Remarks and conclusions concerning vector equations

4.4.1 On the derivation of vector equations

In order to avoid errors[13], we must visibly distinguish the scalar Laplace operator ∇^2 from the vector Laplace operator $\vec{\nabla}^2$, when employing vector equations in the present monograph. These operators are defined as follows:

[13] See Moon & Spencer [159, p. 84].

$$\begin{aligned} &\nabla^2 \stackrel{\text{def}}{=} \operatorname{div}\operatorname{grad}, \\ &\vec{\nabla}^2 \stackrel{\text{def}}{=} \operatorname{grad}\operatorname{div} - \operatorname{curl}\operatorname{curl} \end{aligned} \tag{4.102}$$

and they enter the definitions of more complex scalar and vector operators. We shall list here several typical operators, for which concise notation will be used in this work:

— micropolar centrosymmetric medium (E-N),

$$\begin{aligned} \Box_1 &= (\lambda + 2\mu)\nabla^2 - \rho\partial_t^2, \\ \vec{\Box}_1 &= (\lambda + 2\mu)\vec{\nabla}^2 - \rho\partial_t^2, \\ \Box_2 &= (\mu + \alpha)\nabla^2 - \rho\partial_t^2, \\ \vec{\Box}_2 &= (\mu + \alpha)\vec{\nabla}^2 - \rho\partial_t^2, \\ \Box_3 &= (\beta + 2\gamma)\nabla^2 - 4\alpha - J\partial_t^2, \\ \vec{\Box}_3 &= (\beta + 2\gamma)\vec{\nabla}^2 - 4\alpha - J\partial_t^2, \\ \Box_4 &= (\gamma + \varepsilon)\nabla^2 - 4\alpha - J\partial_t^2, \\ \vec{\Box}_4 &= (\gamma + \varepsilon)\vec{\nabla}^2 - 4\alpha - J\partial_t^2, \end{aligned} \tag{4.103}$$

where μ, λ, α, β, γ, ε, ρ, $J \in \mathbb{R}$;

— micropolar hemitropic medium (A-K),

$$\begin{aligned} &\mathcal{L} = \Box_1\Box_3 - (\kappa + 2\chi)^2\nabla^4, \qquad \nabla^4 = \nabla^2\nabla^2, \\ &L^2 = [(\lambda + 2\mu)(\beta + 2\gamma) - (\kappa + 2\chi)^2]\nabla^2 - 4\alpha(\lambda + 2\mu), \\ &\vec{\mathrm{L}}^2 = [(\lambda + 2\mu)(\beta + 2\gamma) - (\kappa + 2\chi)^2]\vec{\nabla}^2 - 4\alpha(\lambda + 2\mu), \\ &\vec{\mathcal{L}} = \vec{\Box}_1\vec{\Box}_3 - (\kappa + 2\chi)^2\vec{\nabla}^4, \qquad \vec{\nabla}^4 = \vec{\nabla}^2\vec{\nabla}^2, \\ &\vec{\mathcal{D}} = \vec{\Box}_2\vec{\Box}_4 - (\chi + \nu_0)^2\vec{\nabla}^4 + 4\alpha^2\vec{\nabla}^2, \\ &\vec{\mathrm{D}}^2 = [(\mu + \alpha)(\gamma + \varepsilon) - (\chi + \nu_0)^2]\vec{\nabla}^2 - 4\alpha\mu, \\ &\vec{\mathrm{L}}^h = (l_h^2\vec{\nabla}^2 - 1)^2 + \hat{k}\vec{\nabla}^2, \end{aligned} \tag{4.104}$$

where

$$l_h^2 = l^2 - \frac{(\chi + \nu_0)^2}{4\alpha\mu}, \qquad l^2 = \frac{(\gamma + \varepsilon)(\mu + \alpha)}{4\alpha\mu}, \qquad \hat{k} = \frac{\nu_0\mu - \alpha\chi}{\alpha\mu},$$

and μ, α, λ, β, γ, ε, χ, κ, ν_0, ρ, $J \in \mathbb{R}$.

In order to transform the vector equations, one needs to take into account that the following operations hold:

— for a vector field $\mathbf{a}$,

$$\begin{aligned} \operatorname{div} \overrightarrow{\nabla}^2 \mathbf{a} &= \nabla^2 \operatorname{div} \mathbf{a}, \\ \operatorname{curl} \overrightarrow{\nabla}^2 \mathbf{a} &= \overrightarrow{\nabla}^2 \operatorname{curl} \mathbf{a}; \end{aligned} \tag{4.105}$$

— for a scalar field a,

$$\overrightarrow{\nabla}^2 \operatorname{grad} a = \operatorname{grad} \nabla^2 a \tag{4.106}$$

which are carried over to the operators (4.103) and (4.104). In order to obtain vector equations of motion expressed in displacements or rotations, or in displacements and rotations, one should, within the framework of the considered theory, make use of the following equations in the system of curvilinear coordinates: the equations of motion[14], the constitutive relations[15] and the geometric relations. Eliminating the stress tensor (tensors) from the equations of motion by means of the constitutive relations and the geometric relations, we satisfy the compatibility equations[16] and obtain, respectively, the diplacement and rotation equations or the displacement-rotation equations of the given theory.

For instructive purposes, we stressed and analysed in detail in [49] two procedures concerning:

(a) the discussion of the limiting procedure from the results of a more general theory to the results of the limiting theories and limiting problems[17];

(b) the discussion of the limiting procedure from the vector representations of a more general theory to the vector representations in the limiting theories and problems.

In this monograph we paid less attention to procedure **(b)**, hence we shall give now a few other examples of analysis of vector equations and representations.

---

[14] Lysik [147, p. 109] derived equations of motion with a tensor field of couple-stresses in a curvilinear system of coordinates.

[15] The constitutive relations (0.18), written for physical components of stress and deformation fields, are invariant in the sense that they remain valid for spatial systems of coordinates used in this monograph, namely $Ox_1x_2x_3$, Ox_1x_2, (r, θ, z) and (r, θ).

[16] Günther [104] and Kozák & Szeidl [134] defined the geometric relations in the framework of micropolar theory of elasticity, and derived the compatibility equations for a curvilinear system of coordinates.

[17] Considering, for instance, the elastodynamics of a micropolar centrosymmetric medium, by the limiting theories we understand: couple-stress theory, the theory of a hypothetical medium, and the classical theory of elasticity, whereas in the limiting problems we shall include the elastostatics of micropolar theory and two-dimensional (with respect to the position coordinates) problems in elastodynamics (i.e. the problems of plane and axially-symmetric states of strain).

4.4.2 Hemitropic medium. Elastodynamics. Analysis of vector equations

We start with the vector equations (4.1)–(4.6), from which we shall derive special cases according to the diagram in Fig. 4.7.

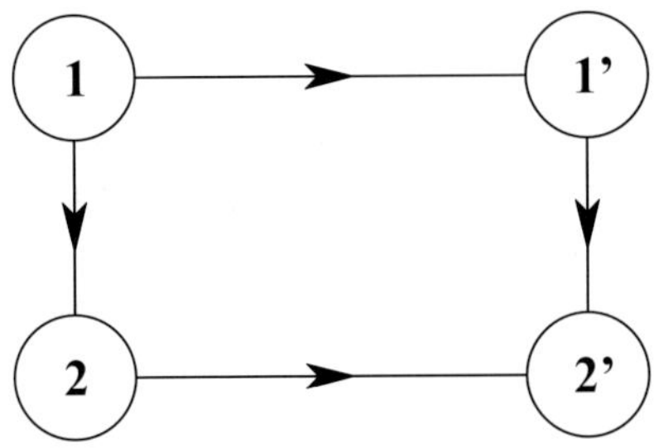

Fig. 4.7. The symbols (1) and (1') in figure refer to elastodynamics and elastostatics of a hemitropic medium, respectively, whereas (2) and (2') denote the elastodynamics and elastostatics of a micropolar centrosymmetric medium, respectively

Case (1) → (1'). Neglecting $\rho\partial_t^2$ and $J\partial_t^2$, we note that dynamical operators are transformed into static operators:

$$
\begin{aligned}
&\Box_1 \mapsto (\lambda+2\mu)\nabla^2, && \vec{\Box}_1 \mapsto (\lambda+2\mu)\vec{\nabla}^2,\\
&\Box_2 \mapsto (\mu+\alpha)\nabla^2, && \vec{\Box}_2 \mapsto (\mu+\alpha)\vec{\nabla}^2,\\
&\Box_3 \mapsto 4\alpha(\nu_1^2\nabla^2-1), && \vec{\Box}_3 \mapsto 4\alpha(\nu_1^2\vec{\nabla}^2-1),\\
&\Box_4 \mapsto 4\alpha(k^2\nabla^2-1), && \vec{\Box}_4 \mapsto 4\alpha(k^2\vec{\nabla}^2-1),
\end{aligned}
\tag{4.107}
$$

where $\nu_1^2 = \dfrac{2\gamma+\beta}{4\alpha}$ and $k^2 = \dfrac{\gamma+\varepsilon}{4\alpha}$.

Equations (4.1) go over into equations (4.10). Moreover, we have

$$
\mathcal{L} \mapsto \nabla^2 L^2 \tag{4.108}
$$

and equation $(4.2)_1$ for the dilatation $e = \operatorname{div}\mathbf{u}$ is transformed into equation $(4.11)_1$. In order to obtain equation $(4.11)_2$ for $z = \operatorname{div}\boldsymbol{\varphi}$, we have to divide equation $(4.2)_2$ by the operator ∇^2. For the vector operators $\vec{\mathcal{L}}$, $\vec{\mathcal{D}}$ we obtain

$$
\vec{\mathcal{L}} \mapsto \vec{\nabla}^2\vec{\mathrm{L}}^2, \qquad \vec{\mathcal{D}} \mapsto \vec{\nabla}^2\vec{\mathrm{D}}^2. \tag{4.109}
$$

Now, taking into account (4.107) and (4.109), we arrive at

$$\begin{aligned}&[\nu_0\overrightarrow{\Box}_2-\alpha(\chi+\nu_0)\overrightarrow{\nabla}^2]\mapsto(\mu\nu_0-\alpha\chi)\overrightarrow{\nabla}^2,\\&\{\overrightarrow{\mathcal{D}}-4[\nu_0\overrightarrow{\Box}_2-\alpha(\chi+\nu_0)\overrightarrow{\nabla}^2]\}\mapsto\overrightarrow{\nabla}^2[\overrightarrow{\mathrm{D}}^2+4(\mu\nu_0-\alpha\chi)],\\&\{\overrightarrow{\mathcal{D}}^2+16[\nu_0\overrightarrow{\Box}_2-\alpha(\chi+\nu_0)\overrightarrow{\nabla}^2]^2\overrightarrow{\nabla}^2\}\\&\qquad\mapsto\overrightarrow{\nabla}^4[\overrightarrow{\mathrm{D}}^4+16(\nu_0\mu-\alpha\chi)^2\overrightarrow{\nabla}^2].\end{aligned}\tag{4.110}$$

We proceed in a similar manner with other operators. Dividing equation (4.4) by $\overrightarrow{\nabla}^2$ and equation (4.6) by $\overrightarrow{\nabla}^4$, we obtain separated static equations (4.13) and (4.15) for the vectors $\mathbf{u}$, $\boldsymbol{\varphi}$, respectively. Disregard the body vectors $\mathbf{X}$, $\mathbf{Y}$ and note that within the theory (1), the vectors $\mathbf{u}$, $\boldsymbol{\varphi}$ satisfy the same equation in $\Omega\times\mathbf{T}_+$:

$$\overrightarrow{\mathcal{L}}\Big\{\overrightarrow{\mathcal{D}}^2+16[\nu_0\overrightarrow{\Box}_2-\alpha(\chi+\nu_0)\overrightarrow{\nabla}^2]^2\overrightarrow{\nabla}^2\Big\}(\mathbf{u},\boldsymbol{\varphi})=(\mathbf{0},\mathbf{0}).\tag{4.111}$$

On the other hand, in the theory (1') the orders of equations for the vectors $\mathbf{u}$ and $\boldsymbol{\varphi}$ are different, and the same equation is satisfied by the vectors $\overrightarrow{\nabla}^2\mathbf{u}$ and $\boldsymbol{\varphi}$ in Ω:

$$\overrightarrow{\nabla}^2[\overrightarrow{\mathrm{D}}^4+16(\nu_0\mu-\alpha\chi)^2\overrightarrow{\nabla}^2]\overrightarrow{\mathrm{L}}^2(\overrightarrow{\nabla}^2\mathbf{u},\boldsymbol{\varphi})=(\mathbf{0},\mathbf{0}).\tag{4.112}$$

Case (1) → (2). Assuming

$$(\chi,\kappa,\nu_0)\mapsto(0,0,0),\tag{4.113}$$

we obtain

$$\mathcal{L}\mapsto\Box_1\Box_3.\tag{4.114}$$

Dividing equation $(4.2)_1$ by the operator $\Box_3$ and equation $(4.2)_2$ by the operator $\Box_1$, we obtain, respectively, in $\Omega\times\mathbf{T}_+$:

$$\begin{aligned}\Box_1 e&=-\operatorname{div}\mathbf{X},\\\Box_3 z&=-\operatorname{div}\mathbf{Y}.\end{aligned}\tag{4.115}$$

Disregard the body vectors $\mathbf{X}$, $\mathbf{Y}$ and note that in the theory (1), $\operatorname{div}\mathbf{u}$ and $\operatorname{div}\boldsymbol{\varphi}$ satisfy the same equation:

$$\mathcal{L}(e,z)=(0,0)\qquad\text{in }\Omega\times\mathbf{T}_+.\tag{4.116}$$

In the theory (2), we obtain the equations

$$\Box_1 e=0,\qquad\Box_3 z=0\qquad\text{in }\Omega\times\mathbf{T}_+.\tag{4.117}$$

For the operators $\overrightarrow{\mathcal{L}}$, $\overrightarrow{\mathcal{D}}$ we get

$$\begin{aligned} \vec{\mathcal{L}} &\mapsto \vec{\Box}_1\vec{\Box}_3, \\ \vec{\mathcal{D}} &\mapsto (\vec{\Box}_2\vec{\Box}_4 + 4\alpha^2\vec{\nabla}^2). \end{aligned} \tag{4.118}$$

Now, removing the operator $\vec{\Box}_3(\vec{\Box}_2\vec{\Box}_4 + 4\alpha^2\vec{\nabla}^2)$ from equation (4.4), we note that the latter is transformed into the following equation in $\Omega \times \mathbf{T}_+$:

$$\begin{aligned} \vec{\Box}_1(\vec{\Box}_2\vec{\Box}_4 + 4\alpha^2\vec{\nabla}^2)\mathbf{u} &= 2\alpha\vec{\Box}_1 \operatorname{curl}\mathbf{Y} \\ &+ [(\lambda+\mu-\alpha)\vec{\Box}_4 - 4\alpha^2] \operatorname{grad}\operatorname{div}\mathbf{X} - \vec{\Box}_1\vec{\Box}_4\mathbf{X}. \end{aligned} \tag{4.119}$$

In turn, dividing equation (4.6) by the operator $\vec{\Box}_1(\vec{\Box}_2\vec{\Box}_4 + 4\alpha^2\vec{\nabla}^2)$, we obtain, in $\Omega \times \mathbf{T}_+$:

$$\begin{aligned} \vec{\Box}_3(\vec{\Box}_2\vec{\Box}_4 + 4\alpha^2\vec{\nabla}^2)\boldsymbol{\varphi} &= 2\alpha\vec{\Box}_3 \operatorname{curl}\mathbf{X} \\ &+ [(\beta+\gamma-\varepsilon)\vec{\Box}_2 - 4\alpha^2] \operatorname{grad}\operatorname{div}\mathbf{Y} - \vec{\Box}_2\vec{\Box}_3\mathbf{Y}. \end{aligned} \tag{4.120}$$

Equations (4.119) and (4.120) refer to the theory (2) and are the separated equations for the vectors $\mathbf{u}$ and $\boldsymbol{\varphi}$, respectively, i.e. equations (1.19) and (1.20). Let us add that equations (4.1) are transformed into equations (1.1), in view of (4.113). When we disregard the body vectors $\mathbf{X}$ and $\mathbf{Y}$, we can see that in the theory (1) the vectors $\mathbf{u}$, $\boldsymbol{\varphi}$ satisfy the same equation (4.111). On the other hand, within the framework of the theory (2), we obtain

$$\begin{aligned} \vec{\Box}_1(\vec{\Box}_2\vec{\Box}_4 + 4\alpha^2\vec{\nabla}^2)\mathbf{u} &= \mathbf{0}, \\ \vec{\Box}_3(\vec{\Box}_2\vec{\Box}_4 + 4\alpha^2\vec{\nabla}^2)\boldsymbol{\varphi} &= \mathbf{0}. \end{aligned} \tag{4.121}$$

Case (1') → (2'). Assuming (4.113), we can see that equations (4.10) go over into equations (1.155). The operator L^2 takes the form

$$L^2 \mapsto 4\alpha(\lambda+2\mu)(\nu_1^2\nabla^2 - 1). \tag{4.122}$$

Removing the operator $(\nu_1^2\nabla^2 - 1)$ in the 4th order equation for the dilatation $e = \operatorname{div}\mathbf{u}$ (equation $(4.11)_1$), we notice that the latter is transformed into

$$\nabla^2 e = -\frac{1}{\lambda+2\mu}\operatorname{div}\mathbf{X}. \tag{4.123}$$

In turn, equation $(4.11)_2$ goes over into

$$(\nu_1^2\nabla^2 - 1)z = -\frac{1}{4\alpha}\operatorname{div}\mathbf{Y}, \tag{4.124}$$

where $z = \operatorname{div}\boldsymbol{\varphi}$ and $\nu_1^2 = \dfrac{2\gamma+\beta}{4\alpha}$.

The vector operators $\overrightarrow{\mathrm{L}}^2$ and $\overrightarrow{\mathrm{D}}^2$ take the form

$$\begin{aligned} \overrightarrow{\mathrm{L}}^2 &\mapsto 4\alpha(\lambda+2\mu)(\nu_1^2\overrightarrow{\nabla}^2-1), \\ \overrightarrow{\mathrm{D}}^2 &\mapsto 4\alpha\mu(l^2\overrightarrow{\nabla}^2-1), \end{aligned} \tag{4.125}$$

where $l^2 = \dfrac{(\gamma+\varepsilon)(\mu+\alpha)}{4\alpha\mu}$.

Dividing equation (4.13) by the operator $(l^2\overrightarrow{\nabla}^2-1)(\nu_1^2\overrightarrow{\nabla}^2-1)$, we obtain the following equation in Ω:

$$\begin{aligned} 2\mu\overrightarrow{\nabla}^2\overrightarrow{\nabla}^2(l^2\overrightarrow{\nabla}^2-1)\mathbf{u} &= \overrightarrow{\nabla}^2\operatorname{curl}\mathbf{Y} - 2\overrightarrow{\nabla}^2(l^2\overrightarrow{\nabla}^2-1)\mathbf{X} \\ &- \mu_0^{-1}\overrightarrow{\nabla}^2\operatorname{curl}\operatorname{curl}\mathbf{X} + (1-\nu)^{-1}(l^2\overrightarrow{\nabla}^2-1)\operatorname{grad}\operatorname{div}\mathbf{X}, \end{aligned} \tag{4.126}$$

where $\nu = \dfrac{\lambda}{2(\mu+\lambda)}$ and $\mu_0 = \dfrac{2\mu}{\gamma+\varepsilon}$.

In turn, dividing equation (4.15) by the operator $(l^2\overrightarrow{\nabla}^2-1)$, we obtain, in Ω:

$$\begin{aligned} 2\mu\overrightarrow{\nabla}^2(\nu_1^2\overrightarrow{\nabla}^2-1)(l^2\overrightarrow{\nabla}^2-1)\boldsymbol{\varphi} \\ = -(2\alpha)^{-1}\mu(l^2\overrightarrow{\nabla}^2-1)\operatorname{grad}\operatorname{div}\mathbf{Y} \\ + (\nu_1^2\overrightarrow{\nabla}^2-1)\operatorname{curl}\mathbf{X} + \alpha_0^{-1}(\nu_1^2\overrightarrow{\nabla}^2-1)\operatorname{curl}\operatorname{curl}\mathbf{Y}, \end{aligned} \tag{4.127}$$

where $\alpha_0 = \dfrac{2\alpha}{\mu+\alpha}$.

Equations (4.126) and (4.127) are the separated equations (1.171) and (1.172), which are satisfied by the vectors $\mathbf{u}$ and $\boldsymbol{\varphi}$, respectively, within the framework of the elastostatics of a centrosymmetric micropolar medium. We found that in the theory (1') (for $\mathbf{X}\equiv\mathbf{0}$ and $\mathbf{Y}\equiv\mathbf{0}$) the vectors $\overrightarrow{\nabla}^2\mathbf{u}$, $\boldsymbol{\varphi}$ satisfy equation (4.112) in Ω. On the other hand, within the model (2') we obtain the following equations in Ω:

$$\begin{aligned} \overrightarrow{\nabla}^2\overrightarrow{\nabla}^2(l^2\overrightarrow{\nabla}^2-1)\mathbf{u} &= \mathbf{0}, \\ \overrightarrow{\nabla}^2(\nu_1^2\overrightarrow{\nabla}^2-1)(l^2\overrightarrow{\nabla}^2-1)\boldsymbol{\varphi} &= \mathbf{0}. \end{aligned} \tag{4.128}$$

Case (2) → (2'). Disregarding the terms $\rho\partial_t^2$, $J\partial_t^2$ in the operators $\Box_\xi$, $\overrightarrow{\Box}_\xi$ ($\xi = 1$–4), we obtain, apart from (4.107), also

$$(\overrightarrow{\Box}_2\overrightarrow{\Box}_4 + 4\alpha^2\overrightarrow{\nabla}^2) \mapsto 4\alpha\mu\overrightarrow{\nabla}^2(l^2\overrightarrow{\nabla}^2-1). \tag{4.129}$$

Now equations (1.1) are transformed into equations (1.155). From equations (4.115) for $\operatorname{div}\mathbf{u}$ and $\operatorname{div}\boldsymbol{\varphi}$ we obtain (4.123) and (4.124), respectively. Finally, transforming the right-hand sides of equations (4.119) and (4.120) appropriately and using equation $(4.102)_2$, we arrive at equations (4.126) and

(4.127). In the sequel, we will derive equations for the limiting problems (i.e. for plane and axially-symmetric strain states) from vector equations (1.1), (4.119), (4.120) and equations (1.155), (4.126), (4.127), referring to (2) and (2'), respectively.

4.4.3 Displacement-rotation equations describing plane and axially-symmetric problems of micropolar theory

The starting point for further considerations is given by vector equations (1.1), i.e. the following equations in $\Omega \times \mathbf{T}_+$:

$$\begin{aligned} &\vec{\Box}_2 \mathbf{u} + (\lambda + \mu - \alpha)\, \mathrm{grad}\,\mathrm{div}\, \mathbf{u} + 2\alpha\, \mathrm{curl}\, \boldsymbol{\varphi} + \mathbf{X} = \mathbf{0}, \\ &\vec{\Box}_4 \boldsymbol{\varphi} + (\beta + \gamma - \varepsilon)\, \mathrm{grad}\,\mathrm{div}\, \boldsymbol{\varphi} + 2\alpha\, \mathrm{curl}\, \mathbf{u} + \mathbf{Y} = \mathbf{0} \end{aligned} \tag{4.130}$$

and the separated equations (4.119) and (4.120). Let us notice a formal similarity between the structures of equations $(4.130)_1$ and $(4.130)_2$, and those of equations (4.119) and (4.120). If we assume the correspondence

$$\begin{gathered} \vec{\Box}_1 \mapsto \vec{\Box}_3 \\ \vec{\Box}_2 \mapsto \vec{\Box}_4 \\ \vec{\Box}_4 \mapsto \vec{\Box}_2 \\ (\lambda + \mu - \alpha) \mapsto (\beta + \gamma - \varepsilon) \\ \mathbf{u} \mapsto \boldsymbol{\varphi} \\ \mathbf{X} \mapsto \mathbf{Y} \end{gathered} \tag{4.131}$$

then equation $(4.130)_1$ is transformed into equation $(4.130)_2$, and equation (4.119) goes over into equation (4.120). It is assumed that the symbols α, $\mathbf{0}$, $\vec{\nabla}^2$, curl, grad div remain unaltered. The given analogy is a tool for additional control in the derivation of equations for plane and axially-symmetric problems. Let us go over to the derivation of the separated equations for u_i, φ_i in the rectangular system of coordinates $Ox_1x_2x_3$. For this purpose, in equations (4.119) and (4.120) for the vector $\mathbf{a} = a_i \mathbf{e}_i$, we should assume

$$\vec{\nabla}^2 \mathbf{a} \stackrel{\mathrm{def}}{=} \mathbf{e}_i \nabla^2 a_i, \tag{4.132}$$

where $\{\mathbf{e}_i\}$ is a basis. The vector Laplace operator $\vec{\nabla}^2$ is equivalent to the triple $(\nabla^2, \nabla^2, \nabla^2)$ of scalar Laplace operators, where $\nabla^2(.) = (.)_{,ii}$. Similarly, taking into account that $\vec{\Box}_\xi \equiv (\Box_\xi, \Box_\xi, \Box_\xi)$, ($\xi$ = 1–4) and that the i-th components of the vectors curl $\mathbf{a}$ and grad div $\mathbf{a}$ take the form $\epsilon_{ijk} a_{k,j}$ and $a_{j,ji}$, respectively, where ϵ_{ijk} is the permutation symbol, equations (4.119) and (4.120) are transformed into the respective equations of (1.29), i.e. into the following equations in $\Omega \times \mathbf{T}_+$:

$$
\begin{aligned}
&\Box_1(\Box_2\Box_4 + 4\alpha^2\nabla^2)u_i = 2\alpha\Box_1\epsilon_{ijk}Y_{k,j} \\
&\qquad + [(\lambda+\mu-\alpha)\Box_4 - 4\alpha^2]X_{j,ji} - \Box_1\Box_4 X_i, \\
&\Box_3(\Box_2\Box_4 + 4\alpha^2\nabla^2)\varphi_i = 2\alpha\Box_3\epsilon_{ijk}X_{k,j} \\
&\qquad + [(\beta+\gamma-\varepsilon)\Box_2 - 4\alpha^2]Y_{j,ji} - \Box_2\Box_3 Y_i.
\end{aligned}
\tag{4.133}
$$

Recall that we are using the index notation subject to the summation convention (comma denotes partial derivative with respect to the position variable, Latin indices i, j, k, ... assume the values 1, 2, 3). Let us go over to the derivation of separated equations for displacements and rotations in two-dimensional problems. This is illustrated by the diagram in Fig. 4.8.

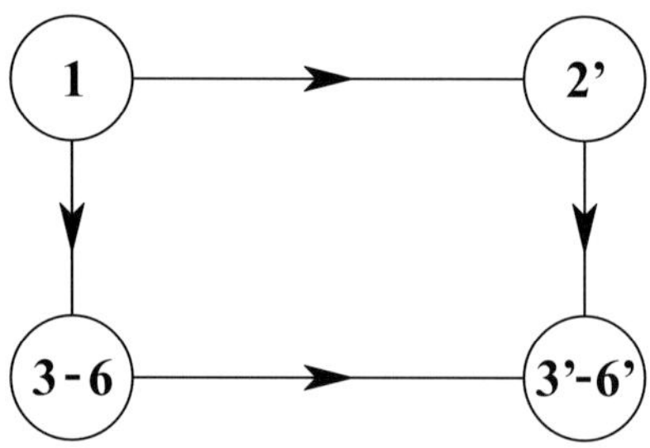

Fig. 4.8. Notation used in the figure is the following: symbols (2) and (2') have the same meaning as in Subsection 4.4.2; symbols (3)–(6) denote, respectively, the first and the second plane problems and the first and the second axially-symmetric problems in elastodynamics. Symbols (3')–(6') refer to elastostatics

Case (2) → ((3), (4)). In micropolar theory, the first and the second plane deformation states are described in the coordinate system of position coordinates Ox_1x_2 by the displacement vector $\mathbf{u}$ and the rotation vector $\boldsymbol{\varphi}$ in the form

$$\mathbf{u} \overset{\text{def}}{=} (u_1, u_2, 0), \qquad \boldsymbol{\varphi} \overset{\text{def}}{=} (0, 0, \varphi_3), \tag{4.134}$$

$$\mathbf{u} \overset{\text{def}}{=} (0, 0, u_3), \qquad \boldsymbol{\varphi} \overset{\text{def}}{=} (\varphi_1, \varphi_2, 0), \tag{4.135}$$

respectively.

In order to transform vector equations (4.119) and (4.120) into equations of the plane problems mentioned, it is necessary to take into account the fact that the spatial equations (4.133), when referred to two-dimensional problems (me mean here the position variables), yield for the components u_3, φ_3 equations of higher orders. The respective equations for those components of diplacements and rotations are degenerate due to the assumption that the derivative $(\partial/\partial x_3)(.)$ vanishes. We obtain the zero component of the vector $\operatorname{grad}\operatorname{div}\mathbf{a}$, where $\mathbf{a} \stackrel{\text{def}}{=} (a_1, a_2, a_3)$. Let us take into account that we have

$$
\begin{aligned}
&\operatorname{grad}\operatorname{div}\mathbf{a} = (a_{,\alpha},\, 0), \qquad a = a_{\beta,\beta},\\
&\operatorname{curl}\mathbf{a} = (\epsilon_{\alpha\beta}a_{3,\beta},\, \epsilon_{\delta\gamma}a_{\gamma,\delta}),\\
&\nabla^2 \to \nabla_1^2, \qquad \nabla_1^2(.) = (.)_{,\alpha\alpha},\\
&\vec{\nabla}^2 \equiv (\nabla_1^2, \nabla_1^2, \nabla_1^2),\\
&\vec{\Box}_\xi \equiv (\Box_\xi, \Box_\xi, \Box_\xi), \qquad \xi = 1\text{–}4,
\end{aligned}
\tag{4.136}
$$

where $\Box_\xi$ — the operator determined by the two-dimensional Laplacian ∇_1^2, $\epsilon_{\alpha\beta}$ — two-indicial permutation symbol, and α, β, γ, δ are indices assuming the values 1, 2.

Now, using (4.136), we obtain from equations (4.119) and (4.120) for $i = \alpha$ the following equations in $\Omega \times \mathbf{T}_+$:

$$
\begin{aligned}
\Box_1(\Box_2\Box_4 + 4\alpha^2\nabla_1^2)u_\alpha &= [(\lambda+\mu-\alpha)\Box_4 - 4\alpha^2]X_{\beta,\beta\alpha}\\
&\quad - \Box_1\Box_4 X_\alpha + 2\alpha\Box_1\epsilon_{\alpha\beta}Y_{3,\beta},
\end{aligned}
\tag{4.137}
$$

$$
\begin{aligned}
\Box_3(\Box_2\Box_4 + 4\alpha^2\nabla_1^2)\varphi_\alpha &= 2\alpha\Box_3\epsilon_{\alpha\beta}X_{3,\beta}\\
&\quad + [(\beta+\gamma-\varepsilon)\Box_2 - 4\alpha^2]Y_{\beta,\beta\alpha} - \Box_2\Box_3 Y_\alpha,
\end{aligned}
\tag{4.138}
$$

respectively.

When $i = 3$, removing the operator $\Box_1$ from equation (4.119) and the operator $\Box_3$ from equation (4.120), we obtain

$$
\begin{aligned}
(\Box_2\Box_4 + 4\alpha^2\nabla_1^2)u_3 &= -\Box_4 X_3 - 2\alpha\epsilon_{\delta\gamma}Y_{\delta,\gamma},\\
(\Box_2\Box_4 + 4\alpha^2\nabla_1^2)\varphi_3 &= -\Box_2 Y_3 - 2\alpha\epsilon_{\delta\gamma}X_{\delta,\gamma},
\end{aligned}
\tag{4.139}
$$

respectively.

Equations (4.137) and $(4.139)_2$ refer to the first plane problem of micropolar elastodynamics and they are equations (3.28). The body loading vectors take the form $\mathbf{X} = (X_1, X_2, 0)$ and $\mathbf{Y} = (0, 0, Y_3)$. In turn, equations (4.138) and $(4.139)_1$ refer to the 2nd plane problem and they coincide with equations (3.171). The triple (Y_α, X_3) refers to body loadings. Let us add that the vector equations (4.130) are transformed into equations (3.1) for the triple (u_α, φ_3) and into equations (3.146) — for the triple (φ_α, u_3).

Case (2') → ((3'), (4')). We begin with the vector equations (1.155), i.e. the following equations in Ω:

$$\begin{aligned}&(\mu+\alpha)\vec{\nabla}^2\mathbf{u}+(\lambda+\mu-\alpha)\,\mathrm{grad}\,\mathrm{div}\,\mathbf{u}+2\alpha\,\mathrm{curl}\,\boldsymbol{\varphi}+\mathbf{X}=\mathbf{0},\\&[(\gamma+\varepsilon)\vec{\nabla}^2-4\alpha]\boldsymbol{\varphi}+(\beta+\gamma-\varepsilon)\,\mathrm{grad}\,\mathrm{div}\,\boldsymbol{\varphi}+2\alpha\,\mathrm{curl}\,\mathbf{u}+\mathbf{Y}=\mathbf{0}\end{aligned}\tag{4.140}$$

and with the separated equations (4.126) and (4.127). In the rectangular system of coordinates $Ox_1x_2x_3$ equations (4.140), (4.126) and (4.127) take the form of (1.190) and (1.191), respectively. Assuming plane deformation state (causes and effects do not depend on the variable x_3), the system of equations (4.140) is transformed (in the coordinate system Ox_1x_2) into two independent systems of equations, (3.65) and (3.217), describing the first and the second plane problems, respectively. Separated equations for the first and the second plane problems are obtained from equations (4.126) and (4.127) by means of (4.136) in the following way. Equation (4.126) for $i=\alpha$ and equation (4.127) for $i=3$ (the latter — after removing the operator $(\nu_1^2\nabla_1^2-1)$) lead to the following equations in Ω:

$$\begin{aligned}&2\mu\nabla_1^2\nabla_1^2(l^2\nabla_1^2-1)u_\alpha=\epsilon_{\alpha\beta}\nabla_1^2Y_{3,\beta}+\mu_0^{-1}\nabla_1^2\epsilon_{\alpha\beta}\epsilon_{\delta\gamma}X_{\delta,\gamma\beta}\\&\qquad-2\nabla_1^2(l^2\nabla_1^2-1)X_\alpha+(1-\nu)^{-1}(l^2\nabla_1^2-1)X_{\beta,\beta\alpha},\\&2\mu\nabla_1^2(l^2\nabla_1^2-1)\varphi_3=-\alpha_0^{-1}\nabla_1^2Y_3-\epsilon_{\delta\gamma}X_{\delta,\gamma},\end{aligned}\tag{4.141}$$

where

$$l^2=\frac{(\gamma+\varepsilon)(\mu+\alpha)}{4\alpha\mu},\qquad \mu_0=\frac{2\mu}{\gamma+\varepsilon},\qquad \alpha_0=\frac{2\alpha}{\mu+\alpha},\qquad \nu=\frac{\lambda}{2(\mu+\lambda)}.$$

They coincide with equations (3.76) for the first plane problem. Let us point out that removing the operator $(\nu_1^2\nabla^2-1)$ can be connected with the constant β in its definition ($\nu_1^2=\frac{2\gamma+\beta}{4\alpha}$), which does not appear in the first plane problem. This remark applies also to elastodynamics and the operator

$$\Box_3=(\beta+2\gamma)\nabla^2-4\alpha-J\partial_t^2,$$

as well as to the first axially-symmetric problem. Equation (4.127) for $i=\alpha$ and equation (4.126) for $i=3$ (the latter — after removing the operator ∇_1^2) lead to the following equations for the second plane problem

$$\begin{aligned}&2\mu\nabla_1^2(\nu_1^2\nabla_1^2-1)(l^2\nabla_1^2-1)\varphi_\alpha=-(2\alpha)^{-1}\mu(l^2\nabla_1^2-1)Y_{\beta,\beta\alpha}\\&\qquad+(\nu_1^2\nabla_1^2-1)\epsilon_{\alpha\beta}X_{3,\beta}+\alpha_0^{-1}(\nu_1^2\nabla_1^2-1)\epsilon_{\alpha\beta}\epsilon_{\delta\gamma}Y_{\gamma,\delta\beta},\\&2\mu\nabla_1^2(l^2\nabla_1^2-1)u_3=\epsilon_{\alpha\beta}Y_{\beta,\alpha}-2(l^2\nabla_1^2-1)X_3+\mu_0^{-1}\nabla_1^2X_3.\end{aligned}\tag{4.142}$$

Case (2) → ((5), (6)). In the cylindrical coordinate system (r,θ,z), the first and the second axially-symmetric problem of micropolar theory are described by the vectors $\mathbf{u}$ and $\boldsymbol{\varphi}$ of the form

$$\mathbf{u} \stackrel{\text{def}}{=} (u_r, 0, u_z), \qquad \boldsymbol{\varphi} \stackrel{\text{def}}{=} (0, \varphi_\theta, 0), \tag{4.143}$$

$$\mathbf{u} \stackrel{\text{def}}{=} (0, u_\theta, 0), \qquad \boldsymbol{\varphi} \stackrel{\text{def}}{=} (\varphi_r, 0, \varphi_z). \tag{4.144}$$

In order to obtain separated equations for the triples $(u_r, \varphi_\theta, u_z)$ and $(\varphi_r, u_\theta, \varphi_z)$ from vector equations (4.119) and (4.120), respectively, one needs to take into account the fact that equations for the components u_θ, φ_θ will be of higher orders. Separated equations for these components are degenerate due to the assumption that the derivative $(\partial/\partial\theta)(.)$ vanishes. We obtain the zero component of the vector grad div $\mathbf{a}$, where $\mathbf{a} \stackrel{\text{def}}{=} (a_r, a_\theta, a_z)$. Therefore we have

$$\begin{aligned} \operatorname{grad}\operatorname{div}\mathbf{a} &= (\partial_r a, 0, \partial_z a), \\ \operatorname{curl}\mathbf{a} &= \left(-\partial_z a_\theta, (\partial_z a_r - \partial_r a_z), \left(r^{-1} + \partial_r\right) a_\theta\right), \end{aligned} \tag{4.145}$$

where $a = \left(r^{-1} + \partial_r\right) a_r + \partial_z a_z$.

The vector Laplace operator $\vec{\nabla}^2$ is equivalent to the triple of scalar operators $(\nabla_0^2, \nabla_0^2, \nabla^2)$:

$$\vec{\nabla}^2\mathbf{a} = (\nabla_0^2 a_r, \nabla_0^2 a_\theta, \nabla^2 a_z), \tag{4.146}$$

where the Laplacian $\nabla^2 = \partial_r^2 + r^{-1}\partial_r + \partial_z^2$ and $\nabla_0^2 = \nabla^2 - r^{-2}$.

Similarly, we have

$$\vec{\Box}_\xi \equiv (\Box_\xi^0, \Box_\xi^0, \Box_\xi), \qquad \xi = 1\text{–}4, \tag{4.147}$$

where the operators $\Box_\xi$, $\Box_\xi^0$ are determined by ∇^2, ∇_0^2, respectively. Note also the following operators:

$$\begin{aligned} \nabla^2\left(r^{-1} + \partial_r\right) &\equiv \left(r^{-1} + \partial_r\right)\nabla_0^2, \\ \partial_r\nabla^2 &\equiv \nabla_0^2\partial_r, \end{aligned} \tag{4.148}$$

which are valid also for the operator $\Box_\xi$. Now, from equations (4.119) and (4.120) (we divide the equation for the component u_θ by the operator $\Box_1^0$ and the equation for φ_θ — by $\Box_3^0$) we obtain separated equations (2.2), (2.3) and (2.249) for the first and the second axially-symmetric problem, respectively.

Case (2') → ((5'), (6')). The starting point is given by the vector equations (4.140), (4.126) and (4.127). In the cylindrical coordinate system (r, θ, z) and under the assumption of axial symmetry, equations (4.140) are transformed into two independent systems of equations, (2.101) and (2.351), for the first and the second axially-symmetric problems, respectively. Separated equations for the triples (4.143) and (4.144) are obtained from equations (4.126) and (4.127) by using (4.145)–(4.148) and taking into account the fact that for the vector $\mathbf{a} \stackrel{\text{def}}{=} (a_r, a_\theta, a_z)$, we have

$$\operatorname{curl}\operatorname{curl}\mathbf{a} = \Big(-\partial_z(\partial_z a_r - \partial_r a_z), -\nabla_0^2 a_\theta, \left(r^{-1} + \partial_r\right)(\partial_z a_r - \partial_r a_z)\Big). \quad (4.149)$$

Now, from equation (4.126) we obtain the following equations:

$$\begin{aligned} 2\mu\nabla_0^2\nabla_0^2(l^2\nabla_0^2 - 1)u_r = & -\partial_z\nabla_0^2 Y_\theta - 2\nabla_0^2(l^2\nabla_0^2 - 1)X_r \\ & + \mu_0^{-1}\nabla_0^2\partial_z(\partial_z X_r - \partial_r X_z) \\ & + (1-\nu)^{-1}(l^2\nabla_0^2 - 1)\partial_r X, \\ 2\mu\nabla^2\nabla^2(l^2\nabla^2 - 1)u_z = & \nabla^2\left(r^{-1} + \partial_r\right)Y_\theta - 2\nabla^2(l^2\nabla^2 - 1)X_z \\ & + (1-\nu)^{-1}(l^2\nabla^2 - 1)\partial_z X \\ & - \mu_0^{-1}\nabla^2\left(r^{-1} + \partial_r\right)(\partial_z X_r - \partial_r X_z), \end{aligned} \quad (4.150)$$

where $X = \left(r^{-1} + \partial_r\right)X_r + \partial_z X_z$,

$$\mu_0 = \frac{2\mu}{\gamma + \varepsilon}, \qquad \nu = \frac{\lambda}{2(\mu + \lambda)}, \qquad l^2 = \frac{(\gamma + \varepsilon)(\mu + \alpha)}{4\alpha\mu}.$$

Removing the operator $(\nu_1^2\nabla_0^2 - 1)$ from equation (4.127), we arrive at

$$2\mu\nabla_0^2(l^2\nabla_0^2 - 1)\varphi_\theta = -\alpha_0^{-1}\nabla_0^2 Y_\theta + (\partial_z X_r - \partial_r X_z), \quad (4.151)$$

where $\alpha_0 = \dfrac{2\alpha}{\mu + \alpha}$.

Equations with the triple $(u_r, \varphi_\theta, u_z)$, (4.150) and (4.151), refer to the first axially-symmetric problem, in which the body loadings of the following form appear:

$$\mathbf{X} = (X_r, 0, X_z), \qquad \mathbf{Y} = (0, Y_\theta, 0). \quad (4.152)$$

From equation (4.127) we obtain

$$\begin{aligned} & 2\mu\nabla_0^2(\nu_1^2\nabla_0^2 - 1)(l^2\nabla_0^2 - 1)\varphi_r = -(2\alpha)^{-1}\mu(l^2\nabla_0^2 - 1)\partial_r Y \\ & \quad - (\nu_1^2\nabla_0^2 - 1)\partial_z X_\theta - \alpha_0^{-1}(\nu_1^2\nabla_0^2 - 1)\partial_z(\partial_z Y_r - \partial_r Y_z), \\ & 2\mu\nabla^2(\nu_1^2\nabla^2 - 1)(l^2\nabla^2 - 1)\varphi_z = -(2\alpha)^{-1}\mu(l^2\nabla^2 - 1)\partial_z Y \\ & \quad + (\nu_1^2\nabla^2 - 1)\left(r^{-1} + \partial_r\right)X_\theta \\ & \quad + \alpha_0^{-1}(\nu_1^2\nabla^2 - 1)\left(r^{-1} + \partial_r\right)(\partial_z Y_r - \partial_r Y_z), \end{aligned} \quad (4.153)$$

where $Y = \left(r^{-1} + \partial_r\right)Y_r + \partial_z Y_z$ and $\nu_1^2 = \dfrac{2\gamma + \beta}{4\alpha}$.
Equation (4.126) leads to the following equation, after division by the operator ∇_0^2:

$$2\mu\nabla_0^2(l^2\nabla_0^2 - 1)u_\theta = (\partial_z Y_r - \partial_r Y_z) - 2(l^2\nabla_0^2 - 1)X_\theta + \mu_0^{-1}\nabla_0^2 X_\theta. \quad (4.154)$$

Equations (4.153) and (4.154) for the triple $(\varphi_r, u_\theta, \varphi_z)$ refer to the second axially-symmetric problem, in which body loadings have the form

$$\mathbf{X} = (0, X_\theta, 0), \qquad \mathbf{Y} = (Y_r, 0, Y_z). \tag{4.155}$$

Case (3)–(6) → (3')–(6'). Here, similarly to the case (2) → (2') discussed in Subsection 4.4.2, one should disregard the inertia terms in the dynamical operators which appear in the equations obtained in (3)–(6), and perform certain transformations on the right-hand sides of the equations, using the definition given by equation $(4.102)_2$. Then one arrives at the respective equations for the problems (3')–(6').

4.4.4 Superposition method. Analysis of equations with the vector ζ

In this subsection we shall present a few remarks pertaining to the vector equations of micropolar elastodynamics and elastostatics, which appear in the superposition method. These remarks will concern the procedure of obtaining from such vector equations the corresponding equations in the superposition method referring to plane and axially-symmetric states of strain. Moreover, from the vector complementary equations in the superposition method we shall derive separated equations for the displacement vector $\mathbf{u}$ and the rotation vector $\boldsymbol{\varphi}$. We start from equations (4.130), to which we introduce the substitution

$$\boldsymbol{\zeta} = \tfrac{1}{2}\operatorname{curl}\mathbf{u} - \boldsymbol{\varphi}. \tag{4.156}$$

The system of equations (4.130) is transformed into a system of equations for the pair $(\mathbf{u}, \boldsymbol{\zeta})$ in $\Omega \times \mathbf{T}_+$, namely:

$$\begin{aligned} &\overset{*}{\overrightarrow{\Box}}_2 \mathbf{u} + (\lambda + \mu)\operatorname{grad}\operatorname{div}\mathbf{u} + \mathbf{X} = 2\alpha\operatorname{curl}\boldsymbol{\zeta}, \\ &\tfrac{1}{2}\overset{*}{\overrightarrow{\Box}}_4 \operatorname{curl}\mathbf{u} + \mathbf{Y} = \overrightarrow{\Box}_4 \boldsymbol{\zeta} + (\beta + \gamma - \varepsilon)\operatorname{grad}\operatorname{div}\boldsymbol{\zeta}, \end{aligned} \tag{4.157}$$

where $\overset{*}{\overrightarrow{\Box}}_2 = \mu\overrightarrow{\nabla}^2 - \rho\partial_t^2$ and $\overset{*}{\overrightarrow{\Box}}_4 = (\gamma + \varepsilon)\overrightarrow{\nabla}^2 - J\partial_t^2$.

We apply the operator $\overset{*}{\overrightarrow{\Box}}_2$ to equation $(4.157)_2$ and arrive at

$$\overset{*}{\overrightarrow{\Box}}_2 \left[\tfrac{1}{2}\overset{*}{\overrightarrow{\Box}}_4 \operatorname{curl}\mathbf{u} - \overrightarrow{\Box}_4 \boldsymbol{\zeta} - (\beta + \gamma - \varepsilon)\operatorname{grad}\operatorname{div}\boldsymbol{\zeta} + \mathbf{Y}\right] = \mathbf{0}. \tag{4.158}$$

Let us assume that for the system of equations (4.130) we have stated an initial-boundary value problem. Then we take the solution for the pair $(\mathbf{u}, \boldsymbol{\varphi})$ in the form

$$\mathbf{u} = \mathbf{u}^0 + \mathbf{u}', \qquad \boldsymbol{\varphi} = \boldsymbol{\varphi}^0 + \boldsymbol{\varphi}' \qquad \text{in } \Omega \times \mathbf{T}_+, \tag{4.159}$$

where (as in Subsection 1.1.1) the pair $(\mathbf{u}^0, \boldsymbol{\varphi}^0)$ constitutes the classical solution, whereas the pair $(\mathbf{u}', \boldsymbol{\varphi}')$ is the complementary solution. This effect is obtained by taking the vector $\boldsymbol{\zeta}$ in the form $\boldsymbol{\zeta} = \boldsymbol{\zeta}^0 + \boldsymbol{\zeta}'$. In the sequel we will not analyse the initial-boundary conditions and we will fix our attention on the field equations. Putting $\boldsymbol{\zeta}^0 = \mathbf{0}$, we derive from (4.157)–(4.159) the classical equations in $\Omega \times \mathbf{T}_+$:

$$\begin{aligned} &\overset{*}{\overrightarrow{\square}}_2 \mathbf{u}^0 + (\lambda+\mu)\,\mathrm{grad\,div}\,\mathbf{u}^0 + \mathbf{X} = \mathbf{0}, \\ &\overset{*}{\overrightarrow{\square}}_2 \boldsymbol{\varphi}^0 + \tfrac{1}{2}\,\mathrm{curl}\,\mathbf{X} = \mathbf{0}, \end{aligned} \tag{4.160}$$

where $\boldsymbol{\varphi}^0 = \frac{1}{2}\,\mathrm{curl}\,\mathbf{u}^0$, from which we obtain the separated equation in $\Omega \times \mathbf{T}_+$:

$$\overrightarrow{\square}_1 \overset{*}{\overrightarrow{\square}}_2 \mathbf{u}^0 = -\overrightarrow{\square}_1 \mathbf{X} + (\lambda+\mu)\,\mathrm{grad\,div}\,\mathbf{X}. \tag{4.161}$$

Moreover, from (4.157)–(4.159) we obtain a system of equations for the pair $(\mathbf{u}', \boldsymbol{\zeta} = \boldsymbol{\zeta}')$ in $\Omega \times \mathbf{T}_+$:

$$\begin{aligned} &\overset{*}{\overrightarrow{\square}}_2 \mathbf{u}' + (\lambda+\mu)\,\mathrm{grad\,div}\,\mathbf{u}' = 2\alpha\,\mathrm{curl}\,\boldsymbol{\zeta}, \\ &\overset{*}{\overrightarrow{\square}}_2 \Big\{ \overset{*}{\overrightarrow{\square}}_4\,\mathrm{curl}\,\mathbf{u}' - 2\big[\overrightarrow{\square}_4 \boldsymbol{\zeta} + (\beta+\gamma-\varepsilon)\,\mathrm{grad\,div}\,\boldsymbol{\zeta}\big] + 2\mathbf{Y} \Big\} \\ &\qquad = \overset{*}{\overrightarrow{\square}}_4\,\mathrm{curl}\,\mathbf{X}. \end{aligned} \tag{4.162}$$

From (4.162) one can derive separate equations for the vector $\mathbf{u}'$ and the vector $\boldsymbol{\zeta}$. Such equations take the following form in $\Omega \times \mathbf{T}_+$:

$$\begin{aligned} &\overset{*}{\overrightarrow{\square}}_2 \overrightarrow{\square}_1 (\overrightarrow{\square}_2 \overrightarrow{\square}_4 + 4\alpha^2 \overrightarrow{\nabla}^2)\mathbf{u}' = 2\alpha \overrightarrow{\square}_1\,\mathrm{curl}\Big(\overset{*}{\overrightarrow{\square}}_2 \mathbf{Y} - \tfrac{1}{2} \overset{*}{\overrightarrow{\square}}_4\,\mathrm{curl}\,\mathbf{X} \Big), \\ &\overset{*}{\overrightarrow{\square}}_2 \overrightarrow{\square}_3 (\overrightarrow{\square}_2 \overrightarrow{\square}_4 + 4\alpha^2 \overrightarrow{\nabla}^2)\boldsymbol{\zeta} \\ &\qquad = \overset{*}{\overrightarrow{\square}}_2 \Big[\alpha \overset{*}{\overrightarrow{\square}}_4 - (\beta+\gamma-\varepsilon) \overset{*}{\overrightarrow{\square}}_2 \Big]\,\mathrm{grad\,div}\,\mathbf{Y} \\ &\qquad\quad + \overset{*}{\overrightarrow{\square}}_2 \overrightarrow{\square}_3 \Big(\overset{*}{\overrightarrow{\square}}_2 \mathbf{Y} - \tfrac{1}{2} \overset{*}{\overrightarrow{\square}}_4\,\mathrm{curl}\,\mathbf{X} \Big). \end{aligned} \tag{4.163}$$

Further, from (4.162) we eliminate the vector $\boldsymbol{\zeta} = \frac{1}{2}\,\mathrm{curl}\,\mathbf{u}' - \boldsymbol{\varphi}'$ and we obtain a system of equations for the pair $(\mathbf{u}', \boldsymbol{\varphi}')$ in $\Omega \times \mathbf{T}_+$:

$$\begin{aligned} &\overrightarrow{\square}_2 \mathbf{u}' + (\lambda+\mu-\alpha)\,\mathrm{grad\,div}\,\mathbf{u}' + 2\alpha\,\mathrm{curl}\,\boldsymbol{\varphi}' = \mathbf{0}, \\ &\overset{*}{\overrightarrow{\square}}_2 \Big[\overrightarrow{\square}_4 \boldsymbol{\varphi}' + (\beta+\gamma-\varepsilon)\,\mathrm{grad\,div}\,\boldsymbol{\varphi}' + 2\alpha\,\mathrm{curl}\,\mathbf{u}' + \mathbf{Y} \Big] \\ &\qquad = \tfrac{1}{2} \overset{*}{\overrightarrow{\square}}_4\,\mathrm{curl}\,\mathbf{X}. \end{aligned} \tag{4.164}$$

From the system of equations (4.164) we derive a separate equation for the vector $\boldsymbol{\varphi}'$:

$$
\begin{aligned}
&\overset{*}{\vec{\Box}}_2 \vec{\Box}_3 (\vec{\Box}_2 \vec{\Box}_4 + 4\alpha^2 \vec{\nabla}^2)\varphi' \\
&\quad = \overset{*}{\vec{\Box}}_2 \left[(\beta + \gamma - \varepsilon)\vec{\Box}_2 - 4\alpha^2 \right] \operatorname{grad} \operatorname{div} \mathbf{Y} \\
&\qquad - \vec{\Box}_2 \vec{\Box}_3 \left(\overset{*}{\vec{\Box}}_2 \mathbf{Y} - \tfrac{1}{2} \overset{*}{\vec{\Box}}_4 \operatorname{curl} \mathbf{X} \right).
\end{aligned}
\tag{4.165}
$$

Now, on the basis of equations (4.162)–(4.165), we shall present special cases of these equations. They will be equations describing plane and axially-symmetric states of strain in the superposition method.

The first plane problem. The systems of functions $(u^0_\alpha, \varphi^0_3)$, $(u'_\alpha, \varphi'_3, \zeta_3)$ and the body loadings X_α, Y_3 are connected with the considered problem. Taking into account the considerations from Subsection 4.4.3 and formulae (4.132) and (4.136), we obtain

$$
\overset{*}{\vec{\Box}}_2 \equiv (\overset{*}{\Box}_2, \overset{*}{\Box}_2, \overset{*}{\Box}_2), \qquad \overset{*}{\vec{\Box}}_4 \equiv (\overset{*}{\Box}_4, \overset{*}{\Box}_4, \overset{*}{\Box}_4), \tag{4.166}
$$

where $\overset{*}{\Box}_2 = \mu \nabla_1^2 - \rho \partial_t^2$, $\overset{*}{\Box}_4 = (\gamma + \varepsilon)\nabla_1^2 - J\partial_t^2$ and $\nabla_1^2(.) = (.)_{,\alpha\alpha}$. Now, from equation $(4.163)_1$ for $i = \alpha$, we obtain an equation for u'_α in $\Omega \times \mathbf{T}_+$, namely

$$
\overset{*}{\Box}_2 \Box_1 (\Box_2 \Box_4 + 4\alpha^2 \nabla_1^2) u'_\alpha = 2\alpha \Box_1 \epsilon_{\alpha\beta} \left(\overset{*}{\Box}_2 Y_3 - \tfrac{1}{2} \overset{*}{\Box}_4 \epsilon_{\gamma\delta} X_{\delta,\gamma} \right)_{,\beta}. \tag{4.167}
$$

Removing the operator $\overset{*}{\Box}_2 \Box_3$ from equation $(4.163)_2$ for $i = 3$, we obtain an equation for ζ_3 in $\Omega \times \mathbf{T}_+$, namely

$$
(\Box_2 \Box_4 + 4\alpha^2 \nabla_1^2)\zeta_3 = \overset{*}{\Box}_2 Y_3 - \tfrac{1}{2} \overset{*}{\Box}_4 \epsilon_{\alpha\beta} X_{\beta,\alpha}. \tag{4.168}
$$

In turn, removing the operator $\Box_3$ from equation (4.165) for $i = 3$, we obtain an equation for φ'_3 in $\Omega \times \mathbf{T}_+$:

$$
\overset{*}{\Box}_2 (\Box_2 \Box_4 + 4\alpha^2 \nabla_1^2) \varphi'_3 = -\Box_2 \left(\overset{*}{\Box}_2 Y_3 - \tfrac{1}{2} \overset{*}{\Box}_4 \epsilon_{\alpha\beta} X_{\beta,\alpha} \right). \tag{4.169}
$$

From the system of equations (4.162) we obtain

$$
\begin{aligned}
&\overset{*}{\Box}_2 u'_\alpha + (\lambda + \mu) u'_{\beta,\beta\alpha} = 2\alpha \epsilon_{\alpha\gamma} \zeta_{3,\gamma}, \\
&\overset{*}{\Box}_2 (\overset{*}{\Box}_4 \epsilon_{\alpha\beta} u'_{\beta,\alpha} - 2\Box_4 \zeta_3 + 2Y_3) = \overset{*}{\Box}_4 \epsilon_{\alpha\beta} X_{\beta,\alpha}.
\end{aligned}
\tag{4.170}
$$

The system of equations (4.164) for the problem considered (without the function ζ_3) takes the following form in $\Omega \times \mathbf{T}_+$:

$$
\begin{aligned}
&\Box_2 u'_\alpha + (\lambda + \mu - \alpha) u'_{\beta,\beta\alpha} + 2\alpha \epsilon_{\alpha\gamma} \varphi'_{3,\gamma} = 0, \\
&\overset{*}{\Box}_2 (\Box_4 \varphi'_3 + 2\alpha \epsilon_{\alpha\beta} u'_{\beta,\alpha} + Y_3) = \tfrac{1}{2} \overset{*}{\Box}_4 \epsilon_{\alpha\beta} X_{\beta,\alpha}.
\end{aligned}
\tag{4.171}
$$

To equations (4.167)–(4.171), concerning the complementary solution, one should add the corresponding classical equations, which can be obtained from equations (4.160) and (4.161). We obtain, in $\Omega \times \mathbf{T}_+$, the following equations:

$$\overset{*}{\Box}_2 u^0_\alpha + (\lambda+\mu) u^0_{\beta,\beta\alpha} + X_\alpha = 0 \tag{4.172}$$

and

$$\Box_1 \overset{*}{\Box}_2 u^0_\alpha = (\lambda+\mu) X_{\beta,\beta\alpha} - \Box_1 X_\alpha, \qquad \overset{*}{\Box}_2 \varphi^0_3 = \tfrac{1}{2}\epsilon_{\beta\alpha} X_{\beta,\alpha}, \tag{4.173}$$

where $\varphi^0_3 = \frac{1}{2}\epsilon_{\alpha\beta} u^0_{\beta,\alpha}$.

The second plane problem. The systems of functions

$$(u^0_3, \varphi^0_\alpha), \qquad (u'_3, \varphi'_\alpha, \zeta_\alpha)$$

and the body loadings X_3, Y_α are connected with this problem. From the system of equations (4.162) we obtain

$$\begin{aligned} &\overset{*}{\Box}_2 u'_3 + 2\alpha\epsilon_{\alpha\beta}\zeta_{\alpha,\beta} = 0, \\ &\overset{*}{\Box}_2 \left[\Box_4 \zeta + (\beta+\gamma-\varepsilon)\zeta_{\beta,\beta\alpha} - \tfrac{1}{2}\overset{*}{\Box}_4 \epsilon_{\alpha\beta} u'_{3,\beta} - Y_\alpha\right] \\ &\quad = \tfrac{1}{2}\overset{*}{\Box}_4 \epsilon_{\beta\alpha} X_{3,\beta}. \end{aligned} \tag{4.174}$$

The system of equations (4.164) for the problem considered takes the following form:

$$\begin{aligned} &\Box_2 u'_3 + 2\alpha\epsilon_{\alpha\beta}\varphi'_{\beta,\alpha} = 0, \\ &\overset{*}{\Box}_2 \left[\Box_4 \varphi'_\alpha + (\beta+\gamma-\varepsilon)\varphi'_{\beta,\beta\alpha} + 2\alpha\epsilon_{\alpha\gamma} u'_{3,\gamma} + Y_\alpha\right] = \tfrac{1}{2}\overset{*}{\Box}_4 \epsilon_{\alpha\beta} X_{3,\beta}. \end{aligned} \tag{4.175}$$

Dividing equation $(4.163)_1$ for $i=3$ by the operator $\Box_1$, we obtain an equation for u'_3, namely

$$\overset{*}{\Box}_2(\Box_2\Box_4 + 4\alpha^2\nabla^2_1) u'_3 = 2\alpha\epsilon_{\gamma\delta}\overset{*}{\Box}_2 Y_{\delta,\gamma} + \tfrac{1}{2}\overset{*}{\Box}_4 \nabla^2_1 X_3. \tag{4.176}$$

In turn, from equation $(4.163)_2$ for $i=\alpha$ we obtain an equation for ζ_α, namely

$$\begin{aligned} \overset{*}{\Box}_2\Box_3(\Box_2\Box_4 + 4\alpha^2\nabla^2_1)\zeta_\alpha &= \left[\alpha\overset{*}{\Box}_4 - (\beta+\gamma-\varepsilon)\overset{*}{\Box}_2\right]\overset{*}{\Box}_2 Y_{\beta,\beta\alpha} \\ &\quad + \overset{*}{\Box}_2\Box_3(\overset{*}{\Box}_2 Y_\alpha - \tfrac{1}{2}\overset{*}{\Box}_4 \epsilon_{\alpha\beta} X_{3,\beta}). \end{aligned} \tag{4.177}$$

From equation (4.165) we obtain for $i=\alpha$ an equation for φ'_α, namely

$$\begin{aligned} \overset{*}{\Box}_2\Box_3(\Box_2\Box_4 + 4\alpha^2\nabla^2_1)\varphi'_\alpha &= \left[(\beta+\gamma-\varepsilon)\Box_2 - 4\alpha^2\right]\overset{*}{\Box}_2 Y_{\beta,\beta\alpha} \\ &\quad - \Box_2\Box_3(\overset{*}{\Box}_2 Y_\alpha - \tfrac{1}{2}\overset{*}{\Box}_4 \epsilon_{\alpha\beta} X_{3,\beta}). \end{aligned} \tag{4.178}$$

The complementary equations (4.174)–(4.178) should be completed with the classical equations. From equations (4.160) and (4.161) we obtain

$$\overset{*}{\square}_2 u_3^0 + X_3 = 0, \qquad \overset{*}{\square}_2 \varphi_\alpha^0 = \tfrac{1}{2}\epsilon_{\beta\alpha} X_{3,\beta}, \tag{4.179}$$

where $\varphi_\alpha^0 = \frac{1}{2}\epsilon_{\alpha\beta} u_{3,\beta}^0$.

The first axially-symmetric problem. With this problem we associate the sets of functions $(u_r^0, \varphi_\theta^0, u_z^0)$, and $(u'_r, u'_z, \varphi'_\theta, \zeta_\theta)$ as well as the body loadings X_r, Y_θ, X_z. Taking into account considerations from Subsection 4.4.3 and formulae (4.145)–(4.148), we also obtain the formula

$$\overrightarrow{\overset{*}{\square}}_2 \equiv (\overset{*}{\square}{}^0_2, \overset{*}{\square}{}^0_2, \overset{*}{\square}_2), \qquad \overrightarrow{\overset{*}{\square}}_4 \equiv (\overset{*}{\square}{}^0_4, \overset{*}{\square}{}^0_4, \overset{*}{\square}_4), \tag{4.180}$$

where

$$\begin{aligned}
&\overset{*}{\square}_2 = \mu\nabla^2 - \rho\partial_t^2, &&\overset{*}{\square}{}^0_2 = \mu\nabla_0^2 - \rho\partial_t^2,\\
&\overset{*}{\square}_4 = (\gamma+\varepsilon)\nabla^2 - J\partial_t^2, &&\overset{*}{\square}{}^0_4 = (\gamma+\varepsilon)\nabla_0^2 - J\partial_t^2,\\
&\nabla^2 = \partial_r^2 + r^{-1}\partial_r + \partial_z^2, &&\nabla_0^2 = \nabla^2 - r^{-2}.
\end{aligned}$$

Now, from equation $(4.163)_1$ we obtain the following equations:

$$\begin{aligned}
&\overset{*}{\square}{}^0_2\square^0_1(\square^0_2\square^0_4 + 4\alpha^2\nabla_0^2)u'_r\\
&\quad = -\alpha\square^0_1\partial_z\left[2\overset{*}{\square}{}^0_2 Y_\theta - \overset{*}{\square}{}^0_4(\partial_z X_r - \partial_r X_z)\right],\\
&\overset{*}{\square}_2\square_1(\square_2\square_4 + 4\alpha^2\nabla^2)u'_z\\
&\quad = \alpha\square_1\left(r^{-1} + \partial_r\right)\left[2\overset{*}{\square}{}^0_2 Y_\theta - \overset{*}{\square}{}^0_4(\partial_z X_r - \partial_r X_z)\right].
\end{aligned} \tag{4.181}$$

Removing the operator $\overset{*}{\square}{}^0_2\square^0_3$ from equation $(4.163)_2$, we obtain an equation for ζ_θ, namely

$$(\square^0_2\square^0_4 + 4\alpha^2\nabla_0^2)\zeta_\theta = \tfrac{1}{2}\left[2\overset{*}{\square}{}^0_2 Y_\theta - \overset{*}{\square}{}^0_4(\partial_z X_r - \partial_r X_z)\right]. \tag{4.182}$$

In turn, removing the operator $\square^0_3$ from equation (4.165), we arrive at the equation for φ'_θ:

$$\overset{*}{\square}{}^0_2(\square^0_2\square^0_4 + 4\alpha^2\nabla_0^2)\varphi'_\theta = -\tfrac{1}{2}\square^0_2\left[2\overset{*}{\square}{}^0_2 Y_\theta - \overset{*}{\square}{}^0_4(\partial_z X_r - \partial_r X_z)\right]. \tag{4.183}$$

The system of equations (4.162) is transformed into:

$$\begin{aligned}
&\overset{*}{\square}{}^0_2 u'_r + (\lambda+\mu)\partial_r e' + 2\alpha\partial_z\zeta_\theta = 0,\\
&\overset{*}{\square}_2 u'_z + (\lambda+\mu)\partial_z e' - 2\alpha r^{-1}\partial_r(r\zeta_\theta) = 0,\\
&\overset{*}{\square}{}^0_2\left[\overset{*}{\square}{}^0_4(\partial_z u'_r - \partial_r u'_z) - 2\square^0_4\zeta_\theta + 2Y_\theta\right] = \overset{*}{\square}{}^0_4(\partial_z X_r - \partial_r X_z),
\end{aligned} \tag{4.184}$$

where $e' = r^{-1}\partial_r(ru'_r) + \partial_z u'_z$.

From the system of equations (4.164) we obtain the following equations for u_r', φ_θ' and u_z':

$$\begin{aligned} &\square_2^0 u_r' + (\lambda+\mu-\alpha)\partial_r e' - 2\alpha\partial_z\varphi_\theta' = 0,\\ &\square_2 u_z' + (\lambda+\mu-\alpha)\partial_z e' + 2\alpha r^{-1}\partial_r(r\varphi_\theta') = 0,\\ &\overset{*}{\square}{}_2^0\big[\square_4^0\varphi_\theta' + 2\alpha(\partial_z u_r' - \partial_r u_z') + Y_\theta\big] = \tfrac{1}{2}\overset{*}{\square}{}_4^0(\partial_z X_r - \partial_r X_z). \end{aligned} \tag{4.185}$$

Equations (4.181)–(4.185) should be completed with the classical equations obtained from equations (4.160) and (4.161), i.e.

$$\begin{aligned} &\overset{*}{\square}{}_2^0 u_r^0 + (\lambda+\mu)\partial_r e^0 + X_r = 0,\\ &\overset{*}{\square}{}_2 u_z^0 + (\lambda+\mu)\partial_z e^0 + X_z = 0, \end{aligned} \tag{4.186}$$

where $e^0 = r^{-1}\partial_r(ru_r^0) + \partial_z u_z^0$, and the equations

$$\begin{aligned} &\overset{*}{\square}{}_2^0\varphi_\theta^0 = \tfrac{1}{2}(\partial_r X_z - \partial_z X_r),\\ &\square_1^0\overset{*}{\square}{}_2^0 u_r^0 = (\lambda+\mu)\partial_r X - \square_1^0 X_r,\\ &\square_1\overset{*}{\square}{}_2 u_z^0 = (\lambda+\mu)\partial_z X - \square_1 X_z, \end{aligned} \tag{4.187}$$

where $\varphi_\theta^0 = \frac{1}{2}(\partial_z u_r^0 - \partial_r u_z^0)$ and $X = r^{-1}\partial_r(rX_r) + \partial_z X_z$.

The second axially-symmetric problem. With this problem we associate the sets of functions $(0, u_\theta', 0)$, $(\varphi_r', 0, \varphi_z')$ and $(\zeta_r, 0, \zeta_z)$ as well as the body loadings $(0, X_\theta, 0)$ and $(Y_r, 0, Y_z)$.

Here, equations (4.164) for the triple $(\varphi_r', u_\theta', \varphi_z')$ take the following form in $\Omega \times \mathbf{T}_+$:

$$\begin{aligned} &\square_2^0 u_\theta' + 2\alpha(\partial_z\varphi_r' - \partial_r\varphi_z') = 0,\\ &\overset{*}{\square}{}_2^0\Big[\square_4^0\varphi_r' + (\beta+\gamma-\varepsilon)\partial_r\kappa' - 2\alpha\partial_z u_\theta' + Y_r\Big] = -\tfrac{1}{2}\overset{*}{\square}{}_4^0\partial_z X_\theta,\\ &\overset{*}{\square}{}_2\Big[\square_4\varphi_z' + (\beta+\gamma-\varepsilon)\partial_z\kappa' + 2\alpha\left(r^{-1}+\partial_r\right)u_\theta' + Y_z\Big]\\ &\qquad = \tfrac{1}{2}\overset{*}{\square}{}_4\left(r^{-1}+\partial_r\right)X_\theta, \end{aligned} \tag{4.188}$$

where $\kappa' = r^{-1}\partial_r(r\varphi_r') + \partial_z\varphi_z'$.

Removing the operator $\square_1^0$ from equation $(4.163)_1$, we arrive at the equation for u_θ':

$$\overset{*}{\square}{}_2^0(\square_2^0\square_4^0 + 4\alpha^2\nabla_0^2)u_\theta' = 2\alpha\Big[\overset{*}{\square}{}_2^0(\partial_z Y_r - \partial_r Y_z) + \tfrac{1}{2}\overset{*}{\square}{}_4^0\nabla_0^2 X_\theta\Big]. \tag{4.189}$$

From $(4.163)_2$ we obtain equations for ζ_r, ζ_z, namely:

$$\begin{aligned}
\overset{*}{\Box}{}^0_2\Box^0_3(\Box^0_2\Box^0_4 + 4\alpha^2\nabla^2_0)\zeta_r &= \left[\alpha\overset{*}{\Box}{}^0_4 - (\beta+\gamma-\varepsilon)\overset{*}{\Box}{}^0_2\right]\partial_r\overset{*}{\Box}_2 Y \\
&\quad + \overset{*}{\Box}{}^0_2\Box^0_3(\overset{*}{\Box}{}^0_2 Y_r + \tfrac{1}{2}\overset{*}{\Box}{}^0_4\partial_z X_\theta), \\
\overset{*}{\Box}_2\Box_3(\Box_2\Box_4 + 4\alpha^2\nabla^2)\zeta_z &= \left[\alpha\overset{*}{\Box}_4 - (\beta+\gamma-\varepsilon)\overset{*}{\Box}_2\right]\partial_z\overset{*}{\Box}_2 Y \\
&\quad + \overset{*}{\Box}_2\Box_3\left[\overset{*}{\Box}_2 Y_z - \tfrac{1}{2}\overset{*}{\Box}_4\left(r^{-1}+\partial_r\right)X_\theta\right],
\end{aligned} \tag{4.190}$$

where $Y = \left(r^{-1}+\partial_r\right)Y_r + \partial_z Y_z$.

From (4.165) we obtain the following equations:

$$\begin{aligned}
\overset{*}{\Box}{}^0_2\Box^0_3(\Box^0_2\Box^0_4 + 4\alpha^2\nabla^2_0)\varphi'_r &= \left[(\beta+\gamma-\varepsilon)\Box^0_2 - 4\alpha^2\right]\partial_r\overset{*}{\Box}_2 Y \\
&\quad - \Box^0_2\Box^0_3(\overset{*}{\Box}{}^0_2 Y_r + \tfrac{1}{2}\overset{*}{\Box}{}^0_4\partial_z X_\theta), \\
\overset{*}{\Box}_2\Box_3(\Box_2\Box_4 + 4\alpha^2\nabla^2)\varphi'_z &= \left[(\beta+\gamma-\varepsilon)\Box_2 - 4\alpha^2\right]\partial_z\overset{*}{\Box}_2 Y \\
&\quad - \Box_2\Box_3\left[\overset{*}{\Box}_2 Y_z - \tfrac{1}{2}\overset{*}{\Box}_4\left(r^{-1}+\partial_r\right)X_\theta\right].
\end{aligned} \tag{4.191}$$

To equations (4.188)–(4.191) one should add the classical equations obtained from equations (4.160) and (4.161). Thus we have

$$\overset{*}{\Box}{}^0_2 u^0_\theta + X_\theta = 0 \tag{4.192}$$

and

$$\begin{aligned}
\overset{*}{\Box}{}^0_2\varphi^0_r &= \tfrac{1}{2}\partial_z X_\theta, \\
\overset{*}{\Box}_2\varphi^0_z &= -\tfrac{1}{2}\left(r^{-1}+\partial_r\right)X_\theta,
\end{aligned} \tag{4.193}$$

where $\varphi^0_z = \frac{1}{2}\left(r^{-1}+\partial_r\right)u^0_\theta$ and $\varphi^0_r = -\frac{1}{2}\partial_z u^0_\theta$.

From our considerations it follows that in the superposition method we can obtain the complementary equations for plane and axially-symmetric problems from the vector equations (4.162)–(4.165). Proceeding in a formal manner, we use the remark on dividing the equations by operators. Namely, in the first plane problem we divide the equation for φ'_3 by $\Box_3$, and in the first axially-symmetric problem we divide the equation for φ'_θ by $\Box^0_3$. In turn, in the second plane problem we divide the equation for u'_3 by $\Box_1$, and in the second axially-symmetric problem — the equation for u'_θ by $\Box^0_1$. This remark is based on the fact that a constant β appears in the definitions of the operators $\Box_3$ and $\Box^0_3$, which is not present in the first plane or axially-symmetric problems. In turn, a constant λ appears in the definitions of $\Box_1$ and $\Box^0_1$, which is not present in the second plane or axially-symmetric problems. An important feature of the vector $\boldsymbol{\zeta}$ is the fact that in the equations for the components ζ_3, ζ_θ, apart from the respective operators $\Box_3$, $\Box^0_3$, one should remove the decoupling operator $\overset{*}{\Box}_2$ (for ζ_3) and $\overset{*}{\Box}{}^0_2$ (for ζ_θ). We will show now that in the superposition method one can derive separated equations for the vectors $\mathbf{u}$, $\boldsymbol{\varphi}$ from the vector equations (4.162)–(4.165). For this purpose we introduce the vectors of reduced body loadings of the form

$$\hat{\mathbf{X}} = \mathbf{0}, \qquad \hat{\mathbf{Y}} = \overset{*}{\overrightarrow{\square}}_2 \mathbf{Y} - \tfrac{1}{2} \overset{*}{\overrightarrow{\square}}_4 \operatorname{curl} \mathbf{X} \tag{4.194}$$

and write the system of equations (4.162) in the following form:

$$\begin{aligned} &\overrightarrow{\square}_2 \mathbf{u}' + (\lambda + \mu) \operatorname{grad} \operatorname{div} \mathbf{u}' + \hat{\mathbf{X}} = 2\alpha \operatorname{curl} \boldsymbol{\zeta}, \\ &\overset{*}{\overrightarrow{\square}}_2 \Big\{ \overset{*}{\overrightarrow{\square}}_4 \operatorname{curl} \mathbf{u}' - 2 \Big[\overrightarrow{\square}_4 \boldsymbol{\zeta} + (\beta + \gamma - \varepsilon) \operatorname{grad} \operatorname{div} \boldsymbol{\zeta} \Big] \Big\} + 2\hat{\mathbf{Y}} = \mathbf{0}. \end{aligned} \tag{4.195}$$

Separated equations (4.163), equations (4.194) and (4.195) being taken into account, take the form:

$$\begin{aligned} &\overset{*}{\overrightarrow{\square}}_2 \overrightarrow{\square}_1 (\overrightarrow{\square}_2 \overrightarrow{\square}_4 + 4\alpha^2 \overrightarrow{\nabla}^2) \mathbf{u}' \\ &\quad = \overset{*}{\overrightarrow{\square}}_2 \Big\{ \Big[(\lambda + \mu - \alpha) \overrightarrow{\square}_4 - 4\alpha^2 \Big] \operatorname{grad} \operatorname{div} \hat{\mathbf{X}} - \overrightarrow{\square}_1 \overrightarrow{\square}_4 \hat{\mathbf{X}} \Big\} \\ &\qquad + 2\alpha \overrightarrow{\square}_1 \operatorname{curl} \hat{\mathbf{Y}}, \\ &2 \overset{*}{\overrightarrow{\square}}_2 \overrightarrow{\square}_3 (\overrightarrow{\square}_2 \overrightarrow{\square}_4 + 4\alpha^2 \overrightarrow{\nabla}^2) \boldsymbol{\zeta} \\ &\quad = \Big[2\alpha \overset{*}{\overrightarrow{\square}}_4 - 2(\beta + \gamma - \varepsilon) \overset{*}{\overrightarrow{\square}}_2 \Big] \operatorname{grad} \operatorname{div} \hat{\mathbf{Y}} \\ &\qquad + \overset{*}{\overrightarrow{\square}}_2 \overrightarrow{\square}_3 (2\hat{\mathbf{Y}} - \overset{*}{\overrightarrow{\square}}_4 \operatorname{curl} \hat{\mathbf{X}}). \end{aligned} \tag{4.196}$$

Eliminating the vector $\boldsymbol{\zeta} = \frac{1}{2} \operatorname{curl} \mathbf{u}' - \boldsymbol{\varphi}'$ from the system of equations (4.195), we arrive at (4.164), which can be written as

$$\begin{aligned} &\overrightarrow{\square}_2 \mathbf{u}' + (\lambda + \mu - \alpha) \operatorname{grad} \operatorname{div} \mathbf{u}' + 2\alpha \operatorname{curl} \boldsymbol{\varphi}' + \hat{\mathbf{X}} = \mathbf{0}, \\ &\overset{*}{\overrightarrow{\square}}_2 \Big[\overrightarrow{\square}_4 \boldsymbol{\varphi}' + (\beta + \gamma - \varepsilon) \operatorname{grad} \operatorname{div} \boldsymbol{\varphi}' + 2\alpha \operatorname{curl} \mathbf{u}' \Big] + \hat{\mathbf{Y}} = \mathbf{0}. \end{aligned} \tag{4.197}$$

A separate equation for $\boldsymbol{\varphi}'$ (4.165) becomes

$$\begin{aligned} \overset{*}{\overrightarrow{\square}}_2 \overrightarrow{\square}_3 (\overrightarrow{\square}_2 \overrightarrow{\square}_4 + 4\alpha^2 \overrightarrow{\nabla}^2) \boldsymbol{\varphi}' &= 2\alpha \overrightarrow{\square}_2 \overrightarrow{\square}_3 \operatorname{curl} \hat{X} \\ &+ \Big[(\beta + \gamma - \varepsilon) \overrightarrow{\square}_2 - 4\alpha^2 \Big] \operatorname{grad} \operatorname{div} \hat{\mathbf{Y}} - \overrightarrow{\square}_2 \overrightarrow{\square}_3 \hat{\mathbf{Y}}. \end{aligned} \tag{4.198}$$

The introduction of the vectors $\hat{\mathbf{X}}$, $\hat{\mathbf{Y}}$ means that, when we assume the formal correspondence

$$\begin{aligned} &\mathbf{u}' \mapsto \mathbf{u}, \qquad &&\boldsymbol{\varphi}' \mapsto \boldsymbol{\varphi}, \\ &\hat{\mathbf{X}} \mapsto \mathbf{X}, \qquad &&\hat{\mathbf{Y}} \mapsto \overset{*}{\overrightarrow{\square}}_2 \mathbf{Y}, \end{aligned} \tag{4.199}$$

and remove the operator $\overset{*}{\overrightarrow{\square}}_2$ from equation $(4.197)_2$, we observe that the system of equations (4.197) is transformed into (4.130), whereas, removing the operator $\overset{*}{\square}_2$ from equations $(4.196)_1$ and (4.198), we find that they go over into equations (4.119) and (4.120), respectively.

Let us pass on to elastostatics. The corresponding field equations with the vectors $\boldsymbol{\zeta}$, $\mathbf{u}'$, $\boldsymbol{\varphi}'$ are given by equations (1.157), (1.163), (1.165), (1.166), (1.168) and (1.169). Let us fix our attention on the equation for the vector $\boldsymbol{\zeta}$ in Ω:

$$\begin{aligned} 4\alpha(\nu_1^2\overrightarrow{\nabla}^2 - 1)(l^2\overrightarrow{\nabla}^2 - 1)\boldsymbol{\zeta} &= -\mu_0^{-1}(\nu_1^2\overrightarrow{\nabla}^2 - 1)\operatorname{curl}\hat{\mathbf{X}} \\ &+ (\nu_1^2\overrightarrow{\nabla}^2 - 1)\overset{*}{\mathbf{Y}} + \left[\tfrac{1}{2}\mu_0^{-1} - (\nu_1^2 - k^2)\right]\operatorname{grad}\operatorname{div}\overset{*}{\mathbf{Y}}, \end{aligned} \tag{4.200}$$

where

$$\nu_1^2 = \frac{2\gamma+\beta}{4\alpha}, \qquad l^2 = \frac{(\gamma+\varepsilon)(\mu+\alpha)}{4\alpha\mu}, \qquad k^2 = \frac{\gamma+\varepsilon}{4\alpha}, \qquad \mu_0 = \frac{2\mu}{\gamma+\varepsilon},$$

as well as on the equations for the vectors $\mathbf{u}'$ and $\boldsymbol{\varphi}'$ in Ω:

$$\begin{aligned} 2\mu\overrightarrow{\nabla}^2\overrightarrow{\nabla}^2(l^2\overrightarrow{\nabla}^2 - 1)\mathbf{u}' &= \overrightarrow{\nabla}^2\operatorname{curl}\overset{*}{\mathbf{Y}} - 2\overrightarrow{\nabla}^2(l^2\overrightarrow{\nabla}^2 - 1)\hat{\mathbf{X}} \\ &- \mu_0^{-1}\overrightarrow{\nabla}^2\operatorname{curl}\operatorname{curl}\hat{\mathbf{X}} + (1-\nu)^{-1}(l^2\overrightarrow{\nabla}^2 - 1)\operatorname{grad}\operatorname{div}\hat{\mathbf{X}}, \\ 2\mu\overrightarrow{\nabla}^2(\nu_1^2\overrightarrow{\nabla}^2 - 1)(l^2\overrightarrow{\nabla}^2 - 1)\boldsymbol{\varphi}' &= -(2\alpha)^{-1}\mu(l^2\overrightarrow{\nabla}^2 - 1)\operatorname{grad}\operatorname{div}\overset{*}{\mathbf{Y}} \\ &+ (\nu_1^2\overrightarrow{\nabla}^2 - 1)\operatorname{curl}\hat{\mathbf{X}} + \alpha_0^{-1}(\nu_1^2\overrightarrow{\nabla}^2 - 1)\operatorname{curl}\operatorname{curl}\overset{*}{\mathbf{Y}}, \end{aligned} \tag{4.201}$$

where $\alpha_0 = \dfrac{2\alpha}{\mu+\alpha}$ and $\nu = \dfrac{\lambda}{2(\mu+\lambda)}$.

In elastostatics, the role of the vectors $\hat{\mathbf{X}}$, $\overset{*}{\mathbf{Y}}$ from elastodynamics is played by the vectors

$$\hat{\mathbf{X}} = \mathbf{0}, \qquad \hat{\mathbf{Y}} = \mathbf{Y} - \mu_0^{-1}\operatorname{curl}\mathbf{X}. \tag{4.202}$$

In order to obtain the equations describing the plane state of strain from equations (4.200) and (4.201), one should take into account the fact that for the vector $\mathbf{a} \overset{\text{def}}{=} (a_1, a_2, a_3)$ in the rectangular coordinate system $Ox_1x_2x_3$ we have

$$\operatorname{curl}\operatorname{curl}\mathbf{a} = (\epsilon_{\alpha\beta}\epsilon_{\delta\gamma}a_{\gamma,\delta\beta}\,, -\nabla_1^2 a_3). \tag{4.203}$$

From equation $(4.201)_1$ we obtain:

— for $i = \alpha$,

$$\begin{aligned} 2\mu\nabla_1^2\nabla_1^2(l^2\nabla_1^2 - 1)u'_\alpha &= \nabla_1^2\epsilon_{\alpha\beta}\overset{*}{Y}_{3,\beta} - 2\nabla_1^2(l^2\nabla_1^2 - 1)\hat{X}_\alpha \\ &- \mu_0^{-1}\nabla_1^2\epsilon_{\alpha\beta}\epsilon_{\delta\gamma}\hat{X}_{\gamma,\delta\beta} + (1-\nu)^{-1}(l^2\nabla_1^2 - 1)\hat{X}_{\beta,\beta\alpha}, \end{aligned} \tag{4.204}$$

where $\hat{X}_\alpha = 0$ and $\overset{*}{Y}_3 = Y_3 - \mu_0^{-1}\epsilon_{\alpha\beta}X_{\beta,\alpha}$;

— for $i = 3$, after removing the operator ∇_1^2,

$$2\mu\nabla_1^2(l^2\nabla_1^2 - 1)u_3' = \epsilon_{\alpha\beta}\overset{*}{Y}_{\alpha,\beta} - 2(l^2\nabla_1^2 - 1)\hat{X}_3 + \mu_0^{-1}\nabla_1^2\hat{X}_3, \tag{4.205}$$

where $\hat{X}_3 = 0$ and $\overset{*}{Y}_\alpha = Y_\alpha - \mu_0^{-1}\epsilon_{\alpha\beta}X_{3,\beta}$.

In turn, from equation $(4.201)_2$ we obtain:

— for $i = \alpha$,

$$\begin{aligned} 2\mu\nabla_1^2(\nu_1^2\nabla_1^2 - 1)(l^2\nabla_1^2 - 1)\varphi_\alpha' &= -(2\alpha)^{-1}\mu(l^2\nabla_1^2 - 1)\overset{*}{Y}_{\beta,\beta\alpha} \\ &+ (\nu_1^2\nabla_1^2 - 1)\epsilon_{\alpha\beta}\hat{X}_{3,\beta} + \alpha_0^{-1}(\nu_1^2\nabla_1^2 - 1)\epsilon_{\delta\gamma}\epsilon_{\alpha\beta}\overset{*}{Y}_{\gamma,\delta\beta}; \end{aligned} \tag{4.206}$$

— for $i = 3$, after removing the operator $(\nu_1^2\nabla_1^2 - 1)$,

$$2\mu\nabla_1^2(l^2\nabla_1^2 - 1)\varphi_3' = \epsilon_{\alpha\beta}\hat{X}_{\beta,\alpha} - \alpha_0^{-1}\nabla_1^2\overset{*}{Y}_3. \tag{4.207}$$

Finally, from equation (4.200) we obtain:

— for $i = \alpha$,

$$\begin{aligned} 4\alpha(\nu_1^2\nabla_1^2 - 1)(l^2\nabla_1^2 - 1)\zeta_\alpha &= -\mu_0^{-1}(\nu_1^2\nabla_1^2 - 1)\epsilon_{\alpha\beta}\hat{X}_{3,\beta} \\ &+ (\nu_1^2\nabla_1^2 - 1)\overset{*}{Y}_\alpha + \left[\tfrac{1}{2}\mu_0^{-1} - (\nu_1^2 - k^2)\right]\overset{*}{Y}_{\beta,\beta\alpha}; \end{aligned} \tag{4.208}$$

— for $i = 3$, after removing the operator $(\nu_1^2\nabla_1^2 - 1)$,

$$4\alpha(l^2\nabla_1^2 - 1)\zeta_3 = -\mu_0^{-1}\epsilon_{\alpha\beta}\hat{X}_{\beta,\alpha} + \overset{*}{Y}_3. \tag{4.209}$$

Equations (4.204), (4.207) and (4.209) refer to the first plane problem (with the triple (u_α, φ_3)) in the superposition method. Equations (4.204) and (4.207) coincide with equations (3.74). To these equations one should add the classical ones, (3.69) and (3.73). Equations (4.205), (4.206) and (4.208) refer to the second plane problem with the triple (φ_α, u_3) in the superposition method. These are equations (3.225) and (3.226), to which one should adjoin the classical equations (3.220) and (3.222). When deriving equations (4.205), (4.207) and (4.209), we divide them by the operators ∇_1^2 and $(\nu_1^2\nabla_1^2 - 1)$, respectively, using the remarks made on vector dynamical equations and taking into account that

$$\begin{aligned} \Box_1 &\mapsto (\lambda + 2\mu)\nabla_1^2, \\ \Box_3 &\mapsto 4\alpha(\nu_1^2\nabla_1^2 - 1). \end{aligned}$$

The purpose of introducing the vectors (4.202) is such that with the correspondence

$$u'_\alpha \mapsto u_\alpha, \qquad \varphi'_3 \mapsto \varphi_3, \qquad \hat{X}_\alpha \mapsto X_\alpha, \qquad \overset{*}{Y}_3 \mapsto Y_3,$$

equations (4.204) and (4.207) are transformed into equations (4.141). In turn, for the correspondence

$$u'_3 \mapsto u_3, \qquad \varphi'_\alpha \mapsto \varphi_\alpha, \qquad \hat{X}_3 \mapsto X_3, \qquad \overset{*}{Y}_\alpha \mapsto Y_\alpha,$$

equations (4.205) and (4.206) go over into equations (4.142). In order to obtain the equations referring to the problems of the axially-symmetric state of strain from equations (4.200) and (4.201), we use formulae (4.145)–(4.149). From equation $(4.201)_1$ we get

$$\begin{aligned}
2\mu\nabla_0^2\nabla_0^2(l^2\nabla_0^2-1)u'_r &= -\nabla_0^2\partial_z\overset{*}{Y}_\theta - 2\nabla_0^2(l^2\nabla_0^2-1)\hat{X}_r \\
&\quad + \mu_0^{-1}\nabla_0^2\partial_z(\partial_z\hat{X}_r - \partial_r\hat{X}_z) + (1-\nu)^{-1}(l^2\nabla_0^2-1)\partial_r\hat{X}, \\
2\mu\nabla^2\nabla^2(l^2\nabla^2-1)u'_z &= \nabla^2\left(r^{-1}+\partial_r\right)\overset{*}{Y}_\theta - 2\nabla^2(l^2\nabla^2-1)\hat{X}_z \\
&\quad - \mu_0^{-1}\nabla^2\left(r^{-1}+\partial_r\right)(\partial_z\hat{X}_r - \partial_r\hat{X}_z) \\
&\quad + (1-\nu)^{-1}(l^2\nabla^2-1)\partial_z\hat{X},
\end{aligned} \tag{4.210}$$

where

$$\hat{X} = \left(r^{-1}+\partial_r\right)\hat{X}_r + \partial_z\hat{X}_z,\ \hat{X}_r = 0,\ \hat{X}_z = 0,\ \overset{*}{Y}_\theta = Y_\theta - \mu_0^{-1}(\partial_z X_r - \partial_r X_z).$$

Moreover, removing the operator ∇_0^2 from equation $(4.201)_1$, we arrive at

$$2\mu\nabla_0^2(l^2\nabla_0^2-1)u'_\theta = \partial_z\overset{*}{Y}_r - \partial_r\overset{*}{Y}_z - 2(l^2\nabla_0^2-1)\hat{X}_\theta + \mu_0^{-1}\nabla_0^2\hat{X}_\theta, \tag{4.211}$$

where $\hat{X}_\theta = 0$, $\overset{*}{Y}_r = Y_r + \mu_0^{-1}\partial_z X_\theta$, and $\overset{*}{Y}_z = Y_z - \mu_0^{-1}\left(r^{-1}+\partial_r\right)X_\theta$.

From equation $(4.201)_2$ we obtain

$$\begin{aligned}
2\mu\nabla_0^2(\nu_1^2\nabla_0^2-1)(l^2\nabla_0^2-1)\varphi'_r &= -(2\alpha)^{-1}\mu(l^2\nabla_0^2-1)\partial_r\overset{*}{Y} \\
&\quad - (\nu_1^2\nabla_0^2-1)\partial_z\hat{X}_\theta - \alpha_0^{-1}(\nu_1^2\nabla_0^2-1)\partial_z(\partial_z\overset{*}{Y}_r - \partial_r\overset{*}{Y}_z), \\
2\mu\nabla^2(\nu_1^2\nabla^2-1)(l^2\nabla^2-1)\varphi'_z &= -(2\alpha)^{-1}\mu(l^2\nabla^2-1)\partial_z\overset{*}{Y} \\
&\quad + (\nu_1^2\nabla^2-1)\left(r^{-1}+\partial_r\right)\hat{X}_\theta \\
&\quad + \alpha_0^{-1}(\nu_1^2\nabla^2-1)\left(r^{-1}+\partial_r\right)(\partial_z\overset{*}{Y}_r - \partial_r\overset{*}{Y}_z),
\end{aligned} \tag{4.212}$$

where $\overset{*}{Y} = \left(r^{-1}+\partial_r\right)\overset{*}{Y}_r + \partial_z\overset{*}{Y}_z$.

Furthermore, removing the operator $(\nu_1^2\nabla_0^2-1)$ from equation $(4.201)_2$, we arrive at

$$2\mu\nabla_0^2(l^2\nabla_0^2-1)\varphi'_\theta = (\partial_z\hat{X}_r - \partial_r\hat{X}_z) - \alpha_0^{-1}\nabla_0^2\overset{*}{Y}_\theta. \tag{4.213}$$

In turn, removing the operator $(\nu_1^2\nabla_0^2-1)$ from equation (4.200), we obtain

$$4\alpha(l^2\nabla_0^2 - 1)\zeta_\theta = -\mu_0^{-1}(\partial_z\hat{X}_r - \partial_r\hat{X}_z) + \overset{*}{Y}_\theta. \tag{4.214}$$

Moreover, without any transformations we get the following equations:

$$\begin{aligned}
&4\alpha(\nu_1^2\nabla_0^2 - 1)(l^2\nabla_0^2 - 1)\zeta_r = \mu_0^{-1}(\nu_1^2\nabla_0^2 - 1)\partial_z\hat{X}_\theta + (\nu_1^2\nabla_0^2 - 1)\overset{*}{Y}_r \\
&\quad + \left[\tfrac{1}{2}\mu_0^{-1} - (\nu_1^2 - k^2)\right]\partial_r\overset{*}{Y}, \\
&4\alpha(\nu_1^2\nabla^2 - 1)(l^2\nabla^2 - 1)\zeta_z = -\mu_0^{-1}(\nu_1^2\nabla^2 - 1)\left(r^{-1} + \partial_r\right)\hat{X}_\theta \\
&\quad + (\nu_1^2\nabla^2 - 1)\overset{*}{Y}_z + \left[\tfrac{1}{2}\mu_0^{-1} - (\nu_1^2 - k^2)\right]\partial_z\overset{*}{Y}.
\end{aligned} \tag{4.215}$$

Equations (4.210), (4.213) and (4.214) refer to the first axially-symmetric problem in the superposition method and are identical to equations (2.110), (2.111) and (2.113), respectively. In turn, equations (4.211), (4.212) and (4.215) refer to the second axially-symmetric problem in the superposition method and are identical to equations (2.359)–(2.361), respectively. In deriving equations (4.211), (4.213) and (4.214), division by the operators ∇_0^2 and $(\nu_1^2\nabla_0^2 - 1)$, respectively, is a conclusion from the discussion of dynamical equations. We have

$$\Box_1^0 \mapsto (\lambda + 2\mu)\nabla_0^2, \qquad \Box_3^0 \mapsto 4\alpha(\nu_1^2\nabla_0^2 - 1)$$

and we take into account the fact that the constants β and λ do not appear in the first and in the second axially-symmetric problems, respectively. Assuming the formal correspondence

$$\begin{aligned}
&u'_r \mapsto u_r, \qquad u'_z \mapsto u_z, \qquad \varphi'_\theta \mapsto \varphi_\theta, \\
&\hat{X}_r \mapsto X_r, \qquad \hat{X}_z \mapsto X_z, \qquad \overset{*}{Y}_\theta \mapsto Y_\theta,
\end{aligned}$$

equations (4.210) and (4.213) are transformed into equations (4.150) and (4.151). In turn, for

$$\begin{aligned}
&\varphi'_r \mapsto \varphi_r, \qquad \varphi'_z \mapsto \varphi_z, \qquad u'_\theta \mapsto u_\theta, \\
&\overset{*}{Y}_r \mapsto Y_r, \qquad \overset{*}{Y}_z \mapsto Y_z, \qquad \hat{X}_\theta \mapsto X_\theta,
\end{aligned}$$

equations (4.211) and (4.212) are transformed into equations (4.153) and (4.154).

Appendix

$$\mathsf{I}^1 = \int_0^\infty \int_0^\infty \xi^1 e^{-\xi x_1} \cos \xi_2 x_2 \cos \xi_3 x_3 \, d\xi_2 \, d\xi_3 = -\frac{\pi}{2} \cdot \frac{R^2 - 3x_1^2}{R^5},$$

$$\mathsf{I}^0 = \int_0^\infty \int_0^\infty \xi^0 e^{-\xi x_1} \cos \xi_2 x_2 \cos \xi_3 x_3 \, d\xi_2 \, d\xi_3 = \frac{\pi}{2} \cdot \frac{x_1}{R^3},$$

$$\mathsf{I}^{-1} = \int_0^\infty \int_0^\infty \xi^{-1} e^{-\xi x_1} \cos \xi_2 x_2 \cos \xi_3 x_3 \, d\xi_2 \, d\xi_3 = \frac{\pi}{2} \cdot \frac{1}{R},$$

$$\mathsf{I}^{-2} = \int_0^\infty \int_0^\infty \xi^{-2} e^{-\xi x_1} \cos \xi_2 x_2 \cos \xi_3 x_3 \, d\xi_2 \, d\xi_3 = -\frac{\pi}{2} \ln(R + x_1),$$

$$\mathsf{J}^1 = \int_0^\infty \int_0^\infty \xi^1 e^{-\xi x_1} \xi_K \sin \xi_2 x_2 \cos \xi_3 x_3 \, d\xi_2 \, d\xi_3 = -\frac{\pi}{2} \cdot 3x_K \cdot \frac{R^2 - 5x_1^2}{R^7},$$

$$\mathsf{J}^0 = \int_0^\infty \int_0^\infty \xi^0 e^{-\xi x_1} \xi_K \sin \xi_2 x_2 \cos \xi_3 x_3 \, d\xi_2 \, d\xi_3 = \frac{\pi}{2} \cdot \frac{3x_1 x_K}{R^5},$$

$$\mathsf{J}^{-1} = \int_0^\infty \int_0^\infty \xi^{-1} e^{-\xi x_1} \xi_K \sin \xi_2 x_2 \cos \xi_3 x_3 \, d\xi_2 \, d\xi_3 = \frac{\pi}{2} \cdot \frac{x_K}{R^3},$$

$$\mathsf{J}^{-2} = \int_0^\infty \int_0^\infty \xi^{-2} e^{-\xi x_1} \xi_K \sin \xi_2 x_2 \cos \xi_3 x_3 \, d\xi_2 \, d\xi_3 = \frac{\pi}{2} \cdot \frac{x_K}{R(R + x_1)},$$

$$\mathsf{J}^{-3} = \int_0^\infty \int_0^\infty \xi^{-3} e^{-\xi x_1} \xi_K \sin \xi_2 x_2 \cos \xi_3 x_3 \, d\xi_2 \, d\xi_3 = \frac{\pi}{2} \cdot \frac{x_K}{R + x_1},$$

$$\mathsf{K}^{-1} = \int_0^\infty \int_0^\infty \xi^{-1} e^{-\xi x_1} \xi_2 \xi_3 \sin \xi_2 x_2 \sin \xi_3 x_3 \, d\xi_2 \, d\xi_3 = \frac{\pi}{2} \cdot \frac{3x_2 x_3}{R^5},$$

$$\mathsf{K}^{-2} = \int_0^\infty \int_0^\infty \xi^{-2} e^{-\xi x_1} \xi_2 \xi_3 \sin \xi_2 x_2 \sin \xi_3 x_3 \, d\xi_2 \, d\xi_3 = \frac{\pi}{2} \cdot \frac{x_2 x_3 (2R + x_1)}{R^3 (R + x_1)^2},$$

$$\mathsf{K}^{-3} = \int_0^\infty \int_0^\infty \xi^{-3} e^{-\xi x_1} \xi_2 \xi_3 \sin \xi_2 x_2 \sin \xi_3 x_3 \, d\xi_2 \, d\xi_3 = \frac{\pi}{2} \cdot \frac{x_2 x_3}{R(R + x_1)^2},$$

$$\mathsf{K}^{-4} = \int_0^\infty \int_0^\infty \xi^{-4} e^{-\xi x_1} \xi_2 \xi_3 \sin \xi_2 x_2 \sin \xi_3 x_3 \, d\xi_2 \, d\xi_3 = \frac{\pi}{2} \cdot \frac{x_2 x_3}{2(R + x_1)^2},$$

$$\mathsf{L}^{-1} = \int_0^\infty \int_0^\infty \xi^{-1} e^{-\xi x_1} \xi_2^2 \cos \xi_2 x_2 \cos \xi_3 x_3 \, d\xi_2 \, d\xi_3 = \frac{\pi}{2} \cdot \frac{R^2 - 3x_2^2}{R^5},$$

$$\begin{aligned}\mathsf{L}^{-2} &= \int_0^\infty \int_0^\infty \xi^{-2} e^{-\xi x_1} \xi_2^2 \cos \xi_2 x_2 \cos \xi_3 x_3 \, d\xi_2 \, d\xi_3 \\ &= \frac{\pi}{2} \cdot \frac{R^2(R+x_1) - x_2^2(2R+x_1)}{R^3(R+x_1)^2},\end{aligned}$$

$$\mathsf{L}^{-3} = \int_0^\infty \int_0^\infty \xi^{-3} e^{-\xi x_1} \xi_2^2 \cos \xi_2 x_2 \cos \xi_3 x_3 \, d\xi_2 \, d\xi_3 = \frac{\pi}{2} \cdot \frac{R(R+x_1) - x_2^2}{R(R+x_1)^2},$$

$$\begin{aligned}\mathsf{L}^{-4} &= \int_0^\infty \int_0^\infty \xi^{-4} e^{-\xi x_1} \xi_2^2 \cos \xi_2 x_2 \cos \xi_3 x_3 \, d\xi_2 \, d\xi_3 \\ &= -\frac{\pi}{2} \left[\ln(R+x_1) + \frac{x_1}{R+x_1} + \frac{x_2^2}{(R+x_1)^2} \right].\end{aligned}$$

References

1. Aero E. L., Kuvshinski E. V., *Continuum theory of asymmetric elasticity. Equlibrium of isotropic body*, Phiz. Tverd. Tela, 1964, **6**, 9, 2689–2699 (in Russian).
2. Aero E. L., Kuvshinski E. V., *Fundamental equations of the theory of elasticity with rotational particle interactions*, Phiz. Tverd. Tela, 1969, **2**, 7, 1399 (in Russian).
3. Al-Hasan M., Dyszlewicz J.,*Stress-temperature equations of motion of Ignaczak type for a three-dimensional problem of micropolar thermoelastisity*, J. of Thermal Stresses, 2001, **24**, 7, 709–722.
4. Al-Hasan M., Dyszlewicz J., *Radiation conditions and integral representations of a solution in coupled micropolar thermoelastisity*, J. of Thermal Stresses, 2001, **24**, 10, 1007–1018.
5. Anthony K. H., *Die Theorie der Disklinationen*, Arch. Rat. Mech. Analysis, 1970, **39**, 43.
6. Ariman T., *Some problems in bending of micropolar plates*, Bull. Acad. Polon. Sci., Sér. Sci. Techn., I–1968, **16**, 7, 259; II–1968, **16**, 7, 301.
7. Ariman T., Kaloni P. N., *Stress concentration effects in micropolar elasticity*, ZAMP, 1967, **18**, 136–141.
8. Askar A., Cakmak A. S., *A structural model of a micropolar continuum*, Int. J. Eng. Sci., 1968, **6**, 10, 583.
9. Babich W. M., Kapilevich M. B., Michlin S. G., Natanson G. N., Riez P. M., Slobodecki L. N., Smirnov M. M., *Linear Equations of Mathematical Physics*, Polish Scientific Publishers, Warsaw 1970 (in Polish).
10. Banks C. B., Sokołowski M., *On certain two-dimensional applications of couple-stress theory*, Int. J. Solids Structures, 1968, **4**, 15–29.
11. Barański W., Wilmański K., Woźniak Cz., *Mechanics of continuous media*, Mech. Teor. i Stos., 1967, **5**, 2, 215 (a review, in Polish).
12. Barański W., Woźniak Cz., *The fibrous body as a simply connected model of multi-hole disc*, Arch. Mech. Stos., 1966, **18**, 3, 273–283.
13. Benedict R. L., Lakes R. S., *Noncentrosymmetry in micropolar elasticity*, Int. J. Eng. Sci., 1982, **20**, 1161–1167.
14. Bieńkowski J., Dyszlewicz J., *The problem of Boussinesq-Mindlin type for elastic half-space with couple-stresses*, manuscript, Institute of Mathematics, Wrocław University of Technology, 1997 (in Polish).
15. Bogy D. B., Sternberg E., *The effect of couple-stresses on singularities due to discontinuous loadings*, Int. J. Solids Structures, 1967, **3**, 757.
16. Boley B. A., Weiner J. H., *Theory of Thermal Stresses*, Wiley, New York, 1960.
17. Brulin O., *Grade Consistent Micropolar Theory of Thermoelasticity*, [in:] *New Problems in Mechanics of Continua*, University of Waterloo Press, 1983, p. 265.

18. Brulin O., Hjalmars S., *Linear grade-consistent micropolar theory*, Int. J. Eng. Sci., 1981, **19**, 1731.
19. Burchuladse T. V., Gegelya T.G., *Development of the method of potentials in elasticity theory*, Tbilisi, 1985, preprint, (in Russian).
20. Carlson D. E., *On Günther's stress functions for couple stresses*, Quart. Appl. Math., 1967, **25**, 2, 139.
21. Chadwick P., Sneddon I. N., *Plane waves in an elastic solid conducting heat*, J. Mech. Phys. of Solids, 1958, **6**.
22. Cosserat E., Cosserat F., *Théorie des Corps Déformables*, A. Hermann, Paris 1909.
23. Cowin S. C., *Stress functions for Cosserat elasticity*, Int. J. Solids Structures, 1970, **6**, 4, 389.
24. Czub M., Dyszlewicz J., *Fundamental solutions for axisymmetric problems of micropolar theory of elasticity with distortion field I.*, Bull. Pol. Ac.: Tech., 1987, **35**, 227–237.
25. Czub M., Dyszlewicz J., *Fundamental solutions for axisymmetric problems of micropolar theory of elasticity with distortion field II.*, Bull. Pol. Ac.: Tech., 1987, **35**, 239–248.
26. Dąbrowski O., *Model investigations of boundary perturbations in a two-curvature convex envelope*, Inż. i Bud., 1962, **19**, 4, 125–128 (in Polish).
27. Dhaliwal R. S., *The steady-state axisymmetric problem of micropolar thermoelasticity*, Archives of Mechanics, 1971, **23**, 5, 705.
28. Ditkin W. A., Prudnikov A. P., *Integral Transformations and Operator Calculus*, Polish Scientific Publishers, Warsaw 1965 (in Polish).
29. Domański Z., Piskorek A., Rojek Z., *On the application of the method of Fischer-Riez-Kupradse to the solution of the first Fourier problem*, Rocz. Pol. Tow. Mat., Ser. 1, Prace Mat. 1972, **16**, 137-147 (in Russian).
30. Doyle J. M., *Singular solutions in elasticity*, Acta Mechanica, 1966, **4**, 1.
31. Drobot S., *On Cosserat Continua*, Appl. Math., 1971, **13**, 3.
32. Dyszlewicz J., *The plane problem of asymmetric elasticity*, Bull. Acad. Polon. Sci., Sér. Sci. Techn., 1971, **19**, 10, 385–392.
33. Dyszlewicz J., *Plane static problem of the micropolar theory of elasticity*, Archives of Mechanics, 1972, **24**, 2, 211–223.
34. Dyszlewicz J., *Stress formulation of the "second" axially-symmetric problem of micropolar theory of elasticity*, Bull. Acad. Polon. Sci., Sér. Sci. Techn., 1973, **21**, 2, 45–56.
35. Dyszlewicz J., *A method of solving static problems of linear asymmetric elasticity*, Mech. Teor. Stos., 1973, **11**, 2, 143–158 (in Polish).
36. Dyszlewicz J., *The static axially-symmetric problem of micropolar elasticity theory and thermoelasticity*, Rozpr. Inż., 1973, **21**, 2, 193-212 (in Polish).
37. Dyszlewicz J., *Axially-symmetric stress problem of micropolar elasticity and thermoelasticity*, Bull. Acad. Polon. Sci., Sér. Sci. Techn., 1973, **21**, 7–8, 321–332.
38. Dyszlewicz J., *Stress functions for a second axially-symmetric problem of micropolar elastostatics*, Bull. Acad. Polon. Sci., Sér. Sci. Techn., 1974, **22**, 10, 515–524.
39. Dyszlewicz J., *Displacement functions for a second axially-symmetric problem of micropolar elasticity*, Bull. Acad. Polon. Sci., Sér. Sci. Techn., 1974, **22**, 10, 525–532.
40. Dyszlewicz J., *The stress and displacement functions for the "second" axisymmetric problem of micropolar elastostatics*, Archives of Mechanics, 1975, **27**, 1, 15–27.

41. Dyszlewicz J., *Three-dimensional problem of micropolar elastostatics*, Bull. Acad. Polon. Sci., Sér. Sci. Techn., I — 1975, **23**, 8, 381–392; II — 1975, **23**, 8, 393–404.
42. Dyszlewicz J., *Functions of displacements for the second axially-symmetric problem of micropolar elastostatics*, Rozpr. Inż., 1975, **26**, 3, 371–379 (in Polish).
43. Dyszlewicz J., *Boundary Value Problems of Asymmetric Elasticity Theory*, dissertation, Institute of Mathematics, Technical University of Wrocław, 1976 (in Polish).
44. Dyszlewicz J., *Selected boundary problems of equations for the plane problems in micropolar theory of elasticity*, Stud. Geotech. et Mech., I – 1980, **2**, 3, 5–20; II — 1980, **2**, 4, 21–36.
45. Dyszlewicz J., *Three-dimensional problem of half-space in micropolar elastostatics*, Bull. Acad. Polon. Sci., Sér. Sci. Techn., 1981, **29**, 1–2, 1–12.
46. Dyszlewicz J., *Three-dimensional problem of half-space in micropolar thermoelastostatics*, Bull. Acad. Polon. Sci., Sér. Sci. Techn., 1981, **29**, 1–2, 13–22.
47. Dyszlewicz J., *Selected boundary-value problems of micropolar elasticity*, Stud. Geotech. et Mech., 1983, **5**, 2.
48. Dyszlewicz J., *Three-dimensional problems in micropolar theory of elasticity*, Stud. Geotech. et Mech., 1983, **5**, 1.
49. Dyszlewicz J., *Boundary Value Problems for Equations of Micropolar Elastostatics and Elastodynamics*, manuscript, Institute of Mathematics, Wrocław University of Technology, 1984 (in Polish).
50. Dyszlewicz J., *Singularity of displacements-rotations and stresses in micropolar semi-space within the framework of elastostatics and thermoelastostatics*, Stud. Geotech. et Mech., 1985, **7**, 4, 19–39.
51. Dyszlewicz J., *Problem of the Boussinesq-Mindlin type for micropolar half-space*, Bull. Pol. Ac.: Tech., 1985, **33**, 113–122.
52. Dyszlewicz J., *Fundamental solutions of micropolar elastostatics*, Bull. Pol. Ac.: Tech., I — 1986, **34**, 179–190; II — 1986, **34**, 191–202.
53. Dyszlewicz J., *Fundamental solutions of micropolar elastostatics. Problem of elastic distortions*, Bull. Pol. Ac.: Tech., 1986, **34**, 203–214.
54. Dyszlewicz J., *Problem of stress concentration in a micropolar medium. Problem of circular hole*, Stud. Geotech. et Mech., 1986, **8**, 4, 25–36.
55. Dyszlewicz J., *On the concentration of stresses in a micropolar medium. Problem of a perfectly stiff circular inclusion*, Stud. Geotech. et Mech., 1987, **9**, 1–2, 59–69.
56. Dyszlewicz J., *Problems of stresses concentration in micropolar medium. Cylindrical inclusion problem*, Bull. Pol. Ac.: Tech., 1987, **35**, 5–14.
57. Dyszlewicz J., *Stress Equations of Motion of Ignaczak Type for Plane and Axially-Symmetric Problems of Micropolar Elastodynamics*, preprint, IPPT-PAN, Warsaw 1987 (in Polish).
58. Dyszlewicz J., *Boundary and Boundary-Initial Value Problems of Micropolar Elastostatic and Elastodynamic Equations*, Wrocław Technical University Press, Wrocław, 1990 (in Polish).
59. Dyszlewicz J., *Representation of Papkovich-Neuber type in micropolar theory of elasticity*, Stud. Geotech. et Mech., 1991, **13**, 3–4, 37–50.
60. Dyszlewicz J., *Three-dimensional problem of the Boussinesq type for micropolar half-space*, Stud. Geotech. et Mech., 1992, 1–2, 33–52.
61. Dyszlewicz J., *Generalized Papkovich-Neuber's representation in the micropolar elasticity*, Bull. Pol. Ac.: Tech., 1994, **42**, 1, 1–12.
62. Dyszlewicz J., *Selected problems of linear asymmetrical thermoelasticity*, J. of Thermal Stresses, 1996, **19**, 2, 185–206.

63. Dyszlewicz J., *Stress equations of motion of Ignaczak type for a second axisymmetric problem of micropolar elastodynamics*, Appl. Math., 1997, **24**, 3.
64. Dyszlewicz J., Kołodziej Cz., *Fundamental solutions of micropolar elastodynamics*, Stud. Geotech. et Mech., 1984, **6**, 4.
65. Dyszlewicz J., Kołodziej Cz., *The Boussinesq problem for the isotropic hemitropic half-space*, preprint, Institute of Mathematics, Wrocław University of Technology, 1984 (in Polish).
66. Dyszlewicz J., Kołodziej Cz., *Displacement-rotation equations of the hemitropic medium*, Bull. Pol. Ac.: Tech, 1985, **33**, 103–112.
67. Dyszlewicz J., Kołodziej Cz., *Fundamental solutions of micropolar elastodynamics. The first axisymmetrical problem. Method of superposition.* Bull. Pol. Ac.: Tech., 1985, **33**, 123–130.
68. Dyszlewicz J., Kołodziej Cz., *Fundamental solutions of micropolar elastodynamics. The second axisymmetrical problem. Method of superposition.* Bull. Pol. Ac.: Tech., 1985, **33**, 131–138.
69. Dyszlewicz J., Kołodziej Cz., *Second dynamic axially-symmetric problem of micropolar theory of elasticity*, Stud. Geotech. et Mech., 1987, **9**, 3–4.
70. Dyszlewicz J., Kołodziej Cz., *Selected one-dimensional problems of elastostatics and thermoelastostatics of the hemitropic micropolar medium*, Bull. Pol. Ac.: Tech., 1987, **35**, 15–26.
71. Dyszlewicz J., Kołodziej Cz., *Axially-symmetric Lamb problems in micropolar theory of elasticity*, Stud. Geotech. et Mech., 1987, **9**, 3–4.
72. Dyszlewicz J., Matysiak S., *Singularity of force-stresses and couple-stresses in a micropolar body due to concentrated loadings*, Mech. Teor. Stos., 1973, **11**, 4, 363–392 (in Polish).
73. Dyszlewicz J., Matysiak S., *Singularity of stresses in linear micropolar medium due to discontinuous loadings*, Mech. Teor. Stos. 1973, **11**, 4, 393-406 (in Polish).
74. Dyszlewicz J., Matysiak S., *Singularity of stresses in a micropolar elastic semi-space due to concentrated load*, Bull. Acad. Polon. Sci., Sér. Sci. Techn., 1973, **21**, 10, 463–476.
75. Dyszlewicz J., Matysiak S., *Singularity of stresses in micropolar elastic semi-space due to discontinuous boundary load*, Bull. Acad. Polon. Sci., Sér. Sci. Techn., 1973, **21**, 12, 605–610.
76. Dyszlewicz J., Rudnicki W., *An axisymmetric contact problem of micropolar elasticity*, Bull. Acad. Polon. Sci., Sér. Sci. Techn., 1972, **20**, 11, 467–476.
77. Dyszlewicz J., Sikora M., *A linear elastic model of the micropolar body of Eringen-Nowacki type. Determination of physical fields in the half-space with an inside heat source*, manuscript, Institute of Mathematics, Wrocław University of Technology, 1997 (in Polish).
78. Dyszlewicz J., Słotwińska B., *Fundamental solutions of micropolar elastodynamics I*, Bull. Pol. Ac: Tech., 1987, **35**, 121–132.
79. Dyszlewicz J., Słotwińska B., *Cerruti problem for elastic semi-space with couple-stresses. Singularity problem for the physical fields*, Stud. Geotech. et Mech., 1995, **16**, 3–4, 61–86.
80. Dyszlewicz J., Słotwińska B., *Elastic semi-space with microstructure. The Boussinesq problem. Singularities of physical fields*, Stud. Geotech. et Mech., 1996, **17**, 1–2.
81. Dyszlewicz J., Słotwińska B., *Elastic micropolar half-space twisted on the boundary*, manuscript, Institute of Mathematics, Wrocław University of Technology, 1997 (in Polish).

82. Dyszlewicz J., Słotwińska B., *Elastic micropolar half-space twisted non-classically on the boundary by a couple-stress. Problem of singularities of physical fields*, manuscript, Institute of Mathematics, Wrocław University of Technology, 1997 (in Polish).
83. Dyszlewicz J., Wytrążek J., *Fundamental solutions of the plane problem of micropolar elasticity with a harmonically varying distortion field*, Eng. Trans. 1992, **40**, 4, 419–434.
84. Dyszlewicz J., Wytrążek J., *Fundamental solutions for second plane problems of micropolar and couple stress theory of elasticity with distortion field*, manuscript, Institute of Mathematics, Wrocław University of Technology, 1997 (in Polish).
85. Dyszlewicz J., Żal P., *Representation of Papkovich-Neuber type in the couple stress theory. Couple-stresses in problems of torsions of half-space*, manuscript, Institute of Mathematics, Wrocław University of Technology, 1997 (in Polish).
86. Eason G., Fulton J., Sneddon I. N., *The generation of waves in an infinite elastic solid by variable body forces*, Phil. Trans. Roy. Soc., 1956, **248**, **A**, 955, 576.
87. Eringen A. C., *Linear theory of micropolar elasticity*, J. Math. Mech., 1966, **15**, 6, 909.
88. Eringen A. C., Suhubi E. S., *Nonlinear theory of simple microelastic solids*, Int. J. Eng. Sci., I — 1964, **2**, 2, 189; II — 1964, **2**, 4, 389.
89. Ewing W. M., Jardetzky W. S., Press F., *Elastic Waves in Layered Media*, McGraw-Hill, New York, 1957.
90. Fischer-Hjalmars I., Hjalmars S., *The coupling between displacement and micropolar rotation estimated from a cubic micro-model*, J. Tech. Phys., 1982, **23**, 23.
91. Frąckiewicz H., *Mechanics of Lattice Media*, Polish Scientific Publishers, Warsaw 1970 (in Polish).
92. Frydrychowicz W., Matysiak S., Olesiak Z., *Mixed Problems of Thermoelasticity for Micropolar Medium*, preprint, IPPT-PAN, Warsaw 1977 (in Polish).
93. Frydrychowicz W., Matysiak S., Olesiak Z., *On thermal stresses in a micropolar layer*, J. of Thermal Stresses, 1979, 3, 437–447.
94. Fung Y. C., *Foundations of Mechanics of Solids*, Polish Scientific Publishers, Warsaw 1966 (in Polish).
95. Gajl O., Suchar M., *The "second" plane distortion problem of the theory of micropolar elasticity*, Bull. Acad. Polon. Sci., Sér. Sci. Techn., 1972, **20**, 9, 305–314.
96. Ganowicz R., *Selected problems of the theory of Reisnner's plates and the theory of three-layer plates*, Mech. Teor. Stos., 1966, **4**, 3 (in Polish).
97. Ganowicz R., *The singular problem in plane medium with couple-stresses*, Rozpr. Inż., 1970, **18**, 3, 407 (in Polish).
98. Gauthier R. D., Jashman W. E., *A quest for micropolar elastic constants*, Part II, Archives of Mechanics, 1981, **33**, 717–737.
99. Gawinecki J., *Existence, uniqueness and regularity of the solution of the first boundary-initial value problem for the equations of linear thermo-microelasticity*, Bull. Pol. Ac.: Tech., 1986, **34**, 487–490.
100. Goodier J. N., *Concentrations of stress around spherical and cylindrical inclusions and flaws*, Trans. Am. Soc. Mech. Engrs., 1933, AMP 55–7, 39–44.
101. Gradstein I. S., Rizhik I. M., *Tables of Integrals, Sums, Series and Products*, Nauka, Moscow, 1971 (in Russian).
102. Green A. E., *Micro-materials and multipolar continuum mechanics*, Int. J. Eng. Sci., 1965, **3**, 5, 533.

103. Grioli G., *Elastica asimetrica*, Annali di Mat. Poura e Applicata, 1960, **4**, 4, 389 (in Romanian).
104. Günther W., *Zur Statik und Kinematik des Cosseratschen Kontinuums*, Abh. der Braunschw. Wiss. Geselschaft, 1958, **10**, 195.
105. Gurtin M. E., *The Linear Theory of Elasticity*, [in:] *Encyclopaedia of Physics*, vol. VIa/2, Springer-Verlag, Berlin, 1972.
106. Gutkowski W., *Cylindrical grid shells*, Archives of Mechanics, 1965, **17**, 3, 481–495.
107. Hadamard J., *Lectures on Cauchy's Problem in Partial Differential Equations*, Yale University Press, 1923.
108. Hjalmars S., *The muscle-relaxed equilibrium of the human spine analyzes by means of an anisotropic beam model*, [in:] *New Problems in Mechanics of Continua*, University of Waterloo Press, 1983, p. 273.
109. Iacovache M., *O extendere a metodei lui Galerkin pentru sistemul ecuatiilor elasticitatii*, Bul. St. Acad. R.P. Române, 1949, Ser. A., **1**, 593–596 (in Romanian).
110. Iesan D., *On the plane coupled micropolar thermoelasticity*, Bull. Acad. Polon. Sci., Sér. Sci. Techn., I — 1968, **16**, 8, 379; II — 1968, **16**, 8, 385.
111. Iesan D., *On the linear theory of micropolar elasticity*, Int. J. Eng. Sci., 1969, **7**, 12, 1213.
112. Iesan D., *Some theorems in a grade consistent micropolar theory of elasticity*, ZAMM, 1982, **62**, 35.
113. Ignaczak J., *Direct determination of stresses from the stress equations of motion in elasticity*, Arch. Mech. Stos., 1959, **5**, 11.
114. Ignaczak J., *A completeness problem for stress equations of motion in the linear elasticity*, Arch. Mech. Stos., 1963, **15**, 2, 225–234.
115. Ignaczak J., *Radiation conditions of Sommerfeld type for elastic materials with microstructure*, Bull. Acad. Polon. Sci., Sér. Sci. Techn., 1970, **18**, 6, 473.
116. Ignaczak J., *Tensorial equations of motion for elastic materials with microstructure*, [in:] *Trends in Elasticity and Thermoelasticity*, Groningen 1971, p. 89.
117. Ignaczak J., *Radiation conditions in asymmetric elasticity*, J. Elast., 1972, **2**, 4, 307.
118. Ignaczak J., *Theorems of Boggio type in asymmetric elastodynamics*, Bull. Acad. Polon. Sci., Sér. Sci. Techn., 1977, **25**, 4, 139–149.
119. Ignaczak J., *Tensorial equations of motion for an elastic medium under initial stresses*, [in:] *New Problems in Mechanics of Continua*, University of Waterloo Press 1983, p. 115.
120. Ignaczak J., Nowacki W., *The Sommerfeld radiation conditions for coupled thermoelasticity*, Archives of Mechanics, 1962, **14**, 1, 3-13.
121. Itou S., *The effect of couple-stresses on the stress concentration around an elliptic hole*, Acta Mech., 1973, **16**, 289.
122. Jemielita G., *Bending of a micropolar rectangular prism*, Bull. Pol. Ac.: Tech., I — 1984, **32**, 625–632; II — 1984, **32**, 633-642.
123. Kaliski S., *On a model of the continuum with essentially nonsymmetric tensor of mechanical stress*, Arch. Mech. Stos., 1963, **15**, 1, 33.
124. Kaliski S., *Thermo-magneto-microelasticity*, Bull. Acad. Polon. Sci., Sér. Sci. Techn., 1968, **16**, 1, 7.
125. Kaliski S., Kapelewski J., Rymarz Cz., *Surface waves on an optical branch in a continuum with rotational degrees of freedom*, Proc. Vibr. Probl., 1986, **9**, 2, 108.
126. Kaliski S., Nowacki W., *Wave type equations of thermo-magnetic-microelasticity*, Bull. Acad. Polon. Sci., Sér. Sci. Techn., 1970, **18**, 4, 277.

127. Kamke E., *On Ordinary Differential Equations*, Nauka, Moscow 1971 (in Russian).
128. Kasprzak W., Lysik B., *Dimensional Analysis in Desigining Experiments*, WASC, Technical University of Wrocław, 1974 (in Polish).
129. Kisiel I., *On the theory of asymmetrically elastic medium (of Taylor-Schneebeli)*, Arch. Hydrotech., I — 1970, **17**, 185-198; II — 1971, **18**, 517–525 (in Polish).
130. Kisiel I., Dmitruk S., Lysik B., *Outline of Rheology of Soils*, Arkady, Warsaw 1969 (in Polish).
131. Koiter W. T., *Couple-stresses in the theory of elasticity*, Koninkl. Nederl. Akad. van Wetenschappen. Proc. Ser. 8, I — 1964, **67**, 1, 17; II — 1964, **67**, 1, 30.
132. Kokot J., Matysiak S., *Singularity of stresses in micropolar half-space in the "second" plane state of deformation induced by concentrated loads*, Bull. Acad. Polon. Sci., Sér. Sci. Techn., 1974, **22**, 9, 727.
133. Kokot J., Matysiak S., *Singularity of stresses in micropolar half-space in the "second" plane state deformation induced by discontinuous loads*, Bull. Acad. Polon. Sci., Sér. Sci. Techn., 1974, **22**, 10, 803.
134. Kozák I., Szeidl Gy., *The field equations and boundary conditions with force stresses and couple stresses in the linearized theory of micropolar elastosotatics*, Acta Techn. Hung., 1980, **91**, 1–2, 57–80.
135. Kunin I. A., *Theory of Elastic Bodies with Microstructure*, Izd. Nauka, Moscow, 1975 (in Russian).
136. Kunin I. A., Kosilova V. G., *Theory of Media with Microstructure*, Institute of Thermal Physics AN SSSR, Novosibirsk 1976.
137. Kupradse V. D., *Methods of Potential Theory in Elastisity*, GIFML, Moscow, 1963 (in Russian).
138. Kupradse V. D., Gegelya T. G., Bashelishili M. O., Burchuladse T. V., *Three-Dimensional Problems of the Mathematical Theory of Elasticity and Thermoelasticity*, Izd. Nauka, Moscow, 1968 and 1976 (in Russian).
139. Kurlandzki J., *Generalized Heaviside-Maxwell equations and Cosserat medium*, Bull. Pol. Ac.: Tech., 1984, **32**, 257.
140. Kuvshinski E. V., Aero E. L., *Continuum theory of asymmetric elasticity*, Phiz. Tverd. Tela, 1963, **5**, 2591 (in Russian).
141. Lakes R. S., Yang J. F. C., *Transient study of couple stress effects in compact bone. Torsion*, J. Biomechanical Engng., 1978, **103**, 275–279.
142. Lamé G., *Leçons sur le Théorie Mathématique de l'Élasticité des Corps Solides*, Paris 1852, Mallet-Bachelier.
143. Lenz J., *Das von einer punktformigen Wärmeguelle erzeugte Verschiebungs und Drenvektorfeld im Cosserat-Kontinuum*, Acta Mech., 1976, **24**, 25–35.
144. Lenz J., *The action of an instataneous heat source in the infinite hemitropic micropolar continuum*, Bull. Acad. Polon. Sci., Sér. Sci. Techn., 1977, **25**, 3, 209–215.
145. Love A. E. H., *A Treatise on the Mathematical Theory of Elasticity*, New York, 1944.
146. Lurie A. I., *Spatial Problems of Elasticity Theory*, Moscow 1955 (in Russian).
147. Lysik B., *Mathematical Foundations of Elasticity Theory*, Technical University of Wrocław, 1970 (in Polish).
148. Majorkowska-Knap K., *Plane Problems of Micropolar Elasticity*, dissertation, Technical University of Warsaw 1972 (in Polish).
149. Majorkowska-Knap K., *Statics of micropolar discs*, Mech. Teor. Stos., 1974, **12**, 4, 577 (in Polish).

150. Matczyński M., Sokołowski M., *Quasi-static problem of a rigidly clamped elastic layer*, Arch. Mech. Stos., 1967, **19**, 6.
151. Matysiak S., *Plane contact problem in asymmetric elasticity theory*, Mech. Teor. Stos., 1975, **13**, 2 (in Polish).
152. Matysiak S., *Certain Problems of Fissures in Asymmetric Elasticity Theory*, dissertation, University of Warsaw 1976 (in Polish).
153. Matysiak S., Olesiak Z., *The axially-symmetric problem of Boussinesq for a heated block in the micropolar theory of elasticity*, Rozpr. Inż., 1979, **27**, 1 (in Polish).
154. Matysiak S., Wachecka-Skowron A., *On a plane contact problem for an elastic micropolar half-space*, Bull. Acad. Polon. Sci., Sér. Sci. Techn., 1974, **22**, 12, 1005.
155. McLachlan N. W., *Bessel Functions for Engineers*, Polish Scientific Publishers, Warsaw 1972 (in Polish).
156. Mikusiński J., *Operational Calculus*, Pergamon, New York 1959.
157. Mindlin R. D., *Influence of couple-stresses on stress-concentrations*, Exper. Mech., 1963, **3**, 1, 1.
158. Mindlin R. D., Tiersten H. F., *Effects of couple-stresses in linear elasticity*, Arch. Rat. Mech. Anal., 1962, **11**, 5, 415.
159. Moon P., Spencer D. E., *Field Theory for Engineers*, D. Van Nostrand Company, Inc. Princeton, New Jersey, Toronto, London, New York, 1961.
160. Mostowski A., Stark M., *Elements of Advanced Algebra*, Polish Scientific Publishers, Warsaw 1975 (in Polish).
161. Muki R., Sternberg E., *The influence of couple-stresses on singular stress concentrations in elastic solids*, ZAMP, 1965, **16**, 5, 611.
162. Muschelishvili N. I., *Selected Basic Problems of the Mathematical Theory of Elasticity*, Izdat. AN SSSR, Moscow 1954 (in Russian).
163. Neuber H., *Über Probleme der Spannungskonzentration im Cosserat-Körper*, Acta Mech., 1966, **2**, 1, 48.
164. Niemish J. N., *The plane problem of elasticity with couple-stress for a region with a circular hole*, Prikl. Mech., 1965, **1**, 5, 52 (in Russian).
165. Nowacki J. P., *The linear theory of dislocations in the Cosserat elastic continuum*, Bull. Acad. Polon. Sci., Sér. Sci. Techn., I — 1973, **21**, 11, 585.
166. Nowacki J. P., *The theory of dislocations in the Cosserat elastic continuum*, Bull. Acad. Polon. Sci., Sér. Sci. Techn., II — 1974, **22**, 7–8, 611.
167. Nowacki J. P., *Some dynamical problems of hemitropic micropolar continuum*, Bull. Acad. Polon. Sci., Sér. Sci. Techn., 1977, **25**, 6, 465–474.
168. Nowacki J. P., *Green functions for a hemitropic micropolar continuum*, Bull. Acad. Polon. Sci., Sér. Sci. Techn., 1977, **25**, 8, 619–625.
169. Nowacki J. P., Nowacki W., *Some problems of hemitropic micropolar continuum*, Bull. Acad. Polon. Sci., Sér. Sci. Techn., 1977, **25**, 4, 297–305.
170. Nowacki W., *Some dynamic problems in thermoelasticity*, Arch. Mech. Stos., 1959, **11**, 2, 259.
171. Nowacki W., *Problems of Thermoelasticity*, Polish Scientific Publishers, Warsaw 1960 (in Polish).
172. Nowacki W., *Green functions for the thermoelastic medium*, Bull. Acad. Polon. Sci., Sér. Sci. Techn., I — 1964, **12**, 6, 315; II — 1964, **12**, 9, 465.
173. Nowacki W., *Couple-stresses in the theory of thermoelasticity*, Bull. Acad. Polon. Sci., Sér. Sci. Techn., I — 1966, **14**, 97; II — 1966, **14**, 203; III — 1966, **14**, 505.
174. Nowacki W., *Dynamical Problems of Thermoelasticity*, Polish Scientific Publishers, Warsaw 1966 (in Polish).

175. Nowacki W., *Mechanics of Structures, III*, Polish Scientific Publishers, Warsaw 1966 (in Polish).
176. Nowacki W., *Green functions for micropolar elasticity*, Bull. Acad. Polon. Sci., Sér. Sci. Techn., 1968, **16**, 11–12, 555.
177. Nowacki W., *Some theorems of asymmetric thermoelasticity*, J. Math. Phys. Sci., 1968, **2**, 2, 111.
178. Nowacki W., *Generalized Love's functions in micropolar elasticity*, Bull. Acad. Polon. Sci., Sér. Sci. Techn., 1969, **17**, 4, 247.
179. Nowacki W., *On the completeness of potentials in micropolar elasticity*, Arch. Mech. Stos., 1969, **21**, 2, 107.
180. Nowacki W., *Theory of Elasticity*, Polish Scientific Publishers, Warsaw 1970 (in Polish).
181. Nowacki W., *Theory of Micropolar Elasticity*, CISM Courses and Lectures, 25, Springer-Verlag, Udine 1970.
182. Nowacki W., *The "second" plane problem of micropolar elasticity*, Bull. Acad. Polon. Sci., Sér. Sci. Techn., 1970, **18**, 11, 899.
183. Nowacki W., *Axially-symmetric problems in micropolar elasticity*, Bull. Acad. Polon. Sci., Sér. Sci. Techn., 1971, **19**, 7–8, 317.
184. Nowacki W., *Plane problems of micropolar elasticity*, Arch. Mech. Stos., 1971, **23**, 5, 587.
185. Nowacki W., *Plane problems of micropolar elasticity*, Bull. Acad. Polon. Sci., Sér. Sci. Techn., 1971, **19**, 6, 427.
186. Nowacki W., *Theory of Asymmetric Elasticity*, Polish Scientific Publishers, Warsaw 1971 (in Polish).
187. Nowacki W., *Distortion problems of micropolar elasticity theory*, Arch. Inż. Ląd., 1972, **18**, 3–4, 375 (in Polish).
188. Nowacki W., *Zweidimensionale Probleme der Mikropolaren Elastostatik*, ZAMM, 1972, **52**, 268.
189. Nowacki W., *Three-dimensional problem of micropolar theory of elasticity*, Bull. Acad. Polon. Sci., Sér. Sci. Techn., 1974, **22**, 5, 363.
190. Nowacki W., *Micropolar Thermoelasticity*, CISM Courses and Lectures, 151, Springer-Verlag, Udine 1978.
191. Nowacki W., *Foundations of Linear Piezoelectricity, Magneto-Thermoelasticity*, CISM Courses and Lectures, 257, Springer-Verlag, 1979.
192. Nowacki W., *Theory of Asymmetric Elasticity*, Polish Scientific Publishers, Warsaw 1981 (in Polish).
193. Nowacki W., *Theory of Asymmetric Elasticity*, Polish Scientific Publishers — Pergamon Press, Warsaw 1986.
194. Nowacki W., Nowacki W. K., *Generation of waves in an infinite micropolar elastic solid*, Bull. Acad. Polon. Sci., Sér. Sci. Techn., I — 1968, **17**, 2, 75; II — 1969, **17**, 2, 83.
195. Nowacki W., Nowacki W. K., *Propagation of monochromatic waves in an infinite micropolar elastic plate*, Bull. Acad. Polon. Sci., Sér. Sci. Techn., 1969, **17**, 29–37.
196. Nowacki W., Nowacki W. K., *The axially-symmetrical Lamb's problem in a semi-infinite micropolar elastic solid*, Proc. Vibr. Probl., 1969, **10**, 2, 97.
197. Nowacki W., Nowacki W. K., *The generation of waves in an infinite micropolar elastic solid*, Proc. Vibr. Probl., 1969, **10**, 2, 170.
198. Olesiak Z., *Stress differential equations of the micropolar elasticity*, Bull. Acad. Polon. Sci., Sér. Sci. Techn., 1970, **18**, 5, 355.
199. Olesiak Z., *On the moment theories of plasticity*, Proc. of the Int. Conf. on Nonlinear Mechanics, October 28–31, 1985 Shanghai, Ed. Science Press, Beijing, China, p. 596–598.

200. Olesiak Z., Wągrowska M., *On shear and rotational yield conditions*, Arch. Mech. Stos., 1985, **37**, 3, 147–155.
201. Palmov V. A., *Basic equations of asymmetric elasticity*, Prikl. Mat. Mech., 1964, **28**, 3, 401 (in Russian).
202. Puri P., *Steady-state thermoelastic problems for the half-space with couple-stresses*, Arch. Mech. Stos., 1970, **22**, 4, 479.
203. Rogula D., *Moment stresses and the symmetry of stress tensor in bodies with no local structure*, Bull. Acad. Polon. Sci., Sér. Sci. Techn., 1970, **18**, 4.
204. Rymarz Cz., *Surface waves in Cosserat's medium*, Bull. Acad. Polon. Sci., Sér. Sci. Techn., 1967, **15**, 3, 177.
205. Rymarz Cz., *Surface waves in a medium with couple-stresses*, Mech. Teor. Stos., 1967, **15**, 3, 177 (in Polish).
206. Sandru N., *On some problems of the linear theory of the asymmetric elasticity*, Int. J. Eng. Sci., 1966, **4**, 1, 81.
207. Savin G. N., *Distribution of Stress Around*, Nauk. Dumka, Kiev 1968 (in Russian).
208. Savin G. N., Niemish J. N., *Investigation of stress concentration in the theory of elasticity with couple-stress*, Prikl. Mech. 1968, **4**, 12, 1 (in Russian).
209. Sawczuk A., *On yielding of Cosserat continuum*, Arch. Mech. Stos., 1967, **19**, 3, 471.
210. Schaefer H., *Das Cosserat-Kontinuum*, ZAMM, 1967, **47**, 8, 485.
211. Sneddon I. N., *The Use of Integral Transforms*, McGraw-Hill Book Co., New York, 1972.
212. Sneddon I. N., Berry D. S., *The Classical Theory of Elasticity*, Springer-Verlag, Berlin, 1958.
213. Sokolnikoff I. S., *Mathematical Theory of Elasticity*, McGraw-Hill, New York, 1956.
214. Sokołowski M., *Couple-stresses in problems of torsion of prismatic bars*, Bull. Acad. Polon. Sci., Sér. Sci. Techn., 1965, **13**, 419.
215. Sokołowski M., *On the Couple-Stress Theory*, Polish Scientific Publishers, Warsaw 1972 (in Polish).
216. Sommerfeld A., *Mechanics of Deformable Bodies*, New York, 1950.
217. Stefaniak J., *A generalization of Galerkin's functions for asymmetric thermoelasticity*, Bull. Acad. Polon. Sci., Sér. Sci. Techn., 1968, **16**, 8, 391.
218. Stefaniak J., *Concentrated Loadings in Asymmetric Thermoelasticity*, Technical University of Poznań, Rozprawy No. 40, 1969 (in Polish).
219. Stefaniak J., *Concentrated loads as body forces*, Revue Roumaine de Mathematiques Pures et Appliquess, 1969, **14**, 4, 119.
220. Stefaniak J., *Solving functions for axisymmetric problems in Cosserat medium*, Bull. Acad. Polon. Sci., Sér. Sci. Techn., 1970, **18**, 9, 203.
221. Suchar M., *Singular solutions of the plane distortion problem of micropolar elasticity*, Bull. Acad. Polon. Sci., Sér. Sci. Techn., 1972, **20**, 7–8, 525.
222. Suchar M., *Stress functions for the "second" plane problem of micropolar elasticity*, Bull. Acad. Polon. Sci., Sér. Sci. Techn., 1972, **20**, 11, 447–456.
223. Suchar M., *On stress and displacement functions for the "second" problem of two-dimensional micropolar elasticity*, Bull. Acad. Polon. Sci., Sér. Sci. Techn., 1972, **20**, 11, 841.
224. Synge J. L., Schild A., *Tensor Calculus*, Polish Scientific Publishers, Warsaw 1966 (in Polish).
225. Szmydt Z., *Fourier Transformation and Linear Differential Equations*, Polish Scientific Publishers, Warsaw 1972 (in Polish).
226. Teodorescu P. P., *Sur la notion de moment massique dans le cas des corps du type de Cosserat*, Bull. Acad. Polon. Sci., Sér. Sci. Techn., 1967, **15**, 1, 57.

227. Tichonov A. N., Samarski A. A., *Equations of Mathematical Physics*, Polish Scientific Publishers, Warsaw 1963 (in Polish).
228. Timoshenko S., Goodier J. N., *Theory of Elasticity*, Arkady, Warsaw 1962 (in Polish).
229. Toupin R. A., *Elastic materials with couple-stresses*, Arch. Rat. Mech. Anal., 1962, **11**, 5, 385.
230. Toupin R. A., *Theories of elasticity with couple-stresses*, Arch. Rat. Mech. Anal., 1964, **17**, 2, 85–112.
231. Truesdell C., Noll W., *The Non-Linear Field Theories of Mechanics, Handbuch der Physik*, III/3, Springer, Berlin-Heidelberg-New York, 1965.
232. Truesdell C., Toupin R. A., *The Classical Field Theories*, [in:] *Encyclopaedia of Physics*, Vol. III/1, Springer-Verlag, Berlin, 1960.
233. Ufland J. S., *Integral Transformations in Problems of Elasticity Theory*, Nauka, Leningrad 1963 (in Russian).
234. Usidus Cz., *Couple-Stresses in a Disk Strip*, Prace IPPT, 1972, **18**, Warsaw (in Polish).
235. Usidus Cz., *Torsion of prismatic bars made of micropolar material*, Mech. Teor. i Stos., 1973, **4**, 4, 53 (in Polish).
236. Usidus Cz., Sokołowski M., *Torsion of cylindrical shafts made of micropolar material*, Bull. Acad. Polon. Sci., Sér. Sci. Techn., 1973, **21**, 1.
237. Voigt W., *Theoretische Studien über die Elastizitätsverhältnisse der Kristalle, I, II*, Abh. der Königl. Ges. der Wiss., Göttingen, 1887, **34**.
238. Wesołowski Z., *On the couple-stresses in an elastic continuum*, Arch. Mech. Stos., 1965, **17**, 219–232.
239. Wickham-Jones T., *Mathematica Graphics*, Springer-Verlag, 1995.
240. Wilmański K., *Dynamics of bodies with microstructure*, Arch. Mech. Stos., 1968, **20**, 6, 705.
241. Wilmański K., Woźniak Cz., *On geometry of continuous medium with microstructure*, Arch. Mech. Stos., 1967, **19**, 5, 715–723.
242. Woźniak Cz., *Thermoelasticity of the bodies with microstructure*, Arch. Mech. Stos., 1967, **19**, 3, 355.
243. Woźniak Cz., *Foundations of Dynamics of Deformable Bodies*, Polish Scientific Publishers, Warsaw 1969 (in Polish).
244. Woźniak Cz., *Lattice-Type Surface Girders*, Polish Scientific Publishers, Warsaw 1970 (in Polish).
245. Zahorski S., *On motion and thermodynamics of non-simple continuum with microstructure*, Arch. Mech. Stos., 1967, **19**, 1, 25.

Author Index

Subject Index

Printing: Saladruck Berlin
Binding: Lüderitz&Bauer, Berlin